OPTIMAL DESIGN OF SWITCHING POWER SUPPLY

OPTIMAL DESIGN OF SWITCHING POWER SUPPLY

Zhanyou Sha, Xiaojun Wang, Yanpeng Wang, and Hongtao Ma

Hebei University of Science and Technology, China

中国电力出版社
CHINA ELECTRIC POWER PRESS

WILEY

This edition first published 2015

Published by John Wiley & Sons Singapore Pte. Ltd., 1 Fusionopolis Walk, #07-01 Solaris South Tower, Singapore 138628, under exclusive license granted by China Electric Power Press for all media and languages excluding Simplified and Traditional Chinese and throughout the world excluding Mainland China, and with non-exclusive license for electronic versions in Mainland China.

For details of our global editorial offices, for customer services and for information about how to apply for permission to reuse the copyright material in this book please see our website at www.wiley.com.

Library of Congress Cataloging-in-Publication Data

Sha, Zhanyou.
Optimal design of switching power supply / Zhanyou Sha, Xiaojun Wang, Yanpeng Wang, and Hongtao Ma.
pages cm
Includes bibliographical references and index.
ISBN 978-1-118-79090-8 (cloth)
1. Switching power supplies – Design and construction. I. Wang, Xiaojun (Electrical engineer) II. Wang, Yanpeng. III. Ma, Hongtao (Electrical engineer) IV. Title.
TK7868.P6S477 2015
621.3815′3 – dc23

2015016439

A catalogue record for this book is available from the British Library.

Cover Image: adventtr/iStockphoto

Typeset in 10/12pt TimesLTStd by SPi Global, Chennai, India

1 2015

Contents

Preface

In recent years, the rapid development of modern power technology and the huge demand of the global market have resulted in higher requirements for the optimal design of switching-mode power supply (SMPS). The so-called optimal design refers to selecting the best design from a variety of design schemes. It should be pointed out that the optimal design is always relative, not absolute. Optimal design is not necessarily the one and only, and there may be several options for designers to choose from. With the development of power technology, the optimal design of SMPS is in endless progress. There are many ways to realize the optimal design of SMPS, which can be approximately grouped into the following aspects: (i) using new devices (e.g., the new single-chip SMPS IC); (ii) using new technologies (e.g., magnetic amplifier, synchronous rectification, active clamp, active power factor correction, and digital power supply); and (iii) using new software (the latest version of the computer-aided design software).

In recent years, with the development and application of green energy-saving power supply, the first author of this book gave 15 lectures on "optimal design of SMPS" and "optimal design of LED drive power supply" in the senior seminar held in Beijing, Shanghai, Suzhou, Ningbo, Hangzhou, and Shenzhen at the invitation of several units such as the China Electronics Standardization Institute under the Ministry of Industry and Information Technology and the China Electronics Enterprises Association. Using these lectures as a basis, the authors put many years of their teaching and scientific research experience into this book for both new and old readers.

The book is scientific, advanced, systematic, and practical and has the following features.

First, it describes the optimal design of SMPS in a comprehensive, in-depth, and systematic manner, including new technologies and applications of SMPS, topology selection of DC/DC converter, peripheral component selection of SMPS, optimal design examples of SMPS, key design points of SMPS, and test technology and protection circuit design.

Second, in view that the SMPS current is moving toward single-chip integration, intellectualization, modularization, and the physical features of short, small, light, and thin, it focuses on the optimal design of single-chip SMPS with the most representative single-chip SMPS IC in the world as an example.

Third, it is novel in content. It introduces many new technologies and their application examples such as the half-bridge LLC resonant converter, synchronous rectification, magnetic amplifier regulation, StackFET™ (stacked FET), suspended high-voltage constant current source, valley fill circuit, active power factor correction, and electromagnetic compatibility design of SMPS.

Fourth, it is easy to understand, delivering high practical value. It not only gives various design examples of SMPS but also describes in detail the key design points of circuits, testing methods, selection of key peripheral components, and the design of protection circuit. It has important reference value for readers to research and develop SMPS.

Fifth, it is informative with wide range of knowledge, easy for readers to comprehend and learn by analogy and apply flexibly.

Professor Zhanyou Sha contributed Chapters 1, 2, and 6–8, and completed the review and integration of the whole book. Associate Professor Hongtao Ma, Professor Xiaojun Wang, and Professor Yanpeng Wang jointly contributed Chapters 3–5, 9, and 10.

We would like to express our sincere thanks to Power Integrations (PI), ON Semiconductor, STMicroelectronics (ST), Texas Instruments (TI), Fairchild, and Philips for their help and support.

We would be very glad if the book can be of use for your SMPS design.

Introduction

This book describes the optimal design of switching-mode power supply (SMPS) in a comprehensive, in-depth, and systematical manner. The book includes 10 chapters. Chapter 1 is the overview on SMPS. Chapter 2 describes the new technologies and applications of SMPS. Chapters 3 and 4 introduce the topology of DC/DC converter and the method of selecting key peripheral components of SMPS, respectively. Chapters 5–8 focus on the power factor correction circuit design of SMPS, the design of high-frequency transformer, the examples of SMPS optimization design, and the key design points of SMPS, respectively. Chapters 9 and 10 introduce the SMPS testing technology and the protection circuit design of SMPS, respectively. This book has important reference value for readers to research and develop new-type SMPS.

This book is rich in content and easy to understand with numerous illustration, delivering high practical value. It is suitable for various technicians in the electronic field, university teachers and students, and fans with a passion for electronics.

Introduction

1

Overview on Switching-Mode Power Supply (SMPS)

1.1 Classification of Integrated Regulated Power Supply

There are tens of thousands of integrated voltage regulators in the market, which can be roughly classified into linear regulators and switching regulators. See Table 1.1 for the classification and characteristics of integrated regulated power supply.

1.1.1 Optimal Design of SMPS

The linear regulator, also known as series regulated integrated regulator, is named for its internal regulating tube, which works in the linear working area and is in series connection with load. It has advantages such as sound voltage regulation performance, low-output ripple voltage, simple circuit, and low cost, while its main disadvantages include relatively bigger voltage drop and high power consumption of the regulating tube with relatively low efficiency of the regulated power supply at about 45%. The linear regulator mainly consists of two types, namely, standard linear regulator using NPN regulating tube, which is also known as NPN linear regulator, and PNP low-dropout (LDO) regulator using PNP regulating tube. Besides, there are quasi low-dropout (QLDO) regulator and very low-dropout (VLDO) regulator. According to the characteristics of output voltage, linear regulators can be divided into different types such as fixed output, adjustable output, positive pressure output, negative pressure output, and multiplexed-output (including tracking output). The efficiency of traditional standard linear regulators is only around 45%, while that of LDO and VLDO can reach 80–90% under low voltage output.

The switching-mode power supply (SMPS) is known as a highly energy-efficient power supply. It leads the development direction of regulated power supply and now has become the leading product of regulated power supply. With the internal key components working under high-frequency switch status, the SMPS consumes quite low energy so that its power efficiency may reach up to 70–90%, twice as high as that of the standard linear regulated power supply. The SMPS integrated circuit mainly consists of the following four types: pulse width modulator (PWM), pulse frequency modulator (PFM), switching regulator, and single-chip SMPS.

According to the circuit principle, regulators can be divided into three types including series regulated linear regulator, shunt regulated linear regulator, and switching regulator.

Optimal Design of Switching Power Supply, First Edition. Zhanyou Sha, Xiaojun Wang, Yanpeng Wang, and Hongtao Ma.

Table 1.1 Classification and Characteristics of Integrated Regulated Power Supply

Integrated regulated power supply	Linear power supply	Standard linear regulator	Fixed type	Three-terminal fixed type	Positive voltage output, negative voltage output
				Multiple-terminal fixed type	Positive voltage output, negative voltage output
			Adjustable type	Three-terminal adjustable type	Positive voltage output, negative voltage output, tracking mode
				Multiple-terminal adjustable type	Positive voltage output, negative voltage output, tracking mode
		Low dropout linear regulator	Low dropout regulator (LDO)	Three-terminal or multiple-terminal fixed/adjustable type	Positive voltage output, negative voltage output, tracking mode
			Quasi low dropout regulator (QLDO)		
			Very low dropout regulator (VLDO)		
	Switching mode power supply (SMPS)	Pulse width modulator (PWM)	With relatively low integration level and complicated peripheral circuit with constitutes high-power SMPS		
		Pulse frequency modulator (PFM)	With relatively low integration level, complicated peripheral circuit, switching frequency may reach over 1 MHz, and high efficiency		
		Switching regulator	With relatively high integration level and power switch tube inside, which needs to be equipped with industrial frequency transformer		
		Single-chip SMPS	With pretty high integration level and simple peripheral circuit, which constitutes SMPS of medium and small power		

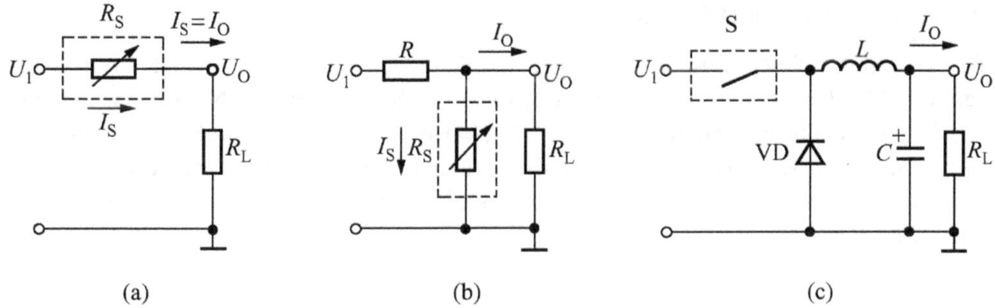

Figure 1.1 Equivalent circuits of the three kinds of regulators: (a) series regulated linear regulator; (b) shunt regulated linear regulator; and (c) switching regulator

Their equivalent circuits are respectively shown in Figure 1.1(a)–(c). In this figure, R_S refers to the equivalent resistance of regulating tube and S the power switch tube. As its output voltage is highly stable and output current is very small, shunt regulated linear regulator is generally used as reference voltage source. The main characteristics of various products are described in the following sections.

1.1.1.1 Three-Terminal Fixed Regulator

Fairchild Semiconductor Corporation firstly launched μ7800 and μA7900 series three-terminal fixed regulators in the beginning of the 1970s. It is a big revolution for integrated circuits of power supply, which greatly simplifies the design and application of power supply. The three-terminal fixed regulator can be placed in the circuits via the simplest way (such as a transistor) and has relatively complete overcurrent protection (OCP), overvoltage protection (OVP), and overheat protection (OTP) functions. At present, μA7800 and μA7900 series three-terminal fixed regulators have become universal ones in the world, with the widest application and largest sale volume. Such a three-terminal fixed regulator is easy to operate and needs no adjustment. It has a simple peripheral circuit and operates reliably and safely, and therefore is applicable to make general or nominal-output regulated power supply. However, it cannot regulate the output voltage or directly output non-nominal voltage, and its voltage is not stable enough.

1.1.1.2 Three-Terminal Adjustable Regulator

Three-terminal adjustable regulator was developed in the late 1970s and the early 1980s, which was the second-generation three-terminal regulator initiated by National Semiconductor (NSC). It not only reserves the advantage of simple structure the three-terminal fixed regulator has but also overcomes the disadvantage that the voltage could not be regulated. Moreover, its voltage stability is increased by one order of magnitude. Therefore, three-terminal adjustable regulator can be used to make laboratory power supply and DC regulated power supply. In addition, it can also be designed as fixed type to replace the three-terminal fixed regulator, to further improve the voltage regulation performance.

1.1.1.3 Low-Dropout (LDO) Regulator

LDO is a high-efficiency linear integrated regulator, whose input–output dropout voltage is about 500 mV, with power efficiency obviously higher than that of NPN linear regulator. VLDO is a new linear integrated regulator developed based on LDO in the beginning of the twenty-first century. VLDO uses the field-effect tube with very low specific on-resistance instead of the power tube that PNP uses, and its input–output dropout voltage can be as low as 45–150 mV.

1.1.1.4 Multiterminal Integrated Regulator

With plenty of pins, the multiterminal integrated regulator is flexible to use but with complicated connections. It can also be classified into fixed type and adjustable type.

1.1.1.5 Tracking Positive–Negative Balanced Output Integrated Regulator

Its characteristic is that when the positive voltage changes for some reasons, the negative voltage output may track automatically and make corresponding change to keep the absolute values of the two equal. The tracking function is particularly important for a precision operational amplifier powered by two supplies, which can prevent the operational amplifier from zero drift rising out of the unbalance between positive voltage and negative voltage.

1.1.1.6 Transformer SMPS without Power Frequency

The SMPS is also known as a low-loss power supply. As its internal parts operate under high-frequency switch status, the SMPS consumes low energy but with power efficiency twice that of ordinary linear regulated power supply.

1.1.1.7 Switching Regulator

A switching regulator is a switch integrated voltage regulator developed in the 1980s and 1990s. With PWM, power output, and protection circuit integrated on the same chip, the switching regulator has efficiency of over 90%. In addition, some switching regulators can even regulate output voltage continuously that can be used to SMPS with dozens to hundreds of wattage.

1.1.1.8 Single-Chip SMPS

With the main circuits (including MOSFET, required analog and digital circuits) of SMPS integrated on chip, the single-chip SMPS with the highest integration level can realize output isolation, PWM, and many other protection functions. A single-chip SMPS fits AC current of 85–265 V and 47–400 Hz via an input rectifier filter; it is thus an AC/DC power converter. The single-chip SMPS integrated circuit has displayed strong vitality since its appearance in the mid-1990s. It has such advantages as high integration level, high cost performance, simplest peripheral circuit, and best performance indicators. Now, single-chip SMPS integrated circuit has become an optimal integrated circuit for the development of medium- and small-power SMPS with wattage below 1000 W, precision SMPS, and SMPS modules. Besides, the development of single-chip SMPS has also created favorable conditions for the optimal design of SMPS.

1.1.1.9 Special SMPS

A special SMPS is characterized by "novelty, uniqueness, and wide application." It is novel in circuit, unique in function, and advanced in performance, with a great variety and wide application. Its varieties include constant voltage/current SMPS, LED driving power for lighting, constant power SMPS, high-voltage pulse power supply, high-power high voltage power supply, battery charger, and so on.

1.2 Characteristics of SMPS

1.2.1 Main Characteristics of SMPS

The SMPS is also called low-loss power supply. With its internal components working in a high-frequency switch status, it consumes low energy, and its power supply efficiency is twice that of the ordinary linear regulated power supply. The integrated circuits for SMPS are classified into two types: one is singled-ended or double-ended output PWM and the other is PFM, both of which can constitute the SMPS without power frequency transformer. Because they realize the voltage transformation and grid isolation with high-frequency transformer of very small volume, they can save power-frequency transformer of cumbersome volume. At present, the work frequency of SMPS has been increased from 20 kHz to hundreds of kilohertz and even above 1 MHz, so has the power efficiency. The output power range includes low power (dozens of wattage), medium power (hundreds of wattage), and high power (thousands of wattage). The disadvantage of SMPS is that the output voltage is not stable enough, and the output ripple is large and its noise is loud, making it inappropriate for making precise regulated power supply. However, it can be used as a pre-voltage regulator, with the standard linear regulator or low dropout linear regulator as the post-voltage regulator to constitute a high efficiency and precise regulated power supply of two stages. Such compound power supply possesses the advantages of SMPS and linear power supply.

Compared with the linear power regulator, although the SMPS is of complex design and some performance indicators are inferior to those of the linear regulator and the noise is loud, the advantages of SMPS mainly lie in the power efficiency, volume, weight, and so on. Especially when it constitutes a high-power regulated power supply, its volume is greatly reduced compared with that of the linear regulated power supply under the condition of the same output power, and the costs also decline significantly.

The efficiency of SMPS generally ranges from 70% to 85% with the maximum of 90%. Equipped with the post-positioned linear regulator and constituting a compound regulated

Table 1.2 Performance comparison of 20 kHz SMPS and linear regulated power supply

Parameter	SMPS	Linear regulated power supply
Power efficiency (%)	70–85	30–40
Output power per unit volume (W/cm^3)	0.12	0.03
Output power per unit mass (W/kg)	88	22
Voltage regulation rate (%)	0.1–1	0.02–0.1
Load regulation rate (%)	1–5	0.5–2
Output ripple voltage (mV, peak–peak value)	50	5
Output noise voltage (mV, peak–peak value)	50–200	Extremely small
Transient response time (μs)	1000	20
Holding time of output voltage after power failure (ms)	20–30	1–2

power supply, it still can achieve a high efficiency ranging from 60% to 65%, while the efficiency of most of the linear regulated power supplies (excluding low dropout linear regulators) only ranges from 30% to 40%. Compared with the linear regulator, the overall size of the traditional 20 kHz SMPS is only 1/4 of that of the linear regulator, and the 100–200 kHz SMPS is 1/8, while the size of new 200 kHz–1 MHz SMPS can be much smaller. After the power outrage, the SMPS can maintain the output voltage for a longer time than the linear regulator, because the latter is generally equipped with a low-voltage input filter capacitor, while the former is equipped with a high-voltage input filter capacitor with the withstand voltage ranging from 200 to 400 V and it saves more charge Q because of the direct proportion between Q and CU_1^2 (U_1 is DC high voltage).

The disadvantage of the SMPS is that it has relatively low-voltage regulation rate and load regulation rate, which takes a long time to respond to the transient state of load change; the output ripple and noise voltage are relatively high so that it is likely to exert electromagnetic interference externally.

1.2.2 Performance Comparison of SMPS and Linear Regulated Power Supply

See Table 1.2 for the performance comparison of 20 kHz SMPS and linear regulated power supply. It can be seen from the table that many technical indicators of the SMPS are superior to those of the linear regulated power supply.

1.3 New Development Trend of SMPS

The SMPS has been developed for decades. The self-excitation push-pull type transistor single transformer DC converter invented in 1955 took the lead in realizing the high-frequency conversion control function; and the one invented in 1957 and the SMPS design without power frequency transformer plan proposed in 1964 forcefully promoted the technological progress of the SMPS. The emerging of the PWM in 1977 and the single-chip SMPS in 1994 paved the way for promotion and popularization of the SMPS. Meanwhile, the frequency of SMPS has also increased from 20 kHz at the beginning to hundreds of thousands of hertz, and even a couple of megahertz. The SMPS is developed to be highly efficient, energy-saving, safe, environment-friendly, short, small, light, and thin. A variety of new technologies, processes, and apparatuses spring up like mushrooms and emerge continuously and the application of SMPS has also been increasingly popular. Next, introduction will be made on the new trend and new technology for the development of SMPS.

1.3.1 New Development Trend of SMPS

1.3.1.1 Green and Energy-Saving SMPS

Many famous integrated circuit manufacturers are making great efforts to develop low-power consumption and energy-saving SMPS integrated circuit. For example, Power Integrations (PI) of the United States adopted the energy-saving technology EcoSmart® to develop single-chip SMPS such as TOPSwitch-HX series. PI announced recently that due to the single-chip SMPS IC EcoSmart®, as it would save electric charge of about USD 3.4 billion for consumers all over the globe. The Green Chip such as TEA1520 series launched by Philips of the Netherlands also attaches great importance to high efficiency and energy saving functions. Besides, the international standards for green and energy-saving power supply have also been widely applied. For instance, the United States has established Energy Star Program in 1992 to reduce the no-load power consumption of the SMPS. The compulsory energy-saving standards established by California Energy Commission (CEC) have been implemented from July 1, 2006, requiring that the standby power consumption and no-load power consumption of electronic products must be reduced substantially. These standards cover all electronic products using external power adapter or charger, including mobile phone, household appliance, portable music player (MP3), handheld game player, electronic toys, and so on.

According to the fourth edition of power saving standard (Code of Conduct) newly published by the European Commission, which came into effect since January 1, 2009, specified that the no-load power consumption of ordinary power supply with rated output power of 0.3–50 W shall not be more than 0.30 W and that of ordinary power supply with rated power of 50–250 W shall not be more than 0.50 W. The new standard raised stricter requirements on the no-load power consumption of mobile phone power supply of 0.3–8.0 W, requiring it not to exceed 0.25 W during January 1, 2009 and December 31, 2010 and 0.15 W from January 1, 2011.

1.3.1.2 Intelligent Digital Power Supply and Programmable SMPS

Digital Power Supply

At present, the SMPS is developed to be intelligent and digitalized. The intelligent digital power system, which emerged at the beginning of the twenty-first century, has drawn great attention of the public for its excellent performance and advanced monitoring functions. The digital power supply, being intelligently adaptive and flexible, is able to directly monitor, process, and adapt to the system condition and meet any complex power requirement. In addition, it also guarantees the reliability of the long-term system operation through remote diagnosis, including fault management, overcurrent protection, and preventing the system from stop.

The promotion of digital power supply created a favorable condition for the optimal design of intelligent power system.

The digital power system has the following features:

1. It is an intelligent SMPS system with a digital signal processor (DSP) or micro controller unit (MCU) as the core and the digital power driver and PWM controller as the control object. The traditional SMPS controlled by MCU (including MCU μP and single-chip machine μC) generally controls only the switch-on and switch-off of the power supply, which is not a digital power supply in real sense.
2. Developed with fusion digital power technology, it realizes the optimal combination of analog element and digital element in the SMPS. For example, the analog element used for the power stage – MOSFET driver can be conveniently connected to the digital power controller and help to manage the power protection and biasing circuit. PWM controller also falls into the category of digital control analog chip.
3. With high integration, it realizes power system on chip, integrating a big number of separated components into a chip or a set of chips.
4. It can make full use of the advantages of DSP and MCU, making the digital power designed reach high technical indicators. For example, the resolution of its PWM can reach 150 ps (or 10^{-12} s), far exceeding the traditional SMPS. The digital power supply can also realize many functions such as multiple phase control, nonlinear control, load share, and fault prediction providing convenience for the research and manufacturing of green and energy-saving SMPS.
5. It provides convenience for building a distributed digital power system.

In March 2005, Texas Instruments (TI) of the United States announced to launch innovative digital power products and displayed the solution of Fusion Digital Power™ that includes the following three types of chips: UCD7K series digital power drivers, UCD8K series PWM controllers, and UCD9K series digital signal processors. Product series have been formed for the above- mentioned chips, supporting both the AC circuit and load power systems. They can be widely applied in telecommunication facilities, computer server, data center power system, and UPS.

Programmable SMPS

The adjustable SMPS changes the output voltage of regulator by manually regulating the resistance value, which is not only precise enough, but also inconvenient for application. Digital potentiometer, which is also called digitally controlled potentiometer (DCP), can replace the adjustable resistance to constitute a programmable SMPS under computer-based control.

The circuit design plan for the programmable SMPS constituted by digital potentiometer is shown in Figure 1.2. Figure 1.2(a) shows replacement of adjustable resistance with DCP, which works in the adjustable resistance model. The adjusted resistance value is R_{DCP}, which, together with R_1, constitutes a sampling circuit$_1$, which is sent to the feedback terminal FB of the switching regulator. The single-chip microcontroller can set the output voltage of adjustable switching regulator by changing the value of R_{DCP}. The second plan is to replace two sampling resistors with R_{DCP} simultaneously, which can save one resistance element, whose simplified circuit is shown in Figure 1.2(b) with other parts being the same with Figure 1.2(a). The third plan is to connect the DCP between R_1 and R_2 in series; the simplified circuit is

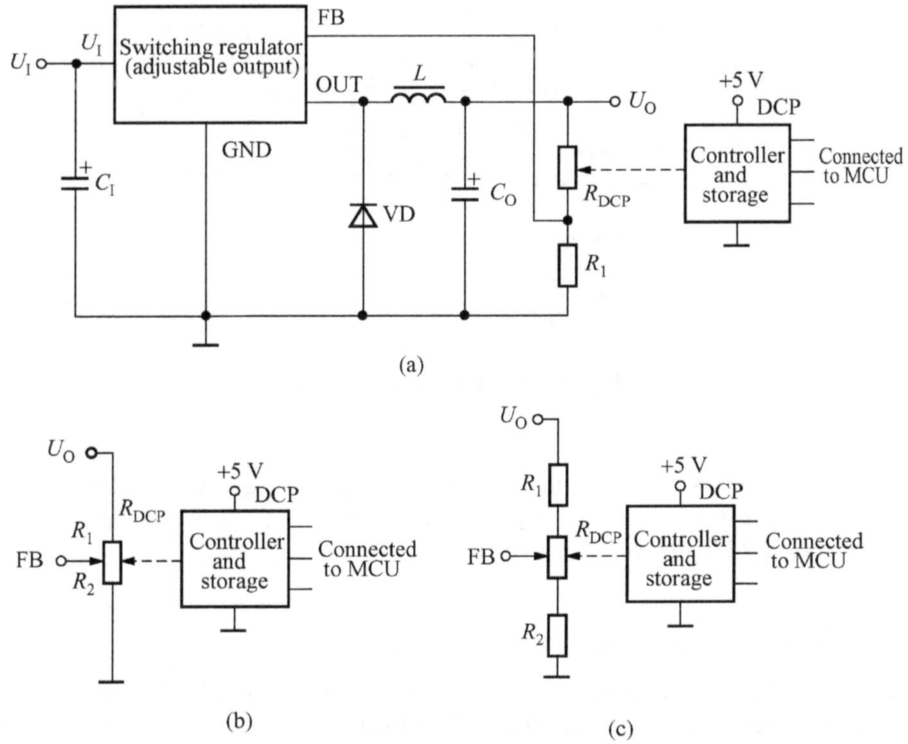

Figure 1.2 Circuit design plan for programmable SMPS constituted by DCP (a) Circuit 1; (b) Circuit 2 (simplified circuit); and (c) Circuit 3 (simplified circuit)

shown in Figure 1.2(c). This circuit is applicable to the fine regulation of output voltage in a small range.

1.3.2 New Technology in the SMPS Field

1.3.2.1 Active Clamp Technology

The function of clamp circuit is to clamp down the peak voltage generated by the SMPS at work time within a certain range in order to protect the power switch tube. The clamp circuits can be divided into passive clamp circuits and active clamp ones. The general R, C, and VDz clamp circuits belong to passive clamp circuits. The advantage of such passive clamp is that it has a simple circuit and can absorb the peak voltage generated by leakage inductance of high-frequency transformers. However, the clamp circuit has larger energy consumption itself, thus reducing the power supply efficiency.

The active clamp circuit invented by VICOR of the United States can significantly reduce power loss of the SMPS. The typical active clamp circuit is shown in Figure 1.3. The active clamp circuit was named for an active power component, MOSFET (V_4), which is used as clamper tube in the circuit. In Figure 1.3, C_c is the clamp capacitor and V_3 is the power switch

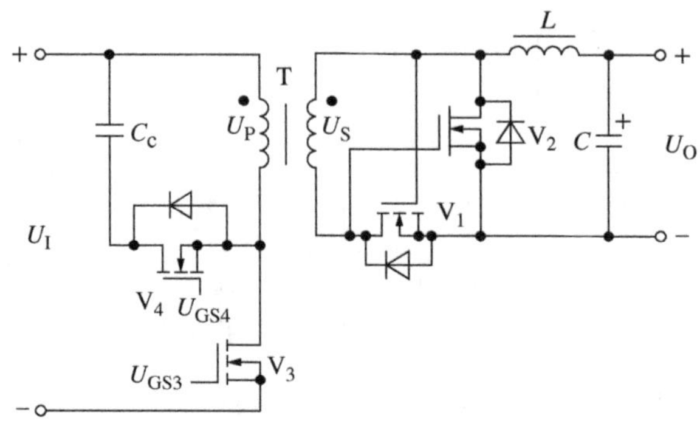

Figure 1.3 Active clamp circuit

tube of SMPS. It can be seen from this figure that U_{GS3} is 0 when V_4 is on, switching V_3 off and U_{GS3} will switch V_3 on when V_4 is off, thus clamping the peak voltage generated by leakage inductance of high-frequency transformers.

1.3.2.2 Synchronous Rectification (SR) Technology

Synchronous rectification (SR) was developed at the end of the twentieth century. It is a new technology of using the special power MOSFET with extremely low on-state resistance instead of rectifier diode to reduce the rectifier loss, which can significantly improve the efficiency of SMPS under low voltage and large current output. SR circuit adopts power MOSFET or Schottky Barrier Diode (SBD) as rectifier tube, requiring that the grid voltage should keep synchronous with the phase of rectified voltage to complete the rectification function. Therefore, it is called SR. For the synchronous rectifier comprising field effect transistor or SBD launched by Onsemi recently, the forward on-state voltage drop generally ranges from 0.2 to 0.4 V and the reverse recovery time is only 100 ns.

1.3.2.3 Soft Switching Technology

The general PWM-type SMPS adopts a "hard switching" technology, with which the voltage or current on VT is not zero when the power switch tube VT is either on or off. Nevertheless, the VT is forced to be on or off when the voltage or current is not zero, thereby increasing the switching loss. The switching loss includes the capacitor loss and the switch overlapping loss of the power switch tube. The capacitor loss, which is also called CU^2f loss, refers to a loss caused by discharging of the distributed capacitance of power switch tube when each switching cycle starts. The switch overlapping loss, caused by the switching time of the power switch tube, increases with the rising of switching frequency, which not only limits the development of high-frequency SMPS, but also easily generates electromagnetic interference.

Soft switching technology shall be introduced to make up the defects of "hard switching" technology. Soft switching refers to zero voltage switching (ZVS) or zero current switching (ZCS). With ZVS and ZCS, the power switch tube can be switched off when voltage and current crossing zero respectively, so as to minimize the switching loss for the purpose of improving the power supply efficiency and protecting the power switch tube.

1.3.2.4 Magnetic Amplifier Regulator Technology

A magnetic amplifier is composed of sampling circuit, reference voltage source, magnetic reset control circuit, controllable magnetic saturation inductor, and PWM. The controllable magnetic saturation inductor acts as a controllable magnetic switch in the voltage regulator circuit, which can accurately regulate the pulse width only by changing the delay time of magnetic reset to achieve accurate voltage regulation. Therefore, the magnetic amplifier is equivalent to an external PWM.

In the positive and negative voltage symmetrical output SMPS, using magnetic amplifier voltage regulator circuit can not only improve the precision of voltage regulation, but also increase the rate of cross loading regulation.

1.3.2.5 New Technologies for Single-Chip SMPS Application

With the increasing popularization of the single-chip SMPS, new technologies are also applied in the circuit design. The following examples are given for illustration.

1. StackFET™ (stack field effect transistor) technology. The single-chip SMPS is integrated with a power field effect transistor MOSFET with a drain-source electrode breakdown voltage of 700 V. When the maximum AC input voltage $U_{I\,(max)}$ is 580 V, the maximum primary voltage of high-frequency transformer reaches nearly 1050 V (including primary induced voltage U_{OR}, which is also called secondary reflected voltage), far above 700 V. To avoid any damage to the internal MOSFET, an MOSFET power field effect tube V can be stacked on its drain electrode as external MOSFET. These are the features of the StackFET circuit.
2. Design of industrial control power supply with ultrawide input range. To ensure that TinySwitch-III can work normally under ultralow AC input voltage, a floating high-voltage constant-current source shall be added exteriorly for the purpose of continuously supplying power to bypass end under low voltage. This technology is applicable for designing industrial control power supply with an ultrawide input range of 18–265 V.
3. PFC circuit. To improve the power factor of SMPS and reduce the total harmonic distortion (THD), AC/DC converter shall be provided with a power factor correction (PFC) circuit. Passive "valley fill circuit" (VFC) can be adopted when high-output ripple voltage (such as the driver composing white light LED lamp) is not required. VFC is used to greatly increase the conduction angle of rectifier diode, changing the input current from peak pulse into a wave form approaching sine wave via filling the valley points. The universal high-power SMPS is generally equipped with a new special chip with active PFC to simplify the circuit design.
4. Single-chip high-power SMPS. In recent years, the maximum output power of single-chip half-bridge LLC resonant converter with PFC, single-chip double-switch forward converter and other chips newly developed by chip manufacturers has reached 600–1000 W, creating favorable conditions for development of cheap high-power SMPS of high quality.
5. LED lighting driving power supply. LED lighting, also called semi-conductor lighting or solid state lighting, falls into the category of energy-saving and environment-friendly

"green lighting" characterized by low power consumption, high luminance, vibration resistance, long service life, small overall dimension, quick response, no pollution to environment, and other noticeable advantages. LED driving power supply is a power unit exclusive for power supply to LED lamps. At present, using new technologies and new processes, chip manufacturers have successively developed a batch of application-specific integrated circuit (ASIC) with advanced performance and unique characteristics. These chips have not only retained the advantages of SMPS chips, such as high efficiency and energy saving, but also are featured by constant current output, dimmable function, and PFC.

1.3.2.6 Highly Reliable Modular Design

It is well-known that integration technology cannot integrate the high-capacity capacitor, inductor, rectifier bridge, potentiometer, and high-power components of 10 A and above into a chip. Therefore, to develop an SMPS, a chip should be selected, the peripheral circuit should be designed, and the printed circuit should be designed as well, which brings inconvenience to users. However, the problems mentioned earlier can be readily solved if a power supply block is used.

The power supply block is a commodity component assembling power supply integrated circuit and miniature electronic components (such as pellet resistance and subminiature electrolytic capacitor) by microelectronic technique to complete a certain specific function. With the structural feature that all components are densely installed on a printed board, so the power supply block is also called secondary integration. The power supply blocks are generally divided into two categories by appearance: totally enclosed and non-removable ones and open ones.

Compared with the traditional whole machine, the whole machine composed of power supply block has the following prominent features: the circuit design can be greatly simplified so that the development cycle of new products can be shortened; with advanced technology and process, the qualified rate and reliability of the whole machine can be improved, and the one-time qualified rate is up to 100%; the volume and weight can be reduced; the machine is easy to install and maintain; and the use of totally enclosed power supply block can prevent forging, so as to protect the rights and interests of manufacturers.

At present, the power supply block is also developed to be intelligent. For example, the intelligent power supply block has achieved higher technical indexes (600 A, 600 V, with various protection functions and the failure self-detection and display function). The mean time between failures (MTBF) of single block has reached 10^7 h with the volume of the block gradually reduced. With the development of surface mount device (SMD) and surface mount technology (SMT), the volume of the power supply block will be further reduced while the performance indicators will be significantly improved.

1.3.2.7 Realization of SMPS Optimal Design Using Software

In recent years, with the development of power supply technology and the popularization of computer application, it has become a new technology in the international power supply field to design the SMPS using computer. At present, software has become the key technology for optimal design of the SMPS. The application of software can give full play to the advantages of high technology and greatly reduce the workload of designers, creating favorable conditions for optimal design of the SMPS.

Besides, computer simulation technology has also become a powerful weapon for developing new SMPS products. With the computer simulation technology, the prototype of SMPS can be created, so that the designers can solve potential technical problems when designing the SMPS before making a model machine, which greatly accelerates the research and development of new SMPS.

1.3.2.8 Anti-Electromagnetic Interference Capacity and Safety

With the increasing popularization of SMPS, higher requirements are put forward for its anti-electromagnetic interference capacity and safety, for which corresponding technical standards have been formulated. For example, the standards in relation to anti-electromagnetic interference include EN55022B, FCC Class B, CE Mark, and VCCI. The standards in relation to safety include IEC950, UL1950, CSA950, TUV-GS (EN60 950), and so on. China has implemented "3C" (China Compulsory Certification), the compulsory product certification, which is also called CCC certification concerning safety, since August 1, 2003.

1.4 Basic Principles of SMPS

1.4.1 Working Mode of SMPS

The SMPS has the following four working modes by control principles:

1. PWM: It is characterized by constant value of switching cycle and achieving voltage regulation by changing duty ratio through pulse width modulation with the core of the PWM.
2. PFM: It is characterized by constant value of pulse width and achieving voltage regulation by changing duty ratio through switching frequency modulation with the core of the PFM.
3. Pulse density modulation (PDM): It is characterized by constant value of pulse width and achieving voltage regulation by changing duty ratio through pulse count modulation. It adopts zero voltage technology and can significantly reduce the loss of power switch tube.
4. Hybrid modulation: It combines the modes referred to in points (1) and (2). Both the switching cycle and the pulse width are not fixed and adjustable. It includes PWM and PFM.

The four working modes discussed in the preceding text are collectively called as "time ratio control" (TRC), in which PWM is used most widely.

It should be noted that PWM can not only be used as an independent integrated circuit (such as UC3842 PWM), but also be integrated into a switching regulator (such as L4960 switching regulator integrated circuit) or an SMPS (such as TOP262E single-chip SMPS integrated circuit), in which the switching regulator belongs to the DC–DC converter and the SMPS is generally an AC–DC converter.

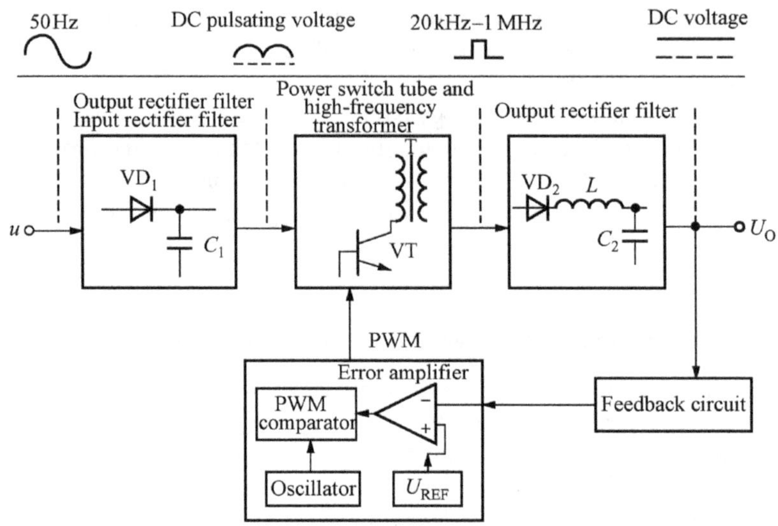

Figure 1.4 Basic composition of the SMPS

1.4.2 Basic Principles of PWM

The circuit of SMPS consisting of five components is relatively complex, which is shown in Figure 1.4. These five components refer to input rectifier filter, including the circuit from AC to input rectifier filter; power switch tube (VT) and high-frequency transformer (T); and control circuit (PWM), including oscillator, reference voltage source (U_{REF}), error amplifier, and PWM comparator. The control circuit can produce PWM signal and its duty ratio is controlled by feedback circuit, output rectifier filter, and feedback circuit. In addition, biasing circuit and protective circuit shall be added. PWM is the core of SMPS.

The operating principle of PWM SMPS is shown in Figure 1.5. u, AC of 220 V, changes into DC voltage U_1 after passing through rectifier filter circuit, then turn into high-frequency square-wave voltage after wave chopping by power switch tube and voltage reduction by high-frequency transformer T, and finally the required DC output voltage V_o through rectifier filter. PWM can produce driving signals with fixed frequency and adjustable pulse width to control the on-off state of power switch tube, so as to adjust the output voltage to achieve voltage regulation.

The sawtooth generator is to provide clock signal, and sampling resistance, error amplifier, and PWM comparator form a closed-ring regulating system. Having been taken as sample by R_1 and R_2, the output voltage U_o is sent to the inverting input terminal of error amplifier to be compared with the reference voltage U_{REF} on the non-inverting input terminal to get the error voltage U_r; then the PWO by PWM comparator is controlled by the amplitude of U_r and finally U_o can remain unchanged by power amplification and buck output circuit. U_J is the output signal of saw-tooth generator.

It should be noted that though the sampling voltage is generally connected to the inverting input terminal of error amplifier, it may also be connected to the non-inverting input terminal, which is related to the polarity of saw-tooth voltage input at the other end of the error amplifier. Generally, the sampling voltage is connected to the inverting input terminal when the input

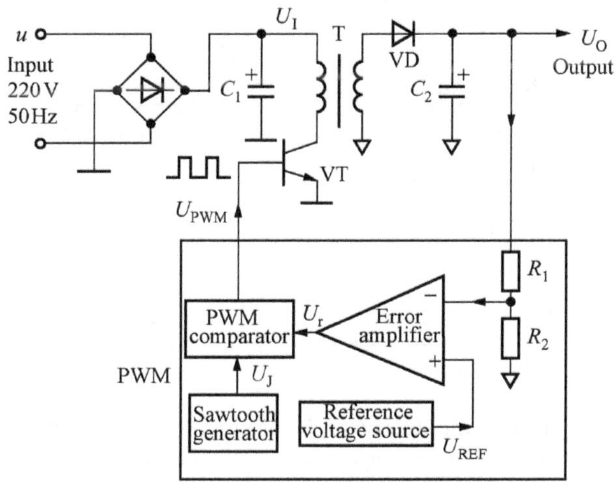

Figure 1.5 Operating principle of PWM SMPS

saw-tooth voltage is of positive polarity and to the non-inverting input terminal when the input saw-tooth voltage is of negative polarity (the same follows).

Given that DC input voltage is U_I, the efficiency of switching regulator is η and the duty ratio is *D*, then the impulse amplitude of power switch tube is $U_P = ⊠ U_I$. The following formula can be worked out:

$$U_o = \eta D U_I \tag{1.1}$$

This indicates that when η and U_I are constant, U_o can be modulated automatically only by changing the duty ratio. When U_o increases because of some reason, $U_r \downarrow \to D \downarrow \to U_O \downarrow$ and when U_o decreases, $U_r \uparrow \to D \uparrow \to U_O \uparrow$

This is the principle of automatic voltage regulation. The wave form of automatic voltage regulation process is shown in Figure 1.6(a) and (b), in which U_J refers to the output voltage of sawtooth generator, U_r is the error voltage, and U_{PWM} is the output voltage of PWM comparator. It can be seen that when U_o decreases, $U_r \uparrow \to D \uparrow \to U_O \uparrow$ and when U_o increases for some reason, $U_r \downarrow \to D \downarrow \to U_O \downarrow$.

1.4.3 Classification of PWM Products

SMPS usually adopts PWM. There are hundreds of PWM integrated circuits. The classification of typical products is shown in Table 1.3. It should be noted that PWM can be divided into doubled-ended output and single-ended output. The former is of push-pull output type that can be used in high-power SMPS from hundreds to thousands of Watt, and the latter, with a simple peripheral circuit, that can be used to make SMPS with medium and low power ranging from dozens to hundreds of wattage. The higher the switching frequency is, the higher the frequency of SMPS is and the smaller its volume is. Generally, the SMPS with switching frequency up to 1MHz is called high-speed SMPS. The products whose models are listed with slashes in Table 1.3 are series products. Take UC1840/2840/3840 for example. The internal circuits and main performance indicators of these products are the same and they only vary in the range of

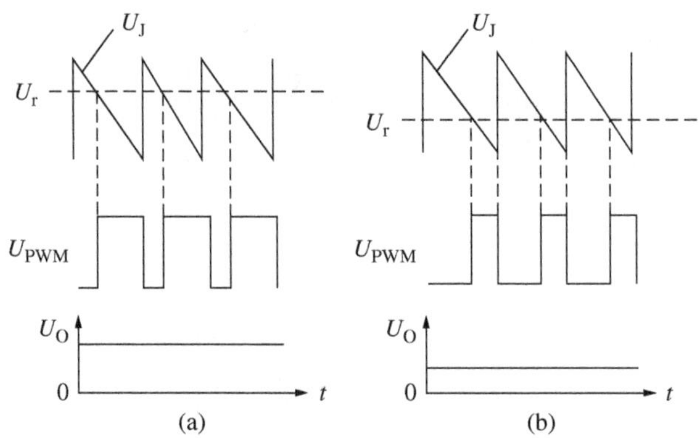

Figure 1.6 Oscillogram of automatic voltage regulation process. (a) The output voltage increases along with the rising of error voltage. (b) The output voltage decreases along with the dropping of error voltage

operating temperature. They respectively fall into Class I Military (−55 to +125 °C), Class II Industrial (−40 to +85 °C), and Class III Civil (0 to +70 °C).

1.5 Control Mode Type of SMPS

SMPS has two types of control modes namely voltage mode control (VMC) and current mode control (CMC) with respective pros and cons. It is hard to say which one is optimal to all applications and they shall be chosen according to actual conditions.

1.5.1 VMC-Type SMPS

VMC is the most common control mode of switching power supply. Take step-down switching regulator (or buck converter), for example. The basic principles and working waveforms of VMC-type buck converters are illustrated respectively in Figure 1.7(a) and (b). The characteristics of VMC-type SMPS are as follows: Firstly, the output voltage is sampled (a sampling resistance divider can be added if necessary), and the sampling voltage U_Q then acts as the input signal of control loop. After that, the sampling voltage U_Q will be compared with the reference voltage U_{REF}, and the comparison result is amplified into the error voltage U_r, which finally is sent to PWM comparator for comparison with sawtooth wave voltage U_J so as to obtain the modulating signal whose pulse width is directly proportional to the error voltage. The oscillator in Figure 1.7 has two channels for output signals: clock signal (in square wave or rectangular wave) and sawtooth signal. C_T refers to the timing capacitor of sawtooth oscillator, T refers to high-frequency transformer, and VT refers to power switch tube. The buck output circuit is composed of rectifier tube VD_1, freewheel diode VD_2, power inductor L, and filter capacitor C_o. R in PWM latch is the reset terminal, S is the set terminal, and Q is the output end of the latch. See Figure 1.7(b) for the output waveform.

The pros of VMC-type SMPS are as follows:

Table 1.3 Classification of typical PWM products

Features	Foreign model	Maximum switching frequency f_{max} (Hz)	Maximum output peak current I_{PM} (A)	Domestic model	Assembling form
Double-ended output, medium speed	MC3520 UC3520	100k	0.1×2	CW3520	DIP-16
	SG3525A	500k	0.4×2	CW3525A	DIP-16
	TL494 UC494A	300k	0.2×2	CW494	DIP-16
Single-ended output, medium speed	UC1840/2840/3840	500k	0.4	CW1840/ 2840/3840	DIP-18
	UC1842/2842/3842	500k	1	CW1842/2842/3842	DIP-8
	UC1841/2841/3841	500k	1		DIP-18
	TEA2018	500k	0.5	CW2018	DIP-8
	μPC1094	500k	1.2		DIP-14
Single-ended output, high speed	UC1823/2823/3823	1M	1.5		DIP-16
	UC1825/2825/3825	1M	1.5		DIP-16
	UC1848/2848/3848	1M	2		DIP-16

UC is developed by Unitrode, which has been incorporated into TI. SG and TL are products of TI, and TEA is a product of ST.

1. It is a closed-loop control system with only one voltage feedback loop (or voltage control loop) and simple circuit design.
2. It can work stably in the process of modulation.
3. With low output impedance, it allows power to be supplied through multiple channels to a load.

The cons of VMC-type SMPS are as follows:

1. Laggard response. Although the current sense resistor R_S is used in the VMC-type circuit, R_S is not connected to the control loop. As a result, when the input voltage changes, the pulse width can be modulated only after the output voltage changes as well. Owing to lag time in the filter circuit, the changes of output voltage can only be seen after multiple cycles. Therefore, the response time of VMC-type SMPS is laggard, which by some degree affects the stability of output voltage.
2. An overcurrent protection circuit shall be additionally designed.
3. Phase compensation of the control loop is quite complex, and closed loop gain will change with input voltage.

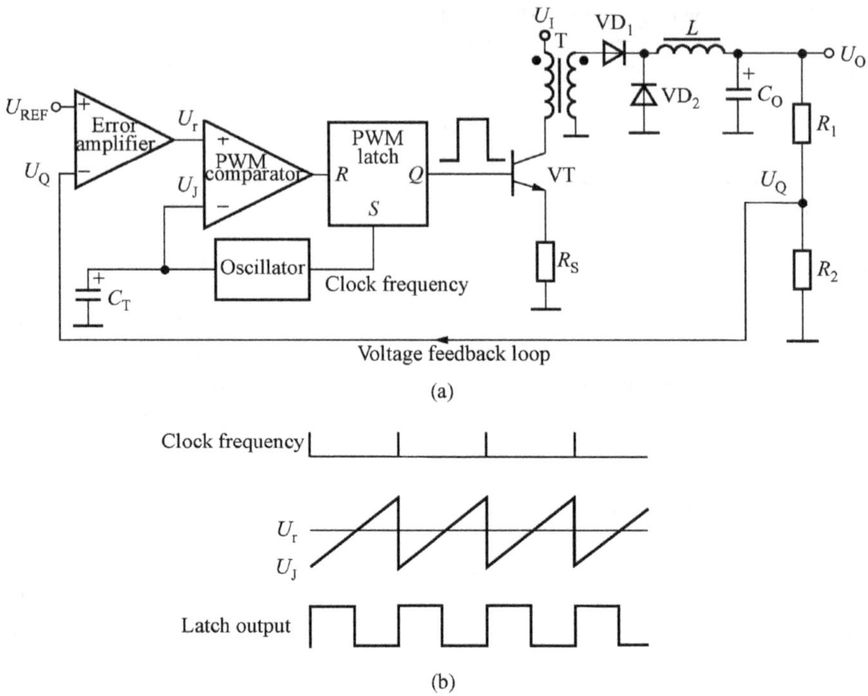

Figure 1.7 (a) Basic principles and (b) working waveforms of VMC-type SMPS

1.5.2 CMC-Type SMPS

CMC-type SMPS contains not only a voltage control loop but also a current control loop. Its basic principles and working waveforms are shown in Figure 1.8(a) and (b). The voltage drop of current sense resistor U_S, and PWM comparator also serves as a current sense comparator.

CMC-type SMPS detects the switching current in the power switch tube by sense resistor and limits the current cyclically, which helps to realize overcurrent protection. Fixed-frequency clock pulse will set PWM latch, driving signals output from Q are high-level signals that will make power switch tube VT-conductive and the primary side current of the high-frequency transformer increase linearly. When the voltage drop on current sense resistor R_s reaches and exceeds U_r, the current sense comparator will turn over, and the output high-level signals will set the latch, turning signals output from Q into low-level ones, which will cut off the power switch tube until the next clock pulse has the PWM latch set.

The pros of CMC-type SMPS are as follows:

1. It is a double closed-loop control system, of which the external loop is composed of voltage feedback circuits while the internal loop is composed of current feedback circuits, which are under the control of voltage feedback circuits. Compared with voltage feedback circuits, the gain bandwidth of current feedback circuits is larger.
2. It can respond to transient changes of input voltage quickly. This means when the input voltage changes, it can rapidly adjust the output voltage to the stable value, because changes of input voltage will lead to changes of the primary side inductive current, which in turn,

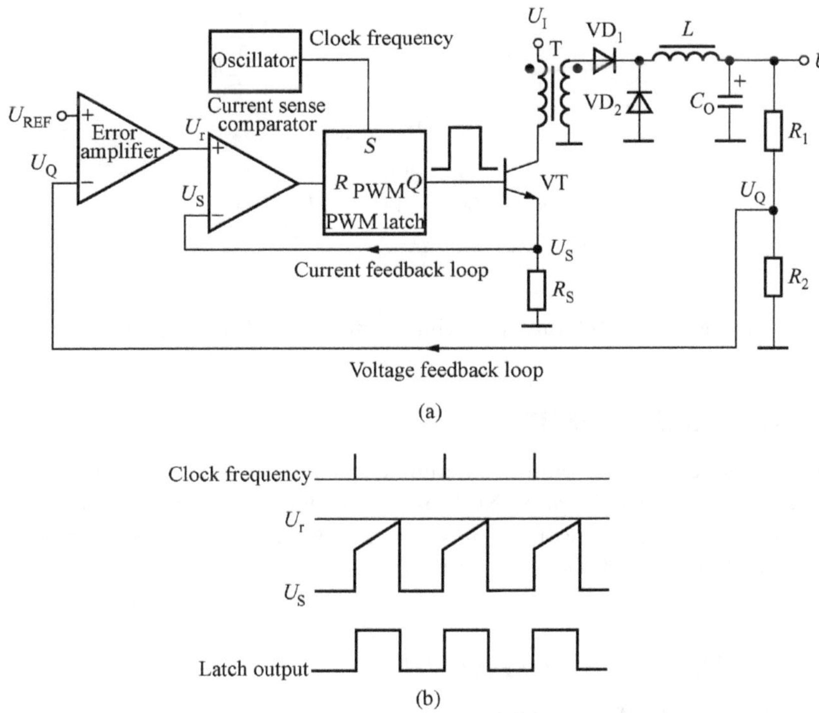

Figure 1.8 (a) Basic principles and (b) working waveforms of CMC-type SMPS

results in changes of U_s. This means that it can change the output pulse's duty ratio directly through the current sense comparator rather than the error amplifier.

3. The joint control of voltage control loop and current control loop may increase the voltage regulation factor.
4. It can simplify the design of error amplifier compensating network.
5. PWM comparator will cut off the power switch tube and maintain output voltage stable as long as the current pulse reaches the set threshold.
6. It is equipped with a limiting current protection circuit. Therefore, the limiting current threshold can be set precisely and simply by changing the value of R_s.

The cons of CMC-type SMPS are as follows:

1. Two control loops make it difficult to design and analyze circuits.
2. A duty ratio of over 50% may lead to instability of the control loops, in which case, a slope compensation circuit needs to be added.
3. It has a poor ability to restrain noises. As the primary side inductor works in the continuous energy storage mode, the rising slope of switching current signal is small. A relatively low noise overlaid to the current signal is likely to cause false operation of the PWM controller, for which a noise suppression circuit will be required.

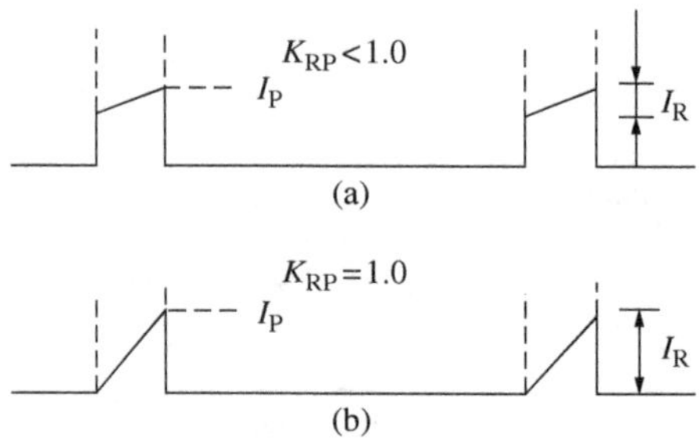

Figure 1.9 Waveforms of switching currents: (a) CUM and (b) DUM

1.6 Working Mode of SMPS

SMPS has two basic working modes: continuous mode (CUM) and discontinuous mode (DUM). Next, taking the single-chip SMPS of TOPSwitch series for example, we introduce the setting methods of the two working modes and then make comparison on power losses under these two working modes so as to derive the conclusion.

1.6.1 Setting Methods of CUM and DUM

1.6.1.1 Characteristics of CUM and DUM

In CUM, the high-frequency transformer starts to work from a non-zero energy storage state in every switching cycle, while in DUM, all energy stored in the high-frequency transformer should be released. Their difference can be seen in Figure 1.9. The switching current in CUM starts from a certain level, goes up to the peak along the slope, and then back to zero rapidly. At this time, the scale factor K_{RP} between primary winding pulse current I_R and peak current I_P is less than 1.0, or

$$I_R = K_{RP} I_P < I_P \tag{1.2}$$

The switching current in DUM starts from zero, goes up to the peak, and then back to zero rapidly. At this time,

$$K_{RP} = 1.0,$$

or

$$I_R = I_P \tag{1.3}$$

1.6.1.2 Working Mode Setting

Using the proportional relationship between I_R and I_P (or the value of K_{RP}), the working modes of SMPS can be quantitatively described. The value range of K_{RP} is 0–0.1. When $I_R = I_P$ and $K_{RP} = 1.0$, the SMPS is set in DUM. When $I_R < I_P$ or $K_{RP} < 1.0$, the SMPS is set in DUM. To be specific, there are two situations: (i) when $0 < I_R < I_P$, or $0 < K_{RP} < 1.0$, it is in CUM; and (ii) ideally, when $I_R = 0$ and $K_{RP} = 0$, it is in absolute CUM, which also can be called extreme CUM. Then the primary winding inductance $L_{p\to}\infty$, and the switching current on primary side appears as a rectangular wave.

As a matter of fact, there is a transition instead of a strict boundary between absolute CUM and DUM. For a given AC input range, a smaller K_{RP} value means a more continuous working mode, larger primary winding inductance, and smaller primary side I_P value and I_{RMS} value. At this time, a lower power TOPSwitch and a larger high-frequency transformer can be used to optimize the design. Otherwise, a larger K_{RP} value means poor continuity and smaller primary winding inductance but relatively larger I_P and primary side RMS current I_{RMS}. At this time, a large power TOPSwitch and a small high-frequency transformer shall be used.

In conclusion, the working mode of SMPS can be set by selection of K_{RP} value. The setting process is $L_p \uparrow\rightarrow \ (I_R < I_P) \rightarrow (K_{RP} < 1.0) \rightarrow$ CUM.

$K_{RP} = 0.4$–1.0 is the best for 100 V/115 V AC power supplies. $K_{RP} = 0.6-1.0$ is preferred for 85–265 V wide-range input or 230 V fixed-input AC power supplies.

1.6.2 Power Consumption Comparison between These Two Working Modes

The following two design cases can illustrate the changes in values of I_P and I_{RMS} corresponding to $K_{RP} = 1.0$ (DUM) and $K_{RP} = 0.4$(CUM) in the wide range input of 85–265 V so that comparison can be made on TOPSwitch power consumptions under these two modes.

1.6.2.1 DUM Design Case

The given working parameters are $K_{RP} = 1.0$, $U_{Imin} = 90\,V$, $D_{max} = 60\%$, $P_O = 30$ W, power efficiency $\eta = 80\%$ and that the primary winding peak current I_P can be expressed either as the function of I_R and K_{RP} or the function of the basic parameters (output power P_O, minimum DC input voltage U_{Imin}, maximum duty ratio D_{max}, and power efficiency η) and I_R; the equations are as follows:

$$I_P = I_R/K_{RP} \tag{1.4}$$

$$I_P = \frac{P_O}{U_{Imin}D_{max}} + \frac{I_R}{2} \tag{1.5}$$

Convert Equation (1.4) to $I_R = K_{RP}I_P$, and put it into Equation (1.5), to calculate the value of I_P:

$$I_P = \frac{2P_O}{U_{Imin}D_{max}\eta(2 - k_{RP})} \tag{1.6}$$

The final RMS current I_{RMS} of primary winding is

$$I_{RMS} = I_P\sqrt{D_{max}\left(\frac{K_{RP}^2}{3} - K_{RP} + 1\right)} \tag{1.7}$$

Put $U_{Imin} = 90V$, $D_{max} = 60\%$, $\eta = 80\%$, $P_O = 30W$, and $K_{RP} = 1.0$ into Equation (1.6) and we will obtain $I_p = 1.39$ A. Put the result into Equation (1.7) to obtain $I_{RMS} = 1.39$ $\sqrt{0.6 \times \left(\frac{1}{3} - 1 + 1\right)} = 0.62$ (A)

1.6.2.2 CUM Design Case

The given working parameters are $K_{RP} = 0.4$, $U_{Imin} = 90V$, $D_{max} = 60\%$, $P_O = 30W$, $\eta = 80\%$. Different from the first case, K_{RP} here is 0.4, which indicates that the working mode is more continuous. Similarly, it can be worked out that $I'_P = 0.87$ A and $I'_{RMS} = 0.54$ A.

It is easy to work out that the peak current in CUM is only 63% of that in DUM, whereas the RMS current in CUM is 87% of that in DUM. Therefore, for the given TOPSwitch chip, the power consumption ratio of the two working modes is

$$\frac{P'_O}{P_O} = \frac{(I'_{RMS})^2 R_L}{(I_{RMS})^2 R_L} = (87\%)^2 = 75.7\%$$

This indicates that, 24.3% of power consumption can be reduced in CUM than in DUM. In other words, under the same output power, CUM allows the use of low-power TOPSwitch, or allows TOPSwitch to work with low power consumption. Besides, when an SMPS is designed in CUM, the AC component on primary side circuit is lower than that of DUM. In addition, it can reduce the skin effect and the power consumption of high-frequency transformers.

1.7 Feedback Type of SMPS

1.7.1 Basic Types of SMPS Feedback Circuit

Consider the Single-Chip SMPS of TOPSwitch series, for example. The SMPS feedback circuits can be grouped into four basic types: (i) basic feedback circuit; (ii) improved basic feedback circuit; (iii) optical coupling feedback circuit with a voltage-regulator tube; and (iv) optical coupling feedback circuit with TL431, whose simplified circuit diagrams are illustrated in Figure 1.10(a)–(d).

Figure 1.10(a) shows a basic feedback circuit. Its advantages include simple circuit, low cost, and applicability to small-scale and economic SMPS and disadvantages include poor voltage regulation performance, voltage regulation $S_V = \pm 1.5\%$ to $\pm 2.5\%$, and load regulation $S_{I\approx} \pm 5\%$.

Figure 1.10(b) shows an improved feedback circuit. Equipped with a voltage-regulator tube VD_{Z2} and a resistance R_1 additionally, the load regulation can reach ±2.5%. The stable voltage of VD_{Z2} is 22 V generally. The number of turns of the bias winding shall be increased correspondingly for a comparatively higher bias voltage U_B to satisfy the requirement of circuit.

Figure 1.10(c) shows an optical coupling feedback circuit with a voltage-regulator tube. The reference voltage U_Z is provided by VD_{Z2}. LED inside the optical coupler may get an error voltage when the output voltage U_O fluctuates. Therefore, this circuit is equivalent to adding an external error amplifier to TOPSwitch. U_O, which can be regulated using both external and internal error amplifiers. This kind of feedback circuit can make the voltage regulation factor lower below ±1%.

Figure 1.10(d) shows an optical coupling feedback circuit with TL431 with a relatively complicated circuit but the best voltage regulation performance. Here an adjustable precision shunt regulator of TL431 type instead of a common regulator is used to form the external error amplifier so that the precision adjustment to U_O can be realized, and the voltage regulation factor and load regulation factor of single-output SMPS can reach $\pm 0.2\%$ and $\pm 0.5\%$ respectively, which can be compared with the linear regulated power supply. This kind of feedback circuit is suitable to constitute a precision SMPS.

In the design of single-chip SMPS, a proper feedback circuit shall be chosen according to actual conditions so that the specified technical indicators can be reached.

1.7.2 Feedback Principle of the Single-Chip SMPS

Taking the basic feedback circuit of TOPSwitch for example, an in-depth analysis is carried out on the feedback principles of CUM and DUM. It should noted that such feedback principle analysis only discusses the interaction between the primary winding and the output circuit, which is different from that of the control circuit constituted by the bias winding and the external circuit. The control circuit is especially used to adjust the duty ratio. Therefore, the discussion in the following text does not relate to the bias winding.

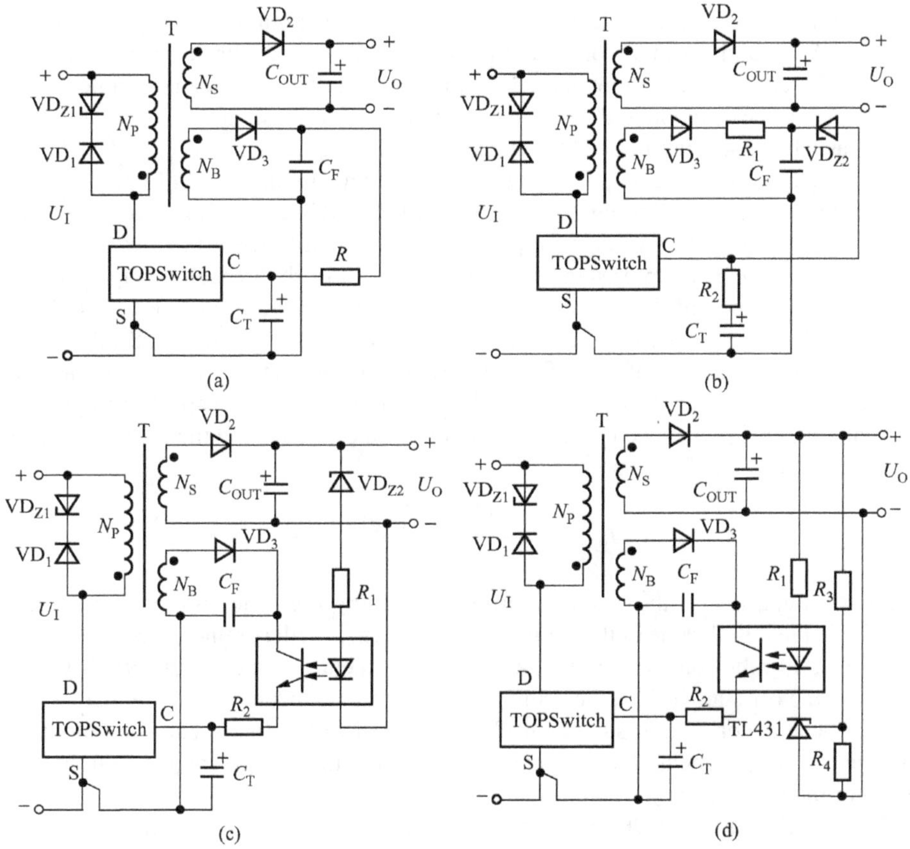

Figure 1.10 Four basic types of feedback circuit: (a) basic feedback circuit; (b) improved basic feedback circuit; (c) optical coupling feedback circuit with a voltage-regulator tube; and (d) optical coupling feedback circuit with TL431

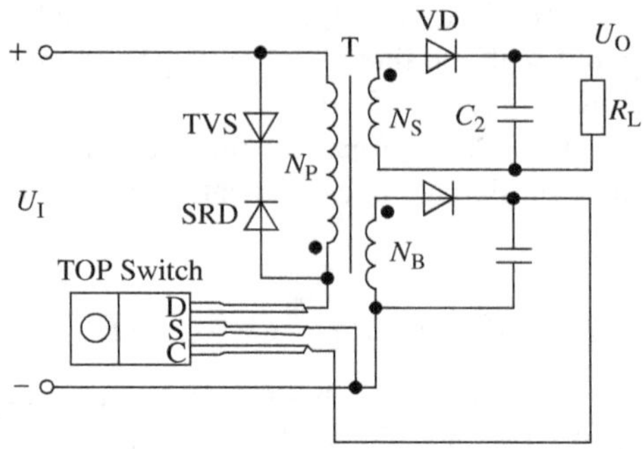

Figure 1.11 Basic feedback circuit of TOPSwitch

1.7.2.1 Basic Feedback Process

Single-chip SMPS of TopSwitch series can be regarded as a single-chip combinational device, which combines the MOSFET and all the needed analog and digital circuits, to perform output insulation, PWM, and various protection functions. The basic feedback circuit of TOPSwitch is shown in Figure 1.11. Being moderately adjusted, the circuit will realize single or multiple output, boost or buck output, and positive or negative voltage output.

In the basic feedback circuit of TOPSwitch, the high-frequency transformer has three major functions of energy storage, output insulation, and voltage regulation. In Figure 1.11 N_P, N_S, and N_B represent primary winding, secondary winding, and bias winding, and respective number of windings. A transient voltage suppressor (TVS) and a super fast recovery diode (SRD) form a drain electrode clamp protection circuit that can absorb the peak voltage generated by the leakage inductance of the primary winding, keeping the drain voltage of MOSFET in a safe range. VD is the output rectifier tube; C_2, the output filter capacitor; RL, the load resistance; and U_O, the output voltage. AC input circuit and the rectifier filter circuit are omitted in Figure 1.11. AC passes through the rectifier bridge and the filter capacitor to produce DC input high voltage U_I when TOPSwitch is on; VD is in off-state, and the primary side current rises along a ramp. The formula is

$$I_{PRI} = I_I + \frac{(U_I - U_{DS(ON)})t_{ON}}{L_P} \tag{1.8}$$

In this formula, I_{PRI} is the primary current containing the peak current I_P and the ripple current I_R; I_I, the initial value of the primary current; $U_{DS\,(ON)}$, the drain-source on-state voltage of MOSFET; t_{ON}, the conducting time; and L_P, the inductance value of the primary winding.

Owing to the off-state of VD, the primary side is insulated from the output load. Therefore, the electric energy originally stored on C_2 is supplied to the load, with the output voltage unchanged. At this moment, the electric energy is stored in the high-frequency transformer in the form of magnetic energy.

During the off-state of TOPSwitch, the magnetic flux in the high-frequency transformer starts to decrease and the polarity of the induced voltage of the secondary winding changes, making the VD connected because of forward bias. As a result, the energy stored in the high-frequency transformer is transferred to the output circuit to power up R_L, and to recharge C_2. The secondary current starts to attenuate from the initial value according to the following

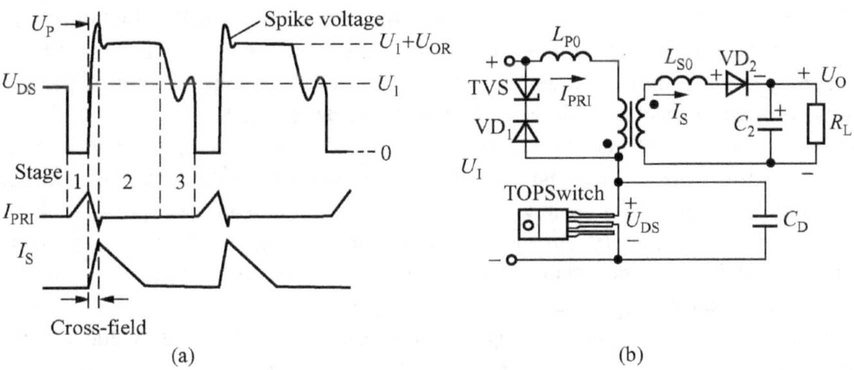

Figure 1.12 Feedback principle of the discontinuous conduction mode in actual situations. (a) Working waveform; and (b) circuit principle

formula:

$$I_S = \frac{I_{PNP}}{N_S} - \frac{(U_O + U_{F1})t_{OFF}}{L_P} \cdot \frac{N_P}{N_S} \geq 0 \tag{1.9}$$

In this formula, I_S is the secondary current; $I_P N_P/N_S$, the initial value of the secondary current; I_P, the peak value before the conduction of TOPSwitch ends; U_{F1}, the forward voltage drop of the output rectifier tube VD; and t_{OFF}, the time period when TOPSwitch is off. During the off-state of TOPSwitch, the output current is provided by C_2 if the secondary current I_S wanes to zero.

TOPSwitch has two working modes, which is decided by the final value of I_S. If I_S wanes to zero during the off-state, the TOPSwitch is working in the DM. Otherwise, it is working in the CM.

1.7.2.2 Feedback Principles of the Two Working Modes in Actual Situations

Ideally, the influence of parasitic elements (including distributed capacitance and leakage inductance) in the feedback circuit can be ignored. However, in actual situations, the impact of distributed capacitance and leakage inductance should be considered; therefore, there exist spike voltage and spike current in the working waveform.

Feedback Principle of the Discontinuous Conduction Mode in Actual Situations

Working waveform and simplified circuit principle of the discontinuous conduction mode in actual situations are illustrated in Figure 1.12(a) and (b) respectively. Figure 1.12(b) indicates that in the discontinuous conduction mode, the period of each switch consists of three stages. In addition, there are three parasitic elements in this actual circuit, namely leakage inductance of the primary winding L_{PO}; leakage inductance of the secondary winding L_{SO}; and distributed capacitance C_D, which is the sum of output capacitance C_{OSS} of TOPSwitch; and distributed capacitance C_{XT} of the primary winding of the high-frequency transformer, or $C_D = C_{OSS} + C_{XT}$. Here we discuss exclusively about the impact of those parasitic elements in the circuit.

In stage 1, C_D would discharge as long as TOPSwitch is conducted. Energy E_D stored on C_D at the end of the previous cycle would be unleashed at the initial stage. As E_D is in direct proportion to $U^2{}_{CD}$, the power efficiency would be significantly declined when C_D has a big volume, and it is even more true when U_I is rather high. It should be noted that because the high-frequency transformer at stage 1 is gathering energy and the current of the secondary winding is zero, the impact of leakage inductance can be ignored.

In stage 2, TOPSwitch is turned off. At the previous stage, energy stored in the high-frequency transformer is transmitted to the secondary winding. At this moment, both leakage inductance L_{PO} and L_{SO} are preventing the current from changing. To be specific, L_{PO} is preventing the primary current I_{PRI} from decreasing, while L_{SO} is preventing the secondary current I_S from increasing. Therefore, a cross-field is formed when I_{PRI} decreases and I_S increases. Finally, I_{PRI} reduces to zero along the diagonal, determined by the leakage inductance L_{PO} and the primary voltage; and I_S climbs to the peak value I_{SP} along the diagonal, determined by the leakage L_{SO} and the secondary voltage. The most important issue is that the primary current in the cross-field must be continued without interruption. The attenuated primary current would charge C_D to U_P when it flows through C_D. The peak voltage U_P generated by leakage inductance L_{PO} would overlay on the waveform of U_{DS}, forming a peak voltage of leakage inductance, which is also called drain-source peak value pulses. The relation is shown herewith:

$$U_{DS} \approx U_I + U_{OR} + U_P \tag{1.10}$$

The actual circuit, drain clamp is usually adopted to protect the circuit, so it is feasible to clamp U_{DS} down below the rated leakage-source breakdown voltage (700 V in general) of TOPSwitch to prevent the chip from damage resulting from increase of U_{DS} caused by U_P.

In stage 3, induced voltage U_{OR} (also called secondary reflected voltage) wanes to zero. Then the high-frequency transformer unleashes all the energy stored at stage 1, reducing the leakage-source voltage from $U_{DS} = U_I + U_{OR}$ at the end of stage 2 to $U_{DS} \approx U_I$. However, this voltage change generates damped oscillation wave that was overlaid on waveform U_{DS} by stimulating the resonance circuit constituted by stray capacitance and the primary inductance, and does not stop oscillating until the TOPSwitch is on power again. Therefore, there are valleys and peaks in waveform U_{DS} at stage 3. Obviously, this damped oscillation wave "modulate" the voltage and energy on C_D and determines the power consumption of transformation in the next cycle of switch.

Feedback Principle of the Continuous Conduction Mode in Actual Situations

There exist the same parasitic elements as those in discontinuous conduction mode in the feedback circuit of continuous current mode in actual situations. Besides, the actual characteristics of the output circuit shall be taken into consideration as well. An ideal rectifier tube needs no time for forward voltage drop and reverse recovery. Time is needed for reverse recovery of junction rectifier tube because there are a few charge carriers passing through the nodes of diode, while that of Schottky diode is brought about by junction capacitance. For single-chip SMPS, Schottky diode that requires a very short time for reserve recovery or super fast recovery diode is recommended to be the output rectifier. Nevertheless, no ordinary low-speed rectifier should be used, for it would not only increase the high-frequency energy consumption or decrease the energy efficiency, but also cause the thermal breakdown of the rectifier.

Working waveform of the continuous conduction mode in actual situations is shown in Figure 1.13. At stage 1, there is still current running at the secondary side when TOPSwitch starts conducting, which means at the moment of power supply the equation $U_{DS} = U_I + U_{OR}$

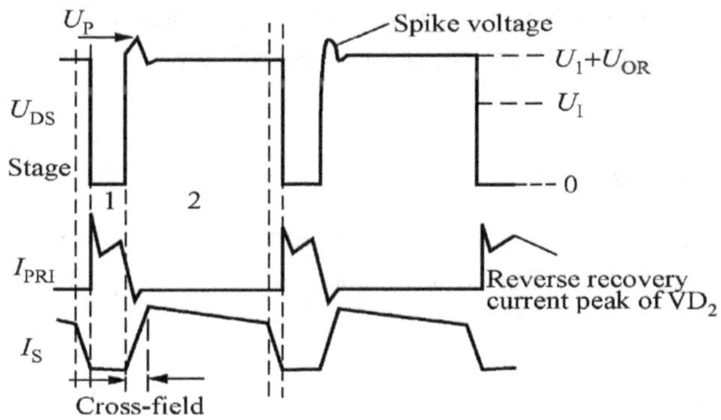

Figure 1.13 Working waveform of the continuous conduction mode in actual situations

rather than $U_{DS} = 0$ makes sense. It turns out that power consumption of TOPSwitch in continuous conduction is higher than that of discontinuous conduction. That's because there is extra energy on distributed capacitance C_D. In addition, it is necessary to charge secondary leakage inductance L_{SO} before turning off the secondary winding output, which generates cross-over current when I_S increases and I_{PRI} wanes. Once L_{SO} is fully charged, the output rectifier would be cut off by a reverse bias and the secondary current I_S would change into zero. Meanwhile, change of I_S would be inducted to the primary winding and form a reverse recovery current peak (spike current) at the leading edge of the primary side current waveform. This spike current leads to a sudden increase of the primary current, which is very likely to cause malfunction of the internal overcurrent protection for the circuit. Therefore, a leading-edge blanking circuit is especially designed in the TOPSwitch, with the purpose of preventing false triggering caused by spike current through blocking the leading edge output by the overcurrent comparator for 180 ns when TOPSwitch is initially conducted.

There is no stage 3 but only stage 2 when TOPSwitch is off. At the moment of turning off TOPSwitch, under the impact of leakage inductance L_{PO} and L_{SO}, the primary current and the secondary current would also form a cross-field, driving U_{DS} up to $(U_I + U_{OR})$. What differs from the discontinuous conduction mode is that induced voltage U_{OR} would exist till the next conduction of TOPSwitch. Therefore, there is no time interval (or stage 3) after U_{OR} decreases to zero.

1.8 Load Characteristics of SMPS

The SMPS supplies power to various loads and each load has its own characteristics. The load characteristics of SMPS can reflect the relation between the output voltage and the load. From a design perspective, there exists a "provide-demand" relationship between the SMPS and the load. On one hand, the SMPS provides stable voltage (or current or power) to the load. On the other hand, the load raises special requirements on the SMPS. Therefore, in order to design a matching SMPS for a load, the designer must know the characteristics of different loads.

Loads of the SMPS generally can be divided into two categories: constant load (also called static load or permanent load) and dynamic load (also named variable load). There are a variety of dynamic loads, such as transient load, constant current load, constant power load, peak power load, inertia load, and low noise load. For dynamic loads, the SMPS shall have the current-limiting protection or current cut-off protection function.

1.8.1 Constant Load

Constant load, usually pure resistor load, is featured by little load current change and can be matched with common SMPSs. This kind of load is ideal but not commonplace in reality.

1.8.2 Transient Load

Transient load is also called high di/dt dynamic load. It is characterized by frequent and transient load current changes and high load current change t rate (di/dt). For instance, the current change rate of high speed logic circuit and radio frequency/microwave transmitter may exceed 100 A/ps. The low-voltage microprocessor lately developed will have an impact on the power supply when switching among different operating modes rapidly, making the supply current vary by several orders of magnitude within nanoseconds. For another example, the power supply voltage of many computers is +3.3 V. When loading data from the database, the power supply shall be able to respond to load current jump of $30\,\text{A}/\mu\text{s}$. Suppose that it takes $1\,\mu\text{s}$ for the load current to change from zero to 5 A, it takes $1/25\,\text{kHz} = 40\,\mu\text{s}$ to complete the change if the bandwidth of SMPS is 25 kHz. Suppose the current rises linearly, the missing quantity of electric charge is $(5\text{A}/2) \times 40\,\mu\text{s} = 100\,\mu\text{C}$. If a fluctuation of 50 mV of +3.3 V voltage is allowed, and the instant energy is provided by the output filter capacitor, then a capacitance of $100\,\mu\text{C}/50\,\text{mV} = 2000\,\mu\text{F}$ is needed to prevent the voltage from dropping below the setting value. It should be noted that rather than using a nominal capacitor, several low capacitance capacitors shall be paralleled with a total capacitance of $2200\,\mu\text{F}$. As the total equivalent series resistance of each output filter capacitor is $\text{RESR} = 50\,\text{mA}/5\,\text{A} = 10\,\text{m}\Omega$, the equivalent series resistance of each filter capacitor of low capacitance is $\text{R}'ESR = 10\,\text{m}\Omega \times n$, where n represents the number of paralleled capacitors. When $n = 4$, $\text{R}'ESR = 40\,\text{m}\Omega$, which may greatly lower the requirement on the output filter capacitor.

Some DC/DC converters adopt broadband and high speed amplifiers so that the switching frequency can reach 2 MHz and the bandwidth 100 kHz, which provides favorable conditions for further improving the transient response of the SMPS and making the power supply smaller.

In order to improve the transient response, AVX tantalum capacitors are recommended as the output capacitor for many new switching regulators. In these capacitors, Tantalum acts as the positive electrode while dilute sulfuric acid as the negative electrode and the oxidation film on the surface of tantalum as the dielectric. It boasts of high insulation resistance, wide-range frequency response, extremely low leakage current, low temperature drift (with the operating temperature ranging from -55 to $+125\,^\circ\text{C}$), small volume, high capacity, stable performance, and long service life. It can be widely used in such high-end technology realms as military, computer, mobile phone, and power controller.

1.8.3 Constant Current Load

Constant current load needs to be provided with constant current. Battery chargers, resistance strain gauge bridges, and constant current transistors are all constant current loads. Common constant voltage/current SMPSs consist of two control loops namely current control loop and voltage control loop. Under normal conditions, the voltage loop works and the SMPS operates in the constant voltage area, and when the output current reaches the limit value, the SMPS enters the constant current area, then the current loop works and the output current remains constant.

1.8.4 Constant Power Load

When the output voltage U_o wanes, the constant power load will increase the output current I_O with the constant power control circuit and vice versa, so as to keep the product I_oU_o of the *U*o and *I*o unchanged and the load power P_L constant. This kind of SMPS can be used to charge battery of laptops as a high-performance, quick, and safe battery charger. The specific property of constant output power is almost a hyperbolic line. When designing the constant-power SMPS, a suitable SMPS integrated circuit shall be chosen based on the maximum load current under the minimum operating voltage so as to ensure enough current for the load even when the voltage is low.

1.8.5 Peak Power Load

For some electronic products, the load needs to be provided with the peak power within a short time, while the output power is required to decrease significantly in non-operating time. It is hard for a common SMPS to meet the requirements mentioned earlier. Such products as ink-jet printers, audio power amplifiers, digital video recorders (DVR), the power supply of data storage device, and direct current motor drives all fall into these electronic products. This kind of SMPS shall be featured by "the maximum peak output/continuous output power ratio" ($P_{OM(pk)}/P_{OM}$). The Peak Switch series and the Tiny Switch-PK series SMPS-integrated circuits lately developed by PI in the US boast excellent peak-power output property.

1.8.6 Inertia Load

Some devices (such as disc drives, fan motors, and motion control systems) need large current in the instant of powering on. If the starting current exceeds the set current value of the SMPS, the output power of the power supply will decrease. As a result, the load takes more time to accelerate or even fails to start. This kind of load is called inertia load, which has similarities with peak power load. The difference between these two loads is that the power of inertia loads only peaks at the starting point, while that of the peak power loads peaks during operation. To meet the requirement of inertia loads, an SMPS with relatively high output power may be chosen to ensure enough extra power.

1.8.7 Low Noise Load

Communication devices such as mobile phones, global positioning systems (GPS), and satellite navigation systems have strict restrictions on power supply noise and radio-frequency interference (RFI) for the noise of the SMPS directly influences the output of the radio frequency power amplifier. A variety of measures can be considered to reduce the noise. On one hand, the device itself shall adopt the circuit of low-noise amplifier; on the other hand, measures such as "frequency jitter" and lowering the switching frequency can be adopted for the SMPS. For example, we can adjust the switching frequency jitter at 250 times per second and the offset $\Delta f = 4\,\text{kHz}$. As the switching frequency changes within a narrow range and is independent of the higher harmonic interference of the central frequency, the noise can be lowered by frequency jittering signals. Some SMPS chips with half-frequency operating modes can reduce the video frequency interference of the SMPS when used as the standby power supply in televisions, DVDs, and video recorder cameras. Besides, hybrid regulated power supplies can be made with switching voltage regulators and low dropout linear regulators to reduce output ripple and noise.

2

New Technology and Its Application of SMPS

2.1 Single-Chip Integration of SMPS

Over the past three decades, the integrated (switching-mode power supply) (SMPS) has been developing toward two directions. One is to achieve the integration of the control circuit, the core unit of SMPS. The controller-integrated circuit of the pulse width modulation (PWM) was developed successfully for the first time in 1977, and such companies as Motorola, Silicon General, and Unitrode of the United States launched a batch of PWM chips one after another, including the typical products MC3520, SG3524, and UC3842. On this basis, high-speed PWM and pulse frequency modulation (PFM) chips (with a switching frequency up to 1 MHz) were developed, including the typical product UC1825.

The other is to achieve single-chip integration for the medium-and low-power SMPS. With all main circuits integrated on one chip, the single-chip SMPS is superior in performance and advance in function. See Table 2.1 for the classification of general single-chip SMPS products.

Taking TOPSwitch-HX series for example, the internal diagram of such products is shown in Figure 2.1. The circuit is mainly composed of 15 parts: (i) shunt regulator/error amplifier; (ii) frequency jitter oscillator; (iii) PWM; (iv) overcurrent comparator; (v) master control, drive and power MOSFET; (vi) over-temperature protection circuit with lag characteristic; (vii)shutdown/automatic restart circuit; (viii) high-voltage current supply; (ix) soft start circuit; (x) under-voltage comparator; (xi) limited current regulator; (xii) circuit tester and comparator; (xiii) shutdown logic; (xiv) trigger; and (xv) leading edge blanking (LEB) circuit. In addition, the 5.8 V reference voltage source and the circuit that can automatically reduce the switching frequency under light load (not indicated in the figure) are also included.

Compared with TOPSwitch-GX, TOPSwitch-HX mainly has the following features: first of all, the maximum output power is increased to 333 from 290 W. Second, a more powerful PWM is used instead of the PWM comparator. Third, two more comparators are added on the basis of the original overcurrent comparator (only for detecting the standard limiting current I_{LIMIT}) to detect the higher limiting current (I_{LIMIT+}) and the lower limiting current (I_{LIMIT-}) respectively. Fourth, the frequency dividing coefficients of the counter in shutdown/automatic restart circuit

Optimal Design of Switching Power Supply, First Edition. Zhanyou Sha, Xiaojun Wang, Yanpeng Wang, and Hongtao Ma.

Table 2.1 Classification of general-purpose single-chip SMPS

The first-generation product	The second-generation product	The third-generation product	The fourth-generation product	The fifth-generation product	
TOPSwitch series TOP100/TOP200 1994	TOPSwitch-II series TOP221–TOP227 1997	TOPSwitch-FX TOP232–TOP234 2000	TOPSwitch-GX TOP242–TOP250 2000–2002	TOPSwitch-HX TOP254–TOP262 2007–2009	TOPSwitch-JX TOP264–TOP271 2011
Three-terminal		Five-terminal	Six/five-terminal	Six/five-terminal	Six-terminal
$P_{OM} \leq 150\,W$		$P_{OM} \leq 75\,W$	$P_{OM} \leq 290\,W$	$P_{OM} \leq 333\,W$	$P_{OM} \leq 177\,W$
Without remote poweroff function		With remote power-off function	With remote power-off function	With remote power-off function	
Switching frequency $f = 100\,kHz$		$f = 130\,kHz/65\,kHz$	$f = 132\,kHz/66\,kHz$	$f = 132\,kHz/66\,kHz$	
AC input voltage range $u = 85–265\,V$ (wide-range input) or $220\,V \pm 15\%$ (fixed input)					
Power supply efficiency η = about 80%					

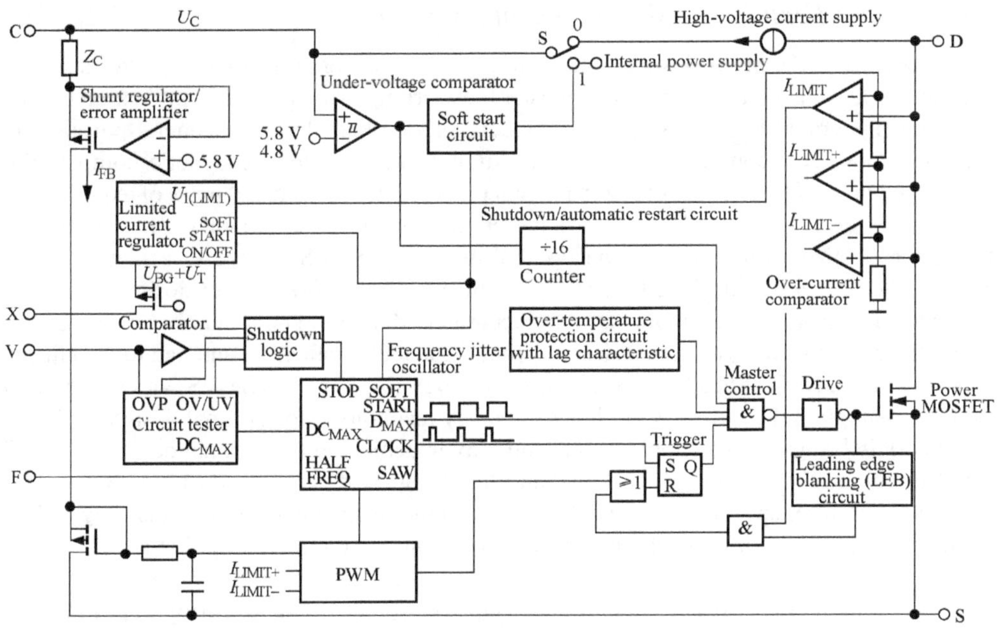

Figure 2.1 Functional block diagram of TOPSwitch-HX

have been changed to ÷16 from ÷8. Fifth, the frequency jitter has been increased that would reduce the cost of electromagnetic interference (EMI) filter. Sixth, an output over-voltage protection circuit is added. Besides, significant improvement has been made to TOPSwitch-HX in other respects. For example, the on-state resistance of power switch tube was further reduced. Therefore, TOPSwitch-HX is superior to TOPSwitch-GX in overall performance.

Furthermore, STMicroelectronics, Philips, and ON Semiconductor have successively developed a number of integrated circuits of single-chip SMPS that can be widely applied in instruments and meters, modern office equipment, industrial control, household appliances, portable battery chargers, consumer electronics, and other fields.

2.2 Computer-Based SMPS Design

In the recent past, with the development of SMPS technology and the popularization of computer application, it has become a new technology to design the SMPS with computer in the international electricity supply field. At present, software has been a key technology for optimal design of SMPS. The application of software gives full play to the high technology and reduces designers' workload, creating favorable conditions for realizing the optimal design of SMPS. Meanwhile, computer simulation technology also has become an important instrument to develop new SMPS. With the computer simulation technology, the designers can solve the technical problems they may encounter in SMPS design by creating an SMPS prototype on the computer instead of a real model, which greatly accelerates the research of new types of SMPS.

2.2.1 Main Features of SMPS Design Software

It is relatively difficult to design a cost-effective SMPS for it involves a wide scope of knowledge. Designers need to not only master the working principles and application circuits of various integrated SMPS products and understand the knowledge of general-purpose and special-purpose semiconductors, analog and digital circuits, electromagnetic compatibility (EMC) and thermodynamics, but also have rich practical experience and mass of experimental data. As the design of SMPS is an iterative process of multiple variables, only continuous adjustment to these variables can the optimal design of SMPS be finally achieved. Traditionally, the SMPS was designed manually, featuring heavy workload for designers and low efficiency. It was hard to estimate variables accurately due to the huge number of them. As a result, the design result was quite different from the actual application and had to be adjusted for many times. Now, it has become a new technology to design the SMPS with computer in the power supply field and software has been a key technology for optimal design of SMPS. This not only brings high technology into full play in achieving the optimal design of SMPS, but also reduces the designers' workload and the design can be finished only within several minutes. Therefore, apart from the experts, any designer is able to design a satisfying SMPS.

The popular SMPS design software in the world now has the following features:

1. They are all PC-based that can determine the best topology circuit of SMPS according to the technical indicators entered by designers and help users to complete the design of low-cost and highly efficient SMPS.
2. They adopt the interactive design mode, guiding users to complete the design via intuitive graphical interface, which makes it simple and convenient to operate.
3. They have the optimal design function.

2.2.2 Classification of Design and Simulation Software of SMPS

See Table 2.2 for the classification of currently developed design and simulation software of SMPS.

2.2.3 Design Process of SMPS Software

2.2.3.1 Design Process of PI Expert 8.5 Software

The simplified design process of PI Expert 8.5 Software is illustrated in Figure 2.2.

2.2.3.2 Design Process of SMPS Design Toolkit 1.6 Software

Design process of the flyback SMPS

SMPS Design Toolkit 1.6 is used to design a flyback SPMS, which is illustrated in Figure 2.3. The design process consists of the following 12 steps (Step 1–Step 12)

Step 1: Determine SMPS specification.

Step 2: Determine the range of input filtering capacitance and DC input voltage.

Table 2.2 Classification of design and simulation software of SMPS

Development company	Software name	Applicable field
PI	PI Expert 8.5	Support the following 20 series of integrated circuits of single-chip SMPS: TOPSwitch-GX (TOP242-TOP250); TOPSwitch-HX (TOP254-TOP261); TOPSwitch-JX (TOP264-TOP271); PeakSwitch (PKS603-PKS607); TinySwitch-II (TNY264-TNY268); TinySwitch-III (TNY274-TNY280); TinySwitch-PK (TNY375-TNY380); DPA-Switch (DPA423-DPA426); LinkSwitch (LNK500, LNK501 and LNK520); LinkSwitch-LP (LNK562-LNK564); LinkSwitch-TN (LNK304-LNK306); LinkSwitch-HF (LNK353-LNK354); LinkSwitch-XT (LNK362-LNK364); LinkSwitch-II (LNK603-LNK606); LinkSwitch-CV (LNK623-LNK364); LinkSwitch-PH (LNK403-LNK409); LinkSwitch-PL (LNK454-LNK460); HiperPLC (PLC810PG); HiperPFS (PFS704-PFS728); HiperTFS (TFS757-TFS764)
Fairchild semiconductor	SMPS Design Toolkit 1.6	Design isolation type, flyback, forward and quasi resonance converters. This software also has the functions of magnetic core database, waveform display and circuit simulation
Micrel	Buck It Designer 1.1.8	Design buck converters
Texas instruments (TI)	LoPwrDC Designer 1.0	Design TPS62XXX series products
	Swift Designer 3.51	Design TPS54XXXX series products
Philips	STARplug design wizard V1.0	TEA1520-TEA1524
American National Semiconductor Corporation (NSC)	SMS3.3	LM2574-LM1577
	SMS4.3	LM2585-LM2588, LM2594-LM2599
	SMS6.24	LM2671-LM2679, LM2594-LM2599, LM2585-LM2587
	SMS6.3	LM2852
	WEBENCH	Large tool software for online design, simulation and test
Linear technology (LT)	Switcher CAD III	SMPS simulation software
ST microelectronics (ST)	VIPer 2.24	SMPS design and simulation software
Ridley engineering	POWER 4-5-6	SMPS design and simulation software
Analog	Saber	Analog and mixed signal simulation software
Microsim	PSpice	General-purpose circuit simulation software

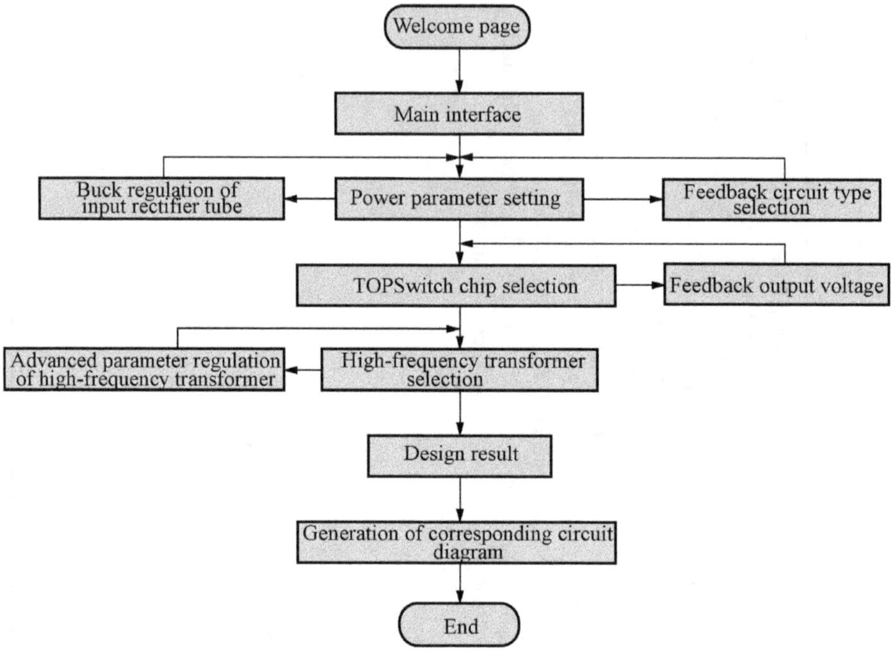

Figure 2.2 Simplified design flowchart of PI Expert 8.5 Software

Step 3: Determine the maximum duty ratio.
Step 4: Determine the inductance value of primary side of high-frequency transformer.
Step 5: Select an applicable chip according to the input power and the drain-source electrode peak value of the power switch tube.
Step 6: Determine the applicable magnetic core and the minimum number of turns of primary winding.
Step 7: Determine the number of turns of each output winding.
Step 8: Determine the wire diameter of each winding, and judge whether the winding window is big enough and whether to change the specification of magnetic core or not.
Step 9: Select an applicable rectifier diode for each output path.
Step 10: Determine the output filtering capacitance.
Step 11: Design RCD-type clamp circuit.
Step 12: Design feedback loop (by means of six sub-steps corresponding to Step 12-1 to Step 12-6).
End of design.

Design process of quasi resonant converter

The basic principle of quasi resonant converter is to connect an external capacitor C_r in parallel between the drain and source electrodes of the power switch tube (MOSFET), with capacitance far more than the distributed capacitance of MOSFET, making the power switch tube be turned off only when the voltage or current waveform is of zero passage, so as to minimize the

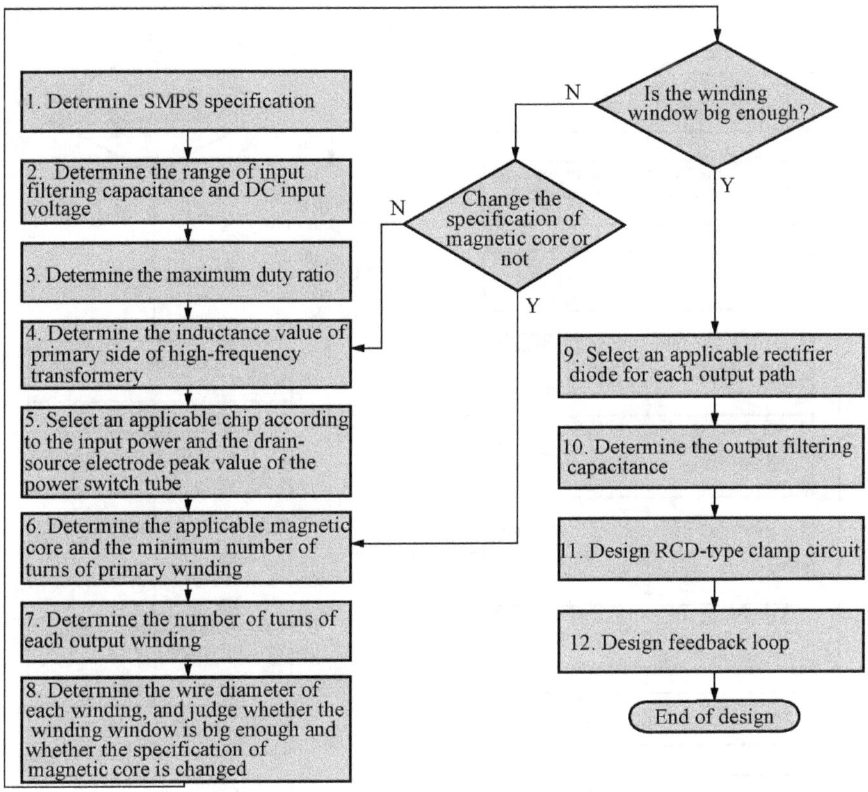

Figure 2.3 Design flowchart of the flyback SMPS

switching losses. SMPS Design Toolkit 1.6 is used to design the quasi resonant converter, the work flow of which is shown in Figure 2.4. The design consists of the following steps (Step 1–Step 14):

Step 1: Determine SMPS specification
Step 2: Determine the range of input filtering capacitance and DC input voltage.
Step 3: Determine the induced voltage.
Step 4: Determine the inductance value of primary side of high-frequency transformer.
Step 5: Select an applicable SMPS chip according to the drain-source electrode peak current of the input power and power switch tube.
Step 6: Determine the applicable magnetic core and the minimum number of turns of the primary winding.
Step 7: Determine the number of turns of output winding and the circuit of auxiliary power.
Step 8: Determine the starting resistance.
Step 9: Determine the wire diameter of each winding.
Step 10: Select the output rectifier diode.
Step 11: Determine the output filtering capacitance.
Step 12: Design the synchronous network.

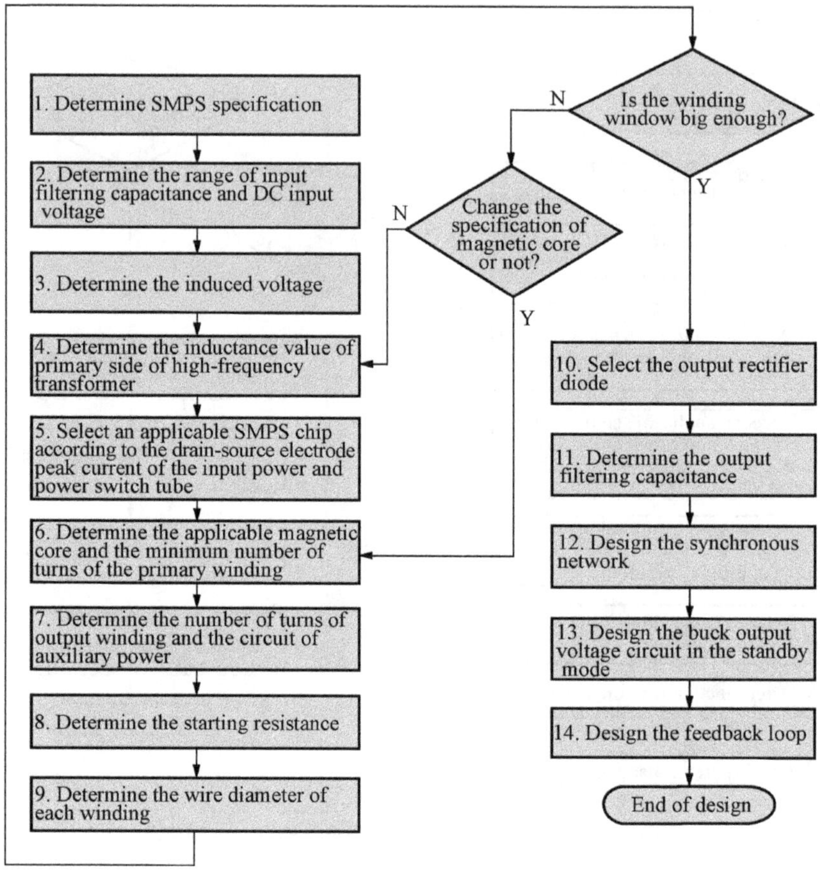

Figure 2.4 Design flowchart of quasi resonant converter

Step 13: Design the buck output voltage circuit in the standby mode.
Step 14: Design the feedback loop (by means of five sub-steps, respectively corresponding to Step 14-1 to Step 14-5).
End of design.

Compared with the flyback SMPS, the design process of quasi resonant converter has been adjusted from 12 steps to 14 steps. Four steps including induced voltage determination, starting resistance determination, synchronous network design, and buck output voltage circuit design in the standby mode have been added, and some design steps have been left out and modified accordingly.

2.2.3.3 Design Process of WEBENCH Online Design and Simulation Software

The design process of the large tool software, namely, WEBENCH online design, simulation and test, is illustrated in Figure 2.5. With WEBENCH tool software, the design can be completed by only four steps:

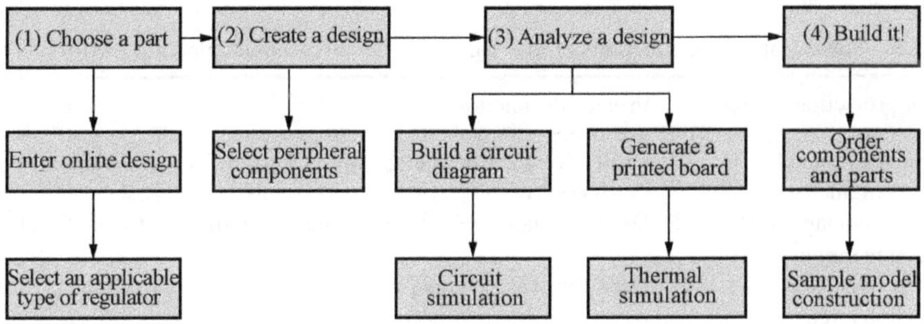

Figure 2.5 Design flowchart of WEBENCH online design and simulation software

Step 1: Choose a part. Select an appropriate type of regulator according to the specification of stabilized power supply (including the input and output voltage range, output current, and other design indexes).

Step 2: Create a design. Select the peripheral components and parameter values.

Step 3: Analyze the design. Evaluate the performance of the design using WEBENCH for circuit simulation and thermal simulation. The performance indexes to be evaluated include the frequency response (cross frequency and phase margin); peak current and voltage; and thermal performance (power-supply efficiency, chip-junction temperature, and temperature of all components and parts). Although evaluation was based on the simulated calculation, the test results are identical with the actual data. The circuit simulation can display the circuit diagram so that users can change components and test the circuit of regulator, including the test of Bode diagram (only for fixed-frequency regulators), initiating stage, stable state, transient input change, and transient output change. However, in each test, both the input voltage and load current might change and the default value may not match with the system. Therefore, users should estimate the expected results in advance. If the simulation results are inconsistent with the expected ones, reasons should be found out.

Step 4: Build it! Complete the design. Order the module, test suite, or evaluation board according to the circuit diagram and printed board diagram generated from the design, so as to apply the final design in reality.

2.3 Internal Protection Circuit of SMPS

There are many different types of protection circuits inside the SMPS. For example, the early TOPSwitch series were designed with over-current protection circuit, over-temperature protection current, power-off and auto restart circuit, and LEB circuit. The protective functions of the new single-chip SMPS are greatly improved with added programmable limited current setting circuit, input under-voltage (UV) protection circuit, input over-voltage (OV) protection circuit, output OV protection circuit, and soft starting circuit.

See Table 2.3 for the classification and functions of internal protection circuits of single-chip SMPS. The internal protection circuit is designed by chip manufacturers, and the external protection circuit is customized by users.

Table 2.3 Classification and functions of internal protection circuit of single-chip SMPS

Name of protection circuit	Protective function
Over-current protection (OCP) circuit	Used to define the limited current I_{LIMIT} of power switch tube MOSFET
Input under-voltage (UV) protection circuit	Used for under-voltage protection when the input voltage is too low
Input over-voltage (OV) protection circuit	Used for over-voltage protection when the input voltage is too high
Output over-voltage protection (OVP) circuit	Used for over-voltage protection when the output voltage is too high
Output overpower protection (OPP) circuit	Used for overload (also called overpower) protection when the output power is too high. Equipped by TOPSwitch-JX series
Output short-circuit protection (SCP) circuit	Rapidly power off the output or reduce the output power substantially by automatically restarting circuit when a short-circuit failure of the power supply occurs
Over-temperature protection (OTP) circuit	When the chip temperature exceeds the maximum junction temperature, the output stage will be turned off and restarted until the temperature returns to a certain value
Adaptive limiting current regulator circuit	In each switching cycle, when the drain current I_D of the power MOSFET reaches I_{LIMIT}, the power MOSFET will be turned off within the rest period of this cycle. If there is a cycle skipping, and the rest cycles will be powered off by the maximum current limitation point, which means that the voltage input is high voltage; the adaptive limited current regulator circuit will reduce the maximum value $I_{LIMIT\ (max)}$ by 10%, so as to reduce the overload output power (PeakSwitch series products)
Programmable limited current setting circuit	Set the value of I'_{LIMIT} externally by changing the resistance value of resistance set by the limited current
Adaptive switching cycle conducting time extension control circuit	Used to extend the conducting period of power MOSFET. As long as the current of primary side is less than I_{LIMIT}, the adaptive switching cycle conducting time control circuit will keep the switching cycle conducted. Even if the maximum duty ratio D_{MAX} is reached, the cycle will not be ended, so as to extend the conducting time, creating a condition for the peak power output (PeakSwitch series products)
Programmable AC voltage detection circuit	It can quickly judge whether the AC input voltage is normal or not. When the AC input voltage is turned off, this circuit can rapidly reset the single-chip SMPS (PeakSwitch series products)
Rapid reset circuit after being re-powered on	Being repowered on, the single-chip SMPS can immediately return to normal operation (PeakSwitch series products)
Open loop fault detection and protection circuit	Once the control loop circuit incurred an open loop fault, the circuit will proceed to the automatic restarting stage to reduce the output power to 6% P_{OM} and return to normal until the fault is eliminated (LinkSwitch-TN series products)
Soft start circuit	When powered on, the output voltage can increase smoothly

(*continued overleaf*)

Table 2.3 (*Continued*).

Name of protection circuit	Protective function
Turn-off/automatic restarting circuit	Once the regulation is out of control, the circuit can restart automatically so that the SMPS can resume normal operation
Leading edge blanking circuit	At the moment when MOSFET is conducted, the circuit blanks the rising edge of the current comparator for some time to avoid the peak current generated by the distribution capacitance of the primary side and the rectifier tube of the secondary side within the reverse recovery time, which may cause early termination of the switching pulse
Frequency jittering circuit	Make the switch frequency jitter within the specified range. The switching frequency changes continuously around the central frequency, and it has no correlation with the higher harmonic disturbance. Therefore, using the frequency jittering signal can reduce the conducted noise
Programmable state controller	The SMPS can be switched between the work state and standby state by means of manual control, microcontroller operation, digital circuit control, and prohibited operation

2.4 Synchronous Rectification (SR) Technology

2.4.1 Brief Introduction to the SR Technology

With the development of power supply technology, the SR (synchronous rectification) technology has been rapidly promoted and widely applied in DC/DC converters with low voltage and high output current in recent years. The loss of DC/DC converter mainly consists of losses of power switch tube, high-frequency transformer, and output rectifier. Under low voltage and high output current, the on-state voltage drop of rectifier diode is relatively higher, and the loss of output rectifier is extremely large. The voltage drop of fast recovery diode (FRD) or super fast recovery diode (SRD) can reach 1.0–1.2 V and even the voltage drop of a low-voltage-drop Schottky barrier diode (SBD) can reach 0.6 V, which lead to increased rectifier losses and lower power-supply efficiency. For example, laptops generally adopt the power supply voltage of 3.3 V or even 1.8, 1.5, and 1.2 V, with the current consumption reaching 20 A, and the rectifier loss of SRD is approaching or even exceeds 50% of the output power. Even with an SBD, the loss of rectifier tube can reach (18–40%)$\cdot P_O$, accounting for over 60% of the total power supply losses. Therefore, the traditional diode rectifier circuits cannot satisfy the demands for high efficiency and small volume of low-voltage and high-current SMPS, a bottleneck restraining the efficiency improvement of DC/DC converters.

SR is a new technology using a special-purpose power MOSFET with an extremely low on-state resistance instead of the rectifier diode to reduce the rectifier loss, which can greatly increase the efficiency of DC/DC converters and there would be no dead-time voltage caused by Schottky barrier voltage. The power MOSFET is a voltage control device, with the on-state volt-ampere characteristics showing a linear relation. When power MOSFET is used as a rectifier, the rectifier function can be realized only if the grid voltage keeps synchronizing with the phase position of the rectified voltage. Therefore, it is called SR.

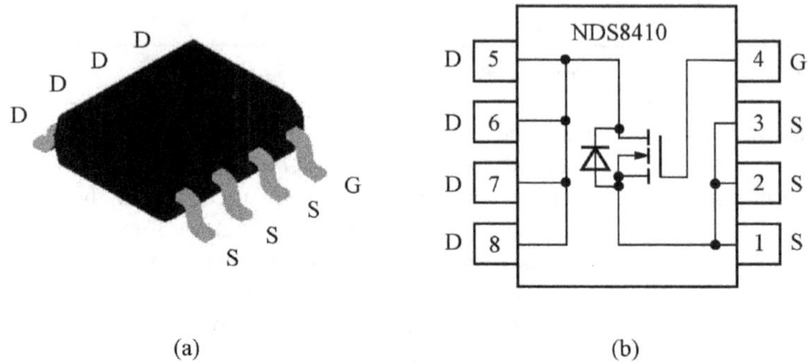

Figure 2.6 Appearance and pin configuration of NDS8410 power MOSFET. (a) Appearance; (b) pin configuration

To satisfy the demands of SR current with high frequency and large capacity, special-purpose power MOSFET products have been developed in recent years. A typical one is NDS8410 N-channel power MOSFET produced by Fairchild semiconductor, whose on-state resistance is 0.015 Ω and appearance and pin configuration are shown in Figure 2.6. In this figure, G refers to the grid, D the drain, and S the source. A fly-wheel diode (FWD) is connected in parallel between the drain and the source of power MOSFET, which will prevent power MOSFET from the damage of back electromotive force when the power switch tube turns off. The main parameters of NDS8410 are as follows: the maximum drain-source voltage: $U_{DSS} = 30\,V$, the maximum grid-source voltage: $U_{GSS} = 20\,V$, the maximum drain current in the continuous conduction mode: 10 A, the maximum drain current when working in the pulse mode: 50 A, the maximum power loss is as low as 1 W, the conduction time and turn-off time are 14 and 56 ns respectively, and operating temperature ranges from −55 to +150 °C.

SI4800 power MOSFET is developed by Philips with TrenchMOS™ technology, its on/off state is controlled by logic level and the drain-source on-state resistance is only 0.0155 Ω. The on-state resistance of the power MOSFET such as IRL3102 (20 V/61 A), IRL2203S (30 V/116 A), and IRL3803S (30 V/100 A) developed by IR is 0.013, 0.007, and 0.006 Ω respectively, and the on-state voltage drop when 20 A current passes is less than 0.3 V. With high input impedance and a short switching time, these special-purpose power MOSFET products have become the first choice for rectifier device in designing low-voltage and high-current power converters.

Besides, IC manufacturers recently released synchronous rectifier integrated circuits (SRICs). For example, the newly released IR1176 is a high-speed CMOS controller specially designed to drive N-channel power MOSFET. IR1176 can run independently without the primary side topology and needs no active clamp, grid drive compensation, and other complicated circuits. IR1176 is applicable to synchronous rectifiers in high-current DC/DC converters with an output voltage of 5 V and below. It greatly simplifies and improves the design of isolated DC/DC converters in broadband network servers. IR1176, equipped with IRF7822 power MOSFET, can increase the efficiency of converters. Given the input voltage of +48 V and the output voltage of +1.8 V and current of 40 A, the efficiency of DC/DC converters will reach 86%. The efficiency can reach 85% even when the output voltage is 1.5 V.

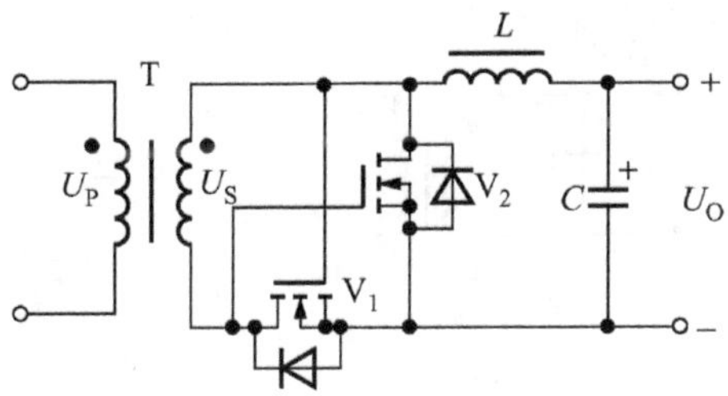

Figure 2.7 The basic principle of single-ended buck synchronous rectifier

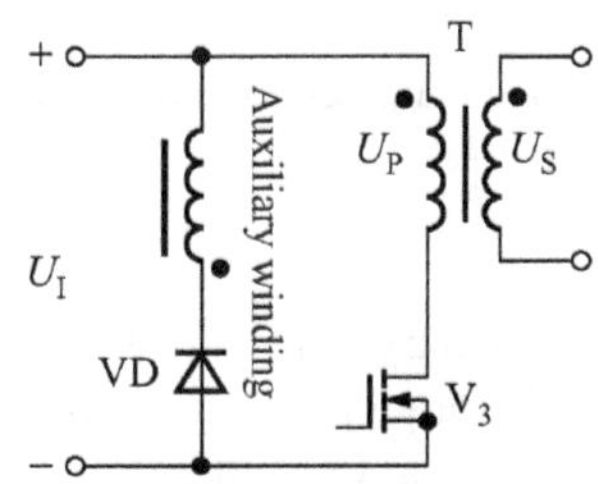

Figure 2.8 The circuit of realizing the magnetic reset of forward DC/DC converter with auxiliary winding

2.4.2 Basic Principle of SR

See Figure 2.7 for the basic principle of single-ended forward and isolated buck synchronous rectifier. V_1 and V_2 are power MOSFET.

In this figure, V_1 plays a rectifying role while V_2 keeps the follow current. The operating principle of such circuit is as follows: at the positive half cycle of secondary voltage, V_1 is on and V_2 is off, and V_1 plays the rectifying role; while at the negative half cycle of secondary voltage, V_1 is off and V_2 is on, and then V_2 keeps the follow current. The power loss of synchronous rectifier circuit mainly consists of the conduction loss and grid drive loss of V_1 and V_2, which constitutes the majority of the power loss given a switching frequency lower and higher than 1 MHz, respectively.

For a forward DC/DC converter, the high-frequency transformer shall be subject to magnetic reset when the power tube is off, so as to prevent the magnetic core of the transformer from saturating. Therefore, a magnetic reset circuit (also called transformer reset circuit) is generally required. Figure 2.8 shows the circuit in which the magnetic reset of forward DC/DC converter is realized with auxiliary winding.

2.5 Active Clamp Technology

The active clamp circuit invented by VICOR in the United States can significantly reduce the power loss of the clamp protection circuit. See Figure 2.9(a) and (b) for the comparison between passive clamp circuit and active clamp circuit. Taking Figure 2.9(a) for example, when the power switch tube MOSFET turns off, U_I is DC input voltage; U_{OR}, the primary

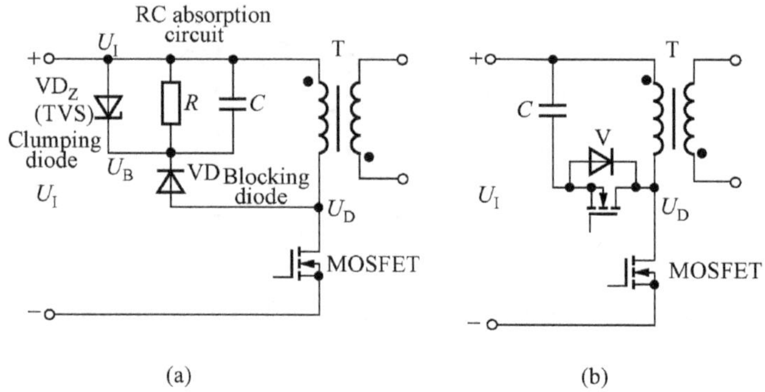

Figure 2.9 Comparison between passive clamp circuit and active clamp circuit. (a) Passive clamp circuit; (b) active clamp circuit

side induced voltage (also called secondary side reflected voltage); U_P, the spike voltage produced by leakage inductance of high-frequency transformer. In this case, the drain voltage is the sum of the above three voltage values. Once the drain voltage U_D exceeds the allowable value, the blocking diode will be conducted immediately, making the clamping diode VD_Z (transient voltage suppressor or TVS) limit U_D within the safe value range rapidly so as to protect MOSFET. R and C constitute the absorption circuit of the spike voltage.

Figure 2.9(b) shows an active clamp circuit, in which an active power device MOSFET (V) with a very low on-state resistance is used as the clamper tube. When MOSFET is off and V is on, the spike voltage can be absorbed by capacitor *C*. As the on-state resistance of V is extremely low and the power loss is quite small, the loss of the clamp circuit can be significantly reduced.

Figure 2.10 shows the simplified circuit diagram of active clamp-mode DC-to-DC converter with SR rectifiers. MOSFET (V_4) is a clamper tube, C_C is a clamp capacitor, and V_3 is the power switch tube of SMPS. It can be seen from this figure that when V_4 is on, V_3 is off because

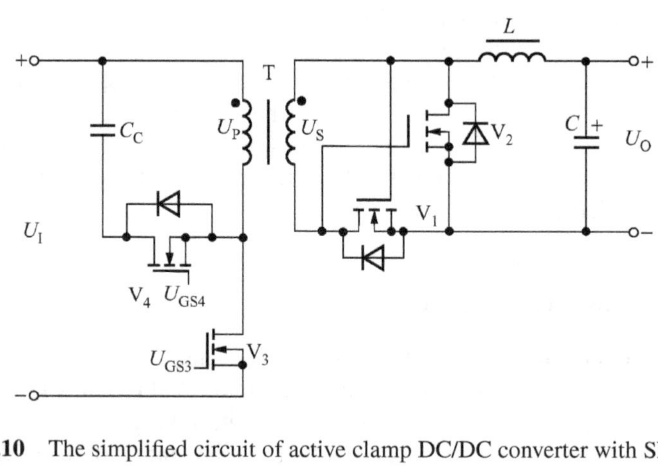

Figure 2.10 The simplified circuit of active clamp DC/DC converter with SR function

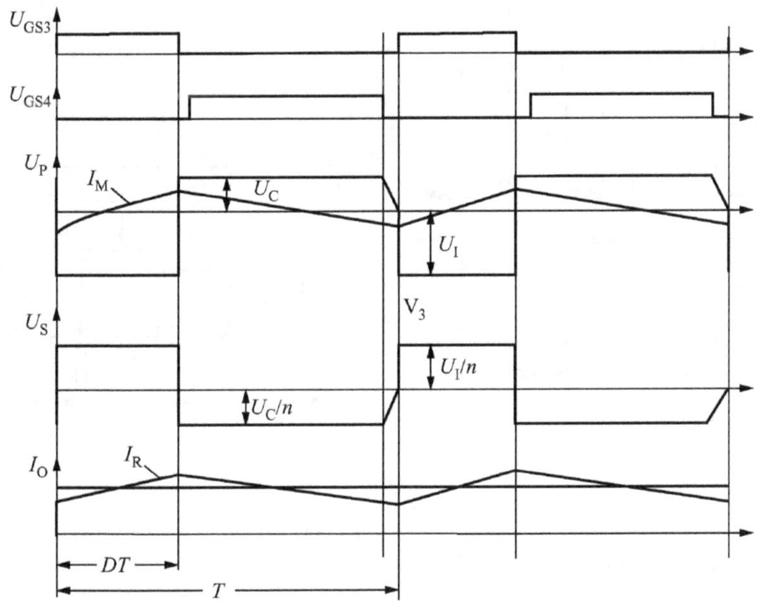

Figure 2.11 Time sequence waveform of the active clamp circuit

$U_{GS3} = 0$. When V_4 is off, V_3 is turned on by U_{GS3}, clamping the spike voltage produced by leakage inductance of the high-frequency transformer. The time sequence waveform of the active clamp circuit is illustrated in Figure 2.11. I_M refers to magnetizing current; I_R, the output ripple current; and U_C, the voltage on clamp capacitor; n is the ratio between the primary and the secondary turns of high-frequency transformer, and DT is the product of duty ratio and switching cycle, or the conduction time of MOSFET.

Taking the DC/DC converter of 3.3 V/20 A active clamp for example, whose typical working waveforms are shown in Figure 2.12. In Figure 2.12(a), A and B refers to drain-source voltage waveforms of V_3 and V_4 respectively, whose voltage amplitudes are both 50 V. Each small grid on the horizontal axis refers to 2.5 μs. Figure 2.12(b) shows the output waveform of synchronous rectifier V_1 with the voltage amplitude of 5 V, which can be filtered to 3.3 V DC output voltage.

2.6 Magnetic Amplifier Regulator Technology

In recent years, the magnetic amplifier voltage regulator circuits made of high-performance amorphous alloy magnetic rings have been widely applied in multi-output SMPS of PCs, particularly for output high current ranging from one ampere to dozens of amperes. Featured with good voltage regulation performance, high efficiency, small volume and low cost, it is worthy of popularization. The magnetic amplifier voltage regulator circuit changes the delay time of magnetic reset with the controllable magnetic saturated inductor and modulates the pulse width precisely to realize the precision voltage regulation. See Figure 8.7 for the typical application circuit.

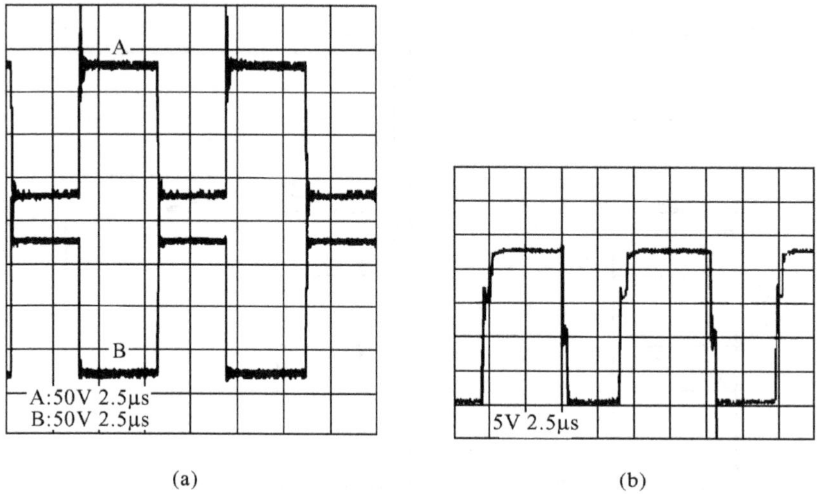

Figure 2.12 The typical working waveforms of the DC/DC converter of 3.3 V/20 A active clamp. (a) Drain-source voltage waveforms of V_3 and V_4; (b) output waveform of synchronous rectifier V_1

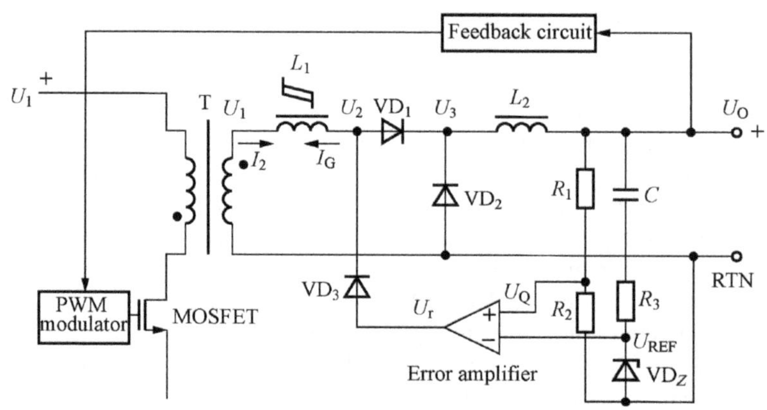

Figure 2.13 The basic principle of the magnetic amplifier voltage regulator circuit

The basic principle of the magnetic amplifier regulator circuit in flyback SMPS is illustrated in Figure 2.13. The output voltage U_O passes the sample resistances R_1 and R_2 and turns into the sampling voltage U_Q which is connected to the inverting input of the error amplifier. The non-inverting input terminal of the error amplifier is connected to the reference voltage U_{REF}. VD_Z refers to voltage regulator tube, and R_3 is biasing resistance. The error amplifier generates error voltage U_r by comparing U_Q and U_{REF}, which passes diode VD_3 and connects the right end of the controllable saturated inductor L_1. VD_1 is the output rectifier and VD_2, the FWD; C, the output filter capacitor; L_2, the magnetic bead restraining switching noises; U_1, U_2, and U_3 symbolize the voltage at left end of L_1, the right end of L_1 and the right end of VD_1, respectively. The upper end of the primary side of the high-frequency transformer connects

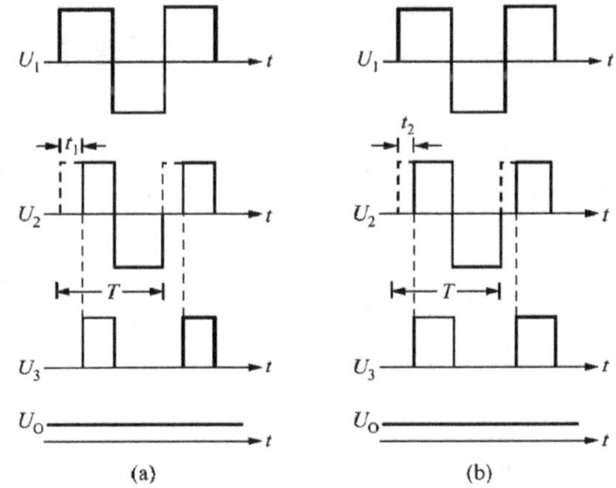

Figure 2.14 Time sequence waveform of the magnetic amplifier. (a) Magnetic reset time is t_1; (b) Magnetic reset time is t_2

the DC input high voltage U_I, while the lower end connects the drain of the power switch tube MOSFET. The feedback signal of output voltage from the feedback circuit is used to adjust the pulse duty ratio. The voltage can be regulated by switching on and off MOSFET.

When MOSFET is on, the energy is stored in the high-frequency transformer, and VD_1 is cut off. When MOSFET is turned off, the energy stored in the high-frequency transformer is transmitted to the secondary side, VD_1 is on, and the magnetic reset current I_G flows leftward through L_1 from the right side to reset the magnetism of L_1. The current direction of the secondary winding I_2 is opposite to that of I_G, so I_2 can only flow through L_2 after neutralizing I_G. This indicates that the secondary side current turns to positive from a negative value and then increases rapidly, making I_2 enter the magnetic saturation state and present low impedance. Obviously, the magnetic reset time is the delay time t_1 for VD_1 starts powering on.

The time sequence waveform of the magnetic amplifier is shown in Figure 2.14(a) and (b) whose corresponding magnetic reset time is t_1 and t_2 respectively. It can be seen from the Figure that the duty ratio of U_2 can be adjusted by changing t_1: $D = t_1/T$. T refers to the switching cycle. To be specific, $D \uparrow \rightarrow U_O \uparrow$ when the magnetic reset time drops from t_1 to t_2. $D \downarrow \rightarrow U_O \downarrow$ when the magnetic reset time increases from t_2 to t_1. Because the magnetic amplifier has the function of "secondary voltage regulation" (the primary voltage regulation is completed by PWM modulator), it can regulate U_O accurately to obtain a high-stability output voltage.

The traditional ferrite core is made of crystalline state materials, the atoms of which are arranged orderly in a three-dimensional space, forming a lattice structure. Amorphous alloy refers to the material whose atoms are arranged in a disordered manner for it fails to crystallize when the substance is sharply cooled down from a liquid (or gas) state. The amorphous alloy can be manufactured with a simple process. Being energy saving, it is a new green and environment-friendly material with advantages of high permeability, high squareness ratio, low core loss, and sound stability in high temperature. This new magnetic material is suitable for making controllable saturated inductors which are used in the ATX power supplies of computers.

Table 2.4 Main parameters of typical products made of common amorphous alloy magnetic rings

Model	Maximum outer diameter D_{max} (mm)	Minimum inner diameter d_{min} (mm)	Maximum height h_{max} (mm)	Magnetic circuit length l_m (cm)	Effective cross-sectional area A_c (cm^2)	Window area W_a (cm^2)	Product of area W_aA_c (cm^4)	Magnetic ring volume V (cm^3)	Squareness ratio	Saturation magnetic flux density B_S (T)	Magnetic flux ϕ (μWB)	Core loss P_D (mW)
MP1005P4AS	10.92	5.59	5.69	2.59	0.060	0.245	0.015	0.155	0.86	0.57	6.82	249
MP1205P4AS	13.84	6.76	6.63	3.14	0.057	0.359	0.020	0.180	0.86	0.57	6.51	289
MP1305P4AS	14.40	7.87	6.71	3.46	0.057	0.487	0.028	0.198	0.86	0.57	6.51	318
MP1405P4AS	15.82	7.87	6.71	3.67	0.083	0.487	0.040	0.304	0.86	0.57	9.43	489
MP1506P4AS	17.12	7.82	8.31	3.86	0.140	0.481	0.067	0.540	0.86	0.57	15.96	869
MP1603P4AS	17.83	11.05	5.13	4.50	0.041	0.959	0.039	0.184	0.86	0.57	4.67	296
MP1805P4AS	20.83	10.80	6.76	4.88	0.108	0.915	0.099	0.529	0.86	0.57	12.36	852
MP2008P4AS	22.23	11.05	10.36	5.15	0.248	0.959	0.238	1.276	0.86	0.57	28.25	2054
MP2303P4AS	24.94	14.86	5.13	6.19	0.081	1.734	0.140	0.501	0.86	0.57	9.23	806
MP2510P4AS	27.79	17.27	11.48	7.01	0.241	2.343	0.565	1.689	0.86	0.57	27.47	2718
MP3210P4AH	34.95	19.86	11.48	8.58	0.388	3.099	1.202	3.330	0.92	0.57	44.24	5359

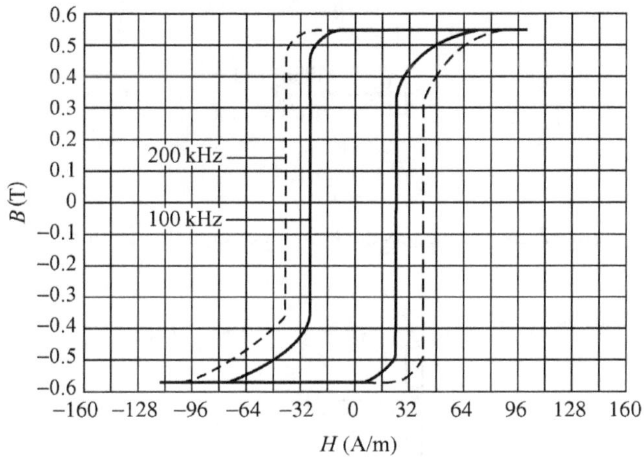

Figure 2.15 The B–H curve of MP1305P4AS

L_1 adopts MP1305P4AS, a high-performance amorphous alloy magnetic ring produced by the American company Metglas, and wound for seven turns evenly with ϕ0.10 mm enameled wire. See Table 2.4 for the main parameters of typical products made of common amorphous alloy magnetic rings. In the model MP1305P4AS, "13" means that the outer diameter is 13 mm long (the nominal value), "5" means that it was 5 mm high (the nominal value). The magnetic circuit is 3.46 cm long; effective cross-sectional area, 0.057 cm^2; weight, 1.50 g; saturation magnetic flux density, 0.57 T; squareness ratio, 0.86; resistance ratio, 0.142 μΩcm; core loss, 318 mW; long-time working temperature, $< 120\,°C$; and Curie point temperature, 225 °C (the hysteresis phenomenon will disappear when the temperature exceeds the point).

The B–H curve of MP1305P4AS is illustrated in Figure 2.15, in which B refers to the magnetic flux density (unit: T) and H refers to the magnetic field strength (unit: A/m). The solid line and dotted line in this figure indicate the switching frequency of 1000 and 200 kHz, respectively.

2.7 Programmable Voltage Regulator Technology

The adjustable SMPS regulator outputs voltage by regulating the resistance value manually, which is not only inaccurate but also inconvenient to use. If the adjustable resistor is replaced with a digital potentiometer, a computer-controlled programmable switch regulator can be developed.

2.7.1 Basic Work Principle of Digital Potentiometers

Digital potentiometers, also called digitally controlled potentiometers (DCP), are new electronic devices with quite a promising development prospect, which can replace the traditional mechanical potentiometer in many fields. They are widely used in instruments and meters,

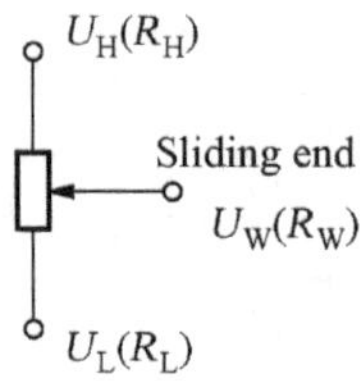

Figure 2.16 The equivalent circuit of digital potentiometers

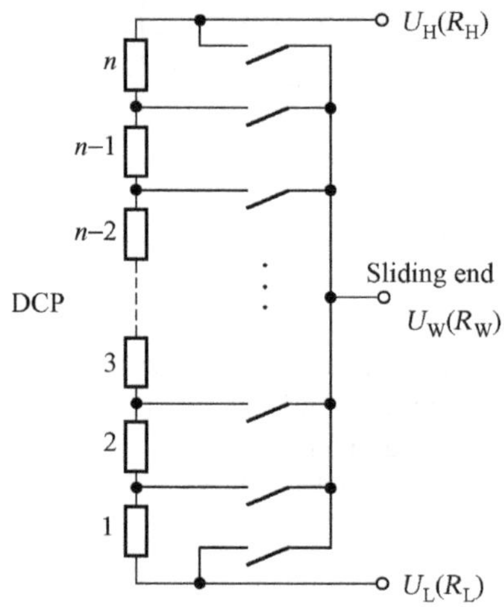

Figure 2.17 The internal simplified circuit diagram of digital potentiometers

computer and communication equipment, household appliance, industrial control, linear and switching regulators.

A digital potentiometer is an integrated three-end variable resistance device, the equivalent circuit of which is shown in Figure 2.16. Its high end, low end, and sliding end are symbolized by U_H, U_L, and U_W, respectively, when used as a voltage divider and by R_H, R_L, and R_W (or H, L, and W) when used as an adjustable resistor.

The internal simplified circuit diagram of digital potentiometers is shown in Figure 2.17. n resistors of the same or different resistance values are connected between U_H end and U_L end (also called R_H end and R_L end) in a series, both ends of each resistor are connected with an analog switch composed of CMOS tube or NMOS tube as the tap of DCP. Such an analog switch is equivalent to a single-pole single-throw switch (SPST switch). Under the control of digital signal, only one analog switch can be closed each time, so as to connect one node of the series resistance to the sliding end.

The schematic diagram of the digital potentiometer is shown in Figure 2.18. Suppose that the digital potentiometer has 16 taps with a stepping quantity of 660 Ω, for every step the sliding end moves, the output resistance increases by 660 Ω. Considering that the sliding end is connected to an analog switch wherever it is, the resistance value of the analog is the sliding resistance and the initial resistance of the digital potentiometer as well. Now assume the sliding

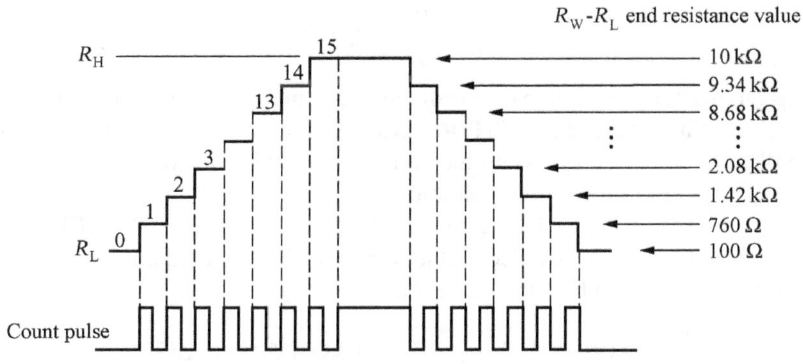

Figure 2.18 The schematic diagram of digital potentiometer

resistance of 100 Ω, when the sliding end moves 15 steps to R_H end, the output resistance between R_W end and R_L end should be $100\,\Omega + 660\,\Omega \times 15 = 10\,\mathrm{k\Omega}$.

2.7.1.1 Basic Configuration Model of Digital Potentiometers

There are two basic configuration modes for the digital potentiometer, one is the adjustable resistor mode and the other is the voltage divider mode, as illustrated in Figure 2.19(a) and (b).

In the adjustable resistor mode, when the digital resistor is configured into the adjustable resistor mode, the resistance of sliding end of the digital potentiometer must be considered. H and L in Figure 2.19(a) refer to high end and low end of the digital potentiometer, respectively, and W refers to the leading end of the sliding end. Suppose the total resistance of H and L ends is R, the sliding end resistance, R_W; the W-H end resistance, R_{WH}; and the W-L end resistance, R_{WL}. Assume that digits of the digital potentiometer is m, the decimal code programming the digital potentiometer is D_n, the range of which is 0 to $(2^m - 1)$. The formulas to calculate R_{WH} and R_{WL} are as follows:

$$R_{WH} = \frac{2^m - D_n}{2^m} \cdot R + R_W \tag{2.1}$$

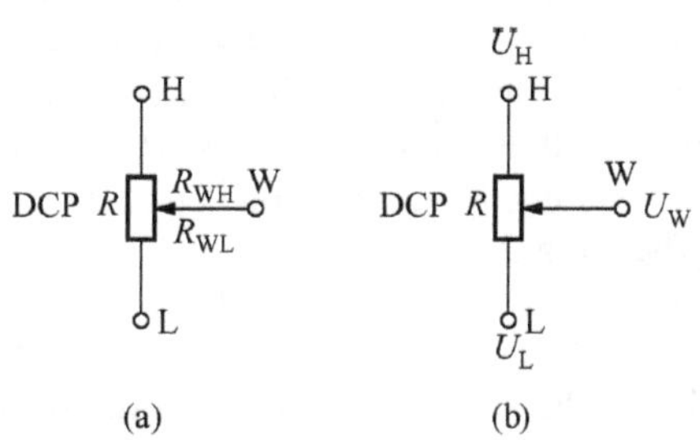

Figure 2.19 The configuration modes of digital potentiometers. (a) The adjustable resistor mode; (b) The voltage divider mode

$$R_{WL} = \frac{D_n}{2^m} \cdot \ R + R_W \tag{2.2}$$

In setting the digital potentiometer, Formulas (2.1) and (2.2) can be used to calculate the resistance value. When the code is 00H (hexadecimal number), the sliding end may be set at the low end L, then $R_{WL} = R_W$. When the code is set at a higher value, the sliding end will move to the high end H of the digital potentiometer.

For example, for the MCP41010 type 256-tap digital potentiometer, $R = 10\,k\Omega$, $R_W = 52\,\Omega$, $m = 8$, $2^m = 256$, D_n = C0H (hexadecimal number) = 192 (decimal number, allowable range 0–255). Substitute them into Formulas (2.1) and (2.2) to get:

$$R_{WH}(C0H) = \frac{256 - 192}{256} \times 10\,k\Omega + 52\,\Omega = 2552\,\Omega$$

$$R_{WL}(C0H) = \frac{192}{256} \times 10\,k\Omega + 52\,\Omega = 7552\,\Omega$$

All the above-mentioned values are typical values, and there will be certain deviation for the actual ones.

In the voltage divider mode when the digital potentiometer is configured into the voltage divider mode, the influence of the resistance at sliding end may not be considered, for the end of the tap is connected to a high impedance circuit. Then, R_{WH} and R_{WL} can be replaced with R_H and R_L, respectively. In Figure 2.19(b), U_H and U_L are, respectively, voltages to ground at the ends of H and L, and U_W is the voltage to ground at the output end. The formula for calculating R_H and R_L is as follows:

$$R_H = \frac{2^m - D_n}{2^m} \cdot R \tag{2.3}$$

$$R_L = \frac{D_n}{2^m} \cdot R \tag{2.4}$$

2.7.1.2 Configuration Mode of Digital Potentiometers with Double Sliding Ends

There is another type of special-purpose digital potentiometers with the structure of "three doubles" (which refers to double interfaces, double circuits, and double sliding ends). The typical product is X9455 produced by XICOR Company, which can realize relatively complex digital control function. X9455 has two serial interfaces, of which one is a two-wire serial interface (SCL and SDA) applicable to reading/writing operations on the digital potentiometer and data transmission, while the other is three-wire boost/buck interface ($U/\overline{D}$, $\overline{CS}$, DS0, and DS1, among which DS0 and DS1 can be taken as one end), which can change the position of the sliding end by boosting/bucking the voltage.

Inside X9455 there is a two-tire serial interface, a three-wire boost/buck interface, circuits of power on reset, interface control and state control, and two digital potentiometers DCP_0 and DCP_1 with two sliding ends and the same structure. Consider the digital potentiometer DCP_0 as an example, whose internal structure is illustrated in Figure 2.20. It includes two resistance arrays, each of which consists of about 255 units of resistance, 256 analog switches, and two independent sliding ends. These analog switches are controlled by two, 8-bit volatile sliding

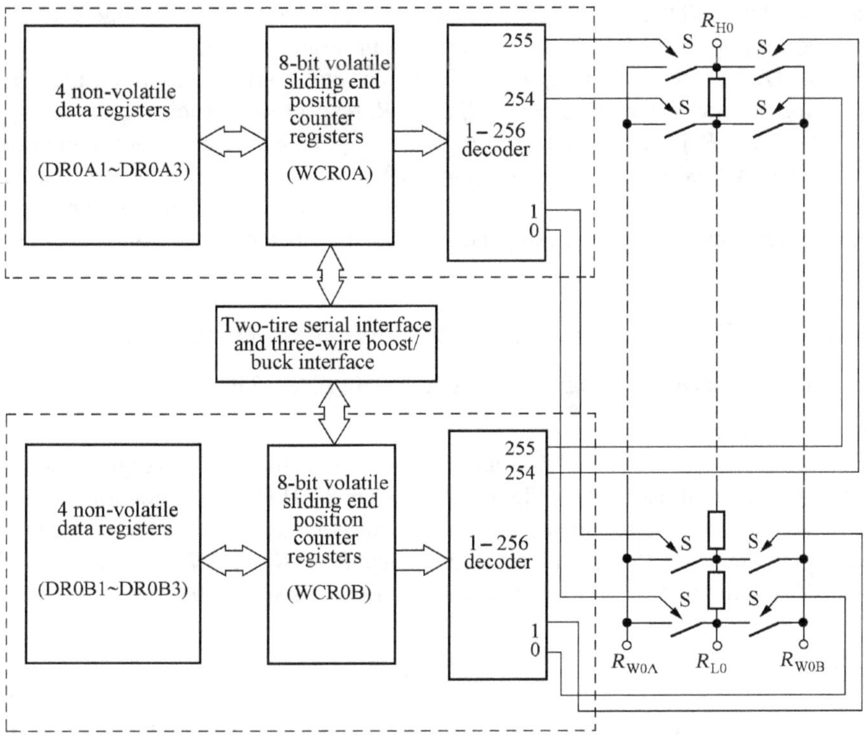

Figure 2.20 The internal structure of the digital potentiometer DCP_0

end position counter registers. One of the switches will be selected and conducted with the 8-bit wave-current ROM (WCR). It should be noted that each sliding end has a specified WCR. When all bits of the WCR are 0, the closest analog switch to R_L end will be selected.

The basic configuration modes of X9455 are illustrated in Figure 2.21(a–c). Figure 2.21(a) shows a 4-end digital potentiometer DPC composed of DCP_0 and DCP_1 of X9455. Figure 2.21(b) shows a 4-end programmable voltage divider composed of only one digital

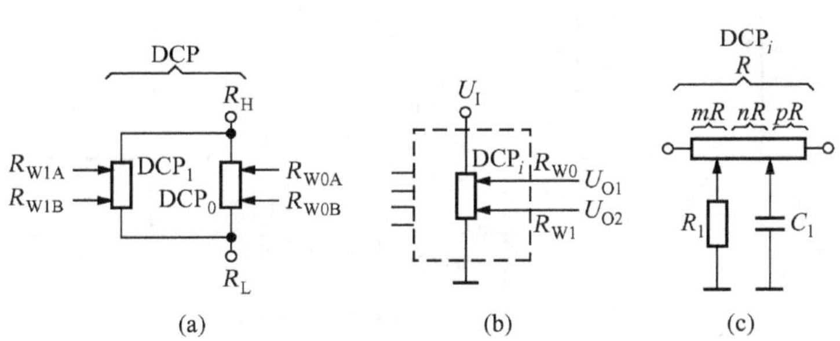

Figure 2.21 The basic method of application of X9455. (a) A 4-end digital potentiometer; (b) a 4-end programmable voltage divider; (c) programmable double T-shaped network

potentiometer DCP_i (DCP_0 or DCP_1) of X9455. U_I is the input voltage and U_{O1} and U_{O2} are two output voltages. Figure 2.21(c) is a programmable double T-shaped network composed of DCP_i. The two sliding ends divide the total resistance R into three parts, with corresponding resistance values of mR, nR, and pR, respectively. Their relational expression is $mR + nR + pR = R$, in which m, n, and p are all proportionality coefficients and satisfy the following relational expression: $m + n + p = 1$ $(0 \leq m,\ n,\ p \leq 1)$. R_1 and C_1 are external resistor-capacitor units. Relatively complex numerical control function can be realized using X9455, which is impossible when using a digital potentiometer with single sliding end.

2.7.2 Circuit Design of Programmable Switching Regulators

2.7.2.1 Typical Circuit of Programmable Switching Regulators

Circuit design scheme

The typical circuit of programmable switching regulators consisting of a digital potentiometer and LM2576-ADJ is illustrated in Figure 2.22. Now a 10 kΩ digital potentiometer DCP is used to replace the sample resistance R_2. The output voltage of the voltage regulator can be set by changing the value of R_{DCP}. For example, when $R_2 = R_{DCP} = 7.76\,\text{k}\Omega$ and $R_1 = 2.0\,\text{k}\Omega$, $U_O = 6.00\,\text{V}$. U_O in the circuit can vary within the range of +1.23 to 6 V.

Instructions

1. As the electric current passing through the digital potentiometer in the switching regulator is low, there is no need to spread its working current in general.
2. For the purpose of improving the stability of output voltage, the metal film resistor with 1% error shall be used for R_1.
3. For the purpose of reducing the output ripple, a stage of LC post-filter can be added at the output stage. Linear integrated voltage regulator can be equipped to LM2576 to constitute a compound regulated power supply which takes LM2576 as the preceding stage and the linear integrated voltage regulator (such as 7805) as the back stage so that respective advantages of the switching regulator and the linear voltage regulator can be brought into full play to design regulated power supply with high efficiency and small output ripple voltage.

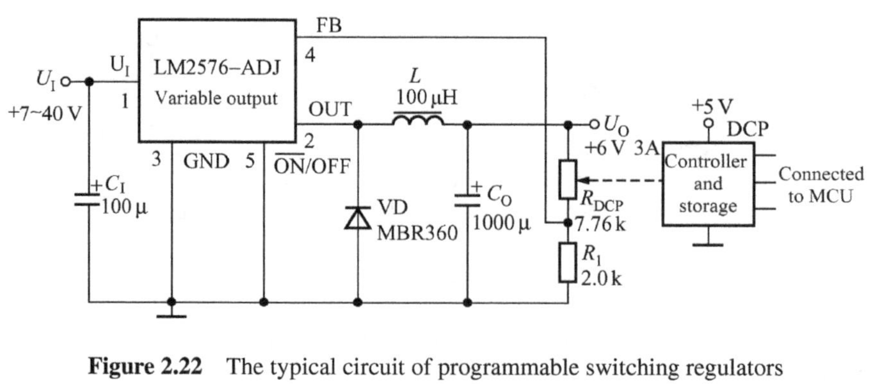

Figure 2.22 The typical circuit of programmable switching regulators

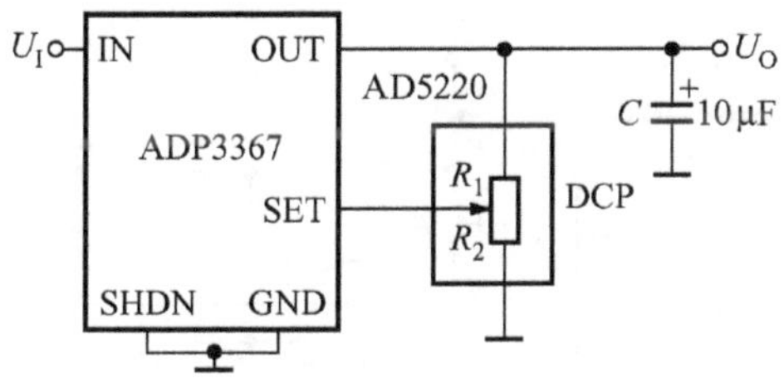

Figure 2.23 The circuit of programmable linear voltage regulator composed of digital potentiometer AD5220

4. When the operating junction temperature of LM2576-ADJ exceeds 110 °C, a bigger radiator must be added to LM2576.
5. The layout of the printed circuit board (PCB) should be scientifically designed, ensuring the minimum ground loop area for the electrical inductance and the shortest leads for C_I, VD, and C_O. Besides, they should be grounded at a single point (which means all the ground terminals should be welded to the same point in the ground wire area). Feedback leads should also be as short as possible. The sample resistance should be as close to the voltage regulator as possible.

2.7.2.2 Programmable Switching Regulators Composed of Digital Potentiometer AD5220

The circuit of programmable linear voltage regulator composed of digital potentiometer AD5220 is illustrated in Figure 2.23. AD5220 is a 128-cap digital potentiometer based on three-wire boost/buck serial interfaces. In the circuit, a low dropout regulator ADP3367 of +5 V fixed or adjustable output is used. When the input–output dropout is 300 mV, the maximum output current may reach 300 mA. The output current is determined by the following formula:

$$U_O = 1.255 \times \left(1 + \frac{R_2}{R_1}\right) \tag{2.5}$$

2.7.2.3 Programmable Switching Regulator Composed of Digital Potentiometer AD5222

The circuit of programmable switching regulator composed of digital potentiometer AD5222 is illustrated in Figure 2.24. AD5222 is a double-circuit, 128-cap digital potentiometer based on three-wire boost/buck serial interfaces, which has four specifications of resistance value: 10 kΩ, 50 kΩ, 100 kΩ, and 1 MΩ. In the circuit, a high-efficiency ADP3020 DC/DC power converter is used, whose switching frequency can reach 300 kHz with conversion efficiency of over 96%. When the specific values of the internal resistance in DCP_1 and DCP_2 are equal, $R_1/R_2 = R_3/R_4$, and the output voltage is expressed as follows:

$$U_O = 1.2 \times \left(1 + \frac{R_1}{R_2}\right) \tag{2.6}$$

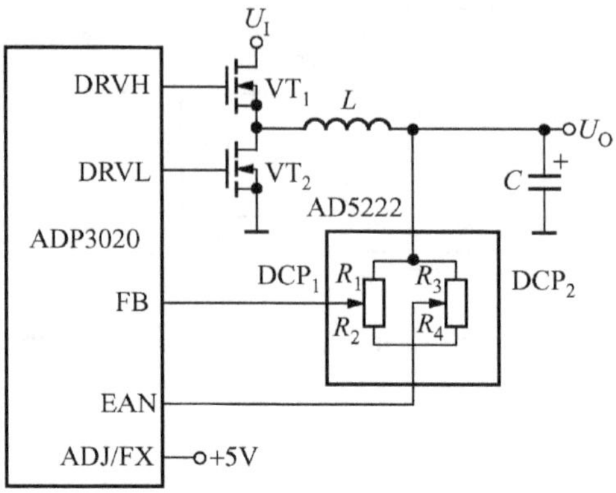

Figure 2.24 The circuit of programmable switching regulator composed of digital potentiometer AD5222

2.7.2.4 Boost DC/DC Power Converter Composed of Digital Potentiometer DS1845

The circuit of boost DC/DC power converters composed of digital potentiometer DS1845 is illustrated in Figure 2.25. Schottky diode is used for VD. DS1845 is a 256-cap nonvolatile digital potentiometer. MAX5025 is selected for boost DC/DC power converters. The rated value of U_O is 32 V with an adjustable range between 27.6 and 36.7 V and a stepping quantity of 35.5 mV.

2.7.2.5 Buck DC/DC Power Converter Composed of Digital Potentiometer DS3903

The circuit of buck DC/DC power converter composed of digital potentiometer DS3903 is illustrated in Figure 2.26(a) and (b). Figure 2.26(a) shows the simplified circuit and Figure 2.26(b) the actual circuit. DS3903 is equipped with two 10 kΩ digital potentiometers and a 90 kΩ one, of which only the latter is used. MAX1776 is a buck DC/DC power converter with switches inside. Its input voltage is 4.5–24 V and output voltage is either fixed or adjustable voltage of 5 V. Its maximum output current is 600 mA. The value of the output voltage can be adjusted by changing the positions of the sliding ends of DS3903 through the two-wire serial bus.

2.8 Digital Power Supply System

Currently, the SMPS is developing to be digital and intelligent. The digital power supply developed at the beginning of the twenty-first century is drawing extensive public attention with its good qualities and advanced monitoring function. The digital power supply provides intellectualized adaptability and flexibility with such power supply management functions as direct monitoring, remote fault diagnosis, and fault handling, which can satisfy the needs of multipurpose power supply systems.

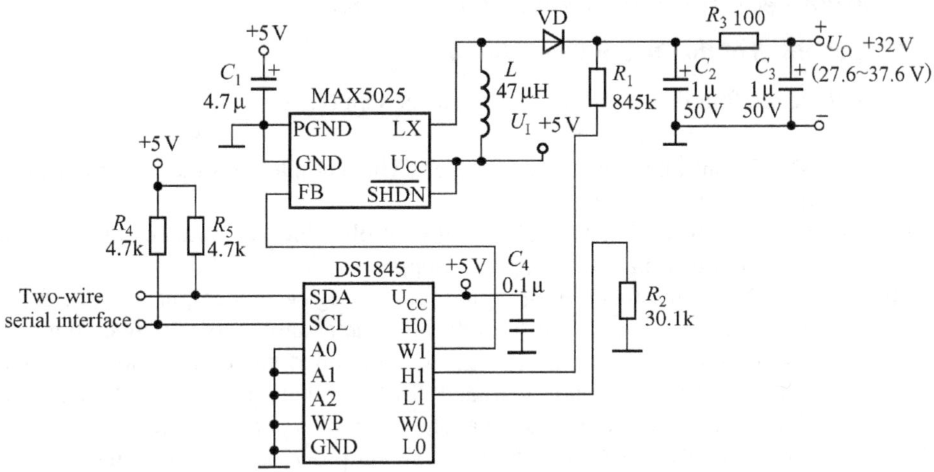

Figure 2.25 The boost DC/DC power converter composed of digital potentiometer ds1845

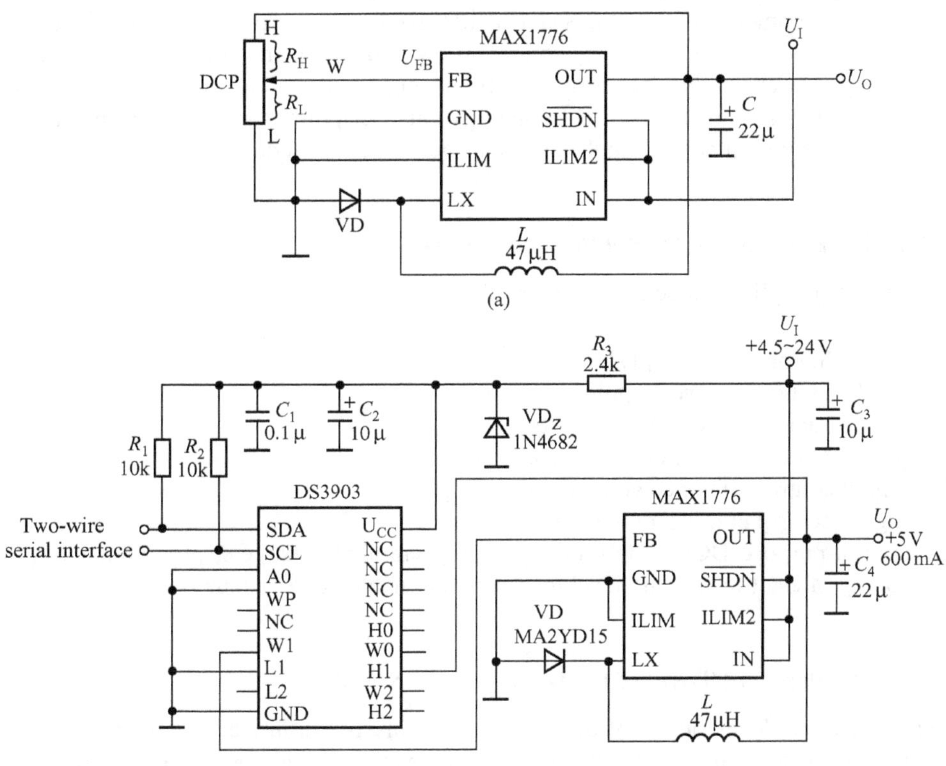

Figure 2.26 The buck DC/DC power converter composed of digital potentiometer DS3903. (a) Simplified circuit; (b) actual circuit

2.8.1 Main Features of the Digital Power Supply

2.8.1.1 Definition of the Digital Power Supply

At present, there are four definitions of digital power supply:

Definition 1: SMPS controlled by digital interface (which emphasizes the "communications" function of digital power supply).

Definition 2: SMPS with the function of digital control (which emphasizes the "numerical control" function of digital power supply).

Definition 3: SMPS with the function of digital monitoring (which emphasizes the function of digital power supply in "monitoring" such parameters as temperature).
The common feature of the above-mentioned three definitions is "analog SMPS → reformed and upgraded SMPS," that emphasizes "power supply control" mainly over the external characteristics of SMPS such as U_O and I_O.

Definition 4: Power supply products with digital signal processor (DSP) or microcontroller unit (MCU) as the core and digital power supply driver and PWM controller as the object of control, which can realize such functions as control, management, and monitoring. They change their external characteristics by setting internal parameters of the SMPS and are designed with the "power supply management" based on "power supply control." Power supply management refers to delivering power supply efficiently to different components of the system so as to reduce the loss maximally. Digital technology must be applied throughout the management of digital power supplies (such as power supply sequencing).

2.8.1.2 Classification of Digital Power Supplies

Digital power supplies can be classified as follows:

1. Standard digital power supply
2. Generalized digital power supply
   a. digital control power (DCP) supply;
   b. programmable switching voltage regulator;
   c. digital control reference voltage source;
   d. digital control current source;
   e. special-purpose DCP supply (such as programmable chargers, programmable power supply of electric welding machines, programmable power adapters).

2.8.1.3 Development Trend of the Digital Power Supply

The research and development of digital power supply began in late 1990s. In 2005, Texas Instruments (TI) took the lead in launching the innovative digital power supply and proposed the solution of Fusion Digital Power™. The solution includes the following three kinds of chips: (i) digital power supply driver (such as UCD7100); (ii) PWM controller (such as UCD8620); and (iii) DSP (such as DSPs UCD9110 and UCD9501 of UCD9K series). Lately, TI launched the third-generation UCD92×× series (including UCD9220, UCD9222,

UCD9240, UCD9246, and UCD9248) integrating digital power supply controller and brand-new plug-in module, which not only supports voltage tracking, tolerance, sequencing, communications and monitoring, but also makes programming more flexible and convenient, further improving the degree of intellectualization of the management of power supply systems.

At present, famous chip manufacturers such as Freescale, Ericsson, Atmel, and Silicon are all developing digital power supply IC, driving the digital power supply to enter a new stage of rapid development. As is predicted by Darnell Group, the price of digital power supply controller IC will be roughly equal to that of the existing analog products. Meanwhile, with the launch of advanced design tool software for digital power supply, the fear of programming of analog power supply designers can be thoroughly eliminated. The tool software can provide graphical user interface (GUI) which needs no programming, thus greatly simplifying power supply design and shortening the R&D cycle. Besides, the CPU of the host computer can communicate with the power supply through interfaces, and the configuration of the power supply can be changed by simply using the software without changing any hardware. This is an advantage over the traditional power supply system. For example, the PC-based full-functional design tool software Digital Power™ Designer, which is provided by IT, can not only simulate, configure, and monitor power supply performance, but also realize circuit simulation and thermal simulation.

Although the digital power supply is still restricted by such factors as weak development capacity of users and high system cost, it is a new technology of great historical significance. The product development of "linear power supply → switching-mode power supply → digital power supply" is an irresistible trend. For example, when the regulated power supply was upgraded from the linear power supply into the SMPS in late 1970s, such problems as high cost and loud output noise also occurred. However, the SMPS has taken the place of the traditional linear power supply in many fields and has become a main-stream product. With the development of technology and expansion of market demand, the upgrading from the analog power supply to the digital power supply will be greatly speeded up. At present, the market sales of digital power supply IC have increased from USD169 million in 2006 to USD796 million in 2011, with an annual average growth rate of 36.4%, which fully pictured the broad development prospect of the digital power supply.

In conclusion, the digital power supply system has such significant advantages as high integration density, high performance-price ratio, advanced power supply management function, simple peripheral circuit and user-friendliness, which created favorable conditions for realizing the optimal design of intelligent power supply systems.

2.8.1.4 Performance Comparison between Digital Power Supplies and Analog Power Supplies

See Table 2.5 for the performance comparison between digital power supplies and analog power supplies. Several points need to be noted: firstly, the traditional digital power supply, which only powers on and off or adjusts the output voltage, is not digital in real sense. Secondly, both the digital power supply and the analog power supply are only habitual titles, between which there is no strict boundary. The PWM of analog SMPS includes such digital circuits as clock and gate circuit, while the digital power supply includes such analog circuits as the reference voltage source and power devices. Therefore, it is

Table 2.5 Performance comparison between digital power supplies and analog power supplies

Digital power supply	Analog power supply
With DSP or MCU (DSP-controlled SMPS may adopt the digital filter with more powerful control, faster response, and better voltage regulation performance)	—
Field programmable (communication, detection, telemetry, and other functions can be realized through software programming)	—
With the control, management, and monitoring functions for complex control	—
Replacement of hardware is not necessary when the performance indexes or power configuration is adjusted	Replacement of hardware is required when the performance indexes are adjusted
Give full play to the advantages of digital signal processor and microcontroller, designing a high-tech digital power supply. For example, the resolution of its pulse width modulation (PWM) can reach the level of 150 ps (1 ps = 10^{-12} s), which is unmatched by traditional SMPS.	Poor performance and reliability
Functions such as multi-phase control, nonlinear control, fuzzy control, load sharing, and failure prediction can be realized, facilitating the development of green energy-saving power supplies	—
Highly integrated, for the convenience of establishing a distributed digital power system	Low integration
Technically complex, and programming is necessary	Technically simple, and no programming is required
Complex internal structure of the digital power system, but with a simple peripheral circuit	Simple internal structure, but with a complex peripheral circuit
High cost	Low cost

not accurate to consider the digital power supply to be completely digital (or fully digital). Thirdly, the analog power supply is a good choice for applications with fewer control parameters and lower cost, while the digital power supply is more applicable to the high-end power systems with more control parameters, faster real-time response, and multiple functions.

2.8.1.5 Typical Products and Applications of Digital Power Supplies

The currently produced digital power chips mainly include four categories: digital power controller, digital power control driver, digital power PWM controller, and digital power transmission module. See Table 2.6 for the new digital power controllers and digital power control drivers produced by TI Company that can provide solutions to designs of AC/DC or DC/DC isolated/non-isolated digital power supplies. The digital power supply is particularly applicable to the high-end power system from AC circuit to load and is widely used in the power

Table 2.6 New digital power controllers and digital power control drivers

Product type	Model	Main features	Encapsulation mode
Digital power controller	F28××Piccolo series	The F28×× Piccolo™ MCU can provide C28× core power, and provide devices with few pins with highly integrated peripheral control equipment	TSSOP-38
	F28×× Delfino	The highly integrated MCU with flash memory is used in the demanding control applications	LQFP-176, BGA-179
	F28×× series	32-bit digital signal controller with ROM	BGA-100, LQFP-100 BGA MICROSTAR-100
	UCD9240	The fully configurable multi-output and multi-phase non-isolated DC/DC PWM controller can control up to 4-way and 8-phase output	VQFN-64, TQFP-80
	UCD9220	The dual-output, multi-phase synchronous buck controller supports the configuration of multiple load points	QFN-48
	UCD9246	The fully configurable multi-output and multi-phase non-isolated DC/DC PWM controller can control up to 4-way and 6-phase output	VQFN (RGC)
	UCD9248	The fully configurable multi-output and multi-phase non-isolated DC/DC PWM controller can control up to 4-way and 8-phase output	TQFP-80
Digital power control driver	UCD7231	Dual high-current driver. Fully compatible with TI fusion digital power controller (such as UCD91xx and UCD92xx series)	QFN-20
	UCD7230A	The operating frequency and duty ratio can be set	QFN-20
	UCD7242	Fully integrated power switch, equipped with the driver for dual synchronous buck converter	QFN (RSJ)
	CSD16325Q5	NexFET™ power MOSFET is designed for the $+10/-8$ V grid-source voltage U_{GS}. It can minimize the drain current loss of power converters, and has been optimized for 5 V grid drivers	SON (DQH)
	CSD16403Q5A	NexFET power MOSFET can minimize the loss of power converter	SON (DQJ)
	CSD16323Q3	NexFET power MOSFET can minimize the loss of power converters, and has been optimized for 5 V grid drivers	SON (DQG)
	CSD16408Q5C	NexFET power MOSFET can minimize the loss of power converter	SON (DQU)

systems for mobile communication equipment, computer servers, and data centers as well as the uninterruptible power supply (UPS).

2.8.2 Basic Constitution of the Digital Power Supply

2.8.2.1 Digital Signal Processor

UCD9501 is a DSP particularly designed by TI Company for digital power systems, and similar products include TMS320F2808 and TMS320F2806, which mainly include the 32-bit CPU of 100 MHz, clock oscillator, three 32-bit timers, watchdog circuit, internal/external interrupt controller, SCI bus, SPI bus, CAN bus, and I^2C bus interface, 12-way PMW signal output, system controller, 16-channel 12-bit ADC, 16K × 16Flash, 6K × 16SARAM, and 1K × 16ROM. It adopts the standard 3.3 V input–output interface, being perfectly compatible with UCD8K series. It is programmable with Power PAD™ HTSSOP and QFN software packages.

2.8.2.2 Digital Power Driver

Both UCD7100 and UCD7201 belong to the digitally controlled power driver chip, which are applicable to UCD9110 and UCD 9501 digital controllers. UCD7100 is a single-ended output and UCD7201 is a double-ended one. However, both of their rated output currents are ±4 A and can drive the MOSFET power switch tube. The main controller can monitor the output current, detect the over-current fault quickly, and power off the power supply promptly with a detection cycle of only 25 ns.

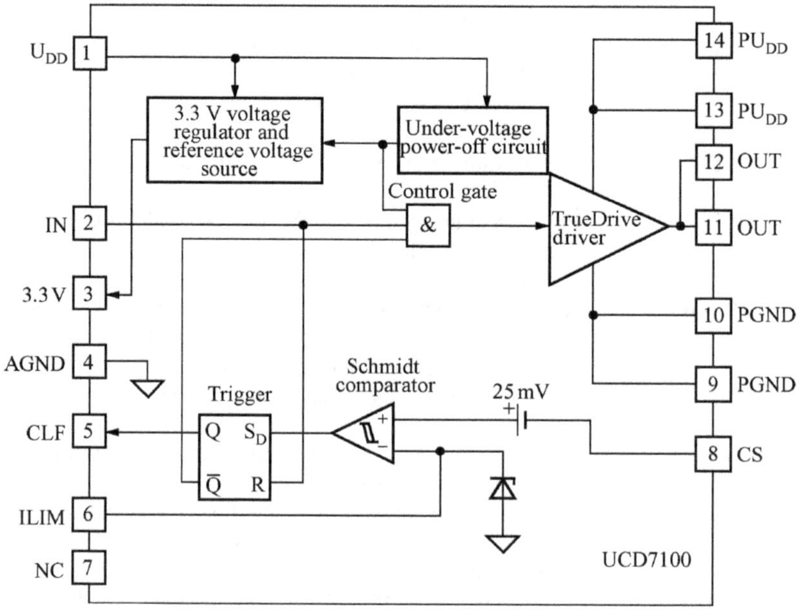

Figure 2.27 The internal block diagram of UCD7100

Consider UCD7100 as an example, whose internal block diagram is shown in Figure 2.27. UCD7100 mainly includes 3.3 V voltage regulator and reference voltage source trigger, Schmidt comparator, under-voltage power-off circuit, control gate, True Drive driver. "True Drive," the proprietary technology of TI Company, is a mixed output stage constituted by creating a pull-up/pull-down circuit with a parallel bipolar transistor and the MOSFET tube. This technology is featured by strong driving capability. It can output normally under low voltage and control the over-voltage and under-voltage protection of external power MOSFET under extremely low output impedance, and the power MOSFET need not be connected with the protective Schottky clamp diode. UCD7100 can provide the grid of MOSFET with a peak current in hundreds of nanoseconds to power on the driver rapidly. And the high-impedance digital input (IN) of UCD7100 can receive the 3.3 V logic level signal with a maximum switching frequency of 2 MHz. The internal circuit can be isolated from the external noise using the Schmidt comparator. UCD7100 can provide the gate pole of MOSFET with a peak current in a few nanoseconds to power on the driver quickly. In case that the PWM output of the controller stops at the high logic level and the over-current failure occurs, the current detection circuit will turn off the output of the driver, and the system can enter the retry mode. The chip can be restarted via the watchdog circuit in the DSP or MCU. The 3.3 V and the 10-mA voltage regulator in the UCD7100 can be used as the power supply of digital controllers.

2.8.2.3 PWM Controller

Both UCD8220 and UCD8620 are double-ended push-pull PWM controllers digitally controlled by DSP or MCU. UCD8220 can be started with a low voltage of 48 V, while UCD8620 is equipped with a 110-V high-voltage starting circuit inside. The internal block diagram of UCD8220 is illustrated in Figure 2.28. UCD8220 mainly includes 3.3 V voltage regulator and reference voltage source, PWM, driving logic, push-pull driver, under-voltage power-off circuit, current limiting circuit, and current detection circuit. UCD8220 and UCD 8620 can run in the peak current mode or the voltage mode. They can not only program the limited current but also output a limited current digital sign under the monitoring of the main controller. The time sequence waveforms of UCD8220 and UCD 8620 are illustrated in Figure 2.29.

2.8.3 *Circuit Design of Digital Power Supply*

2.8.3.1 Circuit Design of Digital Power Supply

Digital power supplies may consist of six parts, including PWM, power driver, DSP, interface circuit, display, and keyboard. The systematic block diagram is illustrated in Figure 2.30 where the DSP UCD9501 is connected with the keyboard and display with the interface chip so that users can not only observe the current power parameters from the display but also modify the power parameters at any time with the keyboard.

To simplify the configuration, the digital power supply may be composed of DSP (UCD9501) and DCP driver (UCD7100), and the circuit is illustrated in Figure 2.31. Passing the rectifier and filter, AC voltage is turned into +36 to 72 V DC input voltage U_I is connected to the primary winding of the high-frequency transformer and then to the analog input terminals AN1 and AN2 of UCD9501, respectively, after passing the dividers R_1 and R_2.

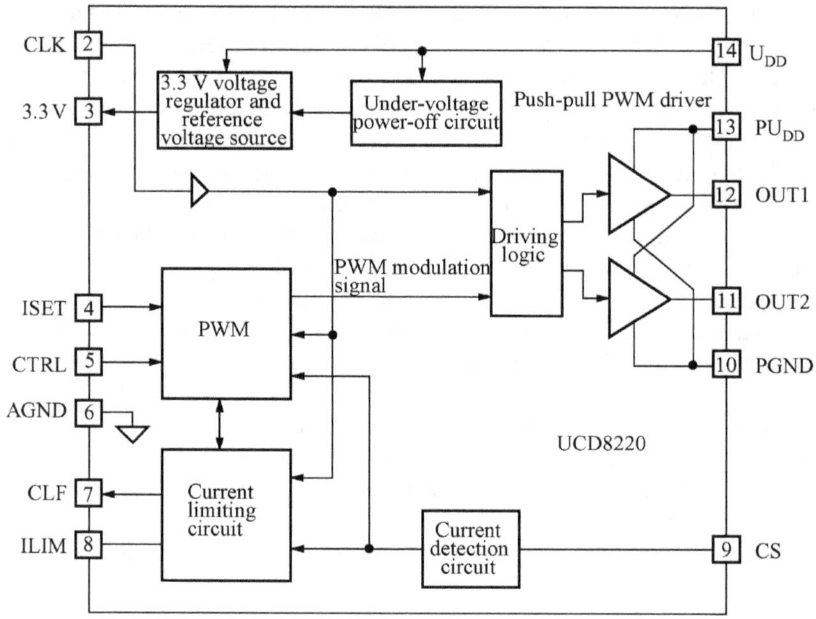

Figure 2.28 The internal block diagram of UCD8220

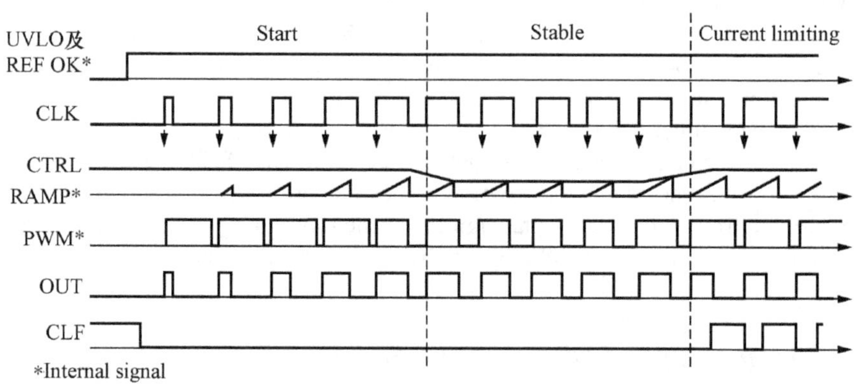

Figure 2.29 The time sequence waveforms of UCD8220/8620

The other end of the primary winding is connected to the power MOSFET. R_3 is the current limiting resistor and R_4, the current sensing resistor. The output voltage of the bias winding is rectified and filtered by VD_1 and C_1 into +12 V DC bias, which is connected to the power terminal U_{DD} of UCD7100. UCD9501 is powered by the 3.3 V voltage from UCD7100. The secondary rectification filter circuit is composed of VD_2, L, and C_2; VD_3 is the FWD and U_O, the DC output voltage. The PWMA from UCD9501 is sent to the IN end of UCD7100. The limited current flag terminal (CLF) of UCD7100 is connected to the interrupt terminal (INT) of UCD9501, and the limited current setting terminal (ILIM) is connected to the GMTR

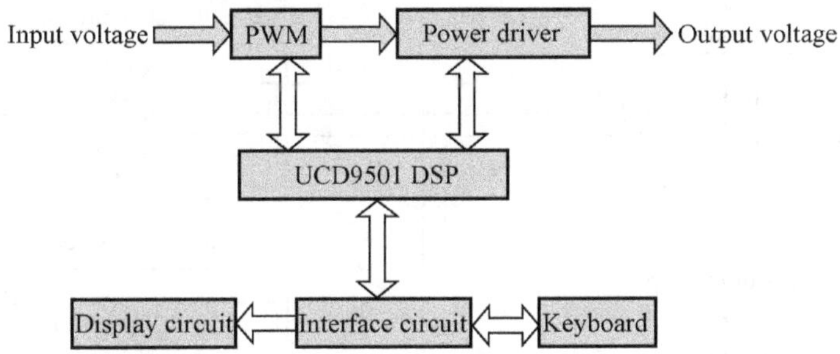

Figure 2.30 The block diagram of digital power supplies

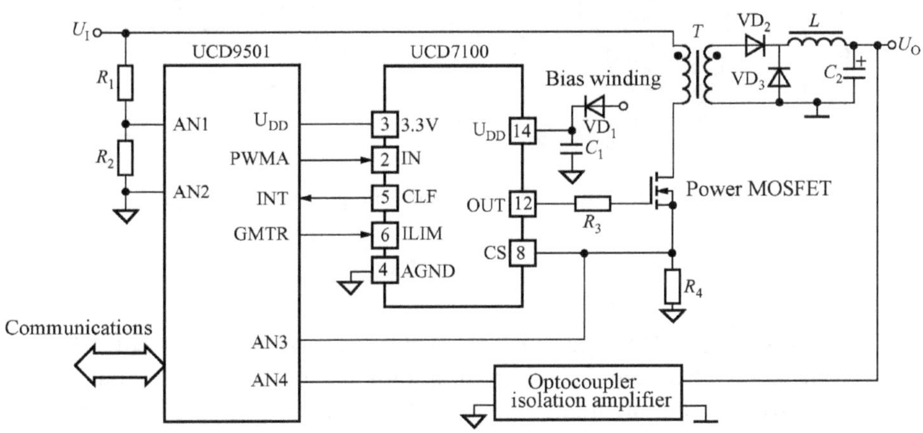

Figure 2.31 The typical circuit of the digital power supply

terminal of UCD9501. The output and input stages can be separated with the optocoupler isolation amplifier.

When $U_{DD} = 12\,V$, the load capacitor of UCD7100 $C_{LOAD} = 10\,nF$, the switching frequency $f = 300\,kHz$, the bias power consumption $P = C_{LOAD}U_{DD}^2f = 10\,nF \times (12\,V)^2 \times 300\,kHz = 0.432\,W$, and the bias current $I = P/U_{DD} = 0.432\,W/12\,V = 0.036\,A$.

If UCD7201 is used, it can drive two external power MOSFETs. In addition, UCD9501 and UCD8620 can be used to constitute digital power system.

2.8.3.2 Basic Structure of Digital Power System

The basic constitution of digital power system is illustrated in Figure 2.32, which mainly includes AC/DC converter, bus interface, multi-, dual- and single-path point of load (POL) converters.

POL converter is called POL converter or POL power for short, which means place an independently buck DC/DC converter near every POL, so as to transform the originally designed "multi-output AC/DC power" into a power system constituting "single-output

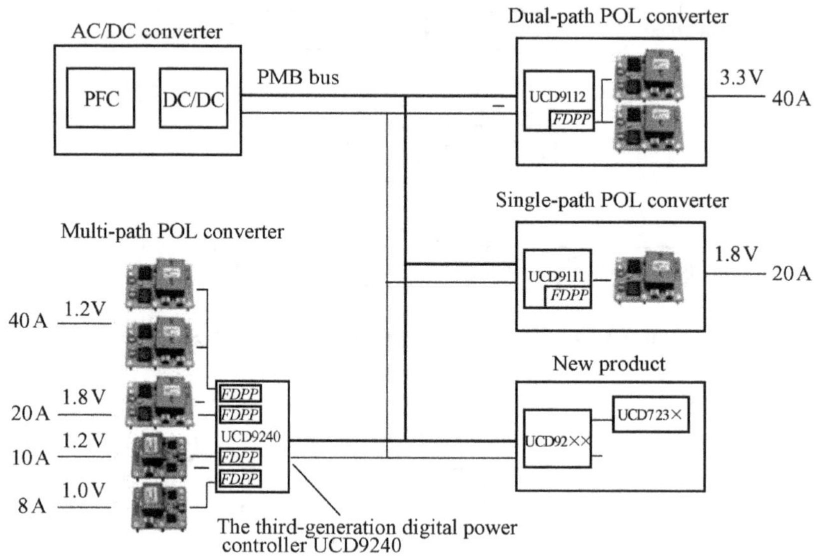

Figure 2.32 The basic constitution of digital power system

AC/DC power + multiple DC/DC converters." This helps to avoid the noise disturbance and the mutual interference between the voltage drop of wires and the different output voltages caused by long wiring from the multi-output AC/DC power to each load, ensuring that the MCU of each load can be powered with accurate, stable, and high-quality supply voltage and meet the demands for low-voltage and high-current output.

2.9 Energy-Saving and Environment-Friendly Technology of SMPS

At present, the integrated circuit of SMPS is developed to be more energy-saving and environment-friendly. Many well-known IC manufacturers attach great importance to energy efficiency and endeavor to develop the energy-saving and environment-friendly integrated circuit of SMPS.

2.9.1 Reduce the Switching Losses with Valley Switching Circuit

The distributed capacitance on the primary winding of the high-frequency transformer is the drain capacitance C_D on the drain pins of MOSFET. The resonant circuit LC composed of C_D and primary winding inductance L_P will form a ringing voltage, which belongs to the disturbance voltage of damped oscillation, and the oscillation frequency of which shall be determined by the formula:

$$f_{\text{ringing}} = \frac{1}{2\pi\sqrt{L_P C_D}} \tag{2.7}$$

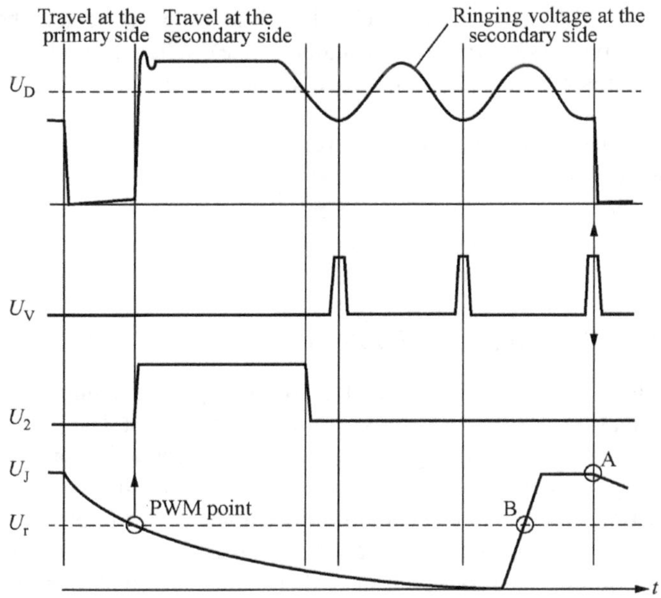

Figure 2.33 Waveforms of valley switching signal with drain voltage and ringing voltage

Obviously, the power loss caused by the ringing frequency during the conducting period of MOSFET is as follows:

$$P_{ON} = \frac{1}{2} C_D U_D^2 f_{ringing} \tag{2.8}$$

To reduce the switching losses, a valley switching circuit is especially equipped in the chip. The waveforms of valley switching signal (U_V) with drain voltage and ringing voltage are illustrated in Figure 2.33. The ringing voltage ($U_{ringing}$) is just overlaid on the waveform of drain voltage. Once the ringing voltage reaches the valley, the valley switching circuit generates a valley switching signal (positive pulse) and cut off the MOSFET, reducing the switching losses. U_2 in the figure refers to the voltage of secondary winding. Point A refers to the starting of a new oscillation cycle with valley switching signal, and point B refers to the beginning of a new oscillation cycle by the traditional PWM.

2.9.2 Reduce the No-Load Power Consumption with EcoSmart Energy-Saving Technology

EcoSmart® is a new energy-saving technology for SMPS launched by PI Company. The single-chip SMPS designed with this technology can significantly reduce the no-load power consumption and improve the power efficiency. At present, PI Company has applied this technology in the integrated circuits of single-chip SMPS, including LinkSwitch, LinkSwitch-XT, LinkSwitch-LP; and LinkSwitch-TN, TinySwitch-III, TOPSwitch-GX, and PeakSwitch series, with the number of the production exceeding one billion.

The theory of EcoSmart energy-saving technology is to design a circuit in the chip for detecting the SMPS in low power state (no-load or standby state). When the no-load state is detected, the following measures will be taken to improve the power efficiency:

1. Reduce the load current by reducing the duty ratio
2. Reduce the output power by "skipping cycles"
3. Reduce the switching losses by automatically reducing the switching frequency, and improve the power efficiency in low-power condition.

2.9.3 Lead-Free Packaging Technology

The lead (Pb) pollutes the environment, and lead poisoning may damage the nervous system, digestive system, and blood circulation system of human beings. Therefore, lead-free packaging has become a critical technology in semiconductor production process. The so-called lead-free packaging aims to eliminate the high energy-consumption and heavy-pollution packaging production process and replace the traditional tin-lead solder with the lead-free one. EU has already developed the restriction of hazardous substances (RoHS) in 2006, prohibiting the use of any hazardous substance that pollutes the environment in the electrical and electronic products and equipment. At present, many integrated circuit manufacturers have applied the lead-free packaging process. For example, the suffix "N" in the model "TOP261EN" of the newly developed single-chip SMPS of PI Company stands for lead-free packaging.

3

Topologies of the DC/DC Converter

3.1 Topologies of the DC/DC Converter

The DC/DC converter mainly includes the following 20 kinds of topologies:

1. Buck converter
2. Boost converter
3. Buck-boost converter
4. Charge pump converter or inverting converter
5. Cuk converter
6. Single-ended primary inductor converter (SEPIC)
7. Zeta converter (ζ converter, similar to SEPIC, but its switch, inductor, and capacitor are located differently in the circuit)
8. Flyback converter
9. Forward converter
10. 2 Switch forward converter
11. Active clamp forward converter
12. Half-bridge converter
13. Full-bridge converter
14. Push-pull converter
15. Phase shift switching ZVT (phase shift switching zero voltage transition)
16. Zero current switching (ZCS) converter
17. Soft switching converter
18. Half-bridge LLC resonant converter
19. Hybrid converter, consisting of the DC/DC (or AC/DC) converter and low dropout (LDO) linear regulator and
20. Programmable converter, consisting of digital potentiometer and DC/DC converter.

See Table 3.1 for 12 commonly used topologies of the DC/DC converter, wherein figures (a)–(l) show different circuit structures, and figures (m)–(x) show corresponding voltage and

Optimal Design of Switching Power Supply, First Edition. Zhanyou Sha, Xiaojun Wang, Yanpeng Wang, and Hongtao Ma.

Table 3.1 The 12 kinds of commonly used topologies of DC/DC converter

Regulator type	Buck	Boost	Buck-boost or inverting converter	SEPIC
Circuit structure	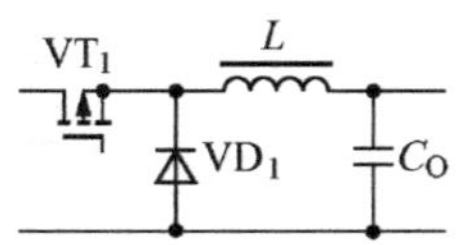 (a)	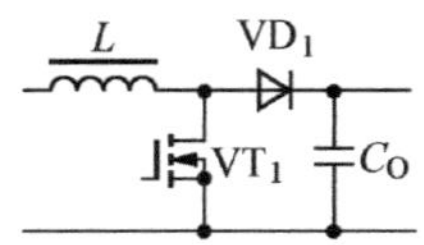 (b)	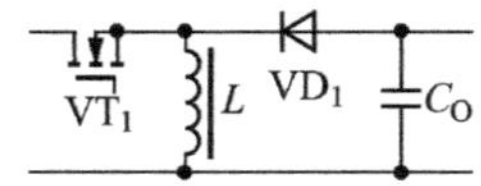 (c)	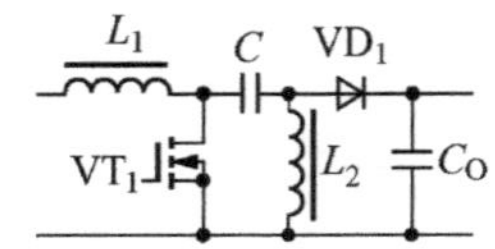 (d)
Ideal transfer function	$\frac{U_O}{U_I} = \frac{t}{T} = D$	$\frac{U_O}{U_I} = \frac{T}{T-t} = \frac{1}{1-D}$	$\frac{U_O}{U_I} = -\frac{t}{T-t} = -\frac{D}{1-D}$	$\frac{U_O}{U_I} = \frac{D}{1-D}$
Maximum drain current	$I_{D\,(max)} = I_O$	$I_{D\,(max)} = \left(\frac{D}{1-D}\right)\ I_O$	$I_{D\,(max)} = \left(\frac{D}{1-D}\right)\ I_O$	$I_{D\,(max)} = \left(\frac{D}{1-D}\right)\ I_O$
Drain voltage	$U_{DS} = U_I$	$U_{DS} = U_O$	$U_{DS} = U_I - U_O$	$U_{DS} = U_I + U_O$
Current on the output rectifier diode	$I_{F1} = (1-D)\ I_O$	$I_{F1} = I_O$	$I_{F1} = I_O$	$I_{F1} = I_O$
Reverse voltage of the output rectifier diode	$U_{R1} = U_I$	$U_{R1} = U_O$	$U_{R1} = U_I - U_O$	$U_{R1} = U_I + U_O$
Voltage and current waveforms	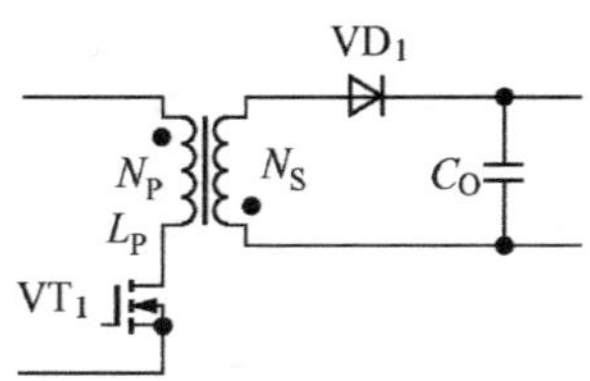 (e)	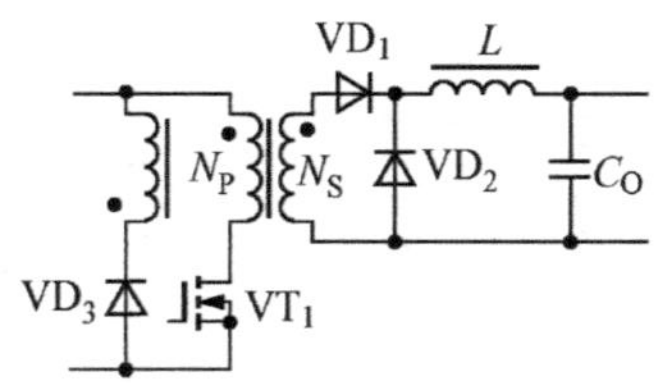 (f)	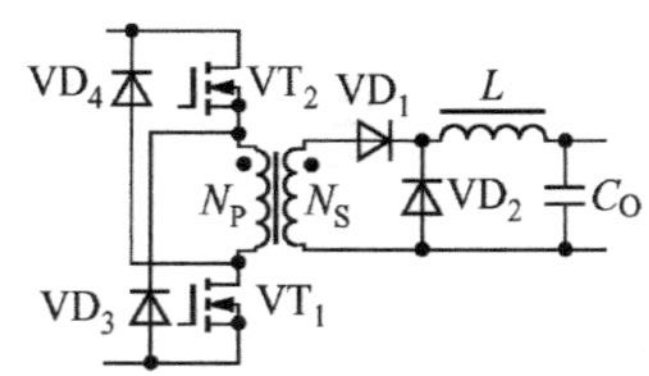 (g)	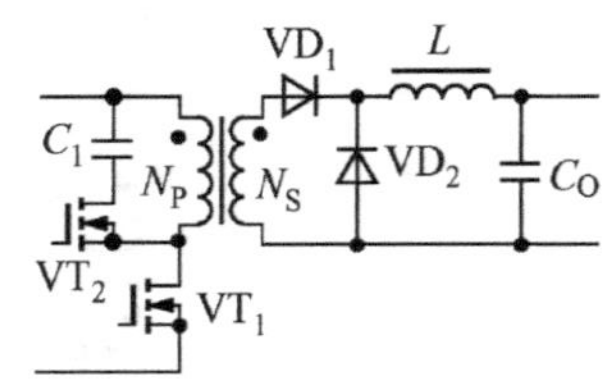 (h)
Main features	$U_O < U_I$	$U_O > U_I$	$U_O < U_I$ or $U_O > U_I$ or $U_O = -U_I$	Adopting two inductors, with high conversion efficiency, suitable for battery-powered portable devices

Typical products	BP2802	LT1937	LTC3452	LM5022
	LM3402	BP1601	LTC3780	AP3031
	CAT4201	XL6004	LTC3453	MAX16807
	SLM2842S	LM3509	SP6686	LTC3783
Regulator type	Flyback	Forward	2-Switch forward	Active clamp forward
Circuit structure	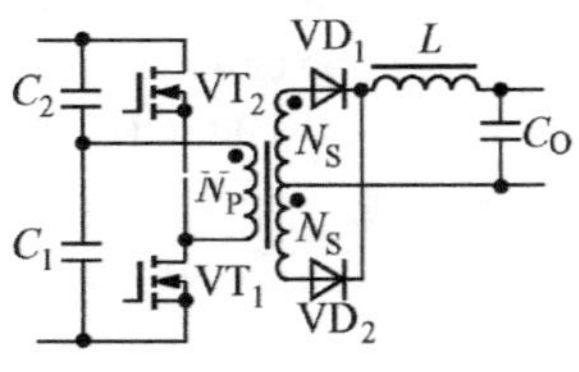 (i)	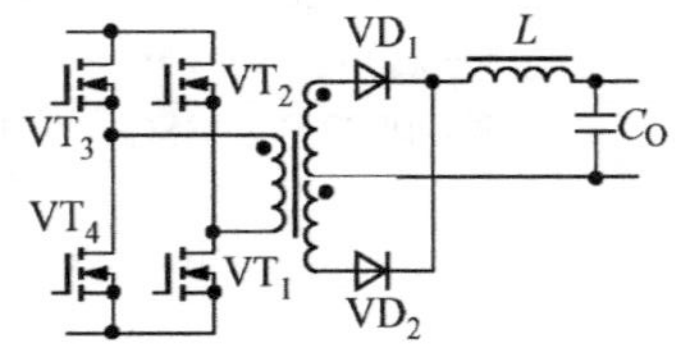 (j)	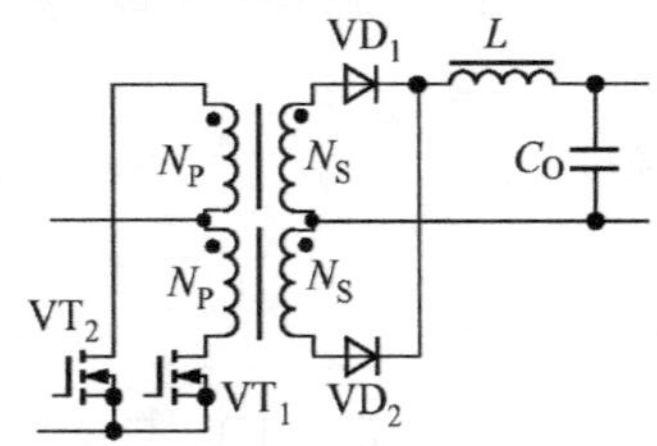 (k)	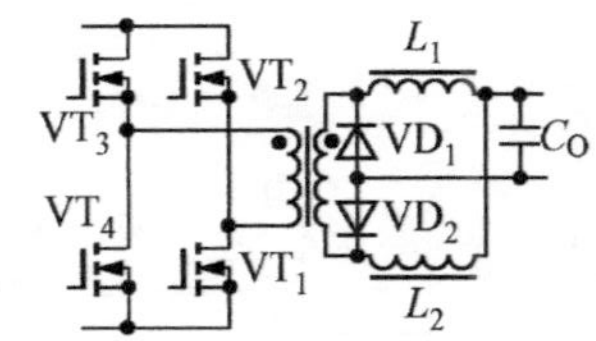 (l)
Ideal transfer function	$\frac{U_O}{U_I} = D\sqrt{\frac{TU_O}{2I_O L_P}}$	$\frac{U_O}{U_I} = \frac{N_S}{N_P} \cdot \frac{t}{T} = \frac{N_S}{N_P} \cdot D$	$\frac{U_O}{U_I} = \frac{N_S}{N_P} \cdot \frac{t}{T} = \frac{N_S}{N_P} \cdot D$	$\frac{U_O}{U_I} = \frac{N_S}{N_P} \cdot \frac{t}{T} = \frac{N_S}{N_P} \cdot D$
Maximum drain current	$I_{D\,(max)} = \frac{U_I t}{L_P}$	$I_{D\,(max)} = \frac{N_S}{N_P} \cdot I_O$	$I_{D\,(max)} = \left(\frac{N_S}{N_P}\right) I_O$	$I_{D\,(max)} = \left(\frac{N_S}{N_P}\right) I_O$
Drain voltage	$U_{DS} = U_I + \frac{N_S}{N_P} \cdot U_O$	$U_{DS} = 2U_I$	$U_{DS} = U_O$	$U_{DS} = \left(\frac{D}{1-D}\right) U_I$
Current on the output rectifier diode	$I_{F1} = I_O$	$I_{F1} = I_O D$	$I_{F1} = I_O D$	$I_{F1} = I_O D$
Reverse voltage of the output rectifier diode	$U_{R1} = U_O + \frac{N_S}{N_P} \cdot U_I$	$U_{R1} = U_O + \frac{N_S}{N_P} \cdot U_I$	$U_{R1} = U_O + \frac{N_S}{N_P} \cdot U_I$	$U_{R1} = U_O + \frac{N_S}{N_P} \cdot \frac{1}{1-D} U_I$

(*continued overleaf*)

Table 3.1 (*Continued*)

Regulator type	Buck	Boost	Buck-boost or inverting converter	SEPIC
Voltage and current waveforms	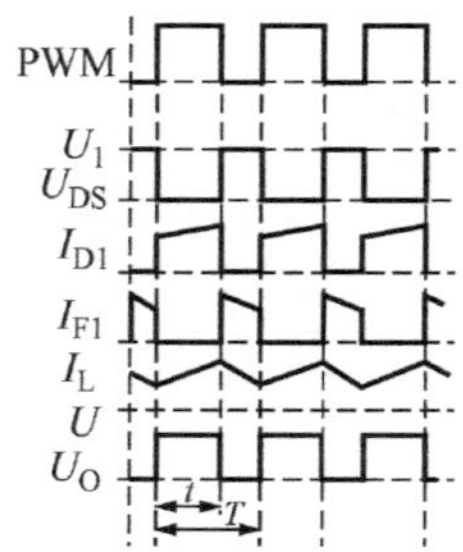 (m)	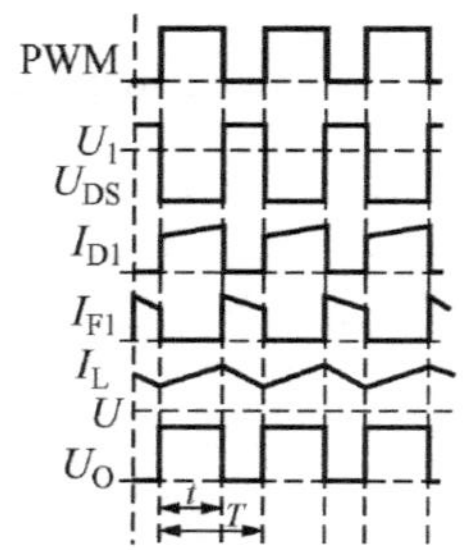 (n)	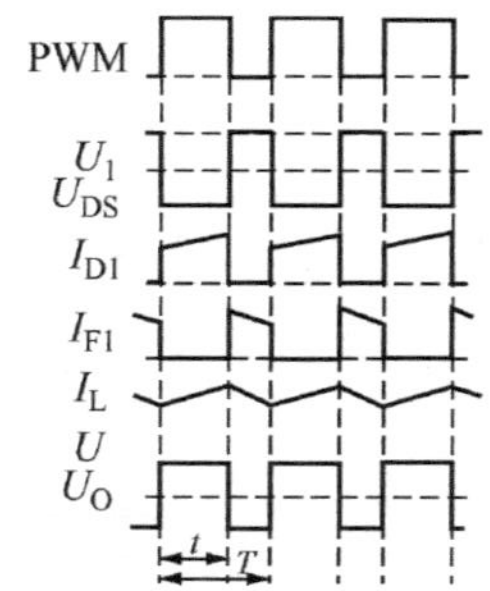 (o)	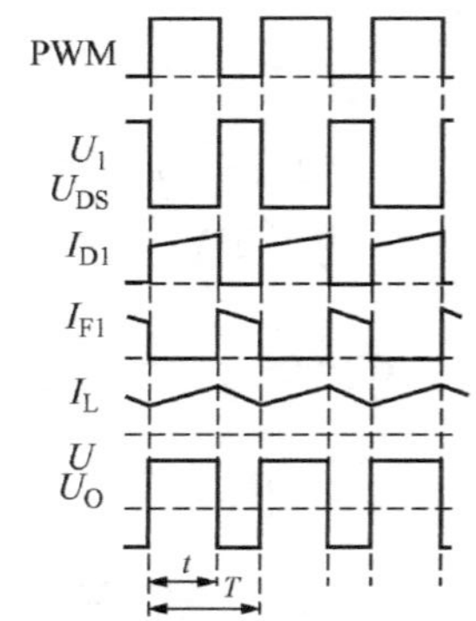 (p)
Main features	When MOSFET is conducted, store electric energy in the primary winding of high-frequency transformer, and when MOSFET is powered off, output electric energy to the secondary side	When MOSFET is conducted, high-frequency transformer transfers energy and stores part of the electric energy in power inductor *L*, and when MOSFET is powered off, the electric energy stored in *L* will supply power to the load through the loop VD_2	Two MOSFETs are used as switching devices in the circuit	Using active clamp forward DC/DC converter
Typical products	NCL30000 UCC28220 UC3842A/3843A	UCC3809-1 UCC38C42 UCC2891	UCC28220 UCC3809-1 UCC38C42	UCC2891 UCC3580-1 UC3824
Regulator type	Half bridge	Full bridge	Push-pull	Phase shift switching ZVT

Circuit structure	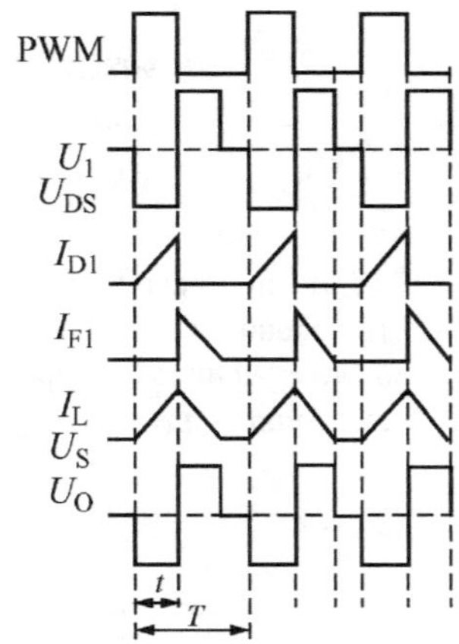 (q)	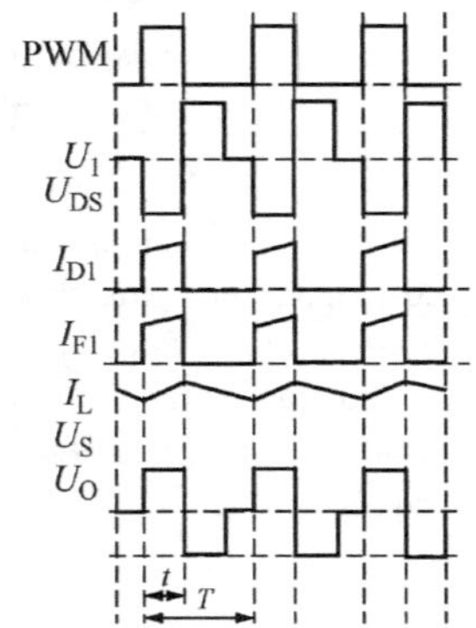 (r)	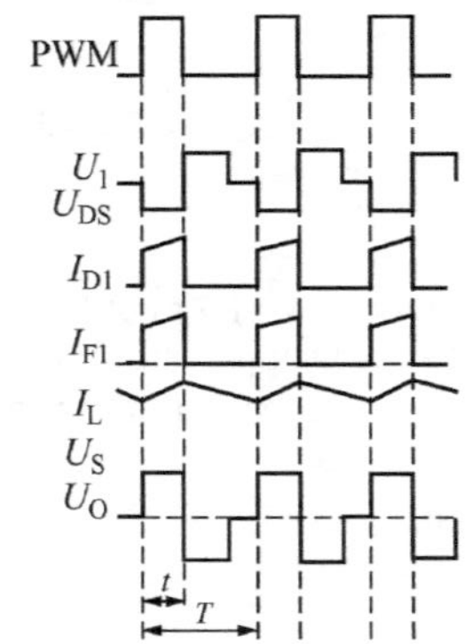 (s)	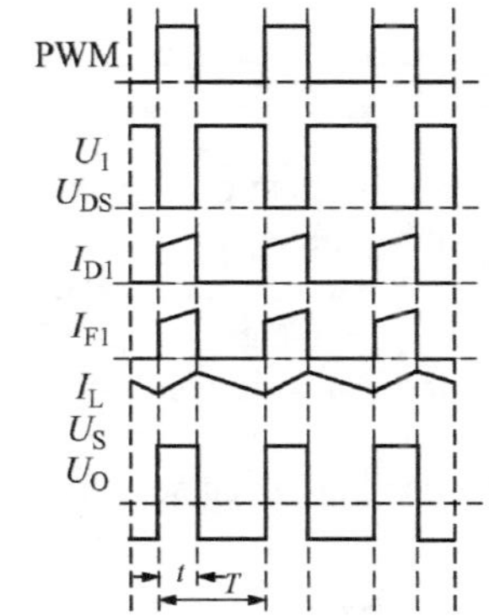 (t)
Ideal transfer function	$\frac{U_O}{U_I} = \frac{N_S}{N_P} \cdot \frac{t}{T} = \frac{N_S}{N_P} \cdot D$	$\frac{U_O}{U_I} = 2 \times \frac{N_S}{N_P} \cdot \frac{t}{T}$ $= 2 \times \frac{N_S}{N_P} \cdot D$	$\frac{U_O}{U_I} = 2 \times \frac{N_S}{N_P} \cdot \frac{t}{T}$ $= 2 \times \frac{N_S}{N_P} \cdot D$	$\frac{U_O}{U_I} = 2 \times \frac{N_S}{N_P} \cdot \frac{t}{T}$ $= 2 \times \frac{N_S}{N_P} \cdot D$
Maximum drain current	$I_{D\,(max)} = \left(\frac{N_S}{N_P}\right) I_O$	$I_{D\,(max)} = \left(\frac{N_S}{N_P}\right) I_O$	$I_{D\,(max)} = \left(\frac{N_S}{N_P}\right) I_O$	$I_{D\,(max)} = \left(\frac{N_S}{N_P}\right) I_O$
Drain voltage	$U_{DS} = U_I$	$U_{DS} = U_I$	$U_{DS} = 2U_I$	$U_{DS} = U_I$
Current on the output rectifier diode	$I_{F1} = DI_O + (1 - 2D)\ \frac{I_O}{2}$	$I_{F1} = DI_O + (1 - 2D)\ \frac{I_O}{2}$	$I_{F1} = DI_O + (1 - 2D)\ \frac{I_O}{2}$	$I_{F1} = \frac{I_O}{2}$
Reverse voltage of the output rectifier diode	$U_{F1} = \frac{U_I}{2} \cdot \frac{N_S}{N_P}$	$U_{F1} = \left(\frac{N_S}{N_P}\right) U_I$	$U_{F1} = \left(\frac{N_S}{N_P}\right) U_I$	$U_{F1} = \left(\frac{N_S}{N_P}\right) U_I$

(*continued overleaf*)

Table 3.1 (*Continued*)

Regulator type	Buck	Boost	Buck-boost or inverting converter	SEPIC
Voltage and current waveforms	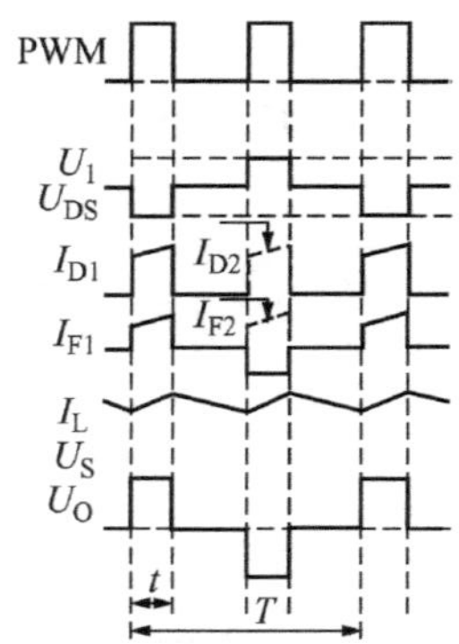 (u)	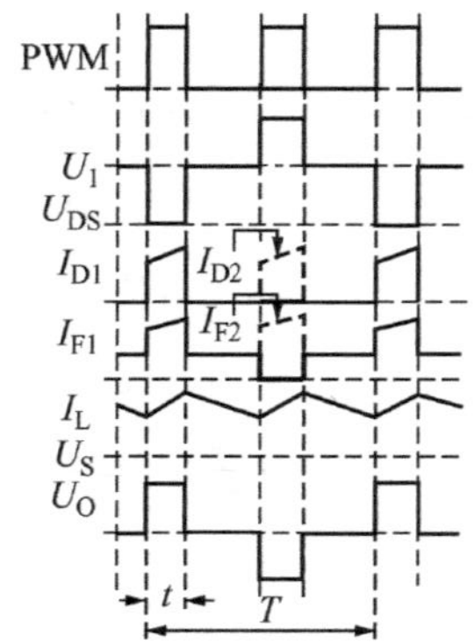 (v)	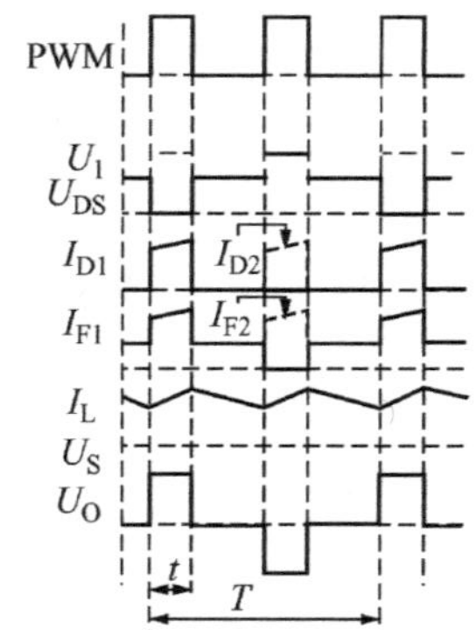 (w)	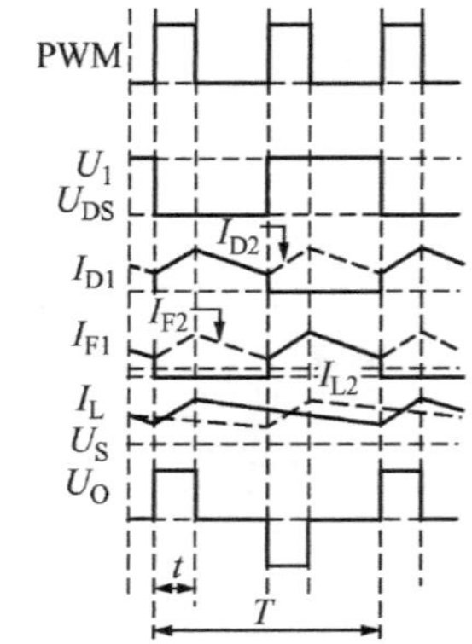 (x)
Main features	Two MOSFETs constitute a half bridge, high-frequency transformer is connected to the diagonal of the bridge	Four MOSFETs constitute a full bridge, capable of generating higher power compared with the half-bridge regulator	The high-frequency transformer with a center tap, two MOSFETs work by turns	Adopting phase shift control and zero-voltage soft-switching technology, greatly reducing the switching losses, and significantly improving the efficiency of the regulator
Typical products	NCP1901 LC810PG UCC3806 UCC3808A	UCC3806 UCC3808A UC3825A	UCC28025 UCC3806 UCC3808A	UCC3895 UC3879 UC3875

Table 3.2 Main technical indicators of typical products of DC/DC converter

Product type	Model	Input voltage U_I (V)	Output voltage U_O (V)	Maximum output current I_{OM} (mA)	Encapsulation mode
Buck	MAX639	5.5–11.5	5	100	DIP-8
	MAX758A	4–16	1.25–U_I	750	DIP-8
	LM2576	≤40 (or ≤60)	3.3, 5, 12, 15 (or adjustable)	3000	TO-220, TO-263
	LM2596	≤40	3.3, 5, 12 (or adjustable)	3000	TO-220, TO-263
Boost	MAX619	2–3.6	5	50	DIP-8
	MAX756	1.1–5.5	3.3 or 5	200	DIP-8
	MAX770	2–16.5	5 (or adjustable)	1000	DIP-8
	MAX773	2–16.5	5, 12, 15 (or adjustable)	1000	DIP-8
	MAX1771	2–16.5	12 (or adjustable)	2000	DIP-8, SO-8
Buck-boost	LTC3441	2.4–5.5	2.4–5.25	1000	DE-12
SEPIC	UC2577-ADJ	3–40	adjustable	3000	TO-263
Inverting converter	ICL7660	1.5–10.5	$-U_I$	20	DIP-8, TO-99
	ICL7662	4.5–20	$-U_I$	20	DIP-8, TO-99
	TC7662B	1.5–15	$-U_I$	20	DIP-8
	MAX764	3–16	–5	250	DIP-8

current waveforms. PWM refers to pulse width modulation waveform; U_I, the DC input voltage; U_{DS}, the drain-source voltage of VT_1 (MOSFET); I_{D1}, the drain current of VT_1; I_{F1}, the working current of VD_1; U_O, the output voltage; I_L, the load current; T, the cycle; t, the time when U_O is high level (or low level); D, the duty ratio, with a relational expression: $D = t/T$, N_P, and N_S that refer to the number of turns of the primary and secondary winding respectively; L_P, the inductance of the primary winding; C_O, the filter capacitor of output terminal; R_L, the load resistance, and so on.

3.1.1 *Main Technical Indicators of the Typical Products of the DC/DC Converter*

At present, there are up to thousands of DC/DC converter models. Table 3.2 shows the main technical indicators of five typical product types of DC/DC converter.

3.2 Basic Principle of Buck Converter

Buck converter is the most commonly used converter. See Figure 3.1 for the basic principle of buck converter. The converter can be equivalent to switch S. When S is switched on, power is supplied to the load while part of the electric energy is stored in L and C. The voltage of L is

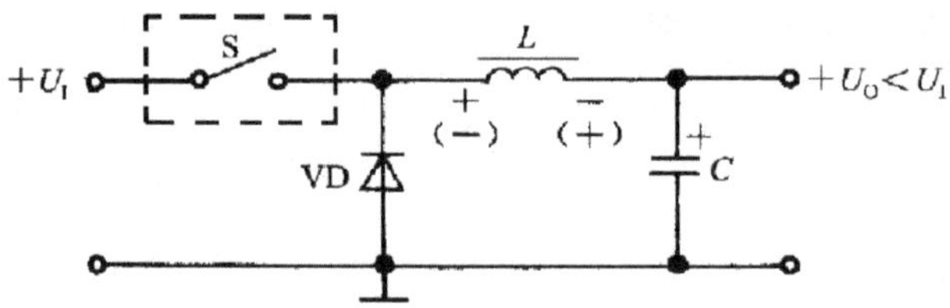

Figure 3.1 Basic principle of buck converter

U_L, and its left is positive and right negative and simultaneously freewheeling diode VD is cut off. When S is switched off, counter electromotive force will be produced on L with the left terminal being negative and the right positive, so that VD is conducted and the electric energy of L is transmitted to the load to maintain a constant output voltage and $U_O < U_I$.

Buck converter has the following features:

1. U_I passes through the switching device, S, first and then the inductor L.
2. $U_I = U_L + U_O$. It is named so just because $U_O < U_I$, and it can reduce voltage.
3. Output and input voltages share the same polarity.
4. The output voltage expression of buck converter is as follows:

$$U_O = \frac{t}{T}U_I \tag{3.1}$$

The typical products of buck converter include LM2576, MAX639, L4960, L4970A, and so on, of which, the peripheral circuit of LM2576 is the simplest.

3.2.1 Simplified Circuit of Buck Converter

Buck converter can convert a DC voltage into a lower one, which is often used in distributed power system. For example, it can convert +24 or +48 V power supply into +15, +12, or +5 V one, with little power loss during the conversion. Buck converter can use one NPN MOSFET (or N-channel MOSFET) as switching device S, making input voltage alternately connect and disconnect the energy-storing inductor L under the control of PWM signal. See Figure 3.2(a) for the simplified circuit of voltage regulation by buck switching, which shows the current paths with the switch on and off.

When the switch is closed, freewheeling diode VD is cut off. As the input voltage U_I is connected with the energy-storing inductor L, input–output voltage difference ($U_I - U_O$) is added to L, thus linearly increasing the current I_L passing through L. During this process, power is supplied to the load and part of electric energy is stored in L and C. The current passing through R_L is I_O, as shown in Figure 3.2(b). When the switch is off, L is disconnected with U_I, but because the inductor current cannot change suddenly in an instant, counter electromotive force is produced on L to maintain constant current passing through the inductor. At this time, freewheeling diode VD is conducted, and the electric energy stored in L is transmitted to the load through the loop constituted by VD to maintain a constant output voltage, as shown in Figure 3.2(c). When the switch is off, C supplies power to the load, which helps to maintain U_O and I_O unchanged.

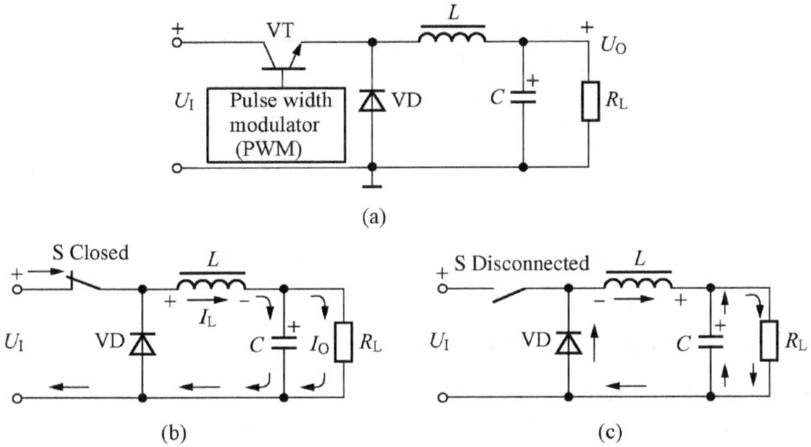

Figure 3.2 Simplified circuit of buck converter. (a) Simplified circuit, (b) current path with the switch on, and (c) current path with the switch off

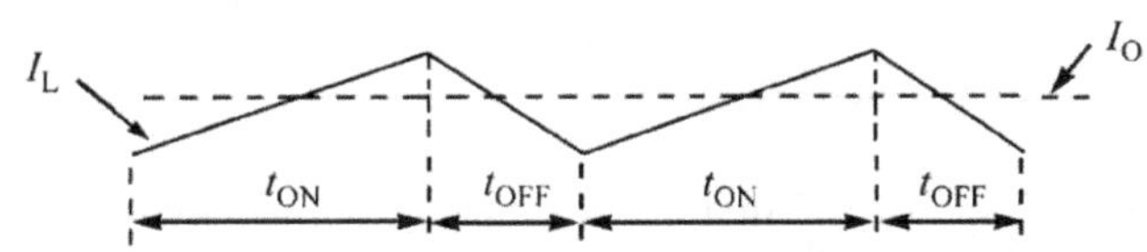

Figure 3.3 Current waveform of the energy-storing inductor in the buck converter

See Figure 3.3 for the current waveform of the energy-storing inductor in the buck converter. This figure shows that the inductor current I_L rises along the ramp when the switch is on (t_{ON}) and falls when the switch is off (t_{OFF}). Therefore, the equivalent load current I_O from the converter is the average of I_L and I_C. The difference between the peaks of inductor current waveform is the inductor ripple current. L should be large enough to ensure that the ripple current is less than 20–30% of the rated DC current.

Finally, a few points need to be noted:

1. Although there is only one inductor in the schematic diagram of buck converter, input and output can also be isolated through input-level transformer and rectifying filter circuit.
2. Buck converter can reduce the input voltage only. If the input voltage is below the desired output voltage, the converter will not work.
3. Buck converter has one output only. If the voltage needs to be changed from +5 to +3.3 V again, the secondary linear converter can be connected to form composite regulated power supply. This design approach is often used in multiple output converters.
4. Buck converter can work in both continuous and discontinuous modes, but its input current is always discontinuous. When MOSFET is off, the input current will be reduced to zero. This requires better performance of the input electromagnetic interference (EMI) filter.

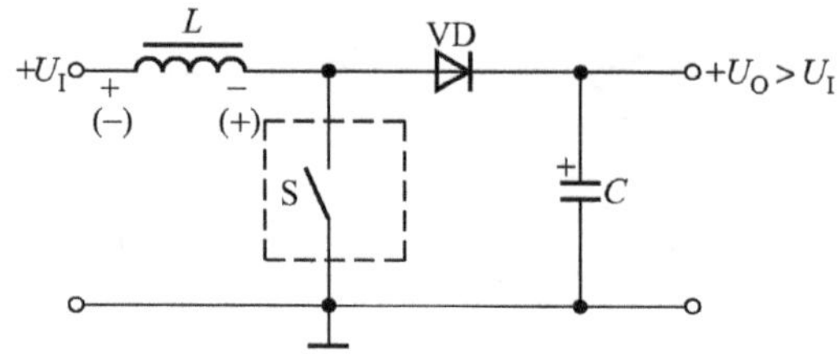

Figure 3.4 Basic principle of boost converter

3.3 Basic Principle of Boost Converter

See Figure 3.4 for the basic principle of boost converter. U_I refers to DC input voltage, U_O, the DC output voltage, and switch S symbolizes a converter. When S is closed, there is current passing through the inductor L to store electric energy. As for the polarity of voltage, the left is positive and the right negative, cutting off the rectifier diode VD, and then C supplies power to the load. When S is off, the counter electromotive force produced on L with the left being negative and the right positive will make VD conducted. The electric charges stored on L supply power to the load through the loop constituted by L, R_L, VD, and charge C. As the switching frequency is high enough, the output voltage U_O can be maintained constant.

Boost converter has the following features:

1. U_I passes through the inductor L first and then the switching device S.
2. $U_O = U_I + U_L - U_D \approx U_I + U_L > U_I$, therefore it is known as boost. It is used to raise voltage to realize $U_O > U_I$. U_L is the voltage drop on the inductor L and U_D is that of freewheeling diode VD, which is usually negligible.
3. The output voltage expression of boost converter is as follows:

$$U_O = \frac{1}{1-D} \cdot U_I \tag{3.2}$$

4. Output and input voltages share the same polarity.

The typical products of boost converter include MAX770, MAX1771, and so on.

3.3.1 Simplified Circuit of Boost Converter

See Figure 3.5(a) for the simplified circuit of boost converter. This figure also shows the current paths under switch on and off. When the switch is closed, the rectifier diode VD is cut off and the input voltage directly returns after passing through the inductor L, linearly increasing the inductor current I_L. At this point, the output filter capacitor C supplies power to the load and the current on the load R_L is I_O, as shown in Figure 3.5(b).

When the switch is off, the inductor current cannot change suddenly in an instant, so counter electromotive force U_L is produced on L to maintain I_L constant. At this point, the rectifier diode VD is conducted, and after U_L is connected to U_I in series, current will be supplied to the load with a voltage over U_I and the output filter capacitor C will be charged, as shown in Figure 3.5(c).

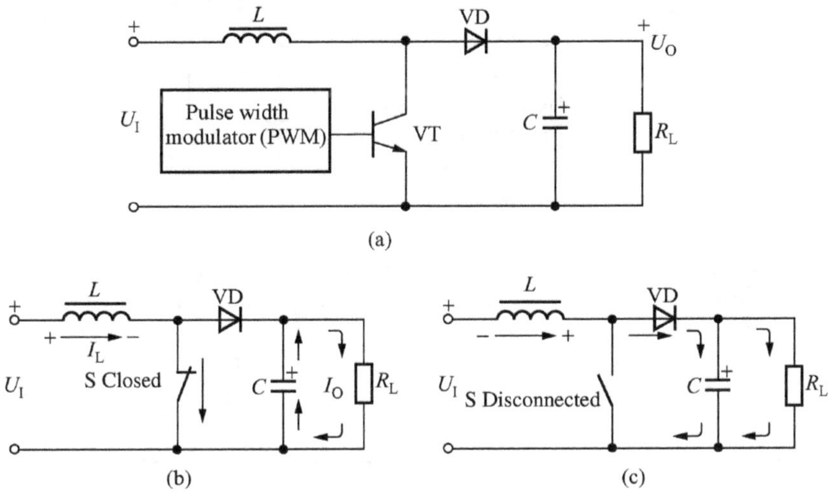

Figure 3.5 Simplified circuit of boost converter. (a) Simplified circuit, (b) current path with the switch on, and (c) current path with the switch off

The total maximum available power of boost converter is equal to the input voltage multiplied by the maximum average input current. As the output voltage of boost converter is higher than the input voltage, the output current must be lower than the input current.

3.4 Basic Principle of Buck-Boost Converter

Buck-boost converter is also known as buck/boost power converter, whose feature is that when the input voltage is higher than the output voltage, the converter works under buck mode, that is, $U_O < U_I$, and when the input voltage is lower than the output voltage, the converter works under boost mode, that is, $U_O > U_I$. Continuous current output can be realized under both operating modes.

See Figure 3.6(a) for the simplified circuit of buck-boost converter. When the switch is closed, the input voltage directly returns through the inductor L with electric energy stored in L. At this point, the output capacitor C supplies current I_O to the load, as shown in Figure 3.6(b). When the switch is off, counter electromotive force is generated on L to make the diode VD become conductive from cutoff, and the inductor current supplies power to the load and charges the output capacitor so as to maintain the output voltage constant, as shown in Figure 3.6(c). Please note that the polarity of the output capacitor C in the buck-boost converter is exactly opposite to that shown in Figure 3.5.

Buck-boost converter mainly has the following features:

1. The buck-boost converter works under discontinuous mode, of which both the input and output current are discontinuous as both of them pass chopper.
2. It has one output only and there is no isolation between the output and the input. The boost output in it cannot be lower than the input voltage. Even though MOSFET is switched off, the output voltage is only equal to the input voltage (with the voltage drop of rectifier diode neglected).

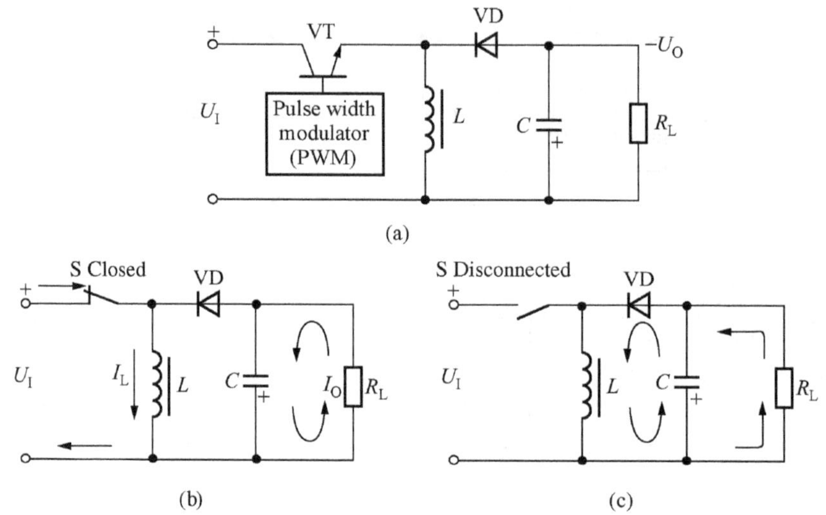

Figure 3.6 Simplified circuit of buck-boost converter. (a) Simplified circuit, (b) current path with the switch on, and (c) current path with the switch off

3. The output voltage expression of buck-boost converter is as follows:

$$U_O = \frac{D}{1-D} \cdot U_I \tag{3.3}$$

4. The polarity of output voltage is always opposite to that of input voltage (note the polarity of capacitor), but the voltage amplitude can be comparatively large or small.

3.5 Basic Principle of Charge Pump Converter

Charge pump converter is also known as switched-capacitor converter or inverting converter, whose feature is that one capacitor is used to transfer energy rapidly under the effect of the switching frequency, making the polarity of output voltage opposite to that of input voltage and the amplitude of the negative output voltage higher or lower than or equal to the input voltage. Therefore, it can also be categorized as a buck-boost converter.

See Figure 3.7 for the circuit principle of charge pump polarity-inverting converter. Take analog switches S_1 and S_2 as one group and S_3 and S_4 as the other; turn on and off the two groups of switches alternately. During the positive half cycle, S_1 and S_2 are closed, while S_3 and S_4 are off, so C_1 is charged to U_{DD}. During the negative half cycle, S_3 and S_4 are closed, while S_1 and S_2 are off, so the positive terminal of C_1 is grounded, and the negative one is connected to U_O. As C_1 and C_2 are connected in parallel, part of the charges in C_1 is transferred to C_2, and negative voltage output is formed on C_2. With the function of the analog switches, C_1 is charged continuously to keep the voltage drop at its both terminals at U_{DD} value. Obviously, C_1 is equivalent to a "charge pump," so it is called pump capacitor, with C_1 and C_2 constituting a pump power. The circuit is a high-efficiency power converter with quite little power loss. The typical products of this power converter include ICL7660 and ICL7662 produced by Itersil and TPS60110, and TPS60111 produced by TI in the United States.

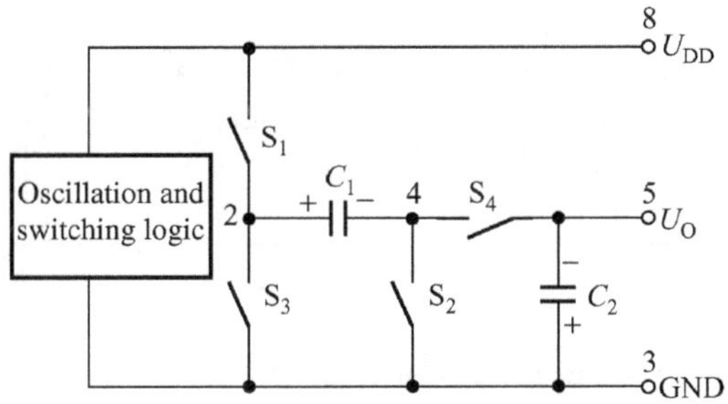

Figure 3.7 Circuit principle of charge pump converter

Charge pump converter has the following features:

1. High power efficiency (up to 90%) and simple peripheral circuit (only two capacitors required), capable of generating voltage-doubling or multiple voltage output.
2. During the switching cycle, charges are stored in the capacitor first, and then transferred to the output terminal. The capacitance of C_1 relates to the switching frequency and the output load current. C_1 and C_2 shall be tantalum capacitors with little drain and stable performance.
3. S_1 and S_2 in the chip can be CMOS analog switch (low current output) or MOSFET (high current output).
4. The output of charge pump converter is not regulated, so a constant voltage control circuit can be added if necessary.

3.6 Basic Principle of SEPIC

The typical products of SEPIC include UC2577-ADJ, BQ24007, and so on. See Figure 3.8 for the simplified circuit of SEPIC. The circuit comprises two inductors L_1 and L_2, two capacitors C_1 and C_2, rectifier VD and switch S (i.e., MOSFET). When S is closed, VD is cut off, the current in L_1 stores energy in L_1 along the loop of $U_I \rightarrow L_1 \rightarrow$ S. At the same time, C_2 stores energy in L_2 through S, and the output capacitor C_2 supplies current I_O to the load, as shown in Figure 3.8(a).

When S is off, counter electromotive force is generated on L_2, making the diode VD become conductive from cutoff. At this point, there are two current paths: one is the inductor current I_{L1} provided by L_1 supplying power to the load R_L along $U_I \rightarrow L_1 \rightarrow C_1 \rightarrow$ VD, and the other is the inductor current I_{L2} provided by L_2 supplying power to R_L along $L_2 \rightarrow$ VD, the total inductor current being $I_{L1} + I_{L2}$, so the output voltage can be kept constant. At the same time, C_1 and C_2 are charged to supplement energy, as shown in Figure 3.8(b).

SEPIC mainly has the following features:

1. The input current is continuous while the inductor current is discontinuous.
2. Two inductors L_1 and L_2 must be used in the circuit, in which L_1 and S play the role of a boost converter, and L_2 and VD serve as a flyback buck-boost converter. Therefore, it is a "boost + buck-boost" converter, allowing the output voltage to be higher (i.e., $U_O > U_I$) or lower (i.e., $U_O < U_I$) than the input voltage, which can be very flexible for use.

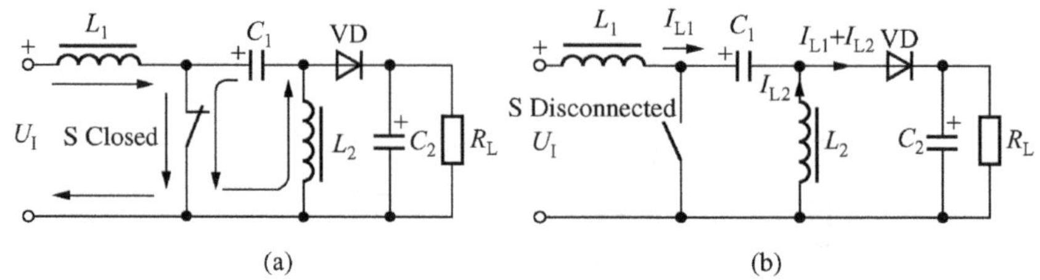

Figure 3.8 Simplified circuit of SEPIC. (a) Current path with the switch on and (b) current path with the switch off

3. L_2 is used to transfer energy to the output end and reset the blocking capacitor C_1.
4. C_1 is not only used as a blocking capacitor, but also equivalent to a "charge pump" for transferring energy. When S is off, C_1 is charged, and when S is closed, C_1 transfers energy to L_2. The capacitor C_1 can absorb the leakage inductance of L_1 through connection to L_1 in series, thereby lowering the requirements for MOSFET.
5. The output voltage expression of SEPIC is as follows:

$$U_O = \frac{D}{1-D} \cdot U_I \tag{3.4}$$

6. It is suitable for the application of a wide range of input voltage (e.g., the battery voltage of automotive electronic equipment) and can also be used as power factor correction circuit. However, the disadvantage is that the circuit is comparatively complex.

3.7 Basic Principle of Flyback Converter

Flyback converter is the most basic topology of switching regulator and SMPS. It is used in a very wide range of applications, and flyback mode is set as default topology by many design softwares.

Those outputting energy to the load during the cutoff of MOSFET are all called flyback converters, which are evolved from the buck-boost converter.

See Figure 3.9(a) and (b) for the basic principle of flyback converter. U_I is the DC input voltage; U_O, the DC output voltage; T, the high-frequency transformer; N_P, the primary winding; and N_S, the secondary winding. V refers to MOSFET, whose gate electrode is connected to the PWM signal and drain electrode (drive side) to the lower end of the primary winding. VD is the output rectifier diode and C, the output filter capacitor. During the positive half cycle of PWM signal, V is conducted and current I_P passes through the primary side to store energy in the primary winding. At this point, the output voltage of the secondary winding is negative at the upper end and positive at the lower end in terms of polarity, so that VD is cut off and has no output, as shown in Figure 3.9(a). During the negative half cycle, V is cut off and no current passes through the primary side. According to the principle of electromagnetic induction, induced voltage U_{OR} will now be generated in the primary winding and voltage U_S generated in the secondary winding with the positive polarity at upper end and the negative polarity at lower end. Therefore, VD is conducted and the output voltage is obtained after rectification and filtering by passing through VD and C, as shown in Figure 3.9(b). Owing to the high

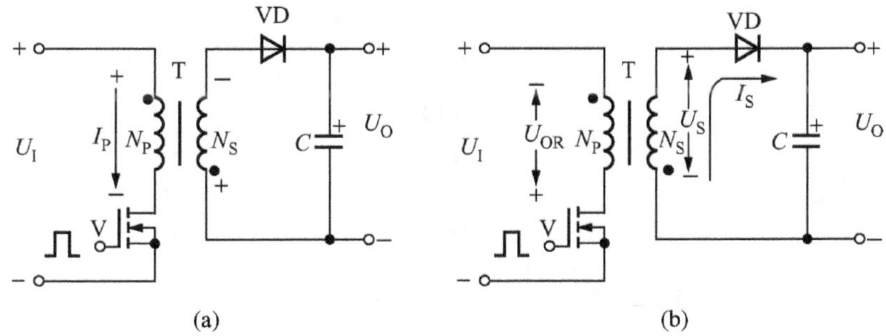

Figure 3.9 Basic principle of flyback converter. (a) Storing energy when MOSFET is conducted and (b) transferring energy when MOSFET is off

switching frequency, the output voltage (i.e., the voltage at both ends of the filter capacitor) is kept basically constant, thereby achieving the purpose of regulating voltage.

Flyback converter has the following main features:

1. The dotted end of the primary winding of high-frequency transformer is opposite to that of the secondary winding in terms of polarity, and the dotted end of the primary winding is connected to the positive terminal of U_I and the other end to the drive side of MOSFET.
2. When MOSFET is conducted, energy is stored in the high-frequency transformer, and when MOSFET is switched off, energy is transferred to the secondary side. A high-frequency transformer is just equivalent to an energy-storing inductor storing and releasing energy continuously.
3. It is capable of working under both continuous (the current of the secondary winding is always greater than zero) and discontinuous (the current of the secondary winding is reduced to zero at the end of each switching cycle) modes.
4. It is capable of constituting either AC/DC converter with AC input or the converter with DC input.
5. The polarity of output voltage can be positive or negative, depending on the winding polarity and the specific connection method of output rectifier.
6. The output voltage can be lower or higher than the input voltage, depending on the turns ratio of high-frequency transformer.
7. The output voltage expression of flyback converter is as follows:

$$U_O = D\sqrt{\frac{TU_O}{2I_O L_P}}U_I \tag{3.5}$$

   wherein U_O/I_O refers to the output impedance of flyback converter.
8. Multiple outputs can be obtained just by adding the secondary winding and associated circuits.
9. For flyback converter, no low-frequency filter inductor (except small magnetic bead inductors, with several microhenry (μH) inductance, designed to suppress high-frequency interference) can be connected in series between the output rectifier diode and the filter capacitor, or it will fail to work normally.

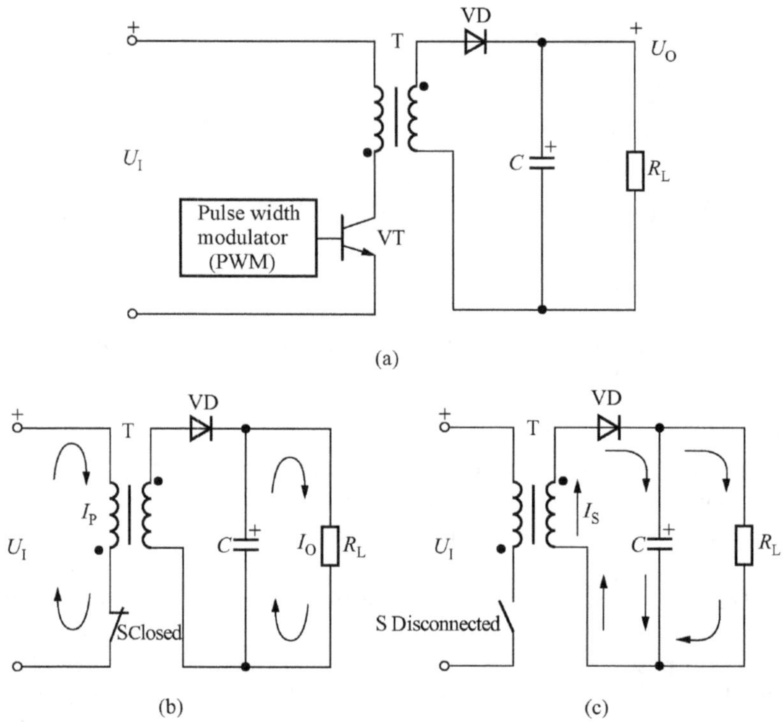

Figure 3.10 Simplified circuit of flyback converter (a) Simplified circuit, (b) Current path with the switch on, and (c) Current path with the switch off

See Figure 3.10(a) for the simplified circuit of flyback converter. In this figure, I_P is the primary current and I_S the secondary current. During the positive half cycle, C supplies power to the load and the load current is I_O. During the negative half cycle, I_S charges C and supplies power to the load to maintain the output voltage constant.

Flyback converter includes two types, the non-isolated and the isolated. See Figure 3.11(a) for the simplified circuit of non-isolated flyback converter. PWM modulation signal is applied from the gate electrode of MOSFET. When MOSFET is switched on, the output voltage is applied to the inductor L, so that the current ramps up and energy is stored in the inductor. When MOSFET is switched off, the inductor current will supply power to the output capacitor C and the load R_L through the rectifier diode VD.

See Figure 3.11(b) for the simplified circuit of isolated flyback converter, whose operating principle is similar to that of Figure 3.11(a). The difference is that it uses the high-frequency transformer to realize isolation and adopts the primary winding instead of inductor L.

3.7.1 Basic Circuit of Multiple Output Flyback Converter

If a high-frequency transformer has multiple secondary windings, then it can be designed into a multiple output flyback converter with all of the outputs isolated from the primary winding. All output voltages can be changed just by adjusting the turns ratio of the primary and the secondary winding.

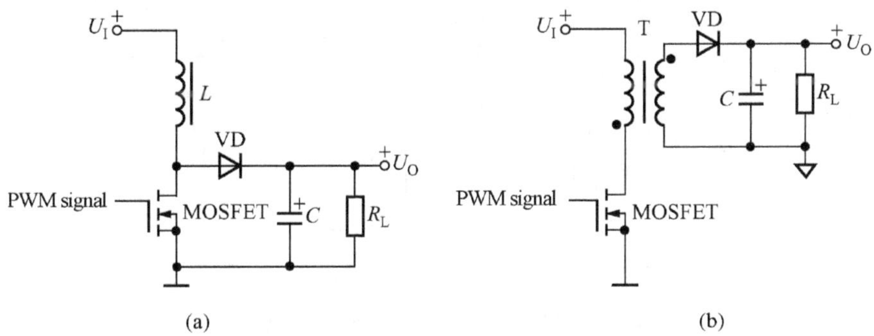

Figure 3.11 The two types of non-isolated and isolated flyback converter. (a) Simplified circuit of non-isolated type and (b) simplified circuit of isolated type

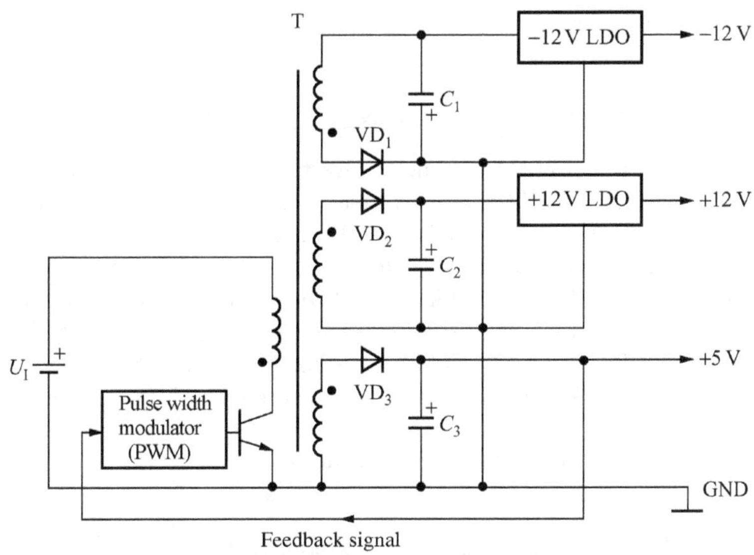

Figure 3.12 Typical circuit of multiple output flyback converter

See Figure 3.12 for the typical circuit of multiple output flyback converter. It is battery-powered and the three paths of output voltage are respectively +5, +12, and −12 V, with high conversion efficiency. The path with the maximum output current (e.g., +5 V) is usually selected as the main output, which can be used to provide feedback signal for the control loop. As the output of this path is adjusted directly, the main output has the best voltage regulation performance. +12 and −12 V are auxiliary output, with load regulation between 5% and 10% in general.

In order to improve the regulation of auxiliary output, an LDO is used in both +12 and −12 V output circuits as post regulator. Because LDO has the advantages of low dropout, low power consumption, and good regulation performance, normal operation can be realized by

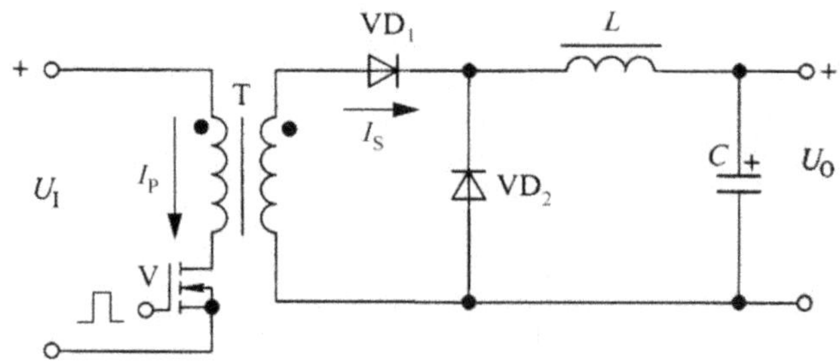

Figure 3.13 Topology of forward converter

ensuring that the output voltage of the secondary side from +12 V is at least 1 V higher than the rated output voltage.

3.8 Basic Principle of Forward Converter

Forward converter can be evolved from buck converter. The difference between them is that high-frequency transformer is added to the former to isolate the secondary and the primary sides. See Figure 3.13 for the topology of forward converter. VD_1 is the rectifier diode; VD_2, the freewheeling diode; and L, the filter inductor with the function of storing energy. Its working principle is as follows: when MOSFET is switched on, VD_1 is conducted, supplying power to the load and storing part of the electric energy in L and C, while VD_2 is cut off. When MOSFET is switched off, VD_1 is cut off, while VD_2 is conducted, and the electric energy stored in L supplies power to the load through the loop constituted by VD_2 to maintain the output voltage constant.

Forward converter mainly has the following features:

1. The polarity of the dotted end in the primary winding is the same as that in the secondary winding, and the other end of the primary winding is connected to the drive side of MOSFET.
2. When MOSFET is conducted, the high-frequency transformer transfers energy with almost no energy stored in it.
3. For the forward converter, a filter inductor must be connected in series between the output rectifier diode and the filter capacitor. The filter inductor can store energy as well, so it is also known as energy storage inductor.
4. The output voltage expression of forward converter is as follows:

$$U_O = \frac{N_S}{N_P} \cdot \frac{t}{T} \cdot U_I = \frac{N_S}{N_P} \cdot DU_I \tag{3.6}$$

5. It is suitable to function as converters with low-voltage output and high current.

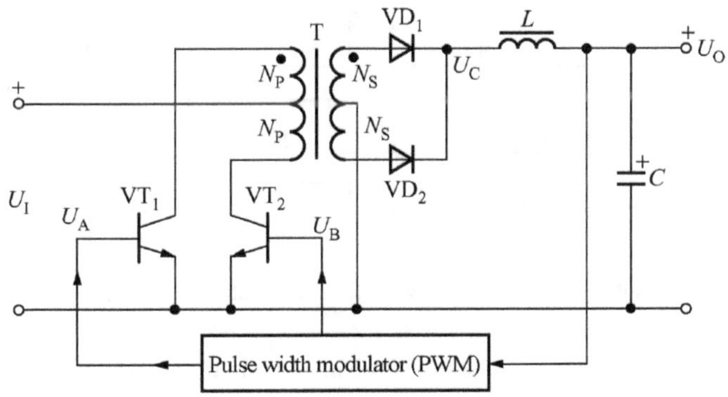

Figure 3.14 Basic principle of push-pull converter

3.9 Basic Principle of Push-Pull Converter

Push-pull converter uses two bipolar power switch tubes (or MOSFET) working alternately to complete the conversion. See Figure 3.14 for its basic principle. This circuit is a forward converter, and both the primary and secondary windings of the high-frequency transformer have a center tap. The two paths of control signal, U_A and U_B, are generated by PWM. When U_A and U_B are high and low level, respectively, MOSFET VT_1 is conducted, and the input voltage U_I passes through the upper half N_P of the primary winding with negative polarity, the voltage polarity being positive at the lower end and negative at the upper end, and at this point VT_2 is cut off. After passing through the high-frequency transformer, the voltage polarity of the lower half N_S of the secondary winding is positive at the lower end and negative at the upper, so that VD_2 is conducted and the current at the secondary side supplies power to the filter capacitor and the load through the output rectifier VD_2. At this point, VD_1 is cut off. Conversely, when U_B and U_A are high and low level, respectively, VD_1 is conducted, while VD_2 is cut off, and the secondary current supplies power to the output filter capacitor and the load through VD_1.

See Figure 3.15 for the timing waveform of push-pull converter, wherein U_A and U_B indicate the waveform of the two input pulses respectively, and U_C refers to the output pulse waveform. It needs to be noted that the pulse frequency of U_C is twice each driving frequency of PWM. For example, when the operating frequency of PWM is 50 kHz, the pulse frequency of U_C is 100 kHz. Push-pull converter is suitable for high-power converter with low-voltage input, such as +12 or +24 V battery-powered systems.

The expression of DC output voltage U_O is as follows:

$$U_O = U_{PK}(T_{ON}/T) \tag{3.7}$$

wherein U_{PK}, the peak voltage of the secondary output pulse, is determined by the following expression:

$$U_{PK} = (U_I - U_{SWITCH})\ N_S/N_P - U_F \tag{3.8}$$

wherein U_{SWITCH} is the saturation voltage of the transistor; N_S/N_P, the turns ratio between the secondary and primary windings; and U_F, the conduction voltage drop of output rectifier. If

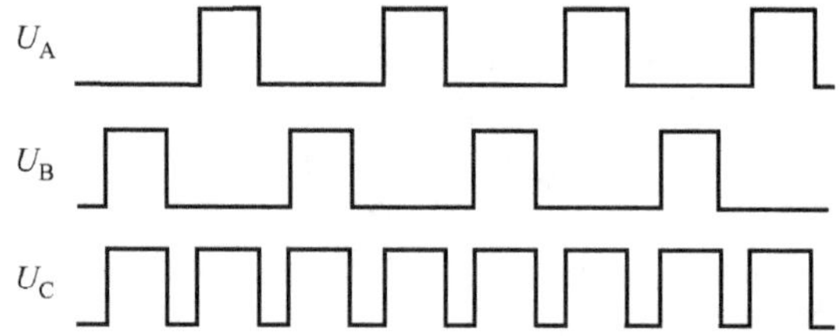

Figure 3.15 Timing waveform of push-pull converter

the bipolar power switch tube is replaced with MOSFET, conversion efficiency can be further improved.

The output voltage expression of push-pull converter is as follows:

$$U_O = 2 \times \frac{N_S}{N_P} \cdot \frac{t}{T} \cdot U_I = 2 \times \frac{N_S}{N_P} \cdot DU_I \tag{3.9}$$

When using push-pull converter, multiple output voltages (including negative voltage output) can be generated by designing more secondary winding, thus providing battery-powered systems with various necessary voltages.

The disadvantage of push-pull converter is that the conduction time of VT_1 and VT_2 must be strictly matched, or the magnetic core of high-frequency transformer will be saturated because of the inconsistent on-off time of them.

3.9.1 Two Types of Push-Pull Converter

Push-pull converter is divided into two topologies of current mode and voltage mode. Their main difference is that the input stage of current mode requires adding a large inductor *L*, but does not need an output filter inductor, while the input stage of voltage mode has no large inductor; however, the output stage must be connected to the filter inductor *L*.

See Figure 3.16(a) for the topology of voltage-mode push-pull converter. It is a forward converter. Two MOSFET power switch tubes V_1 and V_2 are respectively connected to both ends of the primary winding with a center tap and alternately conducted as per the phase difference of 180°.

When V_1 is conducted, positive voltage is applied to the rectifier diode VD_1 to make it conducted. At this point, V_2 is switched off and VD_2 is cut off, and the voltage applied to the drain of V_2 is $2U_I$. This requires MOSFET to withstand a high voltage of at least $2U_I$. For example, if the DC high voltage $U_I \approx +300$ V after a 220-V AC voltage is rectified and filtered, the voltage-withstanding value of MOSFET should be at least $2 \times 300\,V = 600\,V$. In view that there is surge voltage in the electric network, MOSFET withstanding a voltage of 1000 V shall be actually used to avoid tube damage. Furthermore, there shall be a dead time during the conversion of V_1 and V_2 to avoid that the two MOSFETs are conducted simultaneously due to the turn-off delay, which will cause short circuit of the high-frequency transformer and a sudden increase in the current (just with the leakage inductance limiting current in this case), thereby damaging the tube.

See Figure 3.16(b) for the topology of current-mode push-pull converter. A large inductor *L* is connected in series between the input voltage and the transformer. When MOSFET

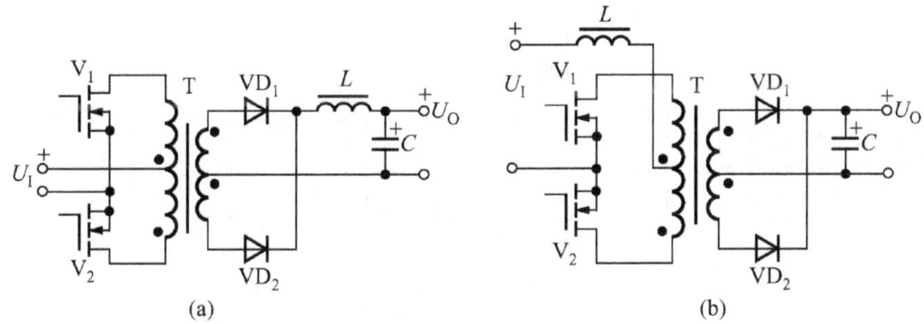

Figure 3.16 Topology of push-pull converter (a) Voltage mode converter and (b) Current mode converter

is conducted, L can be used to reduce the impact current generated during the conduction of MOSFET and rectifier. The disadvantage of the converter is that its output power will be reduced due to the connection of a large inductor.

3.10 Basic Principle of Half/Full Bridge Converter

3.10.1 Basic Principle of Half-Bridge Converter

Half-bridge converter is based on the push-pull converter. It uses two power switch tubes to constitute a half bridge, generally with AC input and applicable to isolated converters with an output power of 500–1500 W. See Figure 3.17 for the basic principle of half-bridge converter, to which a rectifier bridge and a filter capacitor are added at the input stage.

This circuit is basically the same as that in Figure 3.14. The main differences are as follows: (i) The drive circuit must be isolated from the power switch tube and coupled with high-frequency transformer. PWM in Figure 3.17 does provide drive pulses for the power switch tube through one coupling transformer. (ii) The primary winding of high-frequency transformer has no center tap. The timing waveform of half-bridge converter is the same as that of push-pull converter (Figure 3.15); however, its drive circuit is comparatively complicated.

The output voltage expression of half-bridge converter is as follows:

$$U_O = \frac{N_S}{N_P} \cdot \frac{t}{T} \cdot U_I = \frac{N_S}{N_P} \cdot DU_I \tag{3.10}$$

3.10.2 Basic Principle of Full-Bridge Converter

Full-bridge converter requires four power switch tubes to constitute a full bridge. See Figure 3.18 for its basic principle. Among various converters, full-bridge converter has the highest output power, which is suitable for constituting high-power isolated converter with an output power of 1–3 kW. The four power switch tubes are divided into two groups: one is VT_1 and VT_4, and the other is VT_2 and VT_3. When U_B is high level, VT_1 and VT_4 are conducted simultaneously, and when U_A is high level, VT_2 and VT_3 are conducted simultaneously.

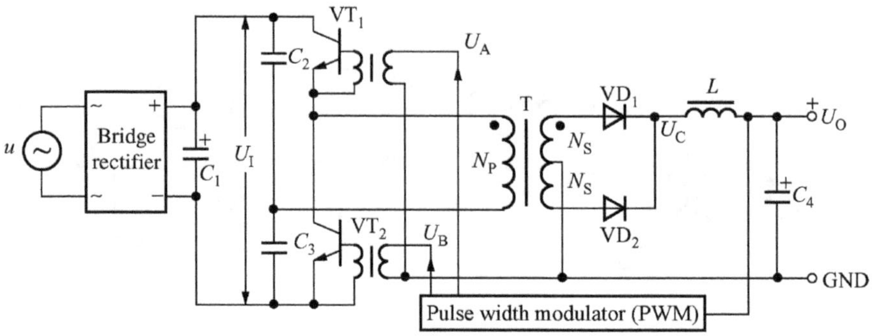

Figure 3.17 Basic principle of half bridge converter

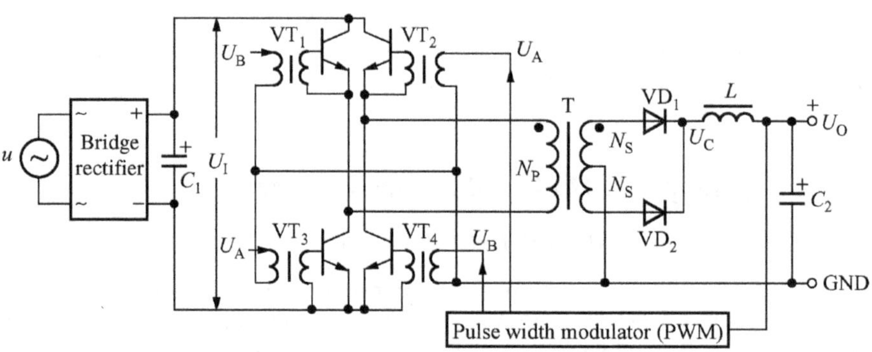

Figure 3.18 Basic principle of full bridge converter

Its timing waveform is similar to that of push-pull converter. Full-bridge converter is also a forward converter.

The DC output voltage expression of full bridge converter is twice that of Formula (3.10), while the peak voltage expression of the secondary pulse of high-frequency transformer is changed into:

$$U_{PK} = (U_I - 2U_{SWITCH})\ N_S/N_P - U_F \tag{3.11}$$

3.11 Basic Principle of Soft Switching Converter

3.11.1 Basic Principle of Resonant Converter

The voltage or current waveform of resonant converter is sinusoidal waveform, which can be achieved with the resonance of inductor and capacitor. The "capacitor" here usually refers to distributed capacitance, whose function is to cut off the power switch tube only when the voltage or current waveform crosses zero so as to minimize the switching loss. Resonant converter is mainly used in high-frequency voltage regulator, in which the turn-off loss exceeds

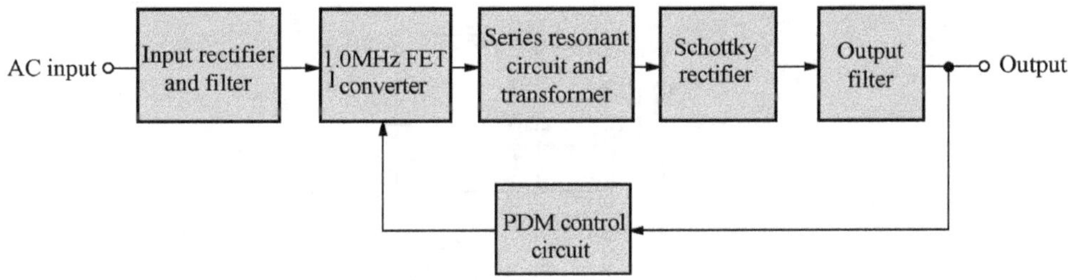

Figure 3.19 Structure diagram of series resonant power supply

the conduction loss. However, the transient state of switching depends on the resonance frequency, and thus the switching frequency of resonant converter is not constant but related to grid voltage and the load.

See Figure 3.19 for the structure diagram of series resonant power supply (SRPS), which mainly consists of an input rectifier and filter, a 1.0-MHz field effect transistor (FET) converter, a series tuned circuit and a transformer, a Schottky rectifier, an output filter, and a PDM control circuit. The switching frequency of SRPS can be up to 1.0 MHz.

The disadvantage of resonant converter is that its switching frequency changes with the load, which brings difficulty to the design of EMI filter. In addition, the distributed capacitance is a part of the resonant circuit and there is high discreteness in the distributed capacitance of power switching tube, which directly affects the switching frequency, thereby further influencing the design of output filter capacitor, and so on. Although connecting an external capacitor in parallel to the power switch tube can significantly reduce the influence of the distributed capacitance; this will increase resonance period, thus making the converter unworkable at high frequencies.

3.11.2 Basic Principle of Quasi Resonant Converter

Quasi resonant converter is named so because it is developed by adding an external capacitor to the resonant converter circuit to make the converter work at constant frequency and it generates resonance only during part of the switching cycle.

See Figure 3.20 for the basic principle of quasi resonant converter. In this circuit, connected in parallel between the drain-source (D-S) of MOSFET is an external capacitor C_1 with a capacitance much larger than the distributed capacitance of MOSFET to eliminate the influence of the distributed capacitance, thereby making the regulator operate at a fixed frequency, which is an important difference of this converter from the resonant converter.

Initially, MOSFET is conducted and the drain voltage is zero. When MOSFET is cut off, the primary inductor L of the high-frequency transformer and the external capacitor C constitute an oscillator circuit. After the first half cycle ($T_1 = T/2$) of resonant waveform, magnetic core is reset. Resonant frequency is determined by L and the value of C. In the latter half cycle ($T_2 = T/2$), because there is no energy stored in the high-frequency transformer, the drain voltage will keep within the amplitude of input voltage U_I until MOSFET is conducted again. See Figure 3.20 for the working waveforms of MOSFET.

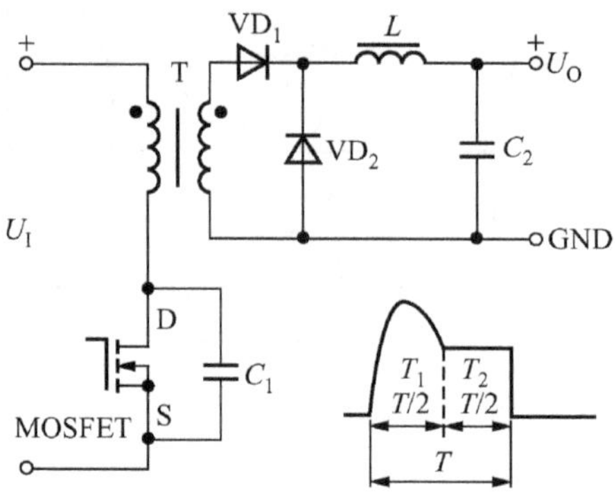

Figure 3.20 Basic principle of quasi resonant converter

The main difference between this converter and the resonant converter is the adoption of PWM and MOSFET as constant frequency switch. The values of L and C must be selected reasonably in designing the circuit. If the values of L and C are too large, the magnetic core will not be reset because the half cycle of resonant waveform exceeds the switching cycle. Conversely, if the values of L and C are too small, the drain voltage will be too high due to the short reset time of magnetic core. Nevertheless, within the operating range of the converter, it is still allowed that the distributed capacitance of MOSFET varies within a comparatively wide range.

When MOSFET is conducted, the energy of C is consumed in MOSFET. In case of low-voltage input, the loss can be negligible as long as the capacity of C is small enough. For example, suppose $C = 100\,\text{pF}$, $U_I = +50\,\text{V}$, and the switching frequency $f = 200\,\text{kHz}$, the loss caused by the capacitor will be only

$$P_C = CU_I^2 f/2 = 100\,\text{pF} \times (50\,\text{V})^2 \times 200\,\text{kHz}/2 = 25\,\text{mW}$$

The disadvantage of soft switching converter is that currently there is no dedicated control chip.

3.11.3 Basic Principle of Full-Bridge Zero Voltage Converter

See Figure 3.21 for the main circuit of full-bridge zero voltage converter. In the circuit, four power switch tubes VT_1–VT_4 constitute a bridge. VD_1 and VD_2 are output rectifiers. The output voltage expression is as follows:

$$U_O = 2 \times \frac{N_S}{N_P} \cdot \frac{t}{T} \cdot U_I = 2 \times \frac{N_S}{N_P} \cdot DU_I \tag{3.12}$$

UC3875 produced by Unitrode can be used as phase-shifted full-bridge PWM controller, which can simultaneously provide four paths of drive signal output to realize phase-shift for the conduction phase of VT_1–VT_4 within 0–80°, thereby avoiding the two power switch tubes on the left and right bridge arms being conducted simultaneously. The adjustment range of duty ratio of UC3875 is 0–100%. See figure (x) in Table 3.1 for the voltage and current waveforms of full-bridge zero voltage converter.

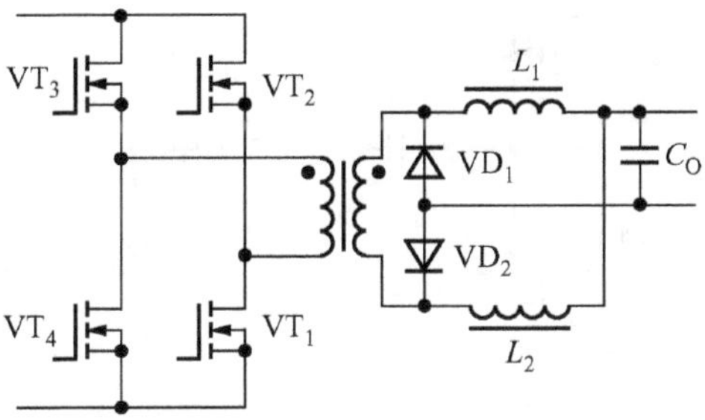

Figure 3.21 Main circuit of full-bridge zero voltage converter

3.12 Basic Principle of Half-Bridge LLC Resonant Converter

With the rapid development of large and extra-large screen LCD TV industry, there are higher requirements for the output power of SMPS. For example, the power range of SMPS required by 32–52 in. LCD TV has been up to 120–300 W, and that of the LCD TV larger than 56 in. can be above 350 W. As the output power of single-ended flyback SMPS is generally no more than 70 W and that of double-ended flyback SMPS is usually 120–180 W, these two types of SMPS cannot meet the above-mentioned requirements. Half-bridge LLC resonant converter, with the advantages of large output power (150–600 W), a small number of required components, high cost-efficiency, high efficiency (up to 99%) and adaptive power factor compensation circuit, is the best choice for producing the SMPS of large and extra-large screen LCD TV. Half-bridge LLC resonant converter is also suitable for such fields as PC power supply, LED street light, communications power supply, industry, medical equipment, and so on. Typical products of integrated circuit of half-bridge LLC resonant converter include PLC810PG from PI, L6599 from ST, NCP1395 and NCP1396 from ON Semiconductor. PLC810PG design can be completed with the application of PI Expert 8.5 software, including automatically selected resonant element values in series and in parallel in accordance with the output load and rated switching frequency defined by users, ensuring zero voltage switching (ZVS) under all load conditions, as well as automatically computing the loss of the main power devices of PFC and LLC and graphically displaying the designed switching frequency and output power. According to actual requirements, zero current switching (ZCS) can be used for output rectifier diode to eliminate the loss of rectifier diode during reverse recovery.

See Figure 3.22(a) and (b) for the basic principle of half-bridge LLC resonant converter. LLC resonant converter is a forward converter. U_I and U_O respectively refer to DC input and output voltage. Half-bridge LLC resonant converter includes a half bridge, two resonant inductors, and one resonant capacitor. In Figure 3.22(a), the half bridge constituted by two N-channel MOSFETs (V_1 and V_2) is driven by LLC controller. In case of different drive circuit, V_1 and V_2 can also use two P-channel MOSFETs. V_1 and V_2 are alternately conducted and cut off at the duty ratio of 50%, and the switching frequency depends on the feedback loop. L_P is the parallel resonant inductance, that is, the inductance of the primary winding of high-frequency transformer. L_S is the series resonant inductance (it can either be the leakage inductance L_{P0} of primary winding or use one separate inductor). The total inductance of both is $L_P + L_S$. C_S is a resonant capacitor, T is a high-frequency transformer, VD_1 and VD_2 are output rectifiers, and C_1 and C_2 are respectively the filter capacitors at the input and output terminals.

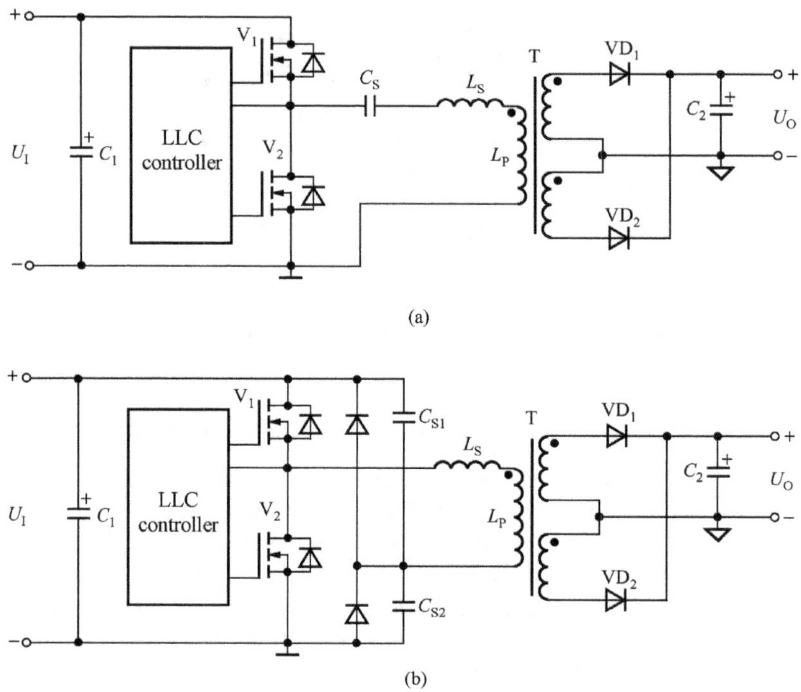

Figure 3.22 Basic principle of half-bridge LLC resonant converter. (a) Single resonant capacitor and (b) double resonant capacitor

The advantages of applying single resonant capacitor are simple wiring and fewer required components. However, the disadvantages are that the ripple and effective value of input current are comparatively high, the effective current passing through the resonant capacitor is comparatively high, and a resonant capacitor withstanding a high voltage up to 600–1500 V is required.

In Figure 3.22(b), two resonant capacitors are used; hence, $C_S = C_{S1} + C_{S2}$. This program can reduce the withstand voltage of each resonant capacitor.

Half-bridge LLC resonant converter has two resonant frequencies: series resonant frequency f_S and parallel resonant frequency f_P, wherein the series resonant frequency is as follows:

$$f_S = \frac{1}{2\pi\sqrt{L_S C_S}} \tag{3.13}$$

and the parallel resonant frequency is as follows:

$$f_P = \frac{1}{2\pi\sqrt{(L_P + L_S)\ C_S}} \tag{3.14}$$

The ratio between the series resonant frequency f_S and the parallel resonant frequency f_P is as follows by dividing Formula (3.13) with Formula (3.14)

$$f_S/f_P = \sqrt{\frac{L_P + L_S}{L_S}} = \sqrt{(L_P/L_S) + 1} \tag{3.15}$$

If $L_P/L_S = k$, then

$$f_S/f_P = \sqrt{k+1} \tag{3.16}$$

Typically, $k = 2–4$, and then the corresponding frequency ratio range is $f_s/f_P = \sqrt{3}–\sqrt{5} =$ 1.732–2.236. The value k can determine the amount of energy stored in the primary winding inductor. The larger the value of k is, the lower the primary current and gain of the converter are, and the greater the operating frequency range is required for regulation.

There is a voltage controlled oscillator (VCO) in the half-bridge LLC resonant converter, which can output two paths of square wave signal with a duty ratio of 50% and phase difference of 180°, and make V_1 and V_2 switched on and off alternately after passing through driver circuit. VCO can adjust operating frequency according to the feedback current of converter and change the voltage gain of half-bridge LLC resonant converter, thus finally realizing the objective of regulating voltage. That is the basic working principle of LLC resonant converter.

3.12.1 The Equivalent Circuit and Voltage Gain Characteristic Curve of Half-Bridge LLC Resonant Converter

See Figure 3.23 for the equivalent circuit of half-bridge LLC resonant converter. Suppose the load resistance is R_L; the secondary AC equivalent resistance, R_{AC}; the quality factor of LLC resonant circuit, Q; the voltage gain, G; and the turns ratio of high-frequency transformer, n $(n = N_P/N_S)$, then the expression is as follows:

$$R_{AC} = \frac{8n^2}{\pi^2} \cdot R_L \tag{3.17}$$

$$Q = \frac{2\pi f_S}{R_{AC}} = \frac{\pi^3 f_S}{4n^2 R_L} \tag{3.18}$$

$$G = 20\lg \frac{U_O}{U_I} \quad \text{(dB)} \tag{3.19}$$

When $k = 3$, the voltage gain characteristic curve of half-bridge LLC resonant converter is shown in Figure 3.23. This figure shows four parallel and four series resonant curves and takes decibels (dB) as the unit of voltage gain G. For example, when $G = 20\lg$ and $(U_O/U_I) = 0$, then $U_O/U_I = 1$; when $G = 20$, $U_O/U_I = 10$; and when $G = -20$, $U_O/U_I = -10$. f_P and f_s

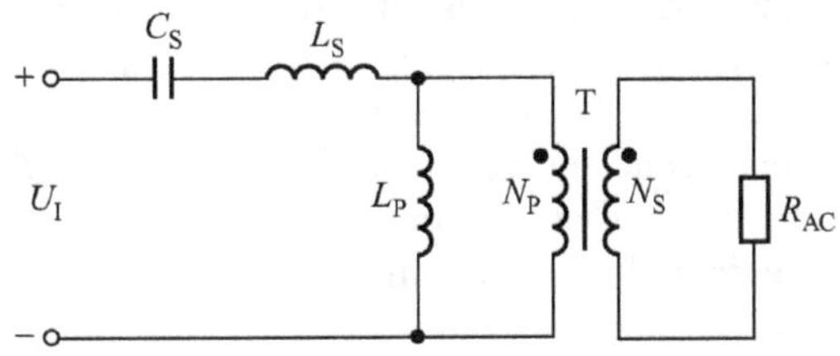

Figure 3.23 The equivalent circuit of half-bridge LLC resonant converter

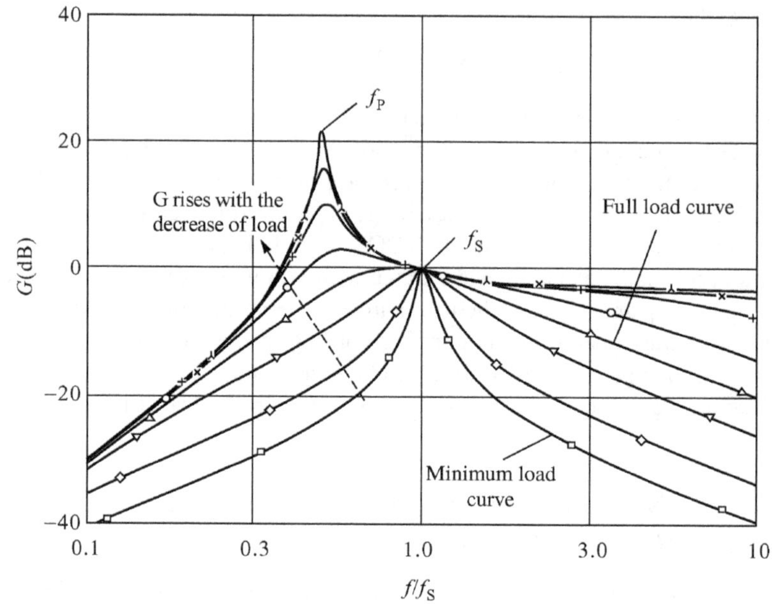

Figure 3.24 Voltage gain characteristic curve of half-bridge LLC resonant converter ($k = 3$)

in the figure respectively correspond to the peak values of parallel and series resonance. f/f_S refers to the ratio of actual working frequency and the series resonant frequency.

Half-bridge LLC converter has the following main features:

1. Half-bridge LLC resonant converter is a kind of frequency-converting converter, whose voltage regulation principles can be summarized as follows: when U_O rises, $f \downarrow \rightarrow G \downarrow \rightarrow U_O \downarrow$ to finally stabilize U_O. Conversely, when U_O drops, U_O will also be stabilized by $f \uparrow \rightarrow G \uparrow \rightarrow U_O \uparrow$. The situation that G gradually increases with the reduction of load is shown by the dotted arrow in Figure 3.24.
2. The series resonant frequency is larger than the parallel resonant frequency, that is, $f_S > f_P$.
3. The quality factor Q is determined by the series resonant frequency f_S and the load resistance R_L. The larger the value of Q is, the wider the working frequency range of the converter will be. If Q is too small, the above-mentioned characteristic curve of gain will be no longer applicable.
4. The half-bridge LLC resonant converter can theoretically work in the following four areas: (i)$f < f_P$, (ii)$f_P < f < f_S$, (iii)$f = f_S$, and (iv)$f > f_S$, but in fact, it can only work in the right area of f_P. Usually, in the case of rated load (full load), the efficiency of converter is the highest when the working frequency is designed as $f = f_S$. When $f \neq f_S$, the output voltage will be reduced with the increase of working frequency. It should be noted that when f is close to f_P, the voltage gain will varies significantly with the load resistance R_L, so it shall be avoided to work in this area.
5. The working frequency f of half-bridge LLC converter depends on the demand for output power P_O. When P_O is comparatively low, the working frequency can be considerably high,

and when P_O is high, the working frequency will be reduced automatically by the control circuit.

6. The following parameters should be focused on when designing the half-bridge LLC converter: the working frequency range required by the output voltage, the voltage regulation range of load, the amount of energy transferred in the resonant loop, and the converter efficiency.

3.13 Basic Principle of the 2-Switch Forward Converter

The 2-switch forward converter, featured by large output power and high conversion efficiency, is suitable for constituting SMPS with large power of hundreds of watts and high efficiency. See Figure 3.25 for the basic principle of the 2-switch forward converter. Two MOSFETs (V_1 and V_2) are used as switching devices in the circuit, which are turned on or off simultaneously

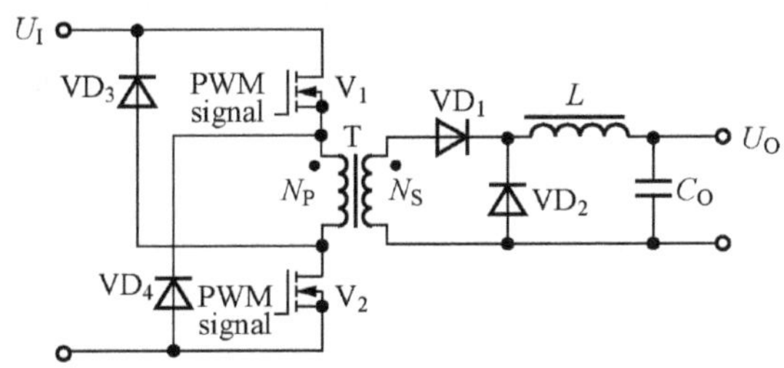

Figure 3.25 Basic principle of the 2-switch forward converter

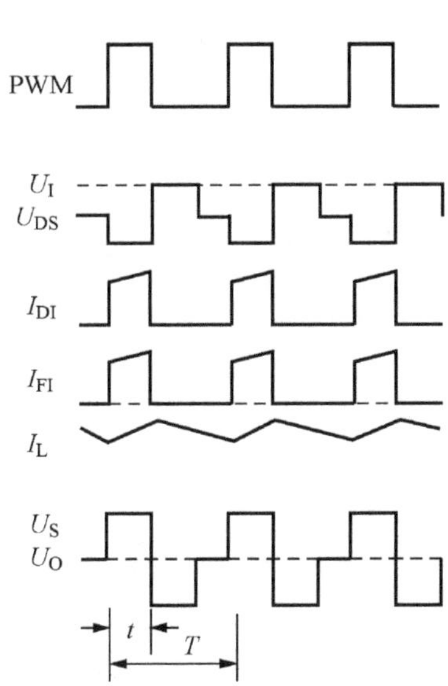

Figure 3.26 Working waveform of the 2-switch forward converter

under the control of PWM signal. VD_3 and VD_4 are the protection diodes of V_1 and V_2, respectively. High-frequency transformer T is used for isolation and voltage transformation. N_P and N_S, respectively, refer to the primary and the secondary windings. VD_1 and VD_2 are the output rectifier and the freewheeling diode, respectively, both of which use ultrafast recovery diode or Schottky diode. L refers to energy-storing inductor. Large-capacity electrolytic capacitor of low equivalent series resistance (ESR) shall be chosen as the output filter capacitor C_O to reduce output ripple voltage.

See Figure 3.26 for the working waveform of 2-switch forward converter. PWM indicates the PWM waveform. U_I is the input voltage. U_{DS} is the drain-source voltage of V_1; I_{D1}, the drain current of V_1; I_{F1}, the working current of VD_1; I_L, the load current; U_S, the output voltage of the secondary winding; U_O, the output voltage (average value); t, the time of high level of U_S; T, the cycle; and the duty ratio, $D = t/T$.

C_O refers to the filter capacitor at the output terminal. The output voltage is determined by the following formula:

$$U_O = \frac{N_S}{N_P} \cdot \frac{t}{T} \cdot U_I = \frac{N_S}{N_P} \cdot DU_I \tag{3.20}$$

4

Method for Selecting Key Peripheral Components of SMPS

4.1 Selection Method for Fixed Resistor

4.1.1 Selection Method for Fixed Resistor

Fixed resistor is the most basic and commonly used electronic component of switched-mode power supply (SMPS). In terms of function, the resistors of SMPS mainly include sampling, current limiting, voltage divider, shunt, bias, feedback, current detecting, bleeder, damping, special [e.g., network resistor, jumper, voltage-sensitive resistor (VSR), NTC thermistor, and fusible resistor] and load resistors, and so on. The resistor, used together with a capacitor, can constitute frequency compensation or phase correction network, and an oscillating circuit, absorption loop, and soft-start circuit. If used together with a clamping diode and capacitor, it can constitute R-, C-, and VD-type clamping diodes. In addition, it can also constitute an EMI filter combined with a capacitor and common-mode inductor.

4.1.1.1 Classification of Resistors

In terms of materials and manufacturing processes, the classification and main features of commonly used fixed resistors are shown in Table 4.1. Common fixed resistors include carbon film, metal film, metal film oxide, wire-wound, chip fixed, network (commonly known as resistor array) resistors, and jumpers [also known as zero ohm jumper (JP)].

Both chip and network resistors are integrated resistors. The wire-wound resistor can further be divided into noninductive (divide winding into two parts and then wind them in opposite directions with the same number of turns to minimize its inductance) and inductive wire-wound resistors. Sometimes, to reduce cost, sampling and current-detecting resistors can be replaced with printed conductors.

Optimal Design of Switching Power Supply, First Edition. Zhanyou Sha, Xiaojun Wang, Yanpeng Wang, and Hongtao Ma.

Table 4.1 Classification and main features of common fixed resistors

Resistor type	Symbol	Power range (W)	Resistance range (Ω)	Relative error (%)	Resistor temperature coefficient (absolute value) (10^{-6}/°C)
Carbon film resistors	RT	0.1–3	1–22 M	±2–10	300–1500
Metal film resistors	RJ	0.1–3	1–5.1 M	±0.1–2	25–100
metal film oxide resistors	RY	0.25–10	0.1–150k	±1–5	100–300
Wire-wound resistors	RX	0.5–10	0.01–10k	±1–10	25–100
Cement wire-wound resistors	RX	2–40	0.01–150k	±1–10	20–300
Power wire-wound resistors	RX	10–1000	0.5–150k	±1–10	20–400
Chip fixed resistors	RC	0.1–5	1–10 M	±0.1–5	100–200
Network resistors	—	0.25–2	10–2.2 M	±0.1–5	50–250
Jumpers	JP	0.125–0.5	0	—	—

4.1.1.2 Nominal Value and Deviation of Resistors

In order to facilitate production and usage, a series of standard values of resistance are developed by the International Electrotechnical Commission (IEC), which are called nominal values of resistors. There are six series of nominal values: E6, E12, E24, E48, E96, and E192. The corresponding permissible deviations are, respectively, ±20%, ±10%, ±5%, ±2%, ±1%, and ±0.5%. Among them, E6, E12, and E24 have 6, 12, and 24 nominal values, respectively, and so on. E48, E96, and E192 series are the precision resistors. The nominal resistance value equals the nominal value of the resistor multiplied by $\ldots\ 10^{-2},\ 10^{-1},\ 10^{1},\ 10^{2},\ 10^{3}\ \ldots$.

The nominal value of the resistors is set according to the principle of error, whose definition formula is as follows:

$$X = \sqrt[y]{10^n} \tag{4.1}$$

wherein

X nominal value of a resistor,

y serial number of the nominal value of the resistor, and

n $n = 0,\ 1,\ 2,\ 3,\ \ldots,\ y - 1$.

For example, suppose the serial number (y) of E24 is 24, the nominal value of the resistor is calculated by the following formula:

$$X = \sqrt[24]{10^n} \quad (n = 0,\ 1,\ 2,\ 3,\ \ldots\ 23) \tag{4.2}$$

If $n = 0$, then $X = 1.0$, if $n = 1$, then $X = 1.1$, and if $n = 3$, then $X = 1.3$ (approximate values).

The nominal value is featured that the positive deviation of a value is connected to the negative deviation of the next value in a series. Consider series E24, for example, $2.0 - 2.0 \times 5\% = 2.0 - 0.1 = 1.9$, while $1.8 + 1.8 \times 5\% = 1.89 \approx 1.9$. In mass production, select first the components with little deviation according to series E24, then series E12, and finally E6 and then mark with specific values after classification. The resistor components that appear in reality are those with special values, such as 2.2, 51, 470 kΩ, and so on, which are obtained by multiplying their nominal values by 10^n. The benefit of setting the nominal value of resistors is to maximize the rate of the qualified products. No matter how large the deviation of resistors produced by factory is, they can be classified into a nominal value of a series according to the standard of the series. Therefore, unless the components are rejected because they are defective in themselves, all the resistor components can be screened for use by certain nominal value and deviation level, no matter what their actual values are, thus greatly simplifying the specification, reducing the cost, and realizing "zero waste." For example, the resistance deviation of series E24 is ±5% and the resistance of a resistor is 1.56 kΩ, as it is the closest to the nominal value 1.6 kΩ, the resistor falls into the specification 1.6 kΩ. Another example is, if the resistance of a resistor is 8.1 kΩ, it can fall into the specification 6.8 kΩ of series E6 and the deviation is +19.1% < +20%, and it can also fall into the specification 8.2 kΩ of series E24 with the deviation of just −1.2%, without exceeding the deviation of ±5% specified for series E24.

The deviation of resistors equals the percentage from dividing the deviation between the actual value and the nominal value by the nominal value, which reflects the accuracy of a resistor. The permissible deviation can be expressed in many ways, such as Roman numerals and percentages: I (±5%), II (±10%), and III (±20%), pure numbers (e.g., 001, 01), or letters and percentages: B (±0.1%), C (±0.25%), D (±0.5%), F (±1%), G (±2%), J(±5%), K (±10%), and M (±20%).

The resistors used for SMPS generally allow the deviation of ±5% to ±10%, but for sampling and current-detecting resistors, the precision resistors with deviation of ±0.5% to ±1% should be adopted. It is worth noting that the resistors with deviation of ±5% to 10% were often used in the past, but the resistors with deviation of ±1% are cheap and easily available now.

4.1.1.3 Rated Power and Voltage

Rated Power. When the ambient temperature is constant under standard atmospheric pressure, the permissible power of a resistor that can be under load continuously for a long term without changing its performance is called rated power, also known as nominal power. For example, the rated power of RT carbon film resistors refers to the permissible power under the ambient temperature of 40 °C. Rated power usually includes 1/16, 1/8, 1/4, 1/2, 1, 2, 5, 10 W, and so on. The higher the rated power is, the larger the area requiring dissipating heat will be under the same temperature rise condition and the bigger the volume of a resistor will be. The price of a resistor mainly depends on its rated power, but it is irrelevant to its resistance, because the costs for producing resistors with high or low resistance are equal under the same process conditions. The calculation formula for rated power is $P = IU = I^2R = U^2/R$.

Rated power relates to the ambient temperature and the atmospheric pressure. For RT resistors, at the ambient temperature above 40 °C (70 °C above for RJ metal film resistors), the rated power is reduced by approximately 1.5% per 1 °C of temperature rise. Under forced air cooling conditions, the service power can be larger than the rated power. When the atmospheric pressure is decreased, the air will be comparatively thin owing to the decrease in its density, worsening the cooling conditions, so the permissible power shall be lower than the rated power.

When selecting the power of resistors, rated power must be derated for use. In general, the actual power consumption of a resistor shall not exceed half of its rated power in order to guarantee a long-term stable operation of the resistor. For example, if the actual power consumption of the resistor is 1/4 W, the resistor of 1/2 W shall be selected to reserve one-time margin of the actual power consumption. However, in a short-term work or under pulsed conditions, the margin can be reduced moderately. In addition, the rated power can also be calculated according to the duty ratio of continuous pulse.

Rated Voltage. The rated voltage of resistors $U = \sqrt{PR}$, while the limiting operating voltage refers to the withstand voltage value of a resistor. When the limiting operating voltage is exceeded, a resistor may be damaged owing to arc discharge. The rated voltage of each resistor can be reduced by connecting resistors in series.

4.1.1.4 Notes for Resistor Selection

1. When the wire-wound resistor works at switching frequency, its distributed inductance shows a comparatively large inductive reactance, and spike voltage will be generated in the case of a sudden change in current. Noninductive wire-wound resistors can be used if necessary.
2. When detecting AC current, current-detecting resistors or current dividers made from manganin wire can be used, because the resistance temperature coefficient of manganin is very low ($\alpha_T < 40 \times 10^{-6}/°C$). The lead of current-detecting resistor should be as short as possible. To reduce costs, the current-sensing resistor sometimes can be replaced with a section of copper foil on the printed circuit board (PCB), the precision of which depends on the size of the copper foil and the resistance temperature coefficient. The resistance temperature coefficient of copper is quite large, about 0.4%/°C. When the thickness of the copper foil is 35 μm, the resistance value at room temperature is determined by the following formula:

$$R = 0.5l/d \tag{4.3}$$

   wherein the unit of R is mΩ, and l and d refer to the length and the width of the copper foil, respectively. For copper foils of thickness 70 μm, the coefficient 0.5 in the above formula should be changed to 0.25.

4.1.2 *Selection Method for Current-Detecting Resistor*

When pulse width modulator (PWM) modulator is used to constitute SMPS, a current-detecting resistor needs to be connected externally to play a protective role when the output current exceeds the limit. The copper conductor on PCB (referred to as PCB conductor) can be used to produce current-detecting resistors, which can reduce volume and cost. However, the power consumption should be settled to prevent the PCB conductor from an increase in the resistance or even burn-through owing to the overheating caused by a too small conductor cross-section of the current-detecting resistor. See Table 4.2 for the parameters of commonly used PCB conductors (English units are generally used for PCB conductors and PCB at present). In designing current-detecting resistors, the following three formulae can be used to calculate the resistance of PCB conductor with the smallest occupation area:

Formula I:

$$\rho(T) = \frac{\rho[1 + \alpha\ (T_A + \Delta T - 20)]}{h}, \tag{4.4}$$

Table 4.2 Parameters of commonly used PCB conductors

Conductor thickness (μm)	Conductor width (in)	Resistance per unit length (mΩ/in)	Conductor thickness (μm)	Conductor width (in)	Resistance per unit length (mΩ/in)
18	0.025	39.3	70	0.025	9.83
	0.050	19.7		0.050	4.91
	0.100	9.83		0.100	2.46
	0.200	4.91		0.200	1.23
	0.500	1.97		0.500	0.49
35	0.025	19.7	106	0.025	6.50
	0.050	9.83		0.050	3.25
	0.100	4.91		0.100	1.63
	0.200	2.46		0.200	0.81
	0.500	0.98		0.500	0.325

wherein

$\rho(T)$ surface resistance (Ω) per unit thickness of PCB conductor at T°C,
ρ resistivity of PCB conductor at 20°C, $\rho = 0.0172\,\Omega \cdot \text{mm}^2/m = 0.0172\ \Omega \cdot \mu\text{m}$,
α resistance temperaturxe coefficient of PCB conductor, $\alpha = 0.00393\,/\,°\text{C}$,
T_A ambient temperature (°C),
ΔT the maximum permissible temperature rise of PCB conductor (°C), and
h the thickness of PCB conductor (μm).

Formula II:

$$b = \frac{I_{OM}}{\sqrt{\frac{\Delta T}{R_{\theta}\rho(T)}}}, \tag{4.5}$$

wherein

b the minimum width (mil) of PCB conductor, 1 mil equals to milli-inch, that is, 1 mil = 0.001 in = 0.0254 mm,
I_{OM} the maximum permissible current (A) when the temperature is ΔT, and
$R_{\theta SA}$ thermal resistance when the CCL of 1 in^2 is in contact with air, 1 in^2 = 645 mm^2. See Figure 8.47 for the relationship curve between the thermal resistance of PCB radiator and the area of cooling copper foil.

Formula III:

$$l = \frac{bR_S}{\rho(T)}, \tag{4.6}$$

wherein

l the length of PCB conductor (mil),
b the width of PCB conductor (mil), and
R_S the expected resistance value (Ω).

See Table 4.3 for the corresponding relationships between the weight and the thickness per unit area of PCB with different specifications, wherein oz/ft^2 refers to "ounce/square foot," 1 ounce = 31.1035 g and 1 square foot = 929.0 cm^2.

Table 4.3 Corresponding relationships between the weight and thickness per unit area of **PCB** with different specifications

Mass per unit area of PCB		Thickness of PCB conductor (μm)
English system of units (oz/ft^2)	International system of units (g/cm^2)	
0.5	0.0167	17.8
1	0.0334	35.6
2	0.0669	71.1
3	0.100	106.7

The following examples describe the method for designing current-detecting resistors. Suppose that the external current-detecting resistance R_S of a SMPS is determined by the following formula:

$$R_S = \frac{0.035\,\text{V}}{I_{\text{LIMIT}}}, \tag{4.7}$$

wherein I_{LIMIT} is the set limiting current value. The designed detecting resistor should occupy the minimum area. According to formula (4.7), when $R_S = 3\,\text{m}\Omega$, the set $I_{\text{LIMIT}} = 11.67\text{A}$.

Given the ambient temperature $T_A = 25\,°\text{C}$, the maximum temperature of PCB conductor $T_M = 100\,°\text{C}$, the maximum permissible temperature rise $\Delta T = 100 - 25\,°\text{C} = 75\,°\text{C}$, the thickness of PCB conductor $h = 35.6\,\mu\text{m}$, then If $I_{\text{OM}} = 10\,\text{A}$, the corresponding $R_S = 3\,\text{m}\Omega$. See Figure 4.1 for the shape of PCB conductor, whose length is l. In order to eliminate the errors caused by the voltage drop of the conductor, it is recommended that the four-wire connection method be applied for the current-detecting resistor. The steps to design a current-detecting resistor are as follows: (i) according to the thickness and the maximum permissible temperature rise of PCB conductor, calculate the surface resistance $\rho(T)$ at 100 °C according to formula (4.4), (ii) according to the maximum current withstood by R_S $I_{\text{OM}} = 10\,\text{A}$, calculate the minimum conductor width b according to formula (4.5), and (iii) according to the required resistance value, calculate the conductor length l according to formula (4.6).

1. Calculate the surface resistance $\rho(T)$ of PCB conductor at 100°C. Put $\rho = 0.0172\,\Omega \cdot \mu\text{m}$, $\alpha = 0.00393/°\text{C}$, $T_A = 25\,°\text{C}$, $\Delta T = 75\,°\text{C}$, and $h = 35.6\,\mu\text{m}$ together into the

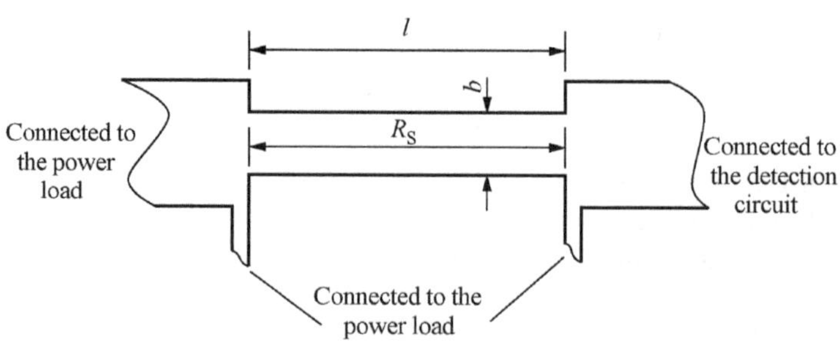

Figure 4.1 The shape of PCB conductor

Table 4.4 **Relationship** of copper foil thickness, width, and permissible current of **PCB** conductor

Copper foil thickness 35 μm		Copper foil thickness 50 μm		Copper foil thickness 70 μm	
Permissible current (A)	Conductor width (mm)	Permissible current (A)	Conductor width (mm)	Permissible current (A)	Conductor width (mm)
0.2	0.15	0.5	0.15	0.7	0.15
0.55	0.2	0.7	0.2	0.9	0.2
0.8	0.3	1.1	0.3	1.3	0.3
1.1	0.4	1.35	0.4	1.7	0.4
1.35	0.5	1.7	0.5	2.0	0.5
1.6	0.6	1.9	0.6	2.3	0.6
2.0	0.8	2.4	0.8	2.8	0.8
3.2	1.0	2.6	1.0	3.3	1.0
2.7	1.2	3.0	1.2	3.6	1.2
3.2	1.5	3.5	1.5	4.2	1.5
4.0	2.0	4.3	2.5	5.1	2.0
4.5	2.5	5.1	2.5	6.0	2.5

formula (4.4), then

$$\rho(T) = \frac{0.0172[10.00393257520]}{35.6} = 635\,\mu\Omega$$

2. Calculate the minimum conductor width b. Put $I_{OM} = 10A$, $\Delta T = 75\,°C$, $R_{\theta SA} = 55\,°C/W$, and $\rho(T) = 635\,\mu\Omega$ together into the formula (4.5), then

$$b = \frac{10}{\sqrt{\frac{75}{55\times 635}}} = 215.8\,\text{mil} \approx 216\,\text{mil}$$

3. Calculate the conductor length l. Put $b = 216\,\text{mil}$, $R_S = 3\,\text{m}\Omega = 3000\,\mu\Omega$, and $\rho(T) = 635\,\mu\Omega$ together into the formula (4.6), then

$$l = \frac{216 \times 3000}{635} = 1020\,\text{mil} = 1.020\,\text{in} = 2.59\,\text{cm}$$

Similarly, when $R_S = 4\,\text{m}\Omega$ (the corresponding $I_{LIMIT} = 8.75A$), it can be worked out that the conductor length $l = 1360.6\,\text{mil} \approx 1361\,\text{mil} = 1.361\,\text{in} = 3.46\,\text{cm}$.

See Table 4.4 for the relationship of copper foil thickness, width, and permissible current of PCB conductor at the room temperature of 25 °C.

4.2 Selection Method for Capacitors

4.2.1 Classification of Common Capacitors for SMPS

Many types of capacitors are used in SMPS, such as input filer, output filter, coupling, bypass, decoupling, noise canceling, oscillating, soft-start, and frequency compensation or phase correction network capacitors as well as safety capacitors of EMI capacitors (including X and Y capacitors). See Section 8.9 for the detailed selection method of EMI capacitors.

In terms of dielectric materials, capacitors mainly include paper, aluminum-electrolytic, tantalum, ceramic (including multilayer ceramic capacitors), film, and chip fixed capacitors. The paper capacitor is generally limited to low-frequency filter circuit owing to its large high-frequency AC loss. Aluminum-electrolytic capacitors are usually used as input or output filters and bypass capacitors in view of their low price. The tantalum capacitor is far superior to the aluminum-electrolytic capacitor in terms of high-frequency characteristics, but it costs comparatively higher, narrow in capacity range (a few hundred microfarads or less), and low in rated voltage. Oscillating and bypass capacitors can be selected from ceramic capacitors. Multilayer ceramic capacitors can be used to replace tantalum capacitors owing to their very low equivalent series resistance (ESR) and a capacity as low as several hundred microfarads. Film capacitors, with plastic as dielectric material, generally include dacron, polyethylene, polystyrene, polypropylene, teflon, polycarbonate capacitors, and so on, among which, the quality of polypropylene capacitor is comparatively better. Chip fixed capacitors are suitable for low-voltage fields owing to their small volumes.

Common nominal values of the capacitor include the following 24 kinds of specifications: 1.0, 1.1, 1.2, 1.3, 1.5, 1.6, 1.8, 2.0, 2.2, 2.4, 2.7, 3.0, 3.3, 3.6, 3.9, 4.3, 4.7, 5.1, 5.6, 6.2, 6.8, 7.5, 8.2, and 9.1.

The DC working voltage of capacitors refers to the highest voltage for a long-term safe work of a capacitor, which is referred to as withstand voltage. Common withstand voltages of capacitors include 6.3, 10, 16, 25, 63, 160, 250, 300, 400, 450, 630, and 1000 V.

There are two kinds of capacitor loss. One is dielectric loss, that is, the energy loss caused by the lagging effect of dielectric conductance and dielectric polarization in insulating materials under the effect of electric field. Dielectric loss can be indicated by loss factor (also known as loss tangent) tgδ, which equals the ratio between the active and reactive powers of a capacitor. tgδ of common electrolytic capacitors is 0.1–0.2, 0.005–0.05 for paper capacitors, and 0.0002–0.002 for mica capacitors. The smaller the value is, the lower the loss will be. Capacitors with small tgδ should be selected for high-frequency circuits. The other kind of capacitor loss is that caused by the ESR of capacitors.

4.2.2 Comparison between Ideal and Actual Capacitors

An ideal capacitor may be considered as a pure capacitor, whose impedance reduces with the increase in frequency. The impedance is expressed as follows:

$$Z = 1/(2\pi fC) \tag{4.8}$$

The actual capacitor contains not only a pure capacitor C and a drain resistance R but also the following two important parameters: one is ESR, indicating the equivalent resistance value R_{ESR} connected with the ideal capacitor in series and reflecting the characteristics of the filter capacitor, and the other is equivalent series inductance (referred to as ESL), indicating the equivalent inductance value L_{ESL} (i.e., distributed inductance) connected with the ideal capacitor in series. The shunting effect of drain resistance R on C can be negligible owing to the large value of the drain resistance, so the total impedance of the actual capacitor is as follows:

$$Z = \sqrt{R_{\mathrm{ESR}}^2 + X^2}, \tag{4.9}$$

wherein X refers to the reactance, $X = X_{\mathrm{L}} - X_{\mathrm{C}} = 2\pi f L_{\mathrm{ESL}} - 1/(2\pi fC)$.

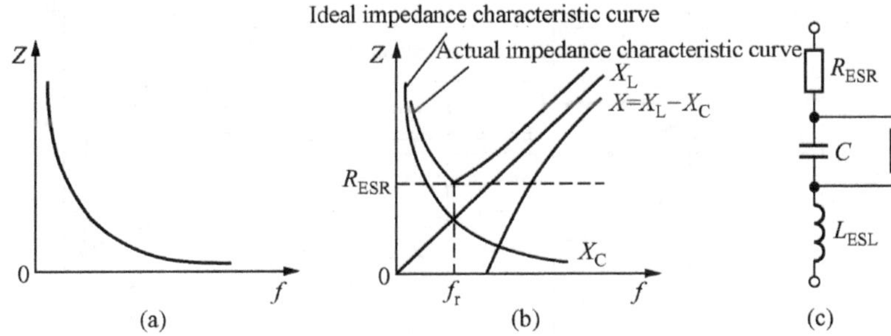

Figure 4.2 Comparison of the impedance characteristics between ideal and actual capacitors. (a) Impedance characteristic of ideal capacitors, (b) comparison of the impedance characteristics between ideal and actual capacitors, and (c) equivalent circuit of an actual capacitor

See Figure 4.2 for the comparison of the impedance characteristics between ideal and actual capacitors. As the figure shows, there is a great difference in the impedance characteristics between them. The impedance characteristic of the former is expressed as hyperbola (in the first quadrant only), while that of the latter is expressed as concavity, with the following features:

1. At the low-frequency stage ($f < f_r$), the capacitor is capacitive, $X \approx X_C = 1/(2\pi fC)$, and the impedance sometimes decreases with the increase in frequency.
2. When the frequency is raised to the self-resonant frequency of equivalent inductance in series (i.e., $f = f_r$), the impedance reaches its minimum $Z = R_{ESR}$.
3. When $f > f_r$, the capacitor is inductive, $X \approx X_L = 2\pi f L_{ESL}$, and simultaneously the impedance increases with the rapid rise of inductive reactance.
4. The higher the self-resonant frequency of a capacitor is, the more suitable for working in high-frequency fields it will be.

4.2.3 Selection Method for Input Filter Capacitor

4.2.3.1 Selection Method for the Capacity of Input Filter Capacitor

The selection of lead styles for input filter capacitors

The input filter capacitor C_I of SMPS, when used together with the bridge rectifier, can convert an AC voltage into a DC high voltage. Input filter capacitors usually adopt aluminum-electrolytic capacitors with a withstand voltage of 400 V. This kind of capacitors include radial (RADIAL, two leads drawn from one end of a capacitor) and axial (AXIAL, two leads, respectively, drawn from both ends of a capacitor) lead types. Figure 4.3 shows the impedance characteristics of three kinds of aluminum-electrolytic capacitors, which can withstand 400 V and use axial and radial leads, respectively.

The numbers in the brackets of the figure refer to the sizes of the capacitors [diameter × length (mm)]. It can be seen from the figure that the impedance of the capacitors using radial lead is very low in the frequency range below 10 MHz, while the capacitors using axial lead begin to show inductive impedance when the frequency is above 1 MHz, whose total impedance increases with the rise of frequency. Therefore, it is recommended to use

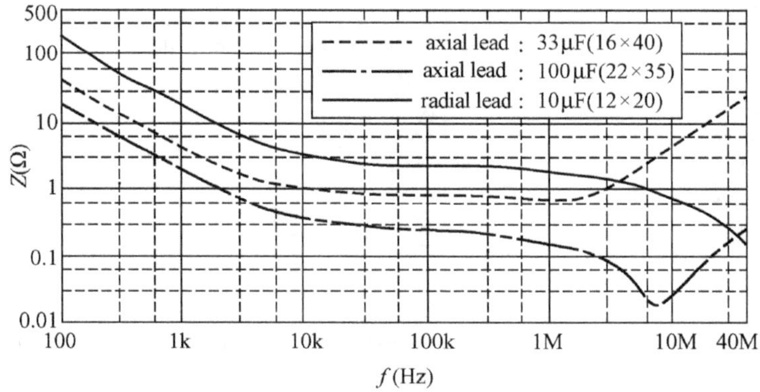

Figure 4.3 Impedance characteristics of three kinds of aluminum-electrolytic capacitors with a withstand voltage of 400 V

the capacitors using radial lead and reduce the lead length as much as possible in installation rather than the capacitors using axial lead, because the lead of the latter is comparatively long (at least equal to the diameter of the capacitor), making the ESL L_{ESL} increase, and thus increasing the total impedance. What needs to be noted is that when the frequency exceeds 1 MHz, compared with the actual impedance of the capacitors using radial lead with a comparatively small capacity, that of the capacitors using axial lead with a comparatively large capacity is higher, which will cause a comparatively large conducted interference current.

The selection of the capacity for input filter capacitor

To reduce the output ripple of rectifier filter, the capacity C_I of the input filter capacitor shall be properly selected. Assume that the scale factor of the capacity (μF) of input filter capacitor required by the output (W) power per unit is k, when the AC voltage $u = 85\text{–}265$ V, then $k = (2\text{–}3)$ μF/W, and when $u = 230\,\text{V}$ $(1 \pm 15\%)$, $k = 1\,\mu\text{F/W}$. See Table 4.5 for the selection method of the capacity of input filter capacitors, wherein P_O refers to the output power of SMPS.

4.2.3.2 Accurate Calculation Method of the Capacity of Input Filter Capacitors

The capacity of input filter capacitors is an important parameter of SMPS. If C_I is too small, U_{Imin} will be greatly reduced, while the input ripple voltage U_R will rise. However, if C_I is too

Table 4.5 Selection method for the capacity of input filter capacitors

u(V)	$U_{I(min)}$ (V)	P_O (W)	k (μF/W)	C_I (μF)
110 (1 ± 15%)	≥90	2–3	(2–3)	≥(2–3) P_O value
85–265	≥90	2–3	(2–3)	≥(2–3) P_O value
230 (1 ± 15%)	≥240	1	1	≥P_O value

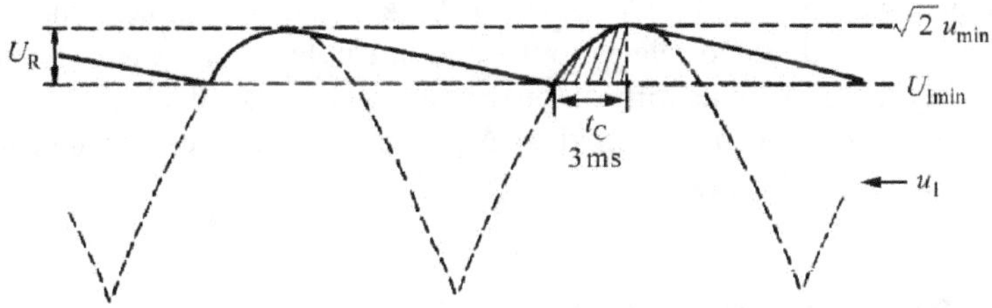

Figure 4.4 Input voltage waveform corresponding to the minimum AC voltage

large, the capacitor cost will increase and there is no remarkable effect of increasing U_{Imin} and reducing the ripple voltage. The following text describes the method of calculating the exact value of C_I.

Assume that the minimum of AC voltage u is u_{min}. u passes through bridge rectification and the filtering of C_I. See Figure 4.4 for the input voltage waveform when $u = u_{min}$. The figure is based on $P_O = P_{OM}$, $f = 50\,Hz$, the conduction time of the rectifying bridge $t_C = 3\,ms$, and $\eta = 80\%$. It can be seen from the figure that a primary ripple voltage with an amplitude of U_R is superposed on the minimum DC high voltage U_{Imin}, which is generated in the process of the charge and discharge of C_I. The accurate value of C_I can be calculated by the following formula:

$$C_I = \frac{2P_O\left(\frac{1}{2f} - t_C\right)}{\eta(2u_{min}^2 - U_{Imin}^2)} \tag{4.10}$$

For example, $u_{min} = 85\,V$ within the wide range of voltage input. Suppose that $U_{Imin} = 90\,V$, $f = 50\,Hz$, $t_C = 3\,ms$, $P_O = 30\,W$, and $\eta = 80\%$, put all of them in the formula (4.10). The result is $C_I = 84.2\,\mu F$ and the scale factor is $C_I/P_O = 84.2\,\mu F/30\,W = 2.8\,\mu F/W$, which is exactly within the permissible range of (2–3) $\mu F/W$.

It is necessary to point out that only when SMPS is idle, the voltage at both ends of the input filter capacitor $U_I \approx \sqrt{2}u$, wherein u refers to the effective value of AC voltage, but when SMPS is with a rated load, $U_I \approx 1.2u$. For example, if the AC input voltage is 85–265 V, the minimum DC input voltage $U_{Imin} = 85\,V \times 1.2 = 102\,V$. That is the "raising" effect of the input filter capacitor on the DC voltage.

4.2.4 Selection Method for Output Filter Capacitor

4.2.4.1 Calculation of the Capacity of the Output Filter Capacitor

Suppose that $C_{O(min)}$ is the minimum of an output filter capacitor, f is the switching frequency, and ΔU_O and ΔI_O are, respectively, the output ripple voltage and the ripple current of the output filter capacitor, then $C_{O(min)}$ can be calculated as follows:

$$C_{O(min)} = \frac{1}{8f\left(\frac{\Delta U_O}{\Delta I_O} - R_{ESR}\right)} \tag{4.11}$$

The ripple current on the load relates to the output ripple voltage and the equivalent load, which is calculated by the formula $\Delta I_L = \Delta U_O / R_L$. ΔI_L is usually much smaller than ΔI_O.

The ripple current on the load relates to the output ripple voltage and the equivalent load and R_{ESR} is the key factor in generating output ripple voltage, its calculation formula is $\Delta I_L = \Delta U_O / R_L$. ΔI_L is usually much smaller than ΔI_O. The approximate formula for the output ripple voltage is $\Delta U_O \approx \Delta I_O R_{ESR}$.

4.2.4.2 Notes for Selecting the Output Filter Capacitor

1. The polarity of the electrolytic capacitor shall not be connected in reverse. The ground terminal of the filter capacitor should be as close to the secondary return terminal (ground) as possible.
2. Electrolytic capacitors shall be derated in usage. Generally, the withstand voltage value should be 1.2–1.5 times of the actual working voltage.
3. In theory, the larger the capacity of a filter capacitor is, the better it is, but actually too much capacity will not significantly improve the filtering effect. That is because the drain resistance rises with the increase of capacity, which results in a corresponding rise of the ESR and inductance.
4. The working life of the electrolytic capacitor also relates to the ripple current, the ambient temperature, and other factors. The larger the ripple current is and the higher the ambient temperature is, the shorter the working life will be. In general, the working life can be doubled approximately with a decline of 10 °C in the ambient temperature. The 105 °C electrolytic capacitor produced by Rubycon of Japan can be used for the SMPS under a high-temperature environment.
5. In addition, to further reduce the ESR, several electrolytic capacitors with the same capacity and low ESR can be connected in parallel to replace an electrolytic capacitor with a large capacity. The ESL L_{ESL} can also be reduced by this method.
6. To reduce output noise, another small capacitor of 0.01–0.1 μF can be connected in parallel to the electrolytic capacitor.
7. To improve the filtering effect, a second-order LC filter can be adopted, its filter inductance can usually be 30–100 μH.
8. For the presence of R_{ESR}, capacitors will cause power loss $I^2 R_{ESR}$ (I refers to the mean square value of current) during charging and discharging, which will result in capacitor fever and the reduction in power supply efficiency. R_{ESR} relates to the frequency, the temperature, and the rated voltage. Assume that the switching frequency of SMPS is 100 kHz, the required output ripple current is 1 A (peak-to-peak value), and the ripple voltage is 50 mV (peak-to-peak value), then the charge variation $\Delta Q = 1\,\text{A} \times (1/100\,\text{kHz}) = 10\,\mu\text{C}$. Without considering R_{ESR}, the required charge will be $C = \Delta Q/U = 10\,\mu\text{C}/50\,\text{mV} = 200\,\mu\text{F}$. Connect two electrolytic capacitors of 100 μF in parallel, assuming that the typical value of R_{ESR} of each capacitor at room temperature is 100 mΩ. In such a case, $R_{ESR} = 50\,\text{mV}/1\,\text{A} = 50\,\text{m}\Omega$ is required in order to make the ripple voltage reduced to 50 mV. Obviously, the requirement can be met just by connecting two electrolytic capacitors of 100 μF in parallel.
9. The filter capacitors with high self-resonant frequency and good temperature characteristic should be used to meet the requirement for filtering of large current with high frequency.

Table 4.6 Service life comparison between solid capacitors and aluminum-electrolytic capacitors

Ambient temperature (°C)	Life of solid capacitors (h)	Life of aluminum-electrolytic capacitors (h)	Life comparison between them
75	60,000	16,000	3.75:1
85	20,000	8000	2.5:1
95	6000	4000	1.5:1

10. Solid capacitors (solid aluminum-electrolytic capacitors) use high molecular polymers with high conductivity as their dielectric instead of electrolyte, featured by a stable working, high-temperature resistance, long life, good high-frequency characteristic, low ESR, energy saving, and environmental protection without the accidents of liquid leakage, explosion, and thermal expansion. The performance of the solid capacitor is far superior to that of the aluminum-electrolytic capacitor, especially suitable for SMPS used under relatively poor working conditions. See Table 4.6 for life comparison between solid capacitors and aluminum-electrolytic capacitors.

4.3 Inductor Characteristics and Selection Method for Magnetic Beads

4.3.1 Inductor Characteristics

A pure inductor can be considered ideal, and its impedance increases linearly with the rise of frequency. The impedance can be expressed as:

$$Z = 2\pi fL \tag{4.12}$$

An actual inductor contains not only a pure inductor but also series DC resistance (R_S) and turn-to-turn capacitance in parallel (C_W, also known as distributed capacitance), so the total impedance of an actual inductor is as follows:

$$Z = \sqrt{R_S^2 + X^2} \tag{4.13}$$

See Figure 4.5 for the comparison of the impedance characteristics between ideal and actual inductors. It can be seen from the figure that the impedance characteristic of the ideal capacitor is expressed as linear shape while that of the actual capacitor as convex shape with the following features:

1. At the low-frequency stage ($f < f_L$), the capacitor is inductive, $X \approx X_L = 2\pi fL$, and the impedance sometimes increases with the rise of frequency.
2. When frequency is raised to the self-resonant frequency f_L determined by the turn-to-turn capacitance C_W, the impedance reaches its maximum.
3. When $f > f_L$, the capacitor is capacitive, $X \approx X_C = 1/(2\pi fC)$, and the impedance decreases with the rapid reduction of inductive reactance.

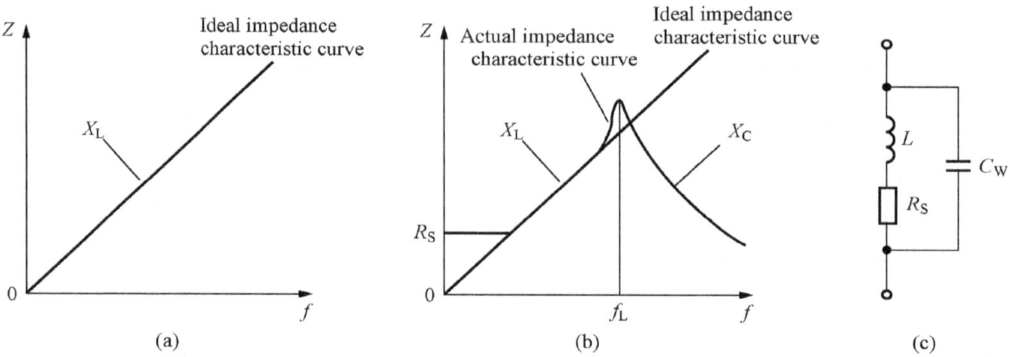

Figure 4.5 Comparison of the impedance characteristics between ideal and actual inductors. (a) Impedance characteristics of the ideal inductor, (b) comparison of the impedance characteristics between actual and ideal inductors, and (c) equivalent circuit of the actual inductor

There are many kinds of inductors used in SMPS, mainly including high-frequency transformers, output filter, EMI filter, magnetic beads, and PFC inductors. See Chapter 6 for the design of the high-frequency transformer, Section 4.4 for the selection method of EMI filter inductors, and Section 5.4 for the selection of PFC inductors.

4.3.2 Selection Method for Magnetic Beads

Owing to the comparatively high switching frequency of SMPS, switching noise will be generated in the reverse recovery after the output rectifier is switched off, and it will be easy to damage the rectifier. Although the RC absorbing circuit composed of series resistor–capacitor units is connected in parallel at both ends of the output rectifier, which can restrain the switching noise to a certain extent, the effect is not ideal and power loss will be caused on resistors. The solution is to connect a magnetic bead in series to the secondary rectifier filter.

4.3.2.1 Performance Features of Magnetic Beads

Ferrite magnetic bead filter (referred to as magnetic bead) is a kind of ultra-compact magnetic material emerged in recent years. It is made through sintering of ferrite materials (or amorphous alloy) and conducting wires at high temperature. Magnetic beads have the following features:

1. Magnetic beads, featured by a large high-frequency loss and high resistivity and permeability (about 100–1500 H/m), can restrain noise interference within a very wide frequency band range. Magnetic beads can be connected in series in signal or power circuits to effectively restrain series-mode noise interference.
2. Magnetic beads can be equivalent to the series circuit of inductors and resistors, whose equivalent inductance and resistance are proportional to the length of them. At DC or low-frequency stage, the inductive reactance of magnetic beads is very low, having no influence on the signal transmission in the data or signal line. At the high-frequency stage

above 10 MHz, however, the inductive reactance is still very small, but the equivalent resistance increases rapidly, causing the total impedance to increase, and thus making high-frequency noise attenuated greatly. In this case, the impedance toward low-frequency signal can be negligible and does not affect the normal work of the circuit. Therefore, the magnetic bead can be equivalent to a low-pass filter, which allows DC current to pass through but filters off high-frequency noise. The performance of this filter is better than that of an ordinary filter inductor. Filter inductors are prone to generate resonance, thus forming a new interference source, but magnetic beads have no such a problem.

3. Magnetic beads are measured according to the impedance generated at a certain frequency with the unit of Ω instead of H. The curve of frequency–impedance characteristics provided in the data sheet of magnetic beads generally takes 100 MHz as a standard. For example, "60R@100 MHz" means that the impedance of a magnetic bead is 60 Ω at frequency 100 MHz, and so on.
4. Magnetic beads can absorb high-frequency components, so it is also called an absorption-type filter. In contrast, EMI filter belongs to the reflection-type filter because the function of its common-mode inductance is to reflect the magnetic interference back to the signal source.
5. Magnetic beads, converting high-frequency energy into eddy current that is dissipated as heat, are referred to as energy dissipation devices. The eddy current loss is directly proportional to the square of noise frequency. By comparison, ordinary inductors are used to store energy, referred to as energy storage elements. The maximum working frequency of a magnetic bead can be up to 1 GHz, while the working frequency of an inductor is generally no more than 50 MHz.
6. Magnetic beads can suppress the switching noise, it actively suppresses high-frequency interference. The reason is that a magnetic bead is connected in the main circuit (i.e., the output circuit) generating spike pulses, whose inductance can be used to reduce the rise rate of spike current, so it is called "active suppression." On the other hand, the EMI filter can only suppress interference passively, so it is called "passive suppression." That is the fundamental difference between them.
7. Multiple magnetic beads may be connected in series or parallel. Usually, magnetic beads should be installed near the interference source.
8. Magnetic beads can be used not only in high-frequency SMPS, electronic measuring instruments, and various circuits with a demanding requirement for noise but also widely in digital household appliances such as mobiles phones, DVD, and digital cameras. Chip magnetic beads can eliminate the radio frequency interference (RFI) in transmission lines. In electromagnetic compatibility (EMC) design, magnetic beads are a common magnetic material used to suppress high-frequency electromagnetic interference.

4.3.2.2 Selection Method for Magnetic Beads

Magnetic beads include tubular, chip, array (commonly known as magnetic bead array) types, and so on. Tubular magnetic beads are further divided into single-hole, double-hole, and porous types to meet different needs. At present, the common dimensions of tubular magnetic beads in the market include $\phi 2.5 \times 3$ (mm), $\phi 2.5 \times 8$ (mm), $\phi 3 \times 5$ (mm), $\phi 3.5 \times 7.6$ (mm), and other specifications. It is tube shaped with leads passing through its core. See Figure 4.6 for

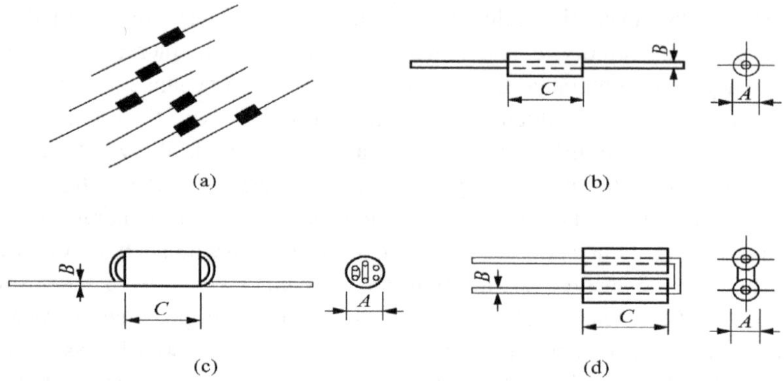

Figure 4.6 Shape and internal structure of typical tubular magnetic beads. (a) Outline drawing of magnetic beads, (b) structures of HT-A62 and HT-B62, (c) structure of HT-S62, and (d) structure of HT-R62

Table 4.7 Technical indicators of typical tubular magnetic beads

Model	Size (mm)			Impedance value (Ω)	
	A	*B*	*C*	25 MHz	100 MHz
HT-A62	3.5 ± 0.2	0.6 ± 0.1	6.0 ± 0.3	50	90
HT-B62	3.5 ± 0.2	0.6 ± 0.1	9.0 ± 0.3	70	120
HT-S62	6.0 ± 0.2	0.6 ± 0.1	10 ± 0.4	320	580
HT-R62	3.5 ± 0.2	0.6 ± 0.1	6.0 ± 0.3	100	130

Table 4.8 Technical indicators of five types of typical chip magnetic beads

Specification	Impedance value (Ω 100 MHz)	Rated current (mA)	Maximum DC resistance (Ω)	Working temperature range (°C)
0402	10–1000	50–500	0.05–1.50	−55 to +125
0603	120–1000	200	0.2–0.7	
0805		200–500	0.15–1.1	−55 to +85, −55 to +125
1206	26–600		0.15–0.90	−55 to +125
1806	80–150		0.1–0.5	

the shape and the structure of typical products of tubular magnetic beads. See Table 4.7 for the technical indicators of four typical products of tubular magnetic beads.

Chip magnetic beads include general purpose, sharp, and high-current (1–6 A) types. See Table 4.8 for the technical indicators of five specifications of chip magnetic beads produced by Murta in Japan. See Figure 4.7 for the dimensions of typical chip magnetic beads. The dimensions of microchip magnetic beads are only 1 mm × 0.5 mm × 0.5 mm (equivalent to 0.04 in × 0.02 in × 0.02 in).

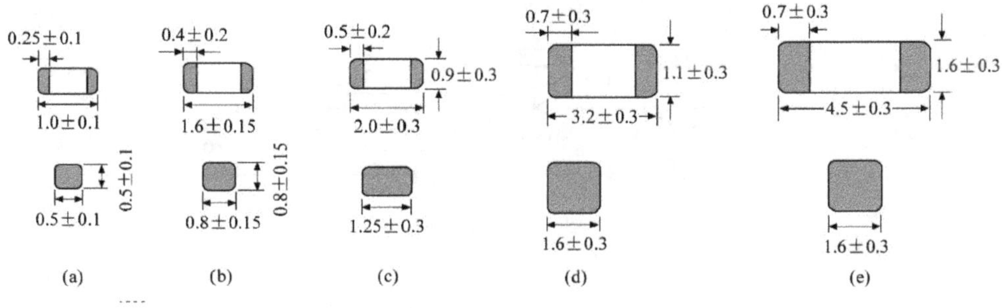

Figure 4.7 Dimensions of typical chip magnetic beads (unit: mm). (a) 0402–BLM10A; (b) 0603–BLM11A; (c) 0805–BLM21A; (d)1206–BLM31A; (e)1806–BLM41A

Magnetic bead array is a kind of integrated chip device with multiple magnetic beads (e.g., 2, 4, 6, and 8) packaged together. For example, BMA2010 magnetic bead array contains four magnetic beads with dimensions of only 2.0 mm × 1.0 mm. Using magnetic bead array can save the PCB area occupied.

Magnetic beads can be equivalent to the series circuit composed of inductance L and loss resistance R. Consider Figure 4.8(a) for the impedance characteristics of BLM03AG700SN1 magnetic beads, for example, wherein Z refers to the impedance, R the loss resistance, X_L the inductive reactance, and $X_L = 2\pi fL$, $Z = \sqrt{R^2 + X_L^2}$. It can be seen from the figure that when frequency f rises from 1 to 1000 MHz, the impedance and the loss resistance of magnetic bead increase rapidly, and when $f = 1000$ MHz, both Z and R reach their maximum, while the inductive reactance increases slowly with the rise of frequency. X_L reaches its maximum when $f = 30$ MHz, and then gradually becomes small till zero when $f = 1000$ MHz.

Some electronic devices can produce spike voltages with high amplitude at a fixed frequency, which are difficult to suppress with ordinary EMI filters. For this case, spike

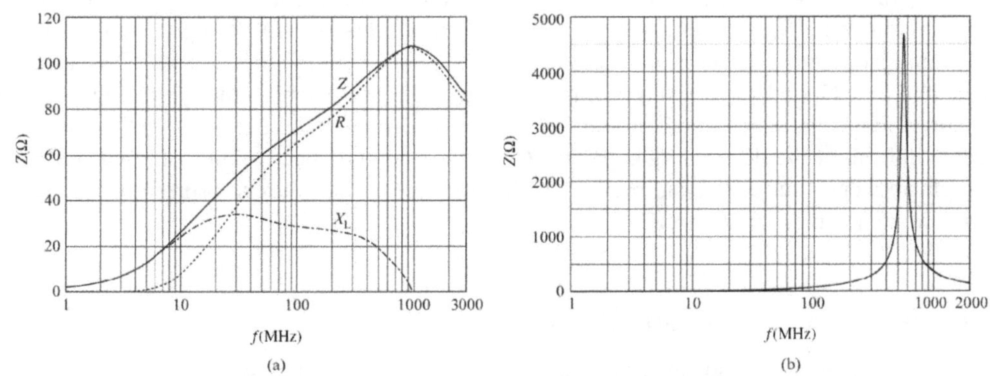

Figure 4.8 Impedance characteristics of two types of magnetic beads. (a) BLM03AG700SN1; (b) BLM15BA750SN1

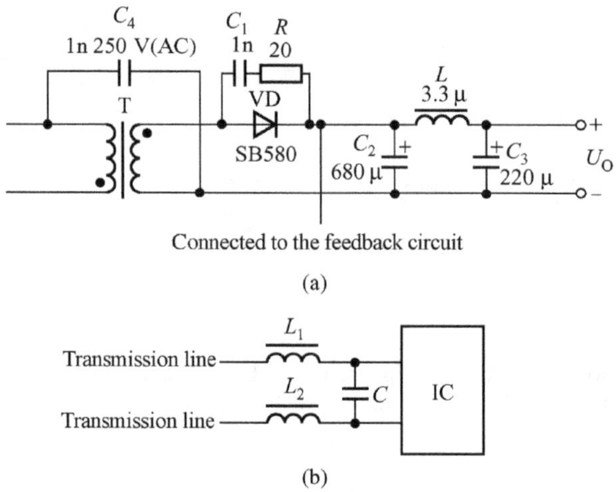

Figure 4.9 Typical applications of magnetic beads. (a) Application circuit (partial) of magnetic beads in SMPS and (b) application circuit of magnetic beads in electronic devices

ferrite magnetic beads are specially designed to solve this problem, and their characteristic curve is a spike shape at a certain frequency. For example, consider Figure 4.8(b) for the impedance–frequency characteristic curve of BLM15BA750SN1 magnetic beads.

4.3.2.3 Application of Magnetic Beads in SMPS

Magnetic beads can be widely used in SMPS. In order to suppress the switching noise of SMPS, magnetic beads can be connected in series in the main circuit (i.e., the output circuit) generating spike pulses to reduce the rise rate of the spike current, thus suppressing the switching noise of SMPS.

See Figure 4.9(a) for the typical application circuit of magnetic beads in SMPS. SB580 Schottky diode (SBD) is used as the secondary output rectifier VD. R and C_1 constitute an electromagnetic interference absorption network. L is a 3.3 μH magnetic bead. C_4 is a safety capacitor, and C_2 and C_3 are output filter capacitors. Figure 4.9(b) is the typical application circuit of magnetic beads in electronic devices, in which magnetic beads L_1 and L_2 are connected in front of the decoupling capacitor C. As the magnetic bead exhibits high impedance at high-frequency current, it can prevent the high-frequency interference current in the transmission line from flowing into IC.

4.4 Selection Method for EMI Filter

EMI filter is a kind of composite device that has been popularized in recent years, it can effectively suppress power network noise and improve the anti-interference ability of electronic devices and the system reliability. Therefore, it is widely used in SMPSs, electronic measuring instruments, computer room equipment, and other fields. EMI filter is composed of capacitors, inductor, and other components, with the advantages of simple structure and low cost, thus facilitating application and popularization.

4.4.1 *Structure Principle and Selection Method of EMI Filter*

4.4.1.1 Basic Circuit of EMI Filter

See Figures 4.10 and 4.11, respectively, for the basic circuit and typical application diagram of EMI filter. The five-terminal device has two input terminals, two output terminals, and one ground terminal. The shell shall be grounded when the filter is in use. The circuit contains a common-mode choke (also known as common-mode inductor) L and filter capacitors C_1–C_4. L has no effect on series-mode interference. However, in the case of common-mode interference, it shows large inductive reactance toward common-mode signals owing to the rapid rise of total inductance after coupling resulted from the same direction of the magnetic flux of the two coils, thus making common-mode signals hard to pass through, so it is called common-mode choke. Its two coils are, respectively, wound on ferrite beads with low loss and high magnetic inductivity. When there is common-mode current,

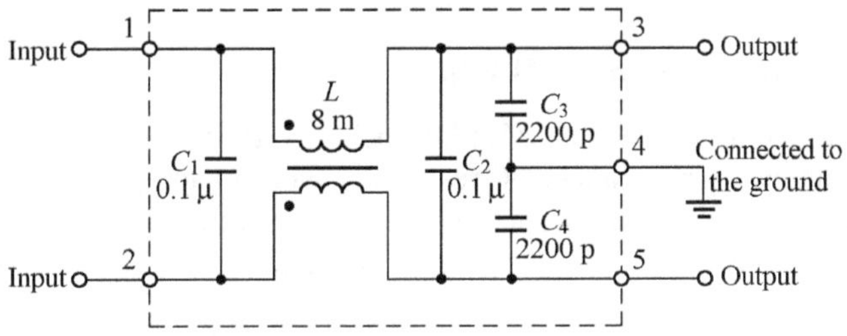

Figure 4.10 Basic circuit of EMI filter

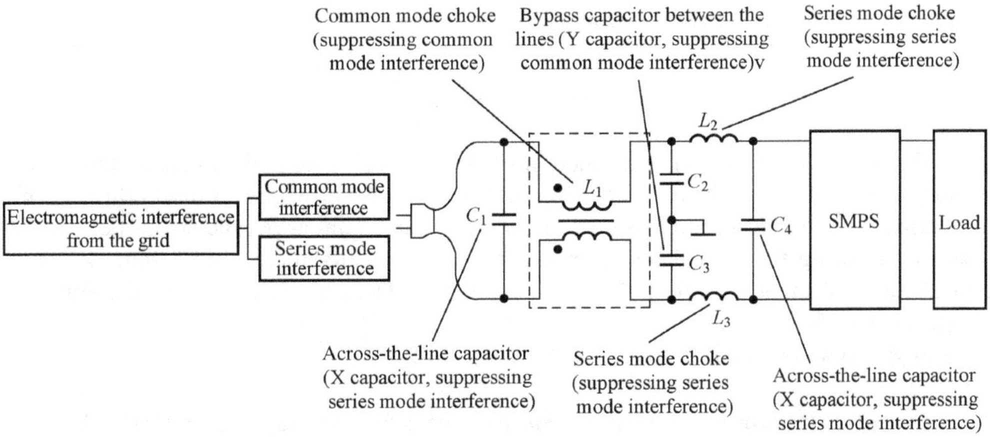

Figure 4.11 Typical application diagram of EMI filter

Table 4.9 The relationship between the inductance range and the rated current

Rated current I (A)	1	3	6	10	12	15
Inductance range L (mH)	8–23	2–4	0.4–0.8	0.2–0.3	0.1–0.15	0.0–0.08

the magnetic fields generated on the two coils will reinforce each other. The inductance of L relates to the rated current I of EMI filter as shown in Table 4.9. It needs to be pointed out that when the rated current is comparatively large, the wire diameter of the common-mode choke should also be increased accordingly to withstand a comparatively large current. In addition, a proper increase of inductance can improve the characteristic of low-frequency attenuation. C_1 and C_2 should be film capacitors with a capacity range of about 0.01–0.47 μF, which are mainly used to filter out the series-mode interference. C_3 and C_4 are connected across the output terminal and make the midpoint of capacitors grounded to effectively suppress common-mode interference. The capacity range of C_3 and C_4 is 2200 pF–0.1 μF. In order to reduce drain current, the capacity of capacitors should be no more than 0.1 μF. The withstand voltage values of C_1–C_4 are 630 V DC or 250 V AC.

4.4.1.2 Main Parameters of the EMI Filter

The main technical parameters of the EMI filter include rated voltage, rated current, drain current, test voltage, insulation resistance, DC resistance, working temperature range, temperature rise in working (T_r), insertion loss (A_{dB}), dimensions, and weight, among which insertion loss is the most important parameter as it is the main indicator to assess the performance of EMI filter.

Insertion loss (A_{dB}), expressed in dB, refers to the ratio of the log of noise voltage on the load before and after inserting EMI filter. The larger the dB value is, the stronger the ability to suppress noise interference will be. Assume that the noise voltages transmitted to the load before and after inserting EMI filter are U_1 and U_2, respectively, and $U_2 \ll U_1$, then the formula of calculating insertion loss at a certain frequency is as follows:

$$A_{dB} = 20\lg\left(\frac{U_1}{U_2}\right) \tag{4.14}$$

As the insertion loss (A_{dB}) is a function of frequency, its theoretical calculation is cumbersome with a comparatively large error, and it is usually provided for users directly after measuring the actual insertion loss corresponding to each point as per the noise spectrum followed by drawing the typical insertion loss curve by manufacturers. See Figure 4.12 for a typical curve. It can be seen from the figure that the product can attenuate the noise voltage of 1–30 MHz by 65 dB.

Finally, several points need to be pointed out as follows:

1. The insertion loss curve of EMI filter can be drawn in another way. $A_{dB} = |20\lg(U_2/U_1)|$ defined by it needs to take the absolute value as A_{dB} itself is negative. The curve shape of this drawing method is in the X-axis symmetry with that in Figure 4.12. Voltage level dB(μV) or

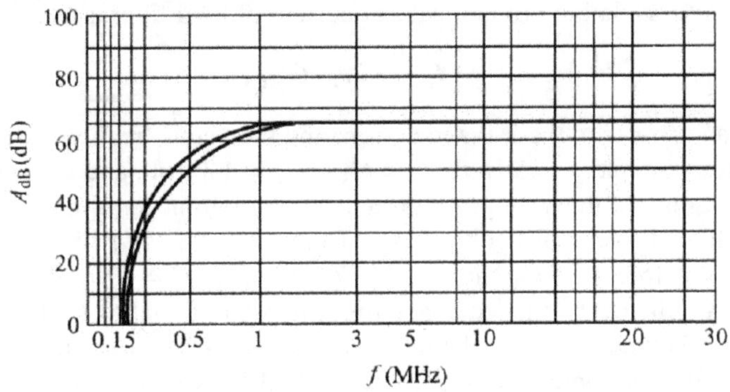

Figure 4.12 A typical insertion loss curve

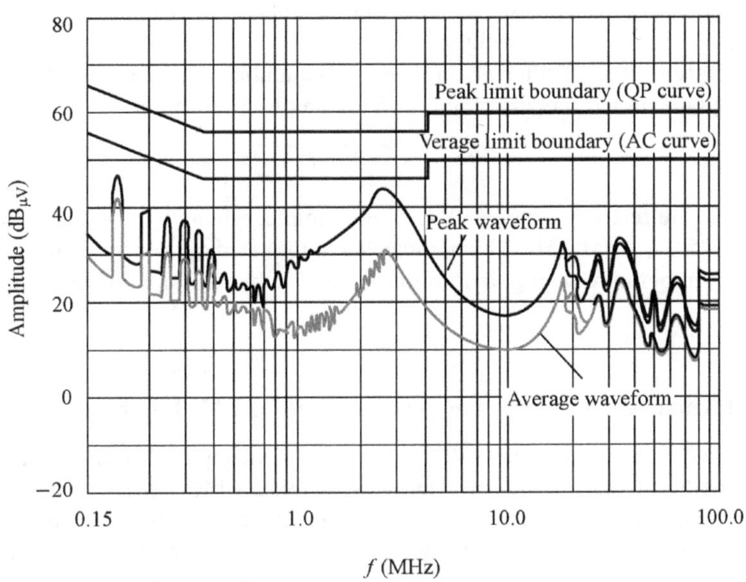

Figure 4.13 EMI waveforms

dB$_{(\mu V)}$ is usually used as the unit of insertion loss. dB $(\mu V) = 20\lg(U_O/1\ \mu V)$, in which the unit of U_O is μV and 1 μV is defined as 0 dB(μV).

2. See Figure 4.13 for the EMI waveforms of typical SMPS, in which the top two lines, respectively, refer to the peak limit boundary (QP curve) and the average limit boundary (AC curve) prescribed by CISPR228/EN55022B international testing standards. It is required that the peak waveform amplitude of the tested SMPS shall not exceed the peak limit boundary and the average waveform amplitude shall not exceed the average limit boundary.

3. The formula for calculating ground leakage current of EMI filter is as follows:

$$I_{\text{LD}} = 2\pi f\ CU_{\text{C}}, \tag{4.15}$$

wherein I_{LD} and f refer to the leakage current and the mains frequency, respectively. Take Figure 4.10 for example, $f = 50\,\text{Hz}$, $C = C_3 + C_4 = 4400\,\text{pF}$, and U_{C} is the voltage drop on C_3 and C_4, that is, the voltage to ground on the output terminal, and suppose $U_{\text{C}} \approx 220\,\text{V}/2 = 110\,\text{V}$. It is not difficult to know from the formula (4.15) that the leakage current now $I_{\text{LD}} = 0.15\,\text{mA}$. If C_3 and C_4 are 4700 pF, then $C = 4700\,\text{pF} \times 2 = 9400\,\text{pF}$ and $I_{\text{LD}} = 0.32\,\text{mA}$. Obviously, the leakage current is proportional to C. The requirement for leakage current is the smaller the better, and it will be safer. The maximum leakage current specified for electronic equipment is 250 μA–3.50 mA, specific values depending on the type of electronic equipment. According to the provisions of international standard IEC950, the maximum leakage current of class II devices (without protective ground wire) is 250 μA, and the leakage current of the handheld device and that of the mobile equipment (excluding handheld devices) in class I devices (with protective ground wire) are 750 μA and 3.50 mA, respectively. However, the requirements for the leakage current of electronic medical equipment are more stringent.
4. The rated current of EMI filter also relates to the ambient temperature T_{A}. For example, some foreign manufacturers give the following empirical formula:

$$I = I_1\sqrt{(85 - T_{\text{A}})/45}, \tag{4.16}$$

wherein I_1 is the rated current at 40 °C. For example, when $T_{\text{A}} = 50\,°\text{C}$, $I = 0.88I_1$, and when $T_{\text{A}} = 25\,°\text{C}$, $I = 1.15I_1$. This indicates that the rated current value increases with the decrease in temperature owing to an improvement in heat dissipation conditions, and vice versa.

4.4.1.3 Performance Comparison of Several Kinds of EMI Filter

In order to reduce cost and size, the SMPS generally uses simple EMI filter, mainly including the common-mode choke L and filter capacitor. See Figure 4.14(a)–(d) for four kinds of simple EMI filter circuits. Take (c) for example, L, C_1, and C_2 are used for filtering out common-mode interference, and C_3 and C_4 are used for filtering out series-mode interference. In the case of common-mode interference, it shows large inductive reactance toward common-mode signals owing to the rapid rise of total inductance after coupling resulted from the same direction of the magnetic flux of the two coils in L, thus making common-mode signals hard to pass through, so it is called common-mode choke. Its two coils are, respectively, wound on the ferrite beads with low loss and high magnetic inductivity. R refers to a bleeder resistor, which can discharge the charges accumulated on C_3 to avoid affecting the filter characteristic resulted from charge accumulation and can, in the case of power-off, make the input terminals L and N uncharged to ensure safety.

See Figure 4.15 for the comparison of interference waveforms before and after connecting the EMI filter. Curve a refers to the waveform (i.e., EMI peak envelope curve) of the transmitting noise of 0.15–30 MHz on SMPS without connecting EMI filter. Curves b and c, respectively, refer to the waveforms after connecting the EMI filter as shown in Figure 4.14(b) and (d), which can attenuate EMI by 50–70 $\text{dB}_{\mu\text{V}}$, and obviously its effect is better.

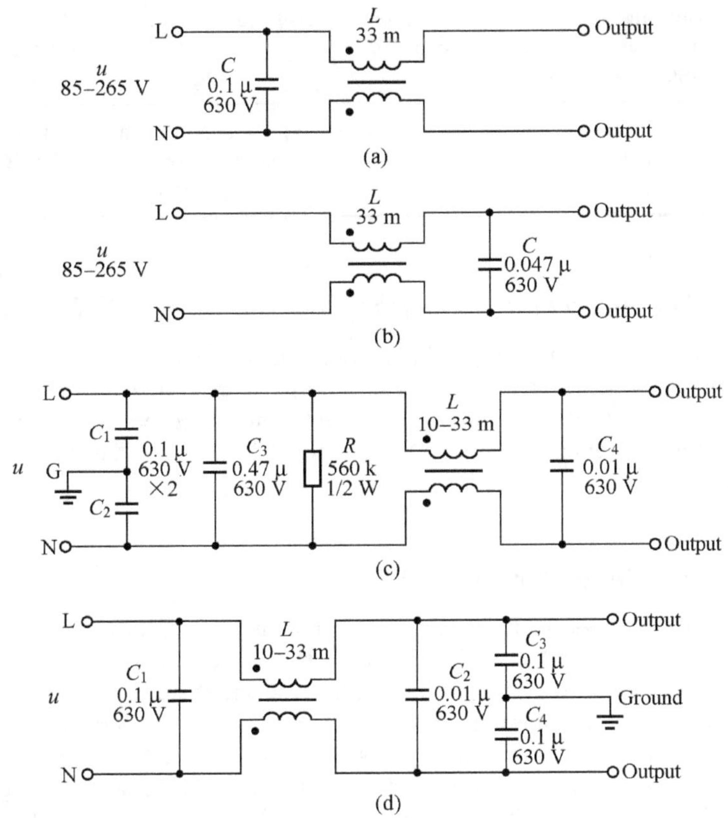

Figure 4.14 Four kinds of common simple EMI filter circuits of SMPS

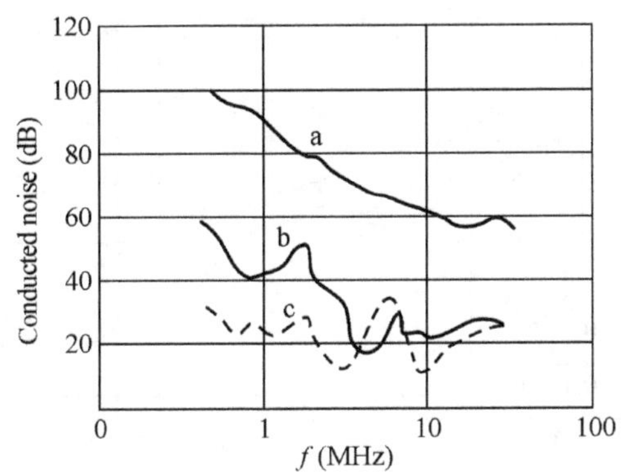

Figure 4.15 Comparison of interference waveforms before and after connecting the EMI filter

Table 4.10 Classification of X capacitors

Classification	Withstandable peak pulse voltage (kV)	IEC-664 insulation class classification	Application field	Withstandable peak pulse voltage U_P before endurance test (kV)
X1	>2.5, ≤4.0	III	High pulse suppression	4 ($C \leq 1.0\,\mu F$)
X2	>2.5	II	General purpose	2.5 ($C \leq 1.0\,\mu F$)
X3	≤1.2	—	General purpose	No requirement

4.4.2 Selection Method for Capacitors and Inductors in EMI Filter

According to the provisions in IEC950 international EMC standard, the capacitors that can filter out the series-mode interference between grid lines are called “X capacitors,” and the capacitors that can filter out the common-mode interference generated by the primary and secondary coupling capacitors are called “Y capacitors.” Both Y and X capacitors are referred to as safety capacitors.

4.4.2.1 Selection Principle for X Capacitors

According to the differences in withstand voltage values and usage, X capacitors can be divided into three types: X1, X2, and X3 capacitors. See Table 4.10 for the classification of X capacitors. X capacitors should only be used in the places where there is no danger of electronic shock for anyone in case of failure of capacitors. X capacitors are generally connected in parallel on AC input terminal and used to suppress series-mode interference in EMI filters. X2 and X3 capacitors are most commonly used in EMI filters. X1 capacitors are generally not used owing to their high costs. When C is, respectively, 0.033, 0.047, 0.1, 0.22, and 0.47 μF, the impedance characteristic curves of X2 capacitors are shown in Figure 4.16. It can been seen from the figure that there is a minimum in the vicinity of the range from several MHz to

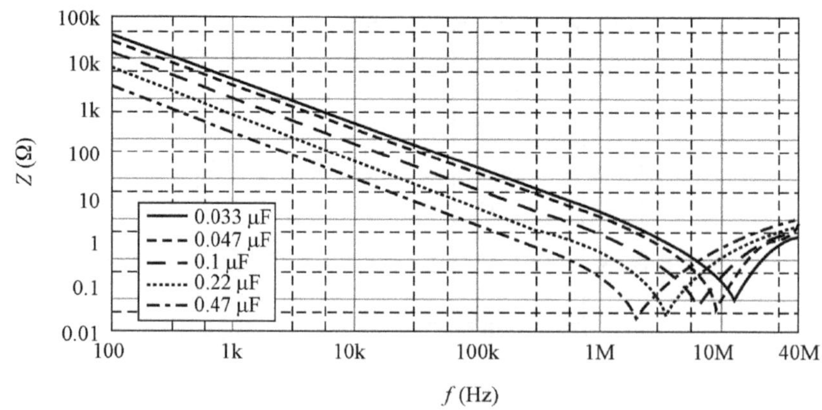

Figure 4.16 Impedance characteristic curve of X2 capacitors

Table 4.11 Classification of Y capacitors

Classification	Insulation type	Rated AC voltage (V)	AC test voltage for QA, periodic and batch testing (V)	Withstandable peak pulse voltage U_P before endurance test (kV)
Y1	Double or reinforced insulation	≤ 250	4000	8.0
Y2	Basic or supplementary insulation	≥ 150, ≤ 250	1500	5.0
Y3	Basic or supplementary insulation	≥ 150, ≤ 250	1500	No requirement
Y4	Basic or supplementary insulation	< 250	900	2.5

10 MHz of the capacitor impedance *Z*. X2 capacitance range used in EMI filters is 1 nF–1 μF, and the best capacitance is generally 0.1–0.33 μF.

4.4.2.2 Selection Principle for Y Capacitors

Y capacitors are divided into four categories, that is, Y1, Y2, Y3, and Y4 capacitors. See Table 4.11 for the classification of Y capacitors, wherein QA refers to quality assurance. Y capacitors should be used in the places where there is a danger of electric shock for people in case of failure occurring to capacitors.

Y capacitors can provide the interference current coupled from the primary side to the secondary side with a return path to prevent the current from being coupled to the ground through the secondary side. In order to prevent the noise of Y capacitors from being coupled to MOSFET source, one terminal of Y capacitors should be connected with primary DC high voltage and the other terminal should be connected to the secondary return terminal RTN [also known as safety extra-low voltage (SELV)], or to the power supply chassis, the shield member or the ground according to the actual situation. See Figure 4.17 for the typical wiring positions of Y capacitors, wherein C_1 refers to Y capacitor. In order to make Y capacitors work efficiently, the

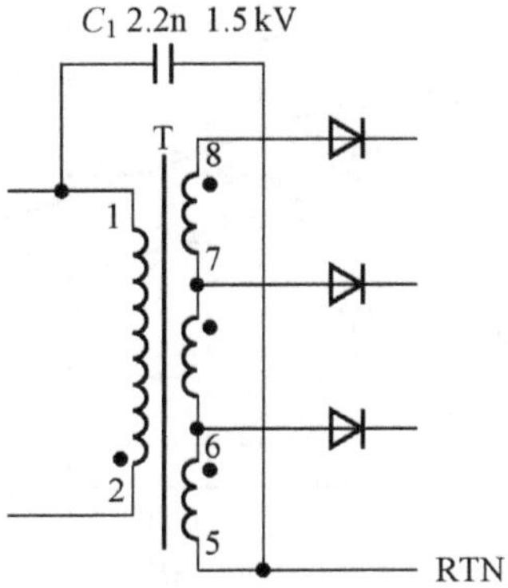

Figure 4.17 Typical wiring positions of Y capacitors

PCB wiring between the pin of high-frequency transformer and it shall be as short as possible and go straight.

The requirements for the maximum permissible drain current are different (usually 0.2–3.5 mA) owing to different AC power network voltage values, so it is necessary to limit the maximum capacity of Y capacitors. For the equipment of class II or two-wire (phase and neutral wires, without ground wire) input conditions, the drain current shall not be larger than 250 μA in case a component fails, so the maximum of Y capacitors should be limited to less than 2.8 nF (i.e., 2800 pF). For equipment of class I or three-wire (phase, neutral, and ground wires) input conditions, the drain current shall be smaller than 3.5 mA in case the ground is disconnected or short circuit is caused by the failure of a component, so the maximum capacity of Y capacitor shall be limited to less than 39 nF. The common capacity range of Y1 capacitor is 1–2.2 nF and the typical value is 1000 pF. It is also advisable to connect two Y2 capacitors of 2200 pF in series to replace one Y1 capacitor of 1000 pF. A proper capacity increase in Y1 capacitor can reduce common-mode EMI noise, but will also increase the drain current to the ground. In the case of two-wire 220 V AC input or AC input with wide voltage range, the AC test voltage of Y1 capacitor connected between the primary DC high voltage and the secondary return terminal RTN is usually 3000 V with the duration of 1 min. However, generally Y1 capacitor is not used for three-wire input, in which case Y2 capacitor can be connected directly between the bridge rectifying output terminal and the ground to safely discharge the fault current to the ground in case of short circuit.

Y capacitors can filter out most of the high-frequency interference within 10–30 MHz frequency range, their resonant frequency should not be less than 40 MHz. However, long lead will reduce resonant frequency to generate interference current, thus causing excessive radiated interference, so the lead of Y capacitors should be as short as possible, which is essential to suppress conducted or radiated interference. When C is 4700, 2200, 1000, 680, and 330 pF, respectively, the impedance characteristic curves of Y2 capacitor are shown in Figure 4.18 correspondingly. It can be seen from the figure that the impedance Z of the capacitor also reduces with the increase of frequency, and the variation rules of impedance and frequency are linear relation within 10 MHz.

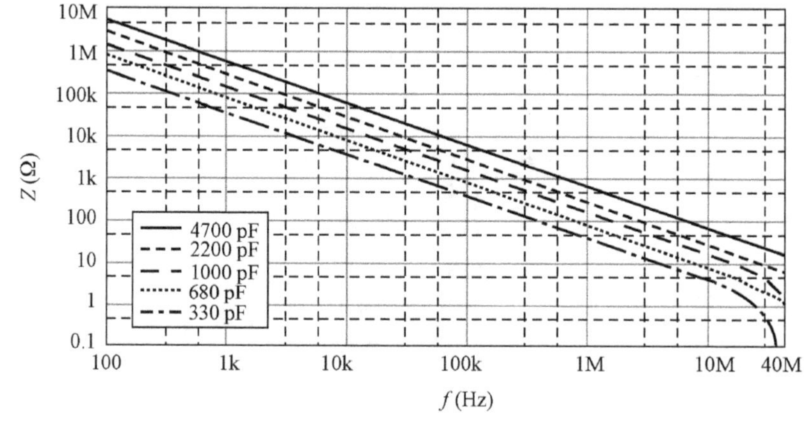

Figure 4.18 Impedance characteristic curves of Y2 capacitor

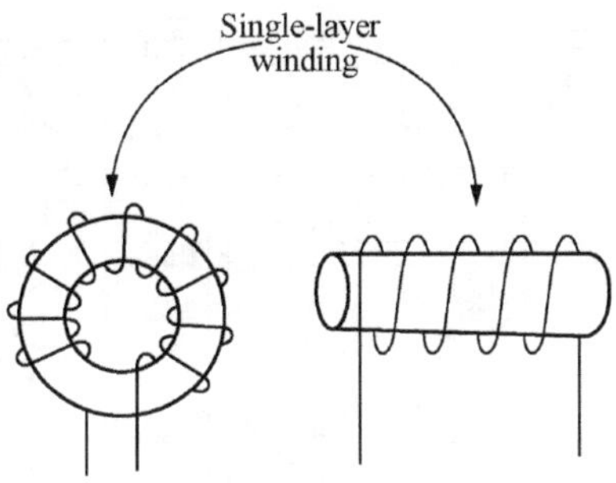

(a) (b)

Figure 4.19 Structure of series-mode choke. (a) The structure with a ferrite magnetic bead and (b) the structure with a solenoid

In order to suppress electromagnetic interference, EMI must be filtered by SMPS to fit the technical indicators specified by EMI standards. There are two methods of reducing EMI: one is to use a simple π filter and one Y capacitor, suitable for SMPSs with small output power of 1–5 W, and the other is to use a filter composed of common-mode choke, and X and Y capacitors, suitable for SMPSs with small and medium power above 5 W.

4.4.2.3 Series-Mode Choke

Series-mode choke is usually wound on a ferrite magnetic bead or solenoid, showing a very high impedance toward series-mode interference. See Figures 4.19 and 4.20, respectively, for the structure and the impedance characteristic curves of series-mode choke. It can be seen from the figures that for the series-mode choke with 1 mH inductance, the series-mode impedance Z reaches the peak when the working frequency is 1 MHz (i.e., 10^6 Hz). The series-mode choke with single-layer winding has the lowest turn-to-turn capacitance and the highest resonant frequency. Series-mode inductance can be selected within the range of 100 μH–5 mH.

4.4.2.4 Common-Mode Choke

Common-mode choke contains two mutually symmetrical coupled inductors. The common-mode inductor winds two separate windings on the same toroid or skeleton-shaped core to ensure good coupling. As the two windings wind the same number of turns in the same direction, the magnetic flux of the series-mode signals from the grid is completely offset in the magnetic core, and the magnetic flux of common-mode signals is mutually reinforced, showing large inductive reactance to common-mode interference and thus making it difficult to pass. In addition, the coupled capacitance between windings can be reduced to the smallest by winding two windings at different positions of the magnetic core. Common-mode inductance is usually 8–33 mH.

See Figure 4.21(a) and (b) for the structure and the equivalent circuit of common-mode choke. Another conspicuous advantage of using common-mode choke is that an equivalent series-mode drain inductor L_0 is connected in series on the common-mode inductor L, and

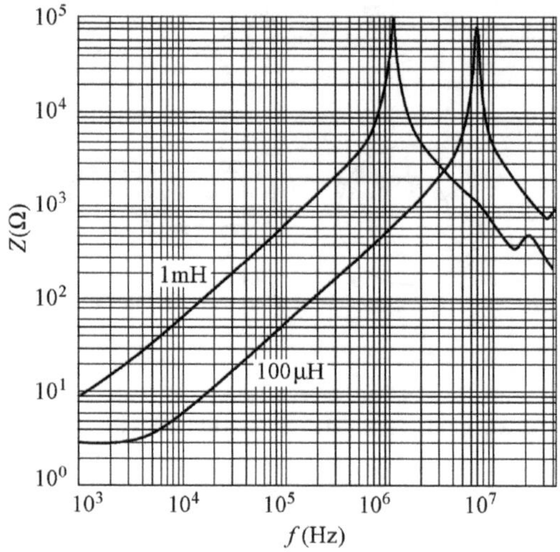

Figure 4.20 Impedance characteristic curve of series-mode choke

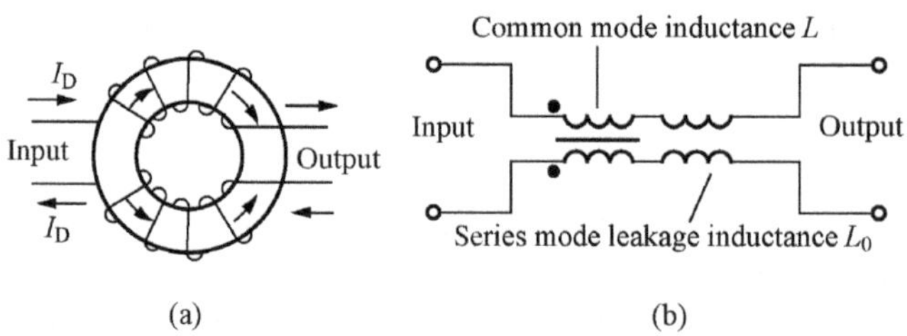

Figure 4.21 Structure of common-mode choke. (a) Structure and (b) equivalent circuit

the former is equivalent to an inherent series-mode choke, capable of suppressing series-mode interference without requiring connecting another discrete series-mode choke. L equals the inductance measured from the other winding after disconnecting one winding. L_0 equals half of the inductance measured from the other winding after making one winding short.

See Figure 4.22(a) and (b), respectively, for the shape of U-type and spool-type common-mode chokes. See Figure 4.23(a) and (b), respectively, for the common-mode impedance characteristic curves of U-type and spool-type common-mode chokes. See Figure 4.24(a) and (b), respectively, for the series-mode impedance characteristic curves of U-type and spool-type common-mode chokes. Figures 4.22 and 4.23, respectively, show the common-mode and the series-mode impedance characteristics of toroidal core common-mode choke of 1 mH. What needs to be noted is that the common-mode and the series-mode impedances of toroidal common-mode choke are significantly lower than those of U-type and spool-type common-mode choke, so an additional series-mode choke is usually required when using toroidal common-mode choke. In view of the reason above, it is not recommended to use toroidal common-mode choke, unless in the case that an additional toroidal common-mode choke is required to suppress high-frequency interference.

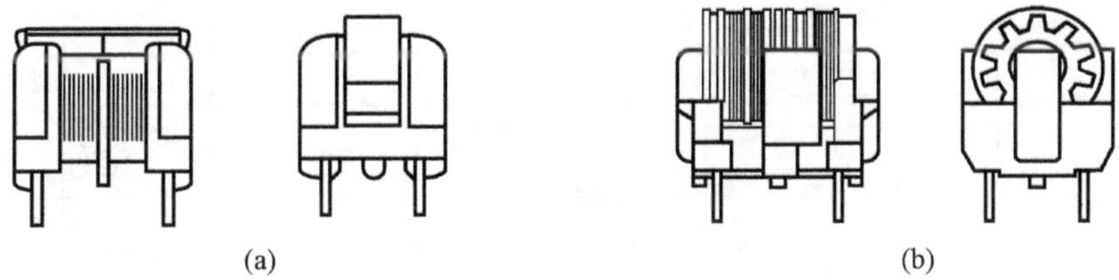

Figure 4.22 The shape of (a) U-type and (b) spool-type common-mode chokes

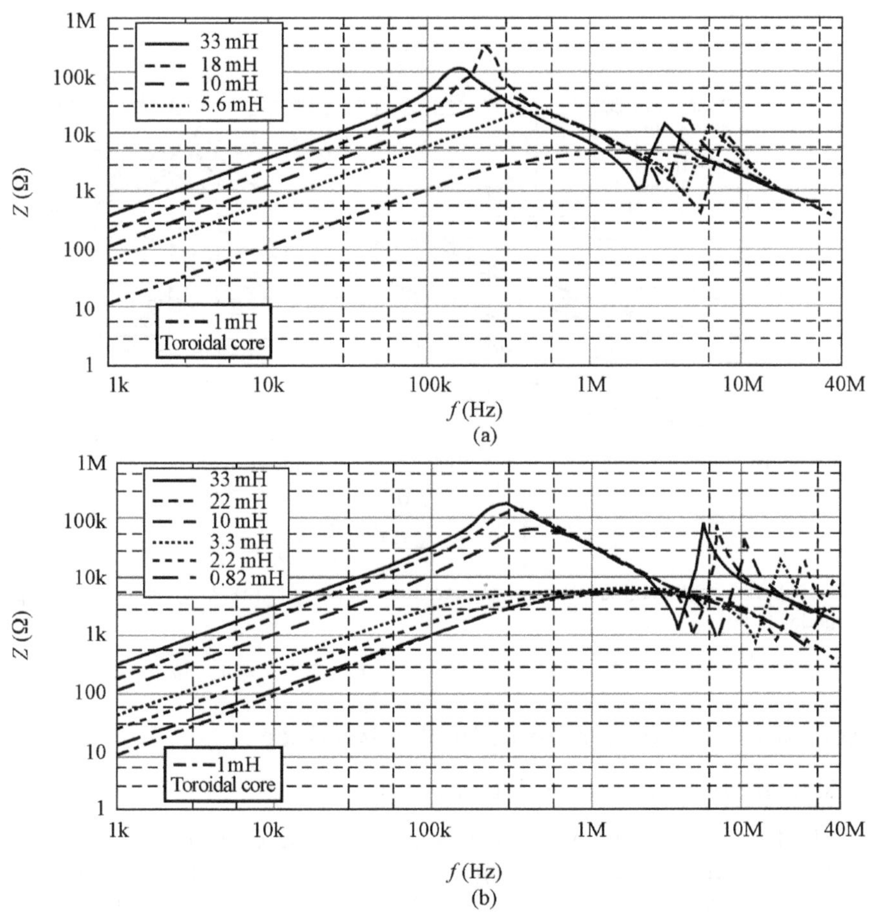

Figure 4.23 Common-mode impedance characteristic curves of (a) U-type and (b) spool-type common-mode chokes

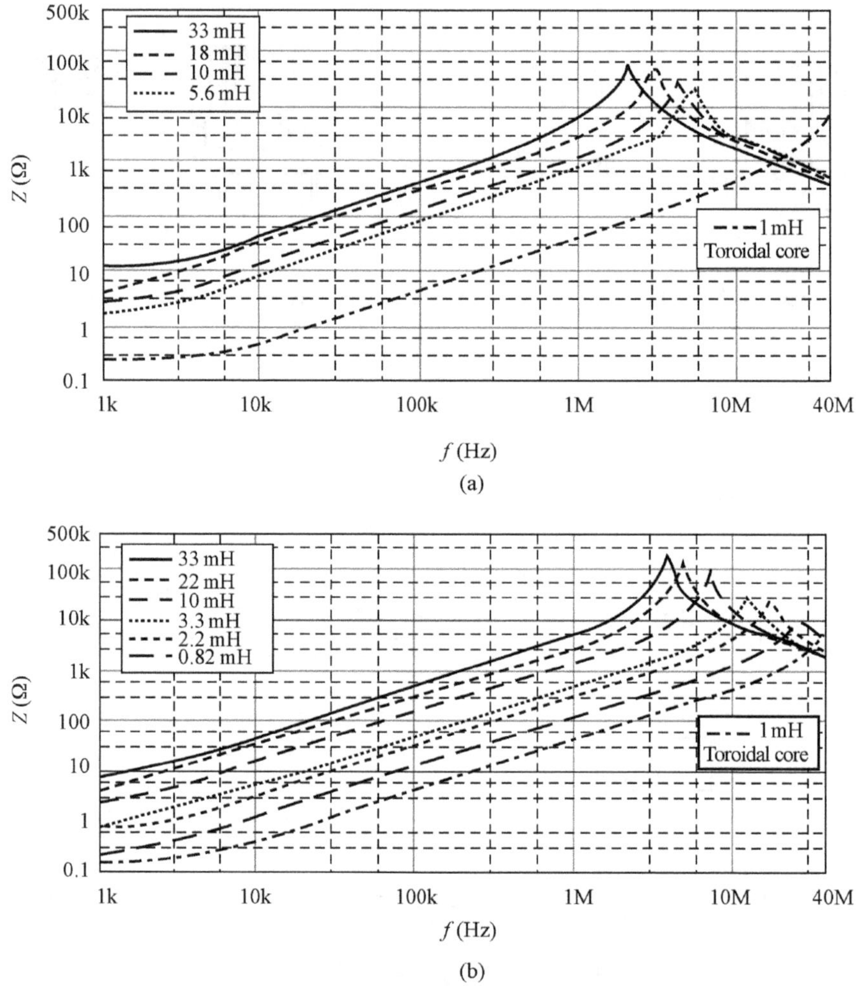

Figure 4.24 Series-mode impedance characteristic curves of (a) U-type and (b) spool-type common-mode chokes

4.5 Selection Method for Input Bridge Rectifier

4.5.1 Selection Method for Bridge Rectifier

Full-wave bridge rectifier (referred to as silicon bridge rectifier) is a kind of semiconductor device forming a bridge with four silicon rectifiers and then enveloped with plastic. It is featured with such advantages as small volume, convenient for use, and good consistency in the parameters of various rectifiers, so that it can be widely used in the rectification circuit of SMPS. Silicon bridge rectifier has four outlet terminals, that is, two AC input terminals and two DC output terminals. See Figure 4.25 for the shapes of several kinds of silicon bridge rectifiers. The maximum average current of silicon bridge rectifier includes the

Figure 4.25 Shapes of several kinds of silicon bridge rectifier

Table 4.12 Main technical indicators of **3KBP005M–3KBP08M** bridge rectifier

Model	3KBP005M	3KBP01M	3KBP02M	3KBP04M	3KBP06M	3KBP08M
U_{RM} (V)	50	100	200	400	600	800
U_F (V)	1.05					
$I_{F(AV)}$ (A)	3.0					
I_{FSM} (A)	80					
$I_{R\ (\mu A)}$	5.0					

specifications of 0.5, 1, 1.5, 2, 3, 4, 6, 8, 10, 15, 25, 35, 40 A, and so on, and the maximum inverse working voltage includes the specifications of 50, 100, 200, 400, 800, 1000 V, and so on. Silicon bridge rectifiers with small power can be welded directly on PCB, but those with large and medium power shall be fixed with screws and installed with a proper heat sink.

The main parameters of bridge rectifier include the peak reverse voltage U_{RM} (V), the forward voltage drop U_F (V), the average rectification current $I_{F(AV)}$ · (A), the peak forward surge current I_{FSM} (A), and the maximum reverse leakage current I_R (μA). The typical products of bridge rectifier include 3KBP005M–3KBP08M produced by VISHAY, whose main technical indicators are shown in Table 4.12. The breakdown reverse voltage U_{BR} of bridge rectifier shall meet the following requirements:

$$U_{BR} \geq 1.25\sqrt{2}u_{max} \tag{4.17}$$

For example, suppose $u_{max} = 265\,\text{V}$, substitute it in the formula (4.17), then it can be calculated that $U_{BR} \geq 468\,\text{V}$. Bridge rectifier with a withstand voltage of 600 V can be adopted.

Bridge rectifier can also be composed of four rectifiers, for example, 1N4007 1A/1000 V silicon rectifiers can be chosen. It needs to be pointed out that, sometimes, in order to reduce the conduction noise below 500 kHz in SMPSs, bridge rectifiers can also be composed of silicon rectifiers and fast recovery diodes (FRDs). In Figure 4.26, two FR106 1A/800 V FRDs (VD_1 and VD_2) and two ordinary 1N4007 silicon rectifiers (VD_3 and VD_4) are used. The reverse recovery time of FR106 is as follows: $t_{rr} \approx 250\,\text{ns}$.

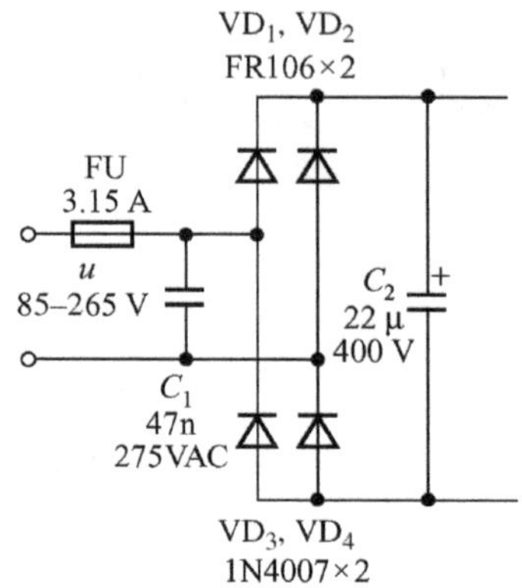

Figure 4.26 Bridge rectifier composed of silicon rectifiers and fast recovery diodes

4.5.2 Conduction Time and Strobing Features of Bridge Rectifier

50 Hz AC voltage becomes pulsating DC voltage u_1 through full-wave rectification, and then is converted into DC high voltage U_I by passing through input filter capacitor. Ideally, the conduction angle of bridge rectifier should be 180° (with a conduction range of 0°–180°), but C is charged by the input current flowing through the bridge rectifier only during the short time when close to AC peak voltage owing to the effect of filter capacitor C. The half cycle of 50 Hz AC is 10 ms, the conduction time of bridge rectifier is $t_C \approx 3$ ms, and its conduction angle is only 54° (with a conduction range of 36°–90°). Therefore, it is the narrow pulse current that actually flows through the bridge rectifier. See Figure 4.27(a) for the principle of bridge rectifying filter circuit and see Figure 4.27(b) and (c), respectively, for the waveforms of the rectified filter voltage and rectified current.

Finally, two points can be summarized:

1. The above characteristics of bridge rectifier can be equivalent to about 30% of input voltage frequency corresponding to the duty ratio.
2. The primary conduction process of rectifier diode can be regarded as a "strobing pulse," whose pulse recurrence frequency is equal to the AC grid frequency (50 Hz).

4.6 Selection Method for Output Rectifier

The output rectifier of SMPS generally uses FRD, superfast recovery diode (SRD), or SBD, which are featured with such advantages as good switching characteristics, short reverse recovery time, large forward current, small size, and easy installation.

4.6.1 Selection Method for Fast and Superfast Recovery Diodes

4.6.1.1 Reverse Recovery Time

Reverse recovery time t_{rr} is defined as the time interval when current passes through zero point, changes from the forward to the reverse, and then from the reverse to the specified low

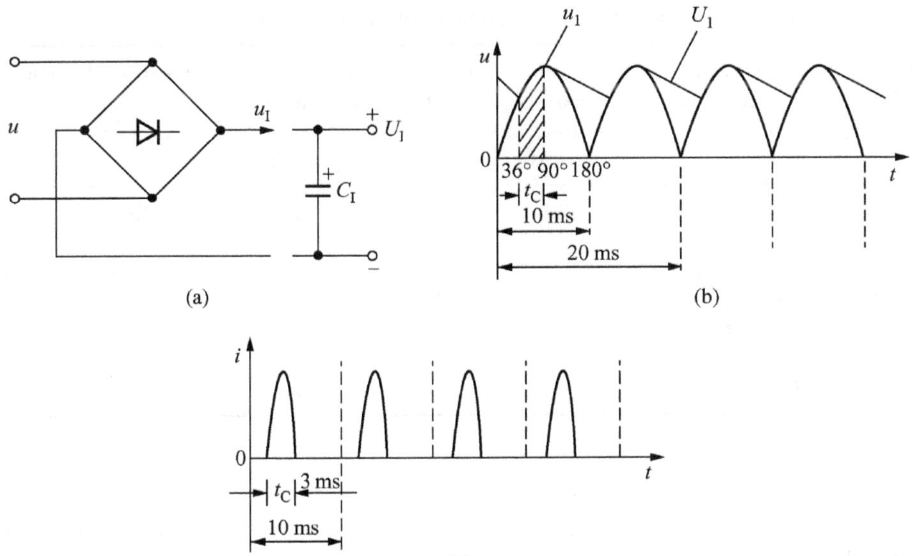

Figure 4.27 Waveforms of rectified filter voltage and rectified current. (a) Filter circuit of bridge rectifier, (b) waveform of rectified filter voltage, and (c) waveform of rectified current

value. It is an important technical indicator to measure the performance of high-frequency rectifying and freewheeling devices. See Figure 4.28 for the waveform of reverse recovery current, wherein I_F refers to forward current, I_{RM} the maximum reverse recovery current, and I_{rr} reverse recovery current, and typically $I_{rr} = 0.1I_{RM}$. When $t \leq t_0$, the forward current $I = I_F$. When $t > t_0$, as the forward voltage of rectifier suddenly becomes reverse voltage, the forward current decreases rapidly and $I = 0$ at the time when $t = t_1$. Then, reverse current I_R flows through the rectifier and increases gradually, reaching the maximum reverse recovery current

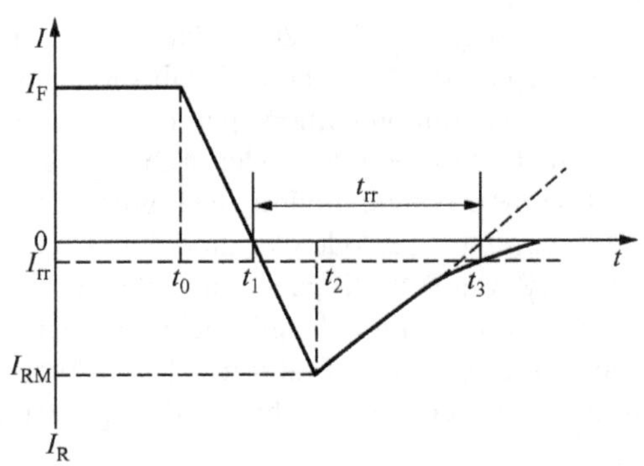

Figure 4.28 The waveform of reverse recovery current

Table 4.13 Main technical indicators of several kinds of FRD and SRD

Model of typical products	Structural features	Reverse recovery time t_{rr} (ns)	Average rectified current I_d (A)	Maximum instantaneous current I_{FSM} (A)	Peak reverse voltage U_{RM} (V)	Packaging form
C20-04	Single diode	400	5	70	400	TO-220
C92-02	Common-cathode pair diode	35	10	50	200	TO-220
MUR1680A	Common-anode pair diode	35	16	100	800	TO-220
EU2Z	Single diode	400	1	40	200	DO-41
RU3A	Single diode	400	1.5	20	600	DO-15

I_{RM} at the time when $t = t_2$. After that, the reverse current decreases gradually under the effect of forward voltage and reaches the specified value I_{rr} when $t = t_3$. The reverse recovery process from t_2 to t_3 is similar to the discharge process of a capacitor. The time interval from t_2 to t_3 is the reverse recovery time t_{rr}.

4.6.1.2 Structural Features of FRD

The internal structure of FRD is different from that of the ordinary diode, in which the base region is added in P- and N-type silicon materials to constitute P–I–N silicon chip. As the base region is very thin with very small reverse recovery charge, not only t_{rr} is greatly reduced but also the transient forward voltage is decreased, enabling the diode to withstand high reverse working voltage. The reverse recovery time of FRD is usually several hundred nanoseconds, the forward voltage drop is about 0.6 V, the forward current is several amperes to several thousand amperes, and the peak reverse voltage is up to several hundred to several thousand volts.

SRD is developed based on FRD, its reverse recovery charge is further reduced and t_{rr} can be as low as tens of nanoseconds.

FRD and SRD below 20 A mostly use TO-220 packaging. In terms of internal structure, they can be divided into two types: single diode and pair diode. A pair diode contains two FRDs or SRDs and is divided into common-cathode pair diode and common-anode pair diode according to the connection method of the two diodes. See Figure 4.29(a) for the outline and the internal structure of C20-04 FRD (single diode). See Figure 4.29(b) and (c) for the outlines and the structures of C92-02 (common-cathode pair diode) and MUR1680A (common-anode pair diode) SRDs. They mainly use TO-220 packaging. See Table 4.13 for the main technical indicators. See Table 4.14 for the models and main parameters of the commonly used SRD_s. FRD and SRD of tens of amperes generally use TO-3P metal-can packaging. Diodes with larger capacity (several hundred to several thousand amperes) use bolt-type or flat-type packaging.

See Figure 4.30 for the typical application of SRD in SMPS. The drain clamp protection circuit in Figure 4.30(a) uses one UF4007 SRD VD_1, and the output rectification

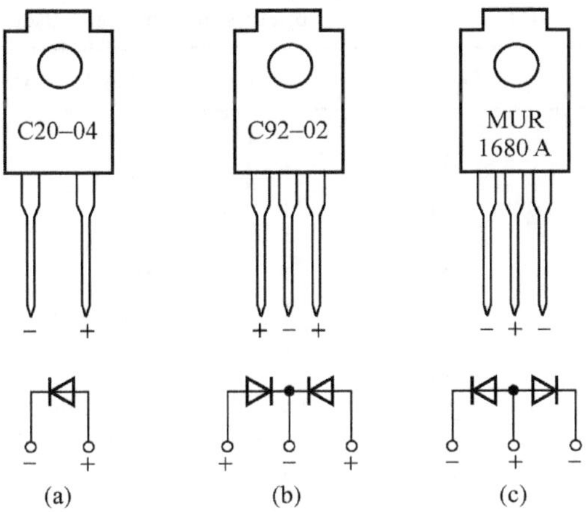

Figure 4.29 Outline and internal structure of three kinds of FRD and SRD. (a) Single diode, (b) common-cathode pair diode, and (c) common-anode pair diode

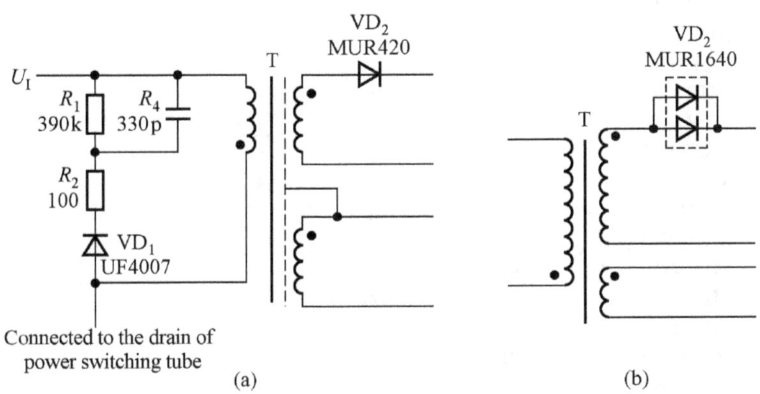

Figure 4.30 Typical application of SRD in SMPS. (a) Application circuit I and (b) application circuit II

circuit uses MBU420 4A/200 V SRD. The output rectifier in Figure 4.30(b) uses one MBU1640 16A/200 V superfast recovery pair diode to meet the need of heavy-current output.

4.6.2 *Selection Method for SBD*

4.6.2.1 Working Principle of SBD

SBD is the metal-semiconductor device that is made by using the rectification characteristics in the barrier formed on the contact surface between the anode made from gold, silver,

Table 4.14 Models and main parameters of commonly used SFD

Model	U_{RM} (V)	I_d (A)	t_{rr} (ns)	Manufacturer
UF4001	50	1	25	GI
UF4002	100	1	25	
UF4003	200	1	25	
UF4004	400	1	50	
UF4005	600	1	30	
UF4006	800	1	75	
UF4007	1000	1	75	
UF5401	100	3	50	
UF5402	200	3	50	
UF5406	600	3	50	
UF5408	1000	3	50	
BYV26A	200	2.3	30	Philips
BYV26B	400	2.3	30	
BYV26C	600	2.3	30	
BYV26D	800	2.3	75	
BYV26E	1000	2.3	75	
BUR130	300	1		Motorola
BUR140	400	1		
BUR150	500	1		
BUR160	600	1		
BUR170	700	1		
BUR180	800	1		
MUR110	100	1		
MUR120	200	1		
MUR410	100	4		
MUR420	200	4		
MUR440	400	40		
MUR610	50	60		
MUR810	100	8		
MUR820	200	8		
MUR1610	100	16	35	
MUR1620	200	16		
BYV27-100	100	2	25	Philips and GI
BYV27-150	150	2	25	
BYV27-200	200	2	25	
BYV32-100	100	20	35	Philips
BYV32-150	150	20	35	
BYV32-200	200	20	35	
BYW29-200	200	8	25	Philips and GI

molybdenum, and other noble metals and the cathode made from N-type semiconducting materials. It is a kind of five-layer device with N-type semiconductor as the substrate of the middle layer, above which is the N^- epitaxial layer with arsenic as dopant. The top is the anode made of the metal material molybdenum. N-type substrate has a very small on-resistance. N^+ cathode layer and cathode metal are sequentially located under the substrate. See Figure 4.31 for the

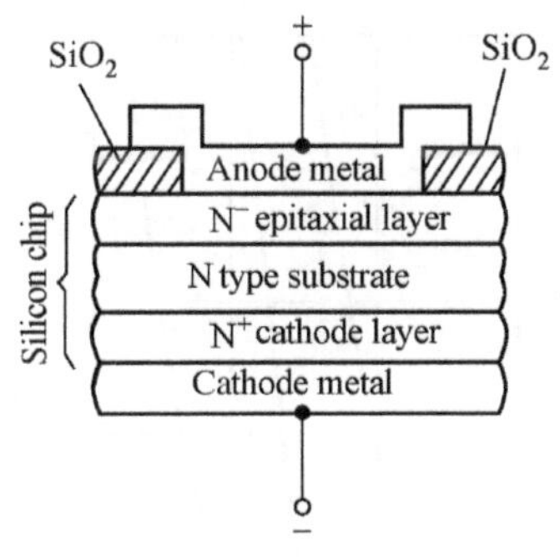

Figure 4.31 Structure of SBD

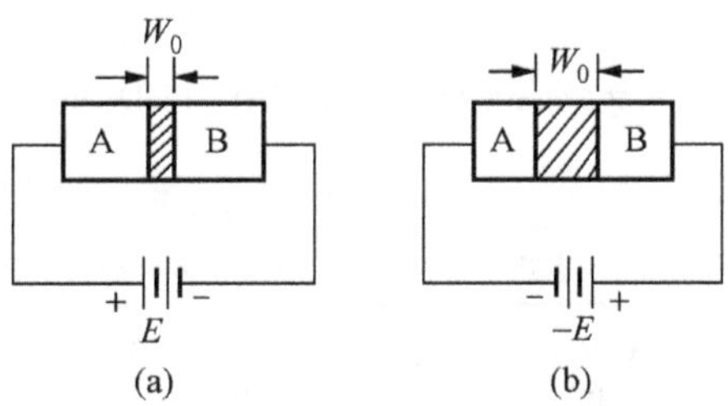

Figure 4.32 Changes in barrier width when bias voltage is applied exteriorly. (a) Positive bias is applied and (b) negative bias is applied

internal structure of typical SBD. Proper Schottky barrier can be formed between the substrate and the anode metal by adjusting structural parameters.

When positive bias E is applied, metal A and N-type substrate B are, respectively, connected to the anode and the cathode of power supply, and the barrier width W_0 becomes narrow. When negative bias $-E$ is applied, the barrier width is increased as shown in Figure 4.32. In recent years, aluminum-silicon SBD manufactured with silicon plane process has made appearance, capable of not only saving noble metals and reducing environmental pollution but also improving the consistency of device parameters. SBD uses only one kind of carrier (electron) for charge transportation without the accumulation of excessive minority carrier in the outside of the barrier, so it has no charge storage effect, thus remarkably improving the switching characteristics. In addition, the reverse recovery time (t_{rr}) can be reduced to less than 10 ns, but its reverse withstand voltage is comparatively low, generally less than 100 V, suitable for working under low voltage and heavy current. The efficiency of low-voltage heavy-current rectification (or freewheeling) circuit can be improved remarkably if its low voltage-drop characteristic is applied.

See the curves of Figure 4.33(a), (b), and (c), respectively, for the typical current–voltage characteristics of SBD, FRD, and high-frequency silicon rectifier. It indicates that the forward conduction voltage drop U_F of SBD is the lowest, and that of FRD is comparatively high, while that of high-frequency silicon rectifier is the highest. Table 4.15 shows the comparison

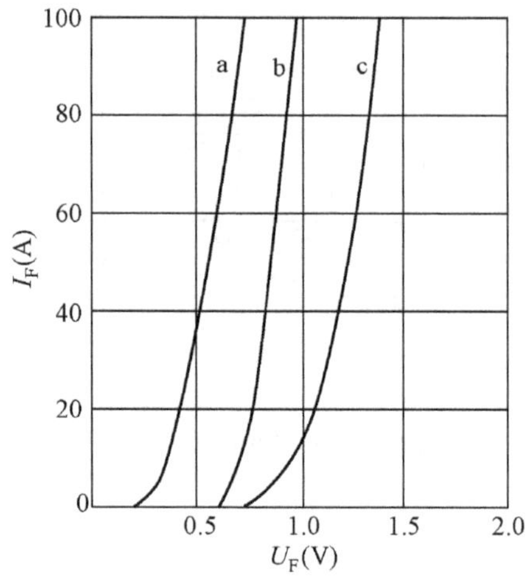

Figure 4.33 Typical current–voltage characteristics of SBD, FRD, and high-frequency silicon rectifier

Table 4.15 Performance comparison between four kinds of typical diode

Semiconductor rectifier diode name	Model of typical products	Average rectified current I_d (A)	Forward conduction voltage drop		Reverse recovery time t_{rr} (ns)	Peak reverse voltage U_{RM} (V)
			Typical value U_F (V)	Maximum U_{RM} (V)		
SBD	16CMQ050	160	0.4	0.8	<10	50
SRD	MUR30100A	30	0.6	1	35	1000
FRD	D25-02	15	0.6	1	400	200
High-frequency rectifier	PR3006	3	0.6	1.2	400	800

between the performance of SBD, SRD, FRD, and high-frequency silicon rectifier. It can be known from the table that the silicon high-speed switching diode is extremely low in t_{rr}, but very small in average rectified current, not suitable for heavy-current rectification. SBDs with small and medium power mostly adopt TO-220 packaging.

4.6.2.2 Typical Application of SBD in SMPS

See Table 4.16 for the SBD models suitable for SMPS output circuit. See Figure 4.34 for the typical application circuit (partial) of SBD in SMPS. In order to reduce the loss of the secondary winding and rectifier, the secondary circuit is composed of two windings, two rectifiers VD_2 and VD_3 connected in parallel, and shares one set of filter in common. All secondary rectifiers use 20 A/100 V Schottky pair diode MBR20100 to reduce the loss of rectifier to the

Table 4.16 Selection of SBD

U_R	SBD		SRD	
	3 A	4–6 A	3 A	4–6 A
20 V	1N5820 MBR320P SR302	1N5823	31DFI HER302	50WF10 MUR410 HER602
30 V	1N5821 MBR330 31DQ03 SR303	50WQ03 1N5824		
40 V	1N5822 MBR340 31DQ04 SR304	MBR340 50WQ04 1N5825		
50 V	MBR350 31DQ05 SR305	50WQ05		
60 V	MBR360 DQ06 SR306	50WR06 50SQ060		

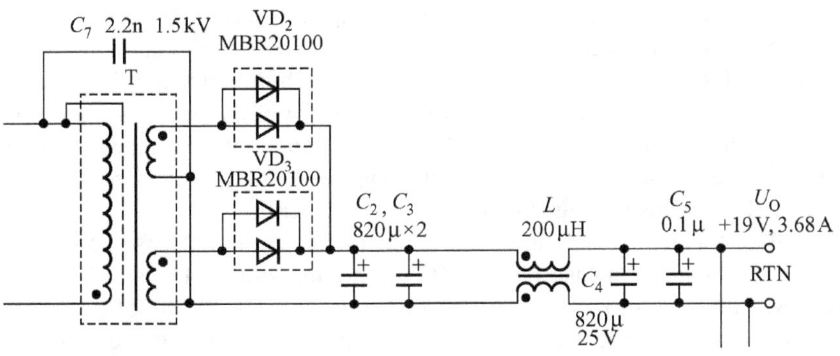

Figure 4.34 Typical application circuit (partial) of SBD in SMPS

minimum. L refers to the common-mode inductor. The output voltage of the SMPS is 19 V and the maximum output current is 3.68A.

4.7 Selection Method for Transient Voltage Suppressor (TVS)

4.7.1 Working Principle of TVS

Transient voltage is caused by sudden release of stored electric energy in a very short time or by large inductance or lightning strike, and so on. See Table 4.17 for the magnitude example of

Table 4.17 Magnitude example of transient voltage source

Type of transient voltage source	Peak voltage U_P (kV)	Peak current I_P	Generating time t_1	Duration t_2
Lightning stroke	25	20 kA	1.2–10 μs	50–1000 μs
Electrostatic discharge	15	30 kA	<1 ns	100 ns
Electromagnetic pulse	1	10 A	20 ns	1 ms

transient voltage source. Suppose that peak transient voltage is U_P, the generating time (t_1) in the table refers to the time when the transient voltage rises from $10\% U_P$ to $90\% U_P$, and duration (t_2) refers to the time when the transient voltage rises from $10\% U_P$ until it reduces to $50\% U_P$; see Figure 4.35(a) and (b), respectively, for the typical test waveform of the lightning stroke and electrostatic discharge transient voltage sources. It can be seen from the figures that the generating time of lightning stroke transient voltage is microseconds order of magnitude while that of electrostatic discharge transient voltage is as low as ns order of magnitude, differing three orders of magnitude between them.

Transient voltage suppressor (TVS) is a new kind of over-voltage protection device. The structural feature of TVS is that PN junction has avalanche effect and its cross-sectional area is much larger than that of the conventional semiconductor diode. Once reverse transient voltage occurs, TVS begins electric conduction and suppresses transient voltage through avalanche effect. It can be used as transient over-voltage protection device owing to its extremely rapid response, stable clamp voltage, capable of withstanding large peak pulse power, small size, and low price. TVS device is divided into the following three types: unidirectional TVS (referred to as unidirectional TVS), bidirectional TVS (referred to as bidirectional TVS), and TVS array. For the TVS devices currently developed abroad, the peak pulse power has reached 60 kW, the maximum peak pulse current can be up to 20 kA, and clamp voltage ranges from 5 V to 3 kV. The latest TVS of SMF series with SMA packaging launched by Littelfuse is 1.1 mm high at most, but capable of withstanding 200 W peak pulse power.

See Table 4.18 for the main performance indicators of commonly used TVS. The parameters listed in the table are measured at 25 °C room temperature. The steady-stage power is generally 5 W. The peak pulse power is divided into 500, 600, 1500, 5000, 15,000 W, and other specifications, related to the duty ratio (D) of interference pulse and ambient temperature (T_A). U_{BM} refers to the maximum value of clamp voltage at high temperature and heavy current.

Take P6KE series for example, there are four models corresponding to P6KE200 (c) A in Table 4.18: P6KE200, P6KE200A, P6KE200C, and P6KE200CA. The suffix in model number without C refers to unidirectional TVS (e.g., P6KE200 and P6KE200A), while that with C refers to bidirectional TVS (e.g., P6KE200C and P6KE200CA), and suffix A indicates that the permissible tolerance of clamp voltage is ±5% (e.g., P6KE200A), while that without A indicates that the permissible tolerance of clamp voltage is ±10% (e.g., P6KE200). The clamping response time of unidirectional TVS is only 1 ps, while that of bidirectional TVS is 1 ns. For P6KE series, the pin near the white ring is positive. TVS can be connected in series or in parallel to increase the peak pulse power, but U_B values of each device in parallel should be equal.

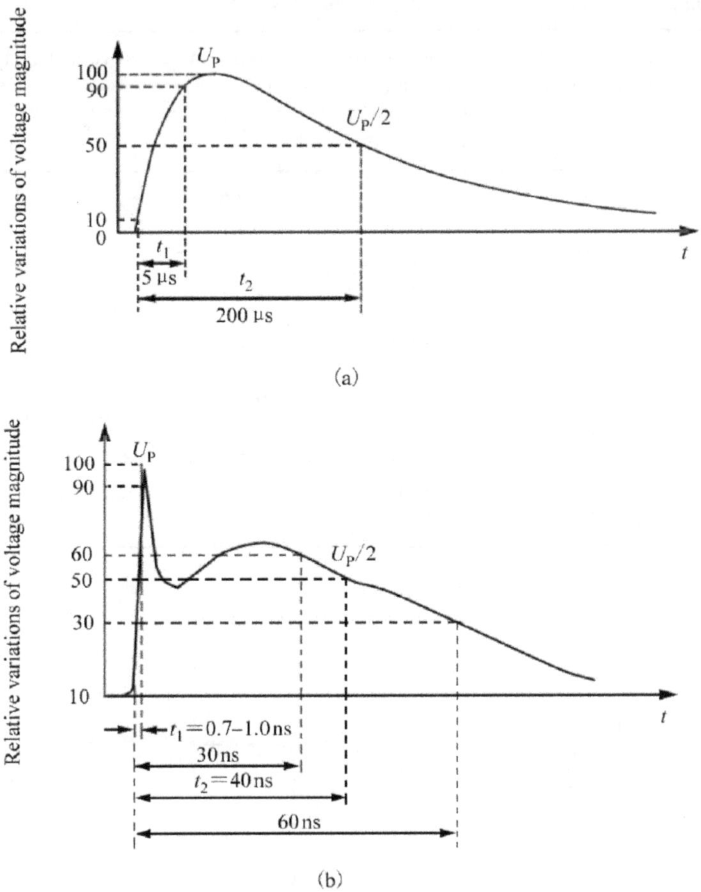

Figure 4.35 Typical test waveforms of two kinds of transient voltage. (a) Typical test waveform of transient voltage at lightning stroke and (b) typical test waveform of transient voltage at electrostatic discharge

See (a), (b), and (c) in Figure 4.36, respectively, for the shape, symbol, and volt–ampere characteristic curve of unidirectional TVS. U_C in Figure 4.36(c) refers to the maximum voltage withstandable within 1 ms. U_R refers to the maximum rated voltage applied to the device before conduction. U_B refers to the clamp voltage with the following relevant formula: $U_R = 0.8U_B$. I_R refers to the leakage current in the case of reverse breakdown, generally less than 10 μA. I_P refers to the maximum peak current withstandable by TVS. Unidirectional TVS can simultaneously suppress interference signals of the same polarity only, applicable to DC circuit.

See (a) and (b) in Figure 4.37, respectively, for the symbol and volt–ampere characteristic curve of bidirectional TVS. This kind of devices can simultaneously suppress interference signals of positive and negative polarities, applicable to AC circuit.

Table 4.18 Main performance indicators of commonly used TVS

Product model	Typical clamp voltage at room temperature U_B (V)	Temperature coefficient of clamp voltage α_T (%/°C)	Maximum clamp voltage U_{BM} (V)	Reverse leakage current I_R (μA)	Peak pulse current I_P (A)
P6KE5(C)A	5	0.057	9.6	5	62.5
P6KE10(C)A	10	0.073	15.0	5	40
P6KE20(C)A	20	0.090	27.7	5	28
P6KE51(C)A	51	0.102	70.1	5	22
P6KE100(C)A	100	0.106	137	5	4.4
P6KE150(C)A	150	0.108	207	5	2.9
P6KE200(C)A	200	0.108	274	5	2.2
P6KE250(C)A	250	0.110	360	5	1.67
P6KE300(C)A	300	0.110	414	5	1.45
P6KE350(C)A	350	0.110	482	5	1.25
P6KE400(C)A	400	0.110	548	5	1.10
1.5KE200(C)A	200	0.108	274	5	5.5

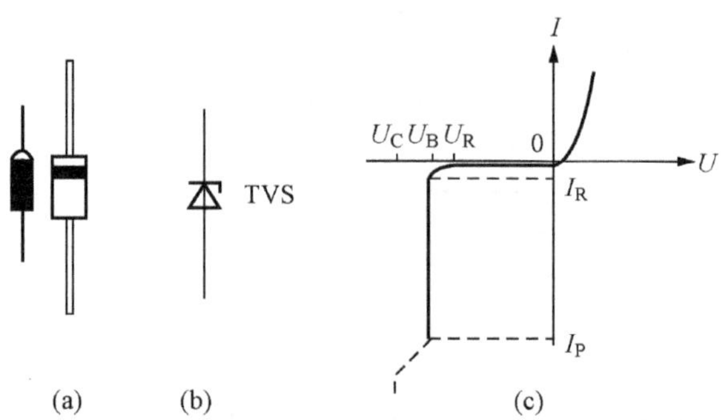

Figure 4.36 Unidirectional TVS. (a) Shape, (b) symbol, and (c) volt–ampere characteristic curve

4.7.2 Selection Method for TVS and Its Typical Application

4.7.2.1 Selection Method for Transient Over-Voltage Protection Circuit

1. The clamp voltage U_B of TVS should be greater than the maximum working voltage U_{max} of the protected circuit by 10–20% in general. If U_B is inappropriate, not only there will be no protection effect but also TVS may be damaged.
2. Unidirectional TVS is generally selected for DC protection, such as the drain clamp protection circuit of power MOSFET. Bidirectional TVS is generally selected for AC protection. TVS array can be selected for multiline protection.
3. TVS series products corresponding to different peak pulse powers are as follows: 500 W (SA series), 600 W (P6KE and SMBJ series), 1500 W (1.5KE series), 5000 W (5KP series), and 15,000 W (15KP series).

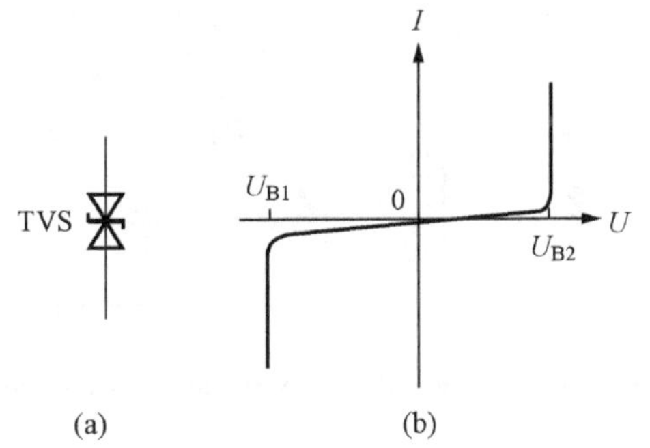

Figure 4.37 Bidirectional TVS. (a) Symbol and (b) volt–ampere characteristic curve

4. The transient pulse withstandable by TVS must be nonrepetitive pulse. However, repetitive pulse may occur in circuit in practical application, so the pulse repetition rate (the ratio between the pulse duration and the interval time) of TVS device is defined as 0.01%, or TVS may be burnt. TVS is very reliable and does not have "aging" problem even under long-term nonrepetitive high-energy shock with large pulses.
5. During the specified pulse duration, the maximum peak pulse power of TVS shall be greater than the peak pulse power that may occur in the protected circuit, and its peak pulse current shall be greater than the transient surge current.
6. The working temperature of TVS is generally −55 to +150 °C. The maximum junction temperature is +175 °C, at which time both the available peak pulse power and peak pulse current reduce to zero.

4.7.2.2 Typical Application of Transient Over-Voltage Protection Circuit

Input protection circuit of AC and DC power supply

See (a) and (b) in Figure 4.38, respectively, for the input protection circuits of AC and DC power supply constituted by TVS. In Figure 4.38(a), two bidirectional TVS, TVS_1 and TVS_2, are, respectively, connected in parallel to the inlet and outlet terminals of isolation transformer to absorb the energy of large transient pulse in positive and negative directions and clamp

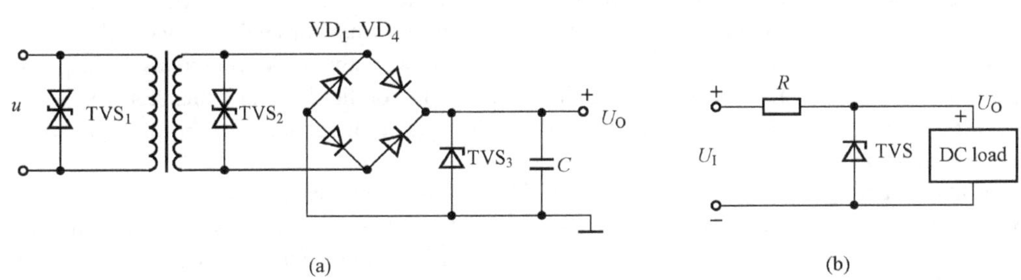

Figure 4.38 Input protection circuits of AC and DC power supply constituted by TVS. (a) Input protection circuit of AC power supply and (b) input protection circuit of DC power supply

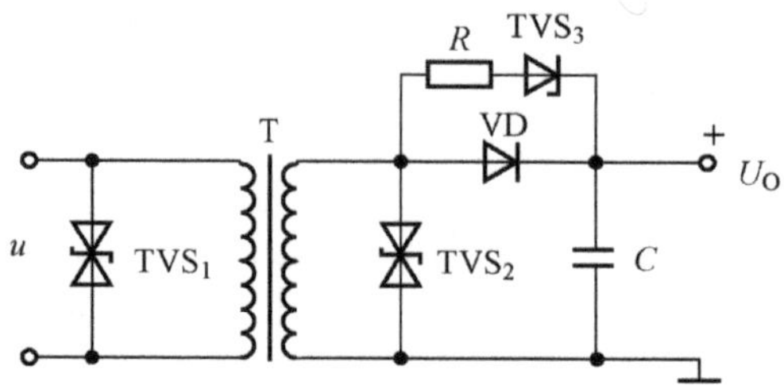

Figure 4.39 Protection circuit of the high-frequency transformer and output rectifier constituted by TVS

the circuit voltage within an allowable range, thus realizing over-voltage protection. As bridge rectifier outputs DC voltage, the outlet terminal of a bridge rectifier should be connected with a unidirectional TVS, TVS_3, in parallel to prevent the load from being shocked by over-voltage. In practice, one to three bidirectional TVSs can be selected according to specific conditions. Figure 4.38(b) is the input protection circuit of DC power supply constituted by unidirectional TVS.

Protection circuit of high-frequency transformer and output rectifier

See Figure 4.39 for the protection circuit of the high-frequency transformer (T) and output rectifier constituted by TVS. For example, when AC input voltage $u = 230\,\text{V}$, 1.5KE440CA bidirectional TVS can be selected for TVS_1 and the clamp voltage is $\pm 440\,\text{V}$ (typical value). When the output voltage of secondary winding is 15 V (AC), 1.5KE24CA can be selected for TVS_2 and the clamp voltage is $\pm 24\,\text{V}$ (typical value). The unidirectional TVS, TVS_3, and the current-limiting resistor R are connected in parallel to both the ends of the output rectifier VD to prevent VD from being broken down by high reverse transient pulse voltage.

4.8 Selection Method for Power Switching Tube

For single-chip SMPSs or switching regulators, the power switching tube is integrated inside the chip. However, if SMPS is constituted by PWM, power switching tube shall be selected. Power switching tube used in SMPS mainly includes three types: the first is bipolar power switching tube, which belongs to bipolar junction transistor (BJT), it was called giant transistor (GTR) at first owing to its high output power and now known as high-power transistors. The second type is metal-oxide-semiconductor field-effect transistor, known as MOSFET or MOS field-effect transistor. The third type is insulated gate bipolar transistor, known as IGBT.

When selecting power switching tube of SMPS, we must pay attention to its conduction voltage drop (or on-state resistance) and switching speed. The conduction voltage drop and the switching speed of power switching tube relate to rated voltage. The higher the rated voltage is, the larger the conduction voltage drop will be and the slower switching speed will be. Therefore, low-voltage power switching tube should be selected under the condition that the rated voltage is 1.2–1.5 times of the actual working voltage.

4.8.1 Selection Method for BJT Power Switching Tube

BJT power switching tube is the BJT, which is characterized by switching and power output. It is called bipolar type owing to two kinds of carrier (electron and hole) flowing through the transistor, different from the field-effect transistor with one kind of carrier only. Bipolar power tube is a current-driven power device, whose withstand voltage is below 1 kV in general use and the working current ranges from several amperes to several hundred amperes. Its advantage is inexpensive, while the disadvantage is low current amplification factor, relatively large drive current, and low switching frequency (below tens of kHz), so it is applicable to the SMPS of small and medium power.

Cautions for using BJT power switching tube:

1. BJT power switching tube has a safe working area with the maximum collector current, the maximum allowable collector power dissipation, secondary breakdown current, and collector–emitter breakdown voltage as the boundary. Either in transient or steady state, the working current and working voltage of transistor shall not exceed the range of safe working area. In addition, the boundary value of safe working area relates to such parameters as ambient temperature and the pulse width. When the ambient temperature rises, the safe working area should be used with decreased rated values.
2. The current amplification factor β of BJT power switching tube is comparatively low, whose minimum value is generally 5–10 times.
3. The collector leakage current will be doubled per 10 °C rise in ambient temperature, which will cause turn-off loss.
4. In order to reduce the conduction loss of power switching tube, it is generally in supersaturation in conduction, which will necessarily increase storage time, thus reducing the switching speed. To reduce storage time, it is necessary to apply reverse voltage to the emitter junction when the power switching tube is turned off. However, if the reverse voltage is too large, the emitter junction will suffer reverse breakdown (the reverse breakdown voltage of emitter junction of silicon power switching tube is about 5–6 V). In order to avoid excessively large breakdown current, resistance can be used to restrict breakdown current to a suitable level.
5. The antisaturation circuit, as shown in Figure 4.40, can be used to turn off the power switching tube quickly. VD_1 and VD_2 are two silicon diodes with conduction voltage drop U_{F1} and U_{F2}, respectively. The collector–emitter saturation voltage of the circuit is as follows: $U_{CE} = U_{F1} + U_{BE} - U_{F2}$. Assume that $U_{F1} = U_{BE} = U_{F2} = 0.7\,V$, $U_{CE} = 0.7\,V + 0.7\,V - 0.7\,V = 0.7\,V$, making excessive drive current pass through the collector

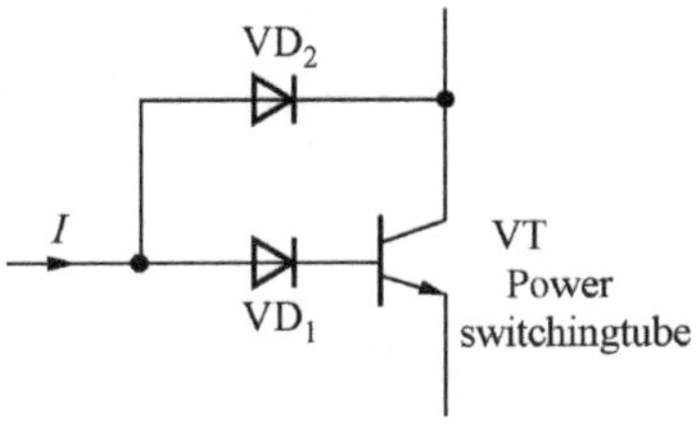

Figure 4.40 Antisaturation circuit

can reduce the saturation depth of power switching tube. To further reduce the saturation depth, another diode VD_3 can be connected in series to VD_1 to make the saturation voltage drop keep at about 1.4 V. In this case, the power switching tube can be in a quasi-saturation state only with a very short storage time, improving the turn-off speed, but the conduction loss will rise.

4.8.2 Selection Method for MOSFET Power Switching Tube

MOSFE is an insulate-gate field-effect transistor. Its main feature is that there is a silicon dioxide insulation layer between the metal gate and the channel, so its input resistance is very high (the maximum can be up to $10^{15}\,\Omega$). It is divided into N-channel and P-channel types. See Figure 4.41 for their symbols. Usually, the substrate and the source electrode S are connected together. According to the mode of conducting, MOSFET can also be divided into enhancement and depletion types. The so-called enhancement type means that the tube is turned off when $U_{GS} = 0$, but when proper U_{GS} ($U_{GS} > 0$ for N-channel tube and $U_{GS} < 0$ for P-channel tube) is applied, the majority carriers will be attracted to the gate to "enhance" the carriers in the region, thus forming a conduction channel. The depletion type means that the channel is formed when $U_{GS} = 0$, and when proper U_{GS} ($U_{GS} < 0$ for N-channel tube) is applied, the majority carriers flow out of the channel, so carriers are "depleted," turning off the tube. VD in Figure 4.41(c) refers to protection diode.

See Figure 4.42 for the two kinds of structure of enhancement-type MOSFET. Take N-channel for example, it forms the source diffusion zone N^+ and the drain diffusion zone N^+ with high doping concentration on P-type silicon substrate, and then, respectively, leads out the source S and the drain D. The source and the substrate are connected internally with equal potential kept between them. The arrow direction of the symbol in Figure 4.42(a) is from outside to inside, indicating pointing from P-type material (substrate) to N-channel. When the drain and the source are connected to the positive and negative poles, respectively, of power supply and $U_{GS} = 0$, the channel current (i.e., drain current) $I_D = 0$. With the gradual rise of U_{GS}, under the attraction of positive voltage of gate, minority carriers with negative charge are induced between the two N^+ diffusion zones to form N-channel from the drain to the source. In addition, the dotted line in the figure refers to the channel. When U_{GS} is greater than the cut-in voltage U_{TN} (usually about +2 V) of the tube, N-channel tube starts conducting, thus forming the drain current I_D.

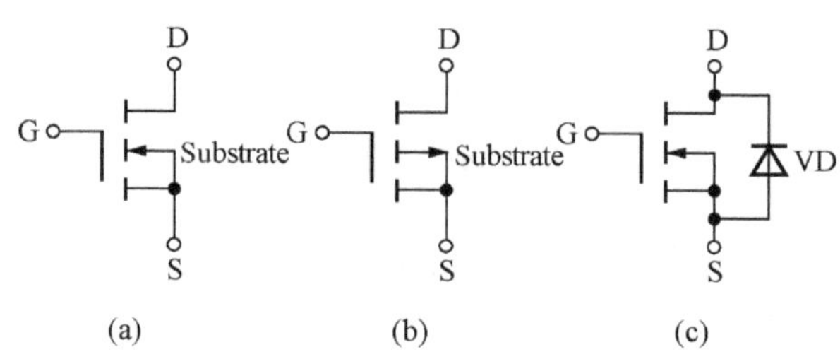

Figure 4.41 The symbols of MOSFET. (a) N-channel tube, (b) P-channel tube, and (c) N-channel tube (with protection diode)

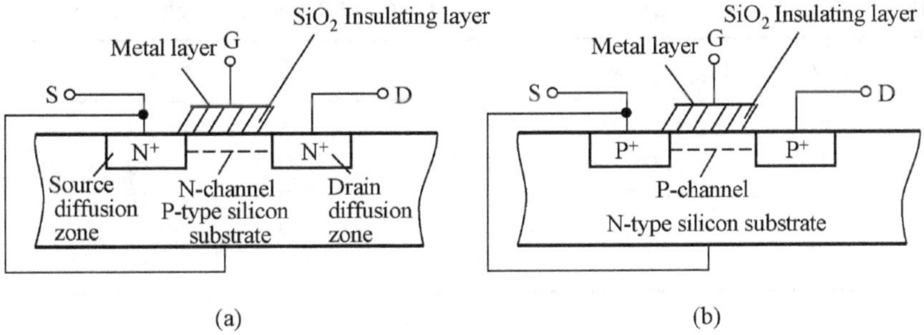

Figure 4.42 The structure of enhancement-type MOSFET. (a) N-channel and (b) P-channel

In comparison with the bipolar power switching tube, MOSFET power switching tube has the following advantages: short turn-on time (several to tens of nanoseconds), suitable for PWM modulators with switching frequency of 100 kHz–1 MHz; using voltage drive, without requiring static drive current; high reliability without secondary breakdown; and a small on-state resistance and low loss. MOSFET power switching tube usually adopts N-channel tube, because the on-state resistance of N-channel tube is smaller than that of the P-channel tube under the same condition, and the switching of the former is faster than that of the latter. For N-channel tube, two-way conduction can be realized as long as a positive voltage is applied between the gate and the source (drain). Therefore, MOSFET may be used for synchronous rectification. The working current of MOSFET power switching tube ranges from a few amperes to several hundred amperes, the output power is from tens of watts to several kilowatts, and the switching frequency is up to several hundred kilohertz to one megahertz or more. Currently, SMPSs with small and medium power mostly use MOSFET as the power switching tube. In addition, there are PMOS, NMOS, and VMOS field-effect transistors.

The DC parameters of MOSFET power switching tube mainly include drain current I_D, drain-source breakdown voltage $U_{(BR)DS}$, drain-source on-state resistance $R_{DS(ON)}$, and drain power dissipation P_D. The AC parameters include turn-on time $t_{d\ (ON)}$, turn-off time $t_{d\ (OFF)}$, input capacitance C_i, output capacitance C_O, and so on. There are mainly two kinds of MOSFET loss in SMPS: conduction and switching losses. Conduction loss occurs on the drain-source on-state resistance R_{ON} when MOSFET is fully turned on. Switching loss refers to the power loss when MOSFET alternates between turn-on and turn-off. In addition, there is the gate loss, that is, the loss resulted from the charge and discharge of MOSFET gate capacitor, but it occurs in the gate resistor or the drive circuit.

See Table 4.19 for the main technical indicators of five kinds of MOSFET power switching tubes produced by IR in the United States. There is a protection diode added between the drain and the source of these tubes.

The following seven key parameters should be focused on in selecting MOSFET power switching tubes:

1. Drain-source breakdown voltage $U_{(BR)DS}$. $U_{(BR)DS}$ must be greater than the maximum voltage $U_{DS(max)}$ that the drain source can withstand, allowing a 10–20% margin left. However, $U_{(BR)DS}$ value shall not be set too high, so as not to increase the cost of devices.

Table 4.19 Main technical indicators of five kinds of MOSFET power switching tube

Model	I_D (A)	$U_{(BR)DS}$ (V)	$R_{DS(ON)}$ (Ω)	P_D (W)	$t_{d(ON)}$ (ns)	$t_{d(OFF)}$ (ns)
IRF840	8	500	0.85	125	14	49
IRFP450	14	500	0.40	190	17	92
IRFP460	20	500	0.27	280	18	110
IRFP250	30	200	0.085	190	16	70
IRFP150	41	100	0.055	230	16	60

2. Maximum drain current $I_{D(max)}$. Generally, $I_{D(max)}$ shall be left a comparatively large margin, which facilitates heat emission and improves the conversion efficiency.
3. Drain-source on-state resistance $R_{DS(ON)}$. Use power MOSFETs with a small $R_{DS(ON)}$ value to reduce transmission loss.
4. Total gate charge Q_G. $Q_G = Q_{GS} + Q_{GD} + Q_{OD}$, wherein Q_{GS} refers to the gate-source charge, Q_{GD} the gate-drain charge (also known as charge on Miller capacitor), and Q_{OD} the excessive charge after the Miller capacitor is fully charged. As the gate charge will cause loss on the drive circuit, Q_G should be as small as possible.
5. Figure of merit (FOM). FOM = $R_{DS(ON)}\ Q_G$, with the unit of $\Omega \cdot nC$. It is an important parameter for assessing power MOSFET. The smaller the FOM value is, the better the performance of a device will be.
6. Output capacitance C_{OSS} (i.e., total drain-source distributed capacitance). An excessive C_{OSS} will increase the switching loss. The reason is that the charge stored on C_{OSS} is released at the beginning of each switching cycle, thus causing loss.
7. Switching time includes turn-on delay time $t_{d(ON)}$ and turn-off delay time $t_{d(OFF)}$. The switching time should be extremely short to reduce the switching loss.

It deserves noting that no indicator should be pursued one-sidedly. For example, some MOSFET power switching tubes have very small drain-source on-state resistance $R_{DS(ON)}$, but have comparatively large output capacitance C_{OSS}, which will increase switching loss, so various indicators should be comprehensively compared before selection.

Take IPP60R099CPA N-channel power MOSFET produced by Infineon in the United States, for example, its main parameters are as follows: $U_{(BR)DS} = 600\,V$, $I_{D(max)} = 19\,A$, $R_{DS(ON)} = 0.09\,\Omega$ (typical value), $Q_G = 60\,nC$, FOM = $R_{DS(ON)}\ Q_G = 0.09\,\Omega \times 60\,nC = 5.4\,\Omega \cdot nC$. $C_{OSS} = 130\,pF$, $t_{d(ON)} = 10\,ns$, and $t_{d(OFF)} = 60\,ns$. The device is applicable to the power switching tube of AC/DC SMPS.

4.8.3 Selection Method for IGBT Power Switching Tube

IGBT is a new kind of high-power power electronic device formed by integrating MOSFET and GTR, which belongs to voltage-controlled high-power device. IGBT takes MOSFET as the input stage (drive circuit) and GTR as the output stage (main circuit). Essentially, IGBT, however, is still an FET, to which just a P-type layer between the drain and the drain zone is added. It integrates the advantages of MOSFET and GTR, featuring in such excellent properties as high input impedance, high withstand voltage, heavy working current, fast speed, good

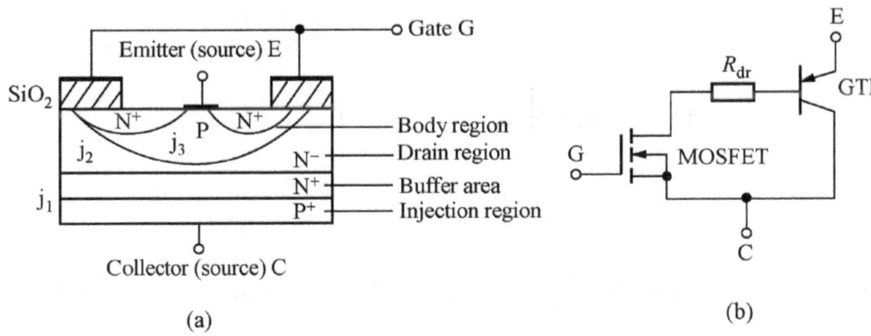

Figure 4.43 Structure and equivalent circuit of IGBT. (a) Sectional view of basic structure and (b) equivalent circuit

thermal stability, and simple circuit. The withstand voltage of IGBT is generally above 500 V, the working current can be up to several hundred amperes, the peak current can be up to several thousand amperes, and the maximum working frequency is 20–30 kHz. IGBT can be mainly applied in such fields as electrical equipment including AC frequency converters and inverters as well as induction cookers.

See Figure 4.43(a) for the structure of IGBT, which is similar to that of MOSFET. The difference is that IGBT adds a P^+ substrate (i.e., the collector of IGBT) on the N^+ substrate (drain) of N-channel MOSFET power switching tube, forming a PN junction j_1, from which the drain is led out. The gate and the source are similar to those of MOSFET. IGBT adopts PNPN four-layer structure, whose output stage can also be regarded as a thyristor constituted by PNP-NPN transistors. However, NPN transistor and the emitter are of short circuit internally, so NPN transistor has no effect. Therefore, IGBT can be regarded as the unidirectional Darlington transistor, which takes N-channel MOSFET as the input stage and PNP transistor as the output stage. It can be seen from the figure that IGBT is equivalent to a thick base region GTR driven by MOSFET, whose simplified equivalent circuit is shown in Figure 4.43(b). In the figure, R_{dr} refers to the diffusion resistance of the thick base region GTR. In view of the idiomatic appellation of bipolar transistor pins, it is specified by the IEC that the gate of IGBT is G, the source outlet terminal is the emitter E, and the drain outlet terminal is the collector C.

N-channel and P-channel IGBTs have two kinds of circuit symbols, respectively, as shown in Figure 4.44(a) and (b). The turn-on and turn-off of IGBT are controlled by the gate voltage. When positive voltage is applied to the gate, channel will be formed inside MOSFET and base current will be provided for PNP transistor to make IGBT turned on. At this time, current will be injected from zone P^+ to zone N^- for conductivity modulation to reduce the resistance R_{dr} of zone N^-, letting the IGBT that can withstand high voltage having a low on-state resistance. When negative voltage is applied to the gate, the channel in MOSFET disappears and the base current of PNP transistor is cut off, thus turning off the IGBT.

The main technical parameters of IGBT power switching tube include the reverse breakdown voltage $U_{(BR)CEO}$, the maximum collector continuous current I_{CM}, the maximum output power P_{OM}, the turn-on time $t_{d(ON)}$, the turn-off time, the gate threshold voltage U_{Ge}, the collector–emitter saturation voltage U_{CE}, and if there is internal protection diode (also known as damping diode). The main foreign manufacturers producing IGBT power switching tube

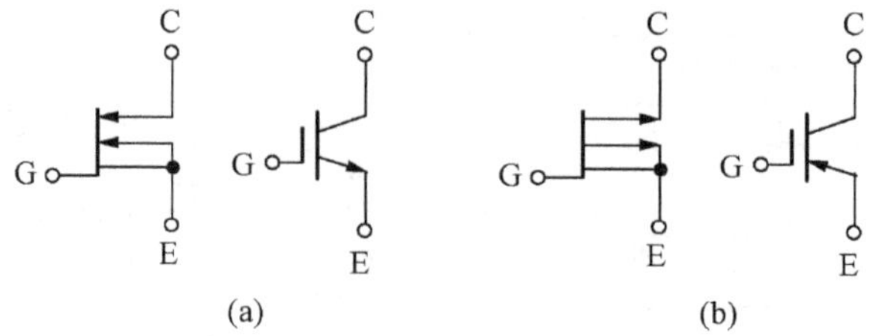

Figure 4.44 Circuit symbols of IGBT. (a) N-channel IGBT and (b) P-channel IGBT

Table 4.20 Main parameters of typical IGBT power switching tubes

Model	$U_{(BR)CEO}$ (V)	I_{CM} (A)	P_{OM} (W)	$t_{d(ON)}$ (ns)	$t_{d(OFF)}$ (ns)	If there is internal protection diode	Packaging form
HGTG20N120CND	1200	63	390	23	200	Yes	TO-247
GT40T101	1500	40	200	700	500	Yes	2–21F2C
SGW25N120	1200	46	313	50	820	Yes	TO-247
SGL40N150	1500	40	200	90	245	No	TO-264

include Fairchild in the United States, Siemens and Infineon in Germany, and Toshiba in Japan. See Table 4.20 for the main parameters of typical IGBT power switching tubes.

For IGBT power switching tube without internal protection diode, a fast recovery damping diode should be connected between C and E, with the positive pole of damping diode connected to C and the negative to E. For example, BY459 fast high-voltage damping diode from Philips can be used, whose reverse withstand voltage is 1500 V, the forward break-over voltage is 0.95 V, the peak repetitive forward current can be up to 100 A, and the reverse recovery time is 250 ns.

The following should be noted in using IGBT:

1. U_{GE} of IGBT module is generally ±20 V. The module will be damaged if the value is exceeded.
2. When the gate is disconnected, it is easy to damage IGBT if a voltage is applied to the main circuit. To prevent such a case, a resistor of about 10 kΩ and a small capacitor should be connected, respectively, in parallel between the gate and the emitter.
3. As IGBT module takes MOSFET as the input stage, it is necessary to take protective measures against static electricity. When IGBT module is used, it is prohibited to touch the gate with hand. If it is necessary to touch the terminals of the module, the static electricity on human body should be released first. The backplane of module should be well grounded. Electric soldering iron or electric welding machine should also be well grounded in welding. The container storing IGBT module shall have no static electricity.
4. To meet the need of large current output, it is allowable to connect multiple IGBT modules in parallel. In this case, the current flowing through each device should be kept balanced.

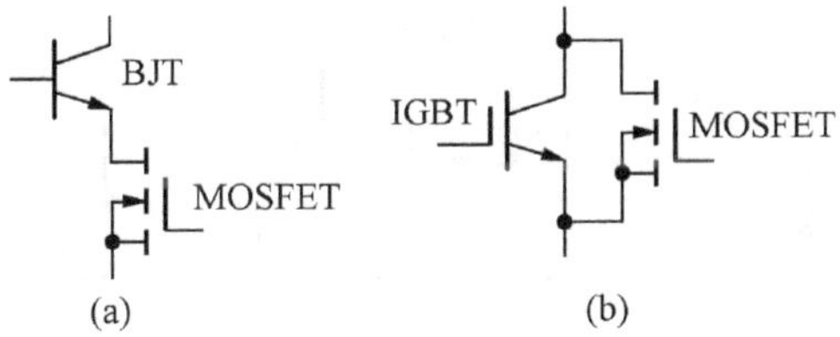

Figure 4.45 Circuit structures of two kinds of multiunit tube. (a) BJT–MOSFET multiunit tube and (b) IGBT–MOSFET multiunit tube

5. In order to improve the switching speed of BJT, BJT and MOSFET can be connected in series to form a multiunit tube, thus enhancing switching speed by making use of the switching characteristics of MOSFET. See Figure 4.45(a) for the circuit structure of BJT–MOSFET multiunit tube. It drives MOSFET power switching tube first when turned on. At the time, BJT is under common base connection, so current is input from the emitter. Owing to the conduction of MOSFET and the reduction of drain voltage, the emitter junction of BJT is forward biased to generate base current and collector current, and saturation conduction of BJT can be realized through positive feedback circuit. At the time of turn-off, MOSFET is turned off first and the emitter junction is reversely biased to turn off BJT immediately. As the frequency characteristic of common base connection is β times that of common emitter connection, turn-off speed can be enhanced greatly. The on-state resistance of low-voltage MOSFET is in the order of magnitude of milliohms only, causing very small conduction loss. The above circuit can be used for double-ended forward SMPSs with large power, with a switching frequency up to 50 kHz. Similarly, MOSFET and IGBT can also be connected in parallel to form a multiunit tube. See Figure 4.45(b) for IGBT–MOSFET multiunit tube. The driving principles of MOSFET and IGBT are as follows: the conduction process is that MOSFET is conducted first under the driving of PWM signals and then IGBT starts to be conducted only when the voltage passes zero. The turn-off process is that IGBT is turned off (at zero voltage) first and then MOSFET is turned off after a delay time. In the period of conduction, the conduction voltage drop of MOSFET is higher than that of IGBT, so most of the current flows through IGBT that bears conduction loss. In addition, switching loss is mainly borne by MOSFET. IGBT–MOSFET multiunit tube can be used for half-bridge or full-bridge topology.

4.9 Selection Method for Optical Coupler

Optical coupler is also known as photoelectric coupler or optical isolator. It transmits electrical signals with light as the medium. Generally, the light projector (infrared emitting diode LED) and the light receiver (phototransistor) are packed in the same shell. When electrical signals are applied to the input terminal, the light projector emits light and the light receiver generates photocurrent after receiving light, which flows out from the output terminal, thus realizing "electricity-light-electricity" conversion. Optical coupler can be widely used in level conversion, signal isolation, interstage isolation, long-distance signal transmission, pulse amplification, solid-state relays, precision SMPSs, instruments, and meters and microcomputer interface.

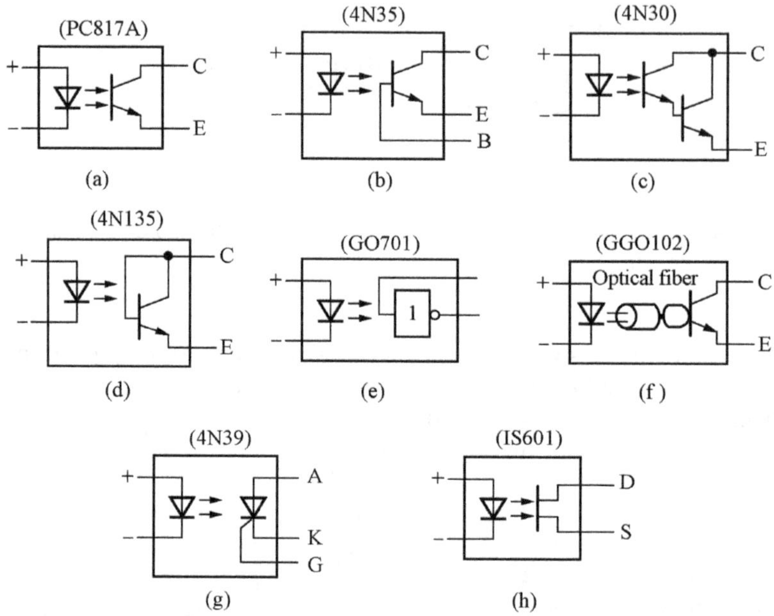

Figure 4.46 Classification and internal circuit of optical coupler. (a) General purpose type (without base lead), (b) general purpose type (with base lead), (c) Darlington type, (d) high-speed type, (e) optical integrated circuit type, (f) fiber type, (g) photosensitive thyristor type, and (h) photosensitive FET type

4.9.1 Basic Principle of Optical Coupler

4.9.1.1 Types of Optical Coupler

See Figure 4.46 for product classification and internal circuit of common optical couplers, in which the words in brackets are the models of eight kinds of typical product. Among them, the general purpose type is medium-speed optical coupler, its current transfer ratio (CTR) is 25–300%. Darlington type optical coupler has a comparatively low speed, but its CTR can be up to 100–5000%. High-speed optical coupler has such advantages as fast speed and good output linearity. The optical coupler constituted by optical integrated circuit belongs to high-speed optical coupler, and its CTR is comparatively large. Fiber-type optical coupler can withstand high voltage, and its insulation voltage is above 100 kV. Photosensitive thyristor type is an optical coupler with large power output, the typical products include 4N39 (containing unidirectional thyristor) and IS607 (containing bidirectional thyristor). Photosensitive FET-type optical coupler features fast speed and dual application of AC and DC. See Table 4.21 for the main technical indicators of typical ordinary optical couplers.

4.9.1.2 Features of Optical Coupler

The main advantages of optical coupler include unidirectional signal transmission, full electrical isolation at input and output terminals, strong anti-interference ability, long transmission distance, long service time, and high transmission efficiency.

Table 4.21 The main parameters of typical ordinary optical couplers

Model	Current transfer ratio CTR (%)	Insulation resistance R (Ω)	Insulation voltage U_{DC} (V)	Maximum forward current I_{FM}(mA)	Reverse breakdown voltage $U_{(BR)CEO}$ (V)	Saturation voltage drop U_{CES} (V)	Dark current I_R (μA)	Maximum power loss P_M (mW)	Packaging form
4N35	>100	—	3550	60	30	0.3	50	—	DIP-6
4N30	>100	10^{11}	1500	60	30	1.0	100	100	DIP-6
GO111	≥ 60	10^{10}	1000	60	≥30	≤0.4	≤10	>5	DIP-6

CTR is an important parameter of optical coupler, usually expressed by direct CTR. When the output voltage is constant, it equals the percentage ratio between the DC output current I_C and the DC input current I_F. The relevant formula is as follows:

$$\text{CTR} = \frac{I_C}{I_F} \times 100\% \tag{4.18}$$

For optical couplers with one phototransistor, CTR range is mostly 20–300%. For example, CTR > 100% for 4N35 optical coupler, while 80–160% for PC817A optical coupler. The CTR range of Darlington optical coupler (e.g., 4N30) can be up to 100–5000%. This indicates that the latter needs a comparatively small input current to obtain the same output current. Therefore, the parameter CTR is similar to h_{FE} of transistor to some extent.

4.9.2 *Selection Method for Linear Optical Coupler*

Linear optical coupler is usually used in optical coupler feedback SMPS, the typical products include PC817A, CNY17-2, and MOC8101. See the dotted and solid lines in Figure 4.47 for the typical CTR-I_F characteristic curves of linear and ordinary optical couplers, respectively. It can be seen from the figure that the CTR-I_F characteristic curve of the ordinary optical coupler is nonlinear, and the nonlinear distortion is particularly severe when I_F is comparatively small, not suitable for analog signal transmission. The CTR-I_F characteristic curve of linear optical

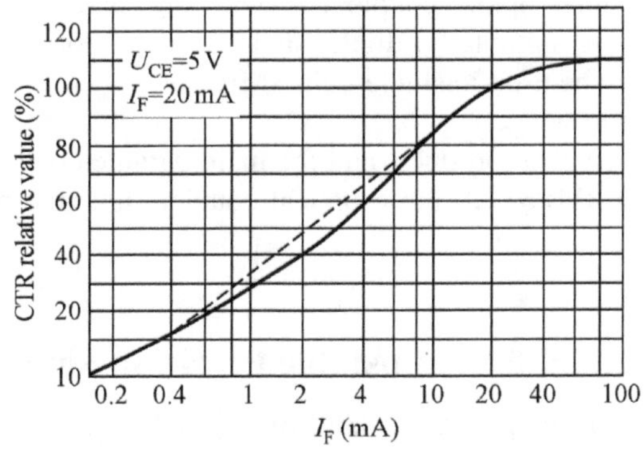

Figure 4.47 CTR-I_F characteristic curves of two kinds of optical coupler

Table 4.22 Typical products and main parameters of linear optical coupler

Product model	CTR (%)	$U_{(BR)CEO}$ (V)	Foreign manufacturer	Packaging form
PC816A	80–160	70	Sharp	DIP-4 (base not led out)
PC817A	80–160	35		
SFH610A-2	63–125	70	Simens	
NEC2501-H	80–160	40	NEC	
CNY17-2	63–125	70	Motorola, Simens, and Toshiba	DIP-6 (base led out)
CNY17-3	100–200	70		
SFH600-1	63–125	70	Simens, Isocom	
SFH600-2	100–200	70		
CNY75GA	63–125	90	Temic	DIP-6 (base not led out)
CNY75GB	100–200	90		
MOC8101	50–80	30	Motorola, Isocom	
MOC8102	73–117	30		

coupler has good linearity, and especially in transmitting small signals, its alternating-current transmission ratio ($\Delta CTR = \Delta I_C/\Delta I_F$) is very close to the CTR value of the direct current transmission ratio, so the linear optical coupler is suitable for analog voltage or current signal transmission and can ensure linear relationship between the input and the output. That is an important feature of it.

Model and parameters of optical coupler must be selected properly in designing optical coupler feedback SMPS. The selection principles are as follows:

1. The allowable range of CTR of optical coupler is 50–200%. This is because when CTR < 50%, LED in the optical coupler needs a comparatively large working current ($I_F > 50\,mA$) to control the duty ratio normally, which will increase the power loss of the optical coupler. If CTR > 200%, false triggering may be caused when the circuit is started up or the load changes suddenly, thus affecting the normal output.
2. It is recommended to use linear optical coupler, featuring that its CTR value can be linearly adjusted within a certain range. Currently, the 4N×× series (e.g., 4N25, 4N26, and 4N35) optical coupler produced by ISOCOM of Britain and Motorola of the United States is widely used. This kind of optical coupler has switching characteristics with poor linearity and uncontrollable CTR, suitable for digital signal transmission (high and low level), so it is not recommended to be used in single-chip SMPS.

See Table 4.22 for the typical products and the main parameters of the linear optical couplers commonly used in SMPS. All these optical couplers take the phototransistor as their receiver tube.

4.10 Selection Method for Adjustable Precision Shunt Regulator

Adjustable precision shunt regulator is a kind of adjustable reference voltage source with current output capability, which possesses excellent performance and low cost, can be widely used in precision SMPS to constitute an external error amplifier, and then form an isolation-type optical coupler feedback circuit together with the linear optical coupler. In addition, it can also constitute voltage comparator, power supply voltage monitor, delay circuit, precision constant current source, and so on. Currently, there are mainly three models of adjustable precision shunt regulators: 2.5 V adjustable precision shunt regulator TL431, 1.24 V low-voltage adjustable precision shunt regulator LMV431, and 0.9 V low-voltage adjustable precision shunt regulator NCP100. In addition, a special type of LT3080 low-dropout linear regulator (LDO) is produced by LT of the United States, which can precisely set the output voltage to be continuously adjustable within the range of 0–10 V just through a resistor. This is the significant difference between LT3080 and other LDOs.

4.10.1 TL431 Adjustable Precision Shunt Regulator

TL431 is the adjustable precision shunt regulator produced by TI and Motorola of the United States, whose output voltage is continuously adjustable within the range of 2.50–36 V.

4.10.1.1 Working Principle of TL431

TL431 mostly uses DIP-8 or TO-92 packaging with the pin array as shown in Figure 4.48(a) and (b), respectively, in which A is the anode to be grounded in use, K the cathode to be connected to the positive power supply through a current-limiting resistor, U_{REF} the setting terminal of output voltage U_O, to be connected to resistance divider externally, and NC the empty pin. See Figure 4.48(c) for the equivalent circuit of TL431, which mainly includes four parts: (i) the error amplifier A, whose noninverting input terminal is connected to the sampling voltage obtained from the resistance divider, while the inverting input terminal is connected to the internal 2.50 V reference voltage U_{ref}. In addition, it is designed that U_{REF} is equivalent to U_{ref} and should be 2.50 V in normal state, so it is also known as the reference terminal. (ii) The internal 2.50 V (the exact value should be 2.495 V) reference voltage source U_{ref}, (iii) NPN transistor VT, which is used to adjust the load current in circuit, and (iv) protection diode VD, which can prevent chip damage caused by the reverse polarity connection between K and A in power supply.

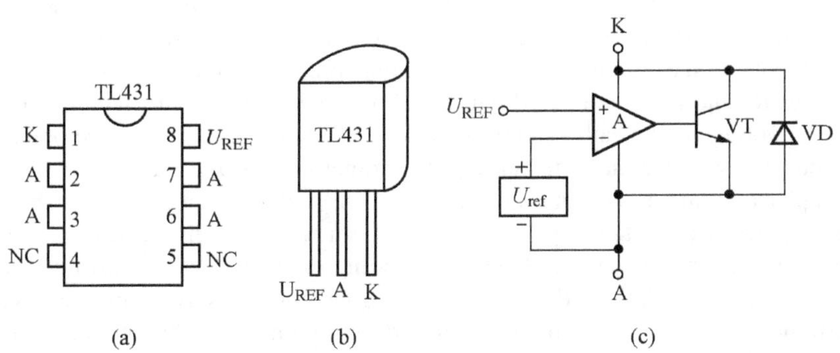

Figure 4.48 Pin array and equivalent circuit of TL431. (a) DIP-8 packaging, (b) TO-92 packaging, and (c) equivalent circuit

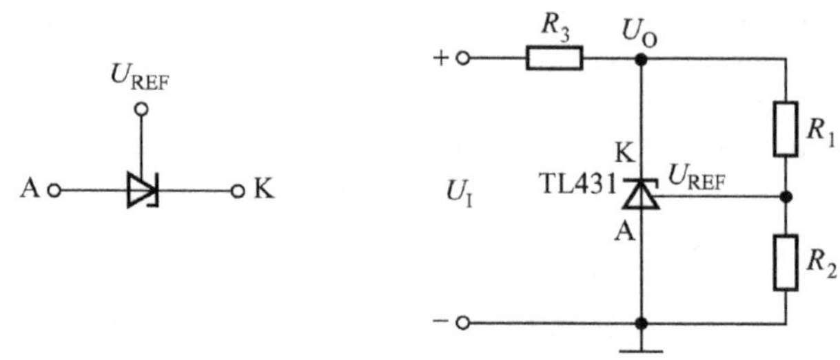

(a) (b)

Figure 4.49 Circuit symbols and basic wiring of TL431. (a) Circuit symbols and (b) basic wiring

See Figure 4.49 for the circuit symbols and basic wiring of TL431. It is equivalent to an adjustable zener diode, whose output voltage is set by the external precision resistances R_1 and R_2. The relevant formula is as follows:

$$U_O = U_{KA} = U_{REF}\left(1 + \frac{R_1}{R_2}\right) = 2.50\ \text{V} - \left(1 + \frac{R_1}{R_2}\right) \tag{4.19}$$

R_3 is the current-limiting resistance of I_{KA}. The selection principle of R_3 is that when the input voltage is U_I, I_{KA} shall be ensured to be within the range of 1–100 mA, so as to make TL431 work normally. The regulation principle of TL431 can be analyzed as follows: when U_O rises for some reason, the sampling voltage U_{REF} will rise correspondingly, making $U_{REF} > U_{ref}$, with high-level output from the comparator. At this time, if VT is conducted, then U_O will decline. Conversely, $U_O \downarrow \rightarrow U_{REF} \downarrow \rightarrow U_{REF} < U_{ref} \rightarrow$ the reversal of comparator leading to low level output $\rightarrow$ VT turn-off $\rightarrow U_O \uparrow$. From the perspective of dynamic equilibrium, U_O will be forced to go to stabilization along with the continuation of the cycle, thus realizing the purpose of voltage regulation and $U_{REF} = U_{ref}$.

4.10.1.2 Typical Application of TL431 in SMPS

TL431 can be widely used in TOPSwitch series single-chip SMPS as the external error amplifier to constitute an optical coupler feedback circuit. See Figure 4.50 for its typical application. When the output voltage U_O fluctuates, the sampling voltage obtained after the voltage division by resistances R_3 and R_4 is compared with the 2.5 V bandgap reference voltage in TL431 to form an error voltage at the cathode, thus making LED working current in the optical coupler change accordingly. After that, the current at terminal C of TOPSwitch is changed through optical coupler to adjust the output duty ratio of TOPSwitch and make U_O unchanged, thus realizing the purpose of voltage regulation. N_P, N_S, and N_F in the figure, respectively, refer to the primary, the secondary, and the feedback winding. VD_Z and VD_1 constitute the primary clamp protection circuit, VD_2 is the secondary rectifier, and C_{OUT} is the filter capacitor at the output terminal. VD_3 and C_F are the rectifier and filter elements at the output terminal. R_1 is the current-limiting resistance of LED.

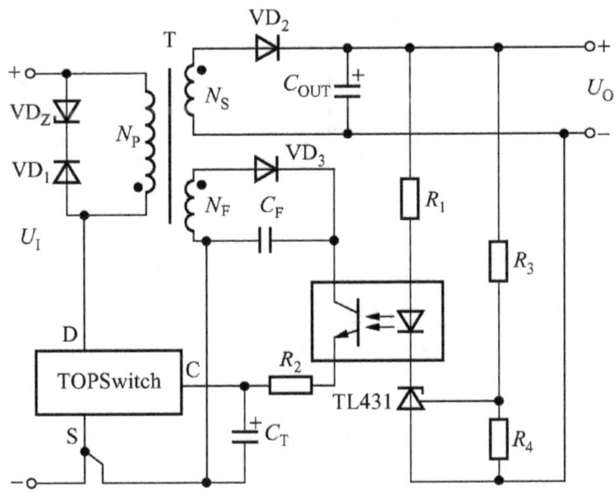

Figure 4.50 Typical application of TL431 in TOPSwitch series single-ship SMPS

4.10.2 NCP100 Low-Voltage Output Adjustable Precision Shunt Regulator

4.10.2.1 Working Principle of NCP100

NCP100 is the low-voltage output adjustable precision shunt regulator produced by ON Semiconductor of the United States, with the output voltage adjustable within the range of 0.9–6.0 V. NCP100 is a kind of three-terminal adjustable device, which can set any reference voltage within the range of 0.9–6.0 V through two external resistors. When the ambient temperature changes from −40 to +85 °C, the variation of cathode working voltage U_{KA} (i.e., the output reference voltage) is only 1.0 mV. The cathode working current $I_{KA} = 0.1$–$20\,mA$. Its dynamic impedance is very low, with a typical value of 0.2 Ω.

NCP100 uses TO-92 or TSOP-5 packaging with pin arrays as shown, respectively, in Figure 4.51(a) and (b), in which A is the anode to be grounded during use, K the cathode to be connected to the positive power supply through the current-limiting resistor, U_{REF} the setting terminal of output voltage U_O, to be connected to the external resistance divider, and NC the empty pin. See Figure 4.51(c) for the equivalent circuit of NCP100, which mainly includes four parts: (i) error amplifier A, whose noninverting input terminal is connected to the sampling voltage obtained from the resistance divider and inverting input terminal is connected to the internal 0.7 V reference voltage U_{ref}, with U_{REF} equivalent to U_{ref} designed and being 0.7 V in normal state, so it is also known as the reference terminal. (ii) Internal 0.7 V (the exact value should be 0.696 V) reference voltage source U_{ref}, (iii) N-channel field-effect transistor V, which is used to adjust the load current in circuit, and (iv) the protection diodes VD_1 and VD_2, of which VD_1 can be used to avoid the error amplifier saturation while VD_2 can be used to prevent chip damage caused by the reverse polarity connection between K and A.

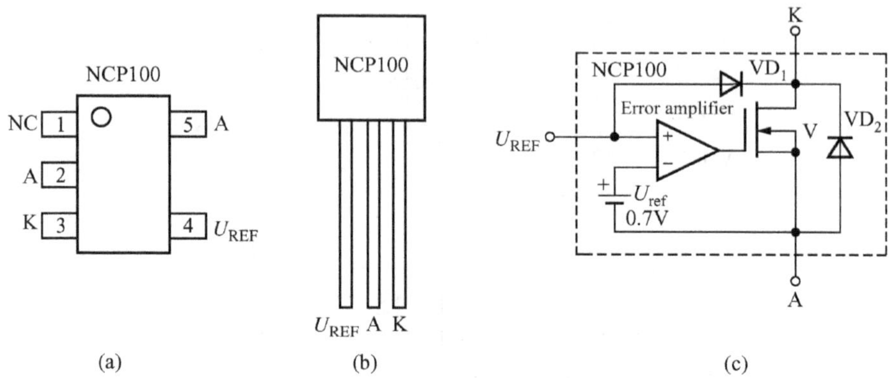

Figure 4.51 Pin array and equivalent circuit of NCP100. (a) SOP-5 packaging, (b) TO-92 packaging, and (c) equivalent circuit

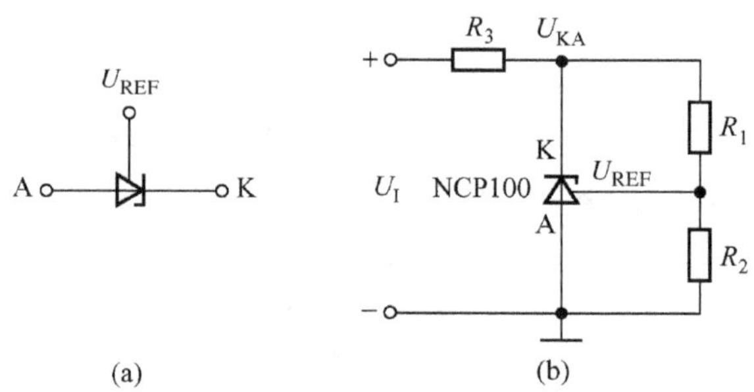

Figure 4.52 (a) Circuit symbols and (b) basic wiring of NCP100

See Figure 4.52(a) and (b), respectively, for the circuit symbols and basic wiring of NCP100. It is equivalent to an adjustable zener diode, whose output voltage is set by the external precision resistances R_1 and R_2. The relevant formula is as follows:

$$U_{KA} = U_{REF}\left(1 + \frac{R_1}{R_2}\right) = 0.7\text{V} \times \left(1 + \frac{R_1}{R_2}\right) \tag{4.20}$$

R_3 is the current-limiting resistance of I_{KA}. When the sampling resistance divider chooses $R_1 = 1.5\,\text{k}\Omega$ and $R_2 = 4.3\,\text{k}\Omega$, it can be gotten that $U_{KA} = 0.94\,\text{V}$ according to formula (4.20).

The regulation principle of NCP100 is as follows: when U_{KA} rises for some reason, the sampling voltage U_{REF} will rise correspondingly, making $U_{REF} > U_{ref}$, and comparator outputs high level. At this time, if V is conducted, U_{KA} will decline. Conversely, $U_{KA} \downarrow \rightarrow U_{REF} \downarrow \rightarrow U_{REF} < U_{ref} \rightarrow$ the reversal of comparator, leading to low level output → V turn-off → $U_{KA} \uparrow$. From the perspective of dynamic equilibrium, U_{KA} will be

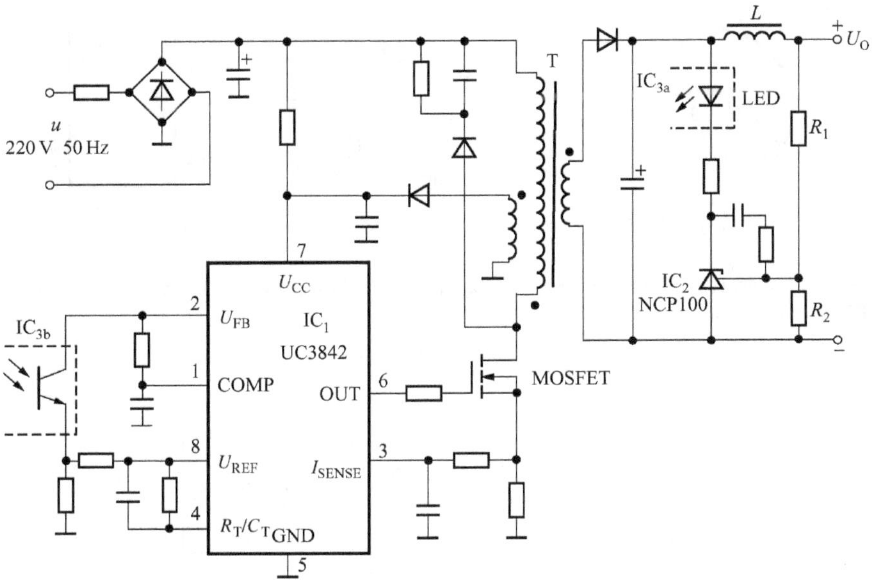

Figure 4.53 Application circuit of NCP100 in SMPS

forced to get to stabilization along with the continuation of the cycle, thus realizing the purpose of voltage regulation and $U_{REF} = U_{ref}$.

4.10.2.2 Typical Application of NCP100 in SMPS

See Figure 4.53 for the application circuit of NCP100 in SMPS, in which UC3842 (IC_1) is the PWM and NCP100 (IC_2) is used as the compensation amplifier to control the feedback circuit. The output voltage is set through R_1 and R_2, which can be lower than that obtained by using TL431. Its minimum output voltage is equal to the sum of the minimum voltage $U_{KA(min)}$ (about 0.9 V) obtained by subtracting the anode voltage from the cathode voltage and the forward conduction voltage drop U_F (about 1.4 V) of LED in the optical coupler (IC_3, including IC_{3a} and IC_{3b}), that is, $U_{O(min)} = U_{KA(min)} + U_F \approx 2.3\,V$.

4.10.3 LMV431 Low-Voltage Adjustable Precision Shunt Regulator

4.10.3.1 Performance Features of LMV431

1. LMV431 is the 1.24 V low-voltage adjustable precision shunt regulator produced by NSC of the United States. It includes three models, LMV431, LMV431A, and LMV431B, their accuracy is 1.5%, 1%, and 0.5%, respectively, and the working temperature range is 0 to + 70 °C and −40 to + 85 °C.

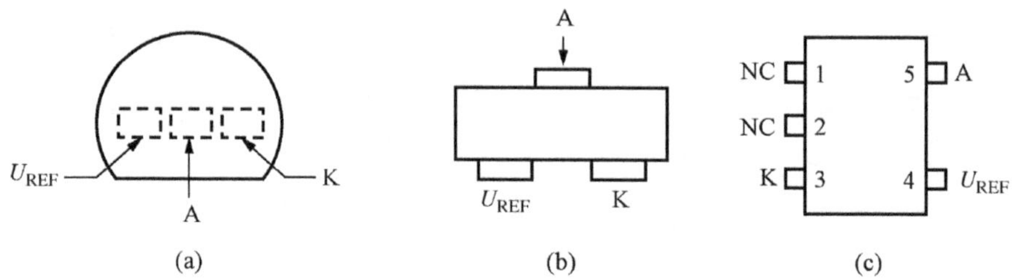

Figure 4.54 Pin arrays of LMV431. (a) TO-92 packaging (bottom view), (b) SOT23-3 packaging, and (c) SOT23-5 packaging

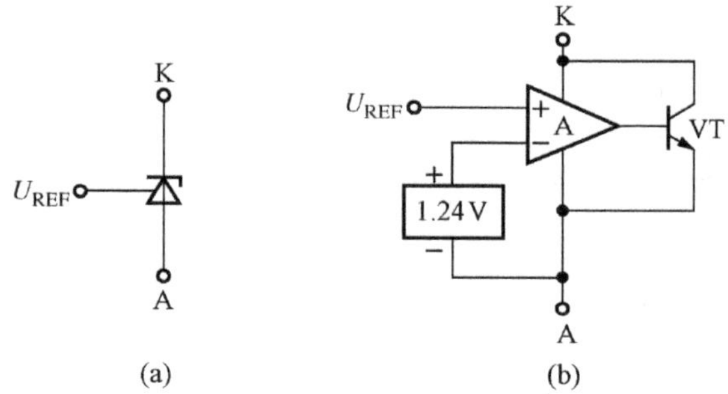

Figure 4.55 (a) Circuit symbols and (b) equivalent circuit of LMV431

2. In order to reduce the power consumption of optical coupler feedback circuit in designing high-efficiency SMPS, conventional 2.5 V adjustable precision shunt regulator can be replaced with LMV431 to reduce the LED forward working current I_{LED} of infrared emission tube in optocoupler from 1 mA to 100 μA.
3. The voltage adjustment range is 1.24–30 V, and the temperature coefficient of voltage within the entire working temperature range is as low as $39 \times 10^{-6}/°C$. In addition, the working current is small (only 55 μA), and the output impedance is low (typically 0.25 Ω).
4. Applicable to isolated precision SMPS, shunt (or series) regulator, constant current source, voltage monitor, and error amplifier.

4.10.3.2 Principle and Application of LMV431

LMV431 uses TO-92, SOT23-3, or SOT23-5 packaging and corresponding pin arrays are as shown in Figure 4.54(a), (b), and (c), respectively. See Figure 4.55(a) and (b) for the current symbols and equivalent circuit, respectively. It contains a 1.24 V bandgap reference voltage source, an error amplifier, and a shunt regulator tube – NPN transistor VT internally.

The connection method for LMV431 is identical with that of TL431. See Figure 4.50 for its typical application. However, the output voltage is changed as follows:

$$U_{KA} = 1.24\text{V} \times \left(1 + \frac{R_1}{R_2}\right) \tag{4.21}$$

4.11 Selection Method for SMPS Protection Elements

Common SMPS protection elements include fuses, fusible resistors, and VSRs.

4.11.1 Selection Method for Fuse

Fuse is indispensable to almost all kinds of electronic equipment. It is commonly known as the protective tube with a circuit symbol of FU. Fuse is made of terne alloy or lead-antimony alloy material, featuring in low fusing point, high resistivity, and fast fusing speed. Under normal circumstances, it is used to connect the input circuit in SMPS. In case of over-load or short-circuit fault, it will be immediately fused when the current flowing through it exceeds the fusing current to disconnect the input circuit, thus realizing over-current protection.

4.11.1.1 Working Principle of Fuse

It is well known that heat will be generated when current flows through conductors owing to certain resistance in the conductors. Heat can be calculated by the following formula: $Q = I^2Rt$, in which Q is the heat quantity (with the unit of J), I the current flowing through a conductor, R the resistance of the conductor, and t the time spent by the current flowing through the conductor. When the current flows through fuse, the temperature of fused mass will rise for the conversion of heat from electric energy. During normal working state, the heat generated by fuse can be emitted into the surrounding air through heat convection, thermal conduction, and so on, to balance heat-producing capability and heat dissipating capacity. If heat-producing capability is larger than heat dissipating capacity, the excess heat will be gradually accumulated in the fuse, so that the temperature will be further increased. When the temperature exceeds the fusing point, the fuse will be melted to cut off the current in the circuit. That is the working principle of fuse.

In fact, whether fuse will be melted or not also depends on heating rate and cooling rate. That involves three cases: first, fuse will not be melted when the heating rate is less than the cooling rate. Second, fuse will not be melted within a long period, when the heating rate is equal to the cooling rate. Third, when the heating rate is greater than the cooling rate, fuse will be melted once the temperature exceeds the fusing point.

Fuse is generally composed of three parts: (i) fuse link, which is the core of fuse and used to cut off the current when fused; (ii) two electrodes, which connect fuse link and circuit, so should have good conductivity and small contact resistance in installation; (iii) clamp and bracket, which should have good mechanical strength, heat durability, and flame retardance and should have no such faults as breakage, deformation, combustion, or short circuit during use. The fuse used in electrical equipment should also have an arc-extinguishing device with quartz sand as commonly used material. In addition, some fuses have a fusing indicating device, which can automatically give light alarm when the fuse is fused. Some fuses even have an indicator.

4.11.1.2 Product Classification of Fuse

See Table 4.23 for detailed product classification of fuse.

Table 4.23 Product classification of fuse

Classification method	Product type
As per rated voltage	High-voltage fuse (e.g., 5 kV fuse used in microwave ovens), low-voltage fuse (e.g., 250 V fuse), and safety-voltage fuse (e.g., 32 V fuse)
As per fusing current	1, 2, 3, 5, 10, 15, 25, 30 A, and so on
As per protection type	Over-current protection fuse, overheating protection fuse (temperature fuse), temperature switch (thermal protector), on-off current fuse, and self-recovery fuse
As per external dimensions	Micro, small, medium, and large fuse Fuse of $\phi 2$, $\phi 3$, $\phi 4$, $\phi 5$, $\phi 6$, and other specifications
As per shape	Flat head tube fuse, pointed tube fuse, screw-type fuse, plug-piece fuse, chip fuse, flat-type fuse, and guillotine-type fuse
As per packaging	Glass, ceramic, chip, and other packaging
As per usage	Fuse used in instruments and household appliances, auto fuse, machine-tool fuse, and electric fuse
As per fusing rate	Special slow fuse (TT), slow fuse (T), medium-speed fuse (M), fast fuse (F), and ultrafast fuse (FF)
As per fusing characteristics	Fast-acting fuse and time-lag fuse
As per adopted safety standards	IEC (standards of China, Europe, etc.), UL (standards of America), and so on

Slow fuse, also known as time-lag fuse, can work properly in the case of nonfault pulse current and provide protection for long-time over-load. Some SMPSs may generate large pulse current which can be up to several times normal working current at the moment of turn-on, which, despite continuing for a very short time, is large enough to make ordinary fuse fused, causing the failure to turn on in SMPS. If fuses with larger capacity are selected, the circuit will not be protected in case of over-load. Slow fuse can absorb energy for its fuse link made with the use of special processes. It can not only resist surge current but also provide effective protection for over-load through adjusting energy absorption ratio properly.

4.11.1.3 Main Parameters of Fuse

1. *Rated Voltage.* Rated voltage refers to the maximum allowable working voltage of fuse under safe working condition. It includes 32, 125, 250, 600 V, and other specifications. It deserves pointing out that whether a fuse is blown or not only depends on the size of current flowing through it, irrelevant to the working voltage. The rated voltage of a fuse is specified only for the safe use of it. A fuse can work safely and reliably only when the working voltage is no more than the rated voltage and there will be no arcing or breakdown when the fuse is blown.
2. *Rated Current.* It refers to the maximum current when a fuse works properly.
3. *Fusing Current.* It is a current value at which a fuse can be blown reliably at rated voltage. Fusing current is equal to the rated current multiplied by the fusing factor that is generally

1.1–1.5. It indicates that even if the current flowing through a fuse is greater than its rated current but does not exceed the fusing current, the fuse will not be blown.

4. *Voltage Drop*. It refers to the voltage drop generated by a fuse when rated current flows through it, reflecting the capacity of the internal resistance of a fuse. The voltage drop of a fuse should be as small as possible to reduce power loss.
5. *Resistance*. The fuse is made of the materials with positive temperature coefficient, whose cold resistance is less than hot resistance.
6. *Ambient Temperature*. The higher ambient temperature is, the higher the working temperature of a fuse will be, the lower its current carrying capacity will be, and the shorter its service life will be. Therefore, working at a comparatively low temperature can extend the service life of a fuse.
7. *Temperature Rise*. It refers to the temperature rise when the current of 1.1 times of rated current flows through a fuse. It is equal to the difference between the measured temperature and the ambient temperature.
8. *Fusing Time*. It refers to the time for fusing a fuse.

4.11.1.4 Notes for Use

1. The rated voltage of the fuse selected should be greater than the input voltage of the circuit protected. For example, when the input voltage of SMPS is AC 220 V, the fuse with a rated voltage of 250 V should be selected.
2. In the actual use of fuse, the rated current should be less than 75% of the nominal value. For example, when the working current in a circuit is 0.75 A, the fuse with a rated current of at least 1 A can be selected.
3. The higher the ambient temperature is, the shorter the service life of a fuse will be. The rated current can be selected according to the temperature influence curve provided by manufacturers. The service life of a fuse will be reduced if the ambient temperature is too high. It is not allowed for the time-lag fuse to work at 150 °C above for a long time. Fast-acting fuse may not work within the range of 175–225 °C for a long time.
4. Fast-acting fuse is applicable to the circuit with a relatively constant working current and a comparatively small surge current. Time-lag fuse is applicable to the circuit with normal surge current only and without any element sensitive to surge current.
5. The rated current will be reduced with the aging of a fuse, easily resulting in error protection, that is, cutting off the circuit in the case of a comparatively small over-load current.
6. The contact resistance between a fuse and its clamp should be as small as possible, generally not exceeding 3 mΩ. When replacing the fuse of SMPS, only the fuse of the same specifications as the original one may be selected.

4.11.2 Selection Method for Fusible Resistor

Fusible resistor has the dual function of resistors and fuses, equivalent to a small resistor in normal operation. When current increases to exceed the fusing current owing to circuit failure, fusible resistor will be blown rapidly to provide the circuit and its elements with over-current protection. Fusible resistor is applicable to the safety device of low-voltage power supply. The

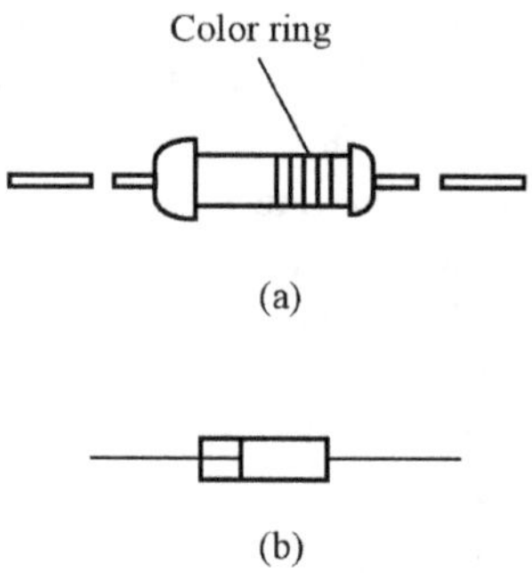

Figure 4.56 Fusible resistor. (a) Shape and (b) national standard symbols

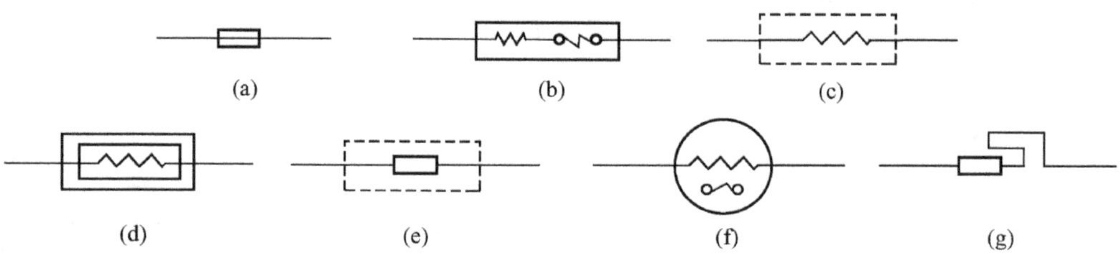

Figure 4.57 Symbols of foreign fusible resistors

advantage of replacing fuses with fusible resistors is that there will be no electric spark or smoke during fusing, safe, and causing no interference.

See Figure 4.56 for the shape and national standard symbols of the fusible resistor. See Figure 4.57 for the symbols commonly used abroad. The power of the fusible resistor is generally 0.125–3 W, and the resistance is from one ohm below to several tens of ohms (the maximum can be up to several thousand ohms). The fusing current is from tens of milliamps to several amps, and the fusing time is from several seconds to tens of seconds. Domestic metal film fusible resistors include RJ90 series. RF10 and RF11 series are fusible resistors with flame retardancy with the technical indicators as shown in Table 4.24 in detail. Fusible resistors are mostly disposable products. When the resistor reaches a certain temperature, the resistive film coated with fusing material will be fused immediately to disconnect the resistor, and the product of the same specification should be used for replacement after fusing. The disposable fusible resistor, when used, should be installed on the PCB in hanging state. Currently, there are recoverable fusible resistors for multiple use. For such resistors, elastic sheet metal or metal wire is welded on the one end with low fusing point solder. In case of overheating, the welding spot will be fused first to disconnect the elastic sheet metal or metal wire from the resistor. So the resistor can continuously be used after repair.

4.11.3 Selection Method for Voltage-Sensitive Resistor

VSR is a kind of over-voltage protection element.

Table 4.24 Technical indicators of RF10 and RF11 series products

Series	Rated power (W)	Resistance range (Ω)	Resistance deviation ± (%)	Stability (%)	Temperature coefficient (10^{-6}/°C)	Withstand voltage (V)	External dimension (mm)
RF10	0.25	0.47–1k	5	5	350	250	$\phi 2.5 \times 7$
	0.5	0.47–1k	5	5	350	250	$\phi 3.9 \times 10.5$
	1	0.47–1k	5	5	350	350	$\phi 5.5 \times 14$
	2	0.47–1k	5	5	350	350	$\phi 6.5 \times 17$
RF11	0.5	0.33–1.5k	5	5	350	1000	$\phi 6.3 \times 13.5$
	1	0.33–1k	5	5	350	1000	$6.5 \times 6.5 \times 14$
	2	0.33–1k	5	5	350	1000	$7.4 \times 7.4 \times 19$
	3	0.33–3.3k	5	5	350	1000	$10.5 \times 10.5 \times 23$

4.11.3.1 Performance Features of Voltage-Sensitive Resistor

VSR is the metal-oxide-semiconductor ceramic element with ZnO or SiC as main materials, whose resistance varies with the terminal voltage. The VSR mainly features in wide working voltage range (6–3000 V, divided into several blocks), fast response to over-voltage pulses (a few nanoseconds to tens of nanoseconds), strong ability to withstand impact current (up to 100 A–20 kA), small leakage current (less than several microamps to tens of microamps), low temperature coefficient of resistance (less than 0.05%/°C), low price, and small size and is ideal for being used as protection element. It can be used to constitute over-voltage protection circuit, lightning protection circuit, spark quench circuit, and surge voltage absorption circuit.

The main parameters of VSR are as follows:

1. Nominal voltage U_{1mA}: the value of the voltage across both the ends of an element when 1 mA DC current flows through. The principle of selecting nominal voltage is as follows: for DC voltage U_{DC}, select $U_{1mA} \geq (1.3–2.6)U_{DC}$. For AC voltage U_{AC}, select $U_{1mA} \geq (1.9–2.2)U_{AC}$. For peak pulse voltage U_P, select $U_{1mA} \geq (1.4–2.0)U_P$.
2. Leakage current: the DC current flowing through an element when the voltage across both the ends of the element is equal to $75\%U_{1mA}$.
3. Discharge current: the maximum pulse current allowed to be passed within the specified time (8/20 μs). Wherein, the time when the pulse current changes from 90% U_P to U_P is 8 μs, and the peak duration time is 20 μs.

See Figure 4.58 for the shape, symbols, and current–voltage characteristic curve of the VSR. VSR itself has no polarity. Its forward and reverse current–voltage characteristic curves are symmetrical and have the function of voltage regulation. Therefore, VSR can be used not only in AC and DC circuits but also as the bidirectional amplitude limit or regulation element in small current (< 1 mA) circuits. The current–voltage characteristic curve of VSR is nonlinear. The relevant formula is as follows:

$$U = CI^{\beta}, \tag{4.22}$$

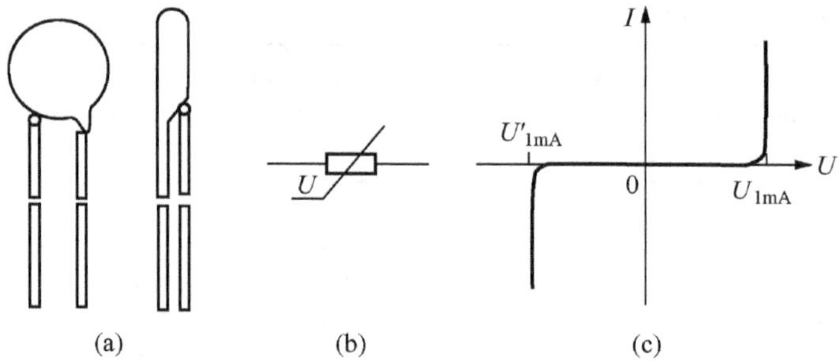

Figure 4.58 Voltage-sensitive resistor. (a) Shape, (b) symbols, and (c) current–voltage characteristic curve

wherein

U the voltage applied across both the ends of VSR
C constant
I the current flowing through VSR, and
β nonlinear coefficient, $\beta < 1$.

4.11.3.2 Product Classification of Voltage-Sensitive Resistor

See Table 4.25 for the model naming of domestic VSRs. For example, MYL1-1 refers to the lightning arrester VSR, and MY31-270/3 refers to the ordinary VSR with a nominal voltage of 270 V and a discharge current of 3 kA. See Table 4.26 for the main parameters of typical VSRs. Each kind of products is further divided into a variety of specifications. The nominal voltage of common VSR includes 6, 18, 22, 24, 27, 33, 39, 47, 56, 82, 100, 120, 150, 200, 216, 240, 250, 270, 283, 360, 470, 850, 900, 1100, 1500, 1800, 3000 V, and others.

4.11.3.3 Typical Application of Voltage-Sensitive Resistor

See Figure 4.59 for the typical application circuit of VSR. Figure 4.59(a) is the input circuit of AC/DC converter with the function of surge voltage protection. Surge voltage protection

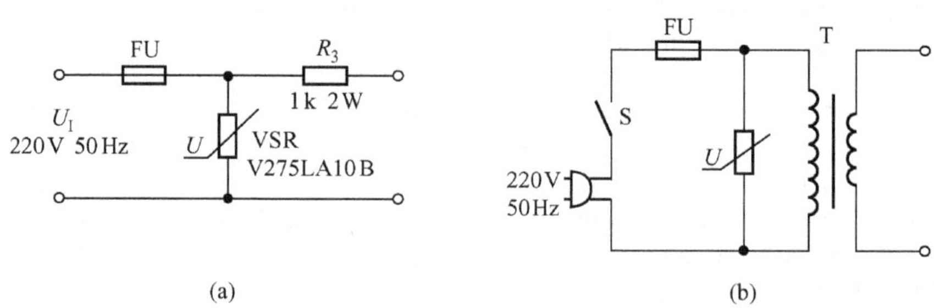

Figure 4.59 Typical application circuit of voltage-sensitive resistor. (a) Surge voltage protection circuit and (b) lightning protection circuit

Table 4.25 Model naming of domestic VSRs

Part I: principal name		Part II: category		Part III: use or feature		Part IV: serial number
Letter	Meaning	Letter	Meaning	Letter	Meaning	
M	Sensitive resistor	Y	Voltage-sensitive resistor	No	Ordinary	The serial number is indicated by numbers. Some serial numbers are followed by the nominal voltage, the discharge current or the resistor diameter, the nominal voltage, the voltage error, and so on
				D	General purpose	
				B	Compensation	
				C	Degaussing	
				E	Noise canceling	
				G	Over-voltage protection	
				H	Arc suppression	
				K	High reliability	
				L	Lightning protection	
				M	Antistatic	
				N	High energy	
				P	High frequency	
				S	Element protection	
				T	Special type	
				W	Voltage regulation	
				Y	Ring type	
				Z	Combination type	

Table 4.26 Main parameters of typical voltage-sensitive resistor

Model	Nominal voltage U_{1mA} (V)	Leakage current (μA)	Discharge current [A/(8/20 μs)]	External dimension (mm)
MYL07DK	22–82	≤10	100	$\phi 10 \times 4.2$
MYL10DK	22–82	≤10	200	$\phi 14 \times 4.3$
MYL14DK	22–82	≤10	500	$\phi 17 \times 4.3$
MYL20D	22–82	≤10	1k	$\phi 23 \times 4.3$
MYL25DK	22–82	≤10	3k	$\phi 28 \times 5$
MYL30DK	82–1500	≤10	5k	$\phi 34 \times 12$
MYL40DK	82–1500	≤10	10k	$\varphi 43 \times 12$

elements use V275LA10B VSR with a nominal voltage of 275 V (effective AC value) and a discharge current of 2500 A, capable of absorbing the surge voltage with a duration of 60 μs and up to 6 kV. R_3 refers to the input resistance of 1 kΩ and 2 W. When the AC input voltage is 120 V, V130LA10B VSR with a nominal voltage of 130 V (effective AC value) can be selected, with R_3 saved. Figure 4.59(b) is the lightning protection circuit added to the user power inlet terminal of 1 + 1 carrier telephone equipment to absorb the over-voltage caused by the induced lightning, thus ensuring personnel and equipment safety.

5

Power Factor Correction Circuit Design of SMPS

5.1 Brief Introduction to Power Factor Correction (PFC)

5.1.1 Power Factor and Total Harmonic Distortion

For undistorted sinusoidal alternating current, the formulae for its input voltage and input current are, respectively, as follows:

$$u = \sqrt{2}\ U_{\cos \omega t} \tag{5.1}$$

$$i = \sqrt{2}\ I\ \cos(\omega t - \phi) \tag{5.2}$$

wherein, u and i are instantaneous values, U and I the root mean squares (RMSs), and ϕ the phase angle. The apparent power of AC input $S = UI$, and the active power $P = UI \cos \phi$. $P = UI = S$ only when $\cos \phi = 1$.

Power factor is referred to as PF, with a GB symbol of λ. Power factor is defined as the ratio between the active power and the apparent power. Relevant formula is as follows:

$$\lambda = \frac{P}{S} = \frac{UI \cos \phi}{UI} = \cos \phi \tag{5.3}$$

The power factor of AC power supply equipment is defined in the case when there is no current waveform distortion. Power factor is reduced for two reasons: one is the phase distortion of the AC input current waveform and the other is the distortion of the AC input current waveform. Phase distortion is usually caused by the load property (inductive or capacitive) of power supply. In this case, the power factor is analyzed in a relatively simple way and is generally calculated as per the formula $\cos \phi = P/(UI)$. However, if AC input current waveform is not sinusoidal wave (e.g., full-wave rectified current waveform), the formula (5.3) is no longer

Optimal Design of Switching Power Supply, First Edition. Zhanyou Sha, Xiaojun Wang, Yanpeng Wang, and Hongtao Ma.

applicable owing to a lot of harmonic wave contained in the AC input current waveform. At this time,

$$i\ (t) = \sqrt{2}\ I_1 \cos(\omega t - \phi_1) + \Sigma I_n \cos(\omega t - \phi_n) \tag{5.4}$$

wherein, I_1 is fundamental current, I_n the nth harmonic current, $\Sigma I_n = \sqrt{I_0^2 + I_1^2 + I_2^2 + \cdots + I_n^2}$, and I_0 the DC component of the current, and it is equal to 0 for pure AC power supply. The power factor redefined should be as follows:

$$\lambda = \frac{I_1}{I_n}\cos\phi_1 = \frac{I_1}{\sqrt{I_0^2 + I_1^2 + I_2^2 + \cdots + I_n^2}}\cos\phi_1 \tag{5.5}$$

wherein, $\cos\phi_1$ is the displacement power factor (DPF, also known as the fundamental power factor). The current distortion component $I_{\text{dis}} = \sqrt{I^2 - I_1^2}$.

At present, all the SMPSs (switching-mode power supply) using AC/DC converters are connected to the grid through rectification circuit. Their input rectifier filters are generally constituted by the bridge rectifier and the filter capacitor, both of which belong to nonlinear components, making SMPS show nonlinear impedance toward the mains. The angle of flow of rectifier diode becomes very narrow owing to the filter capacitor with large capacity, which can be conducted only in the vicinity of the peak AC input voltage, thus causing serious distortion of AC input current and turning it into spike pulse. This kind of current waveform contains a lot of harmonic components, which not only pollute the grid but also significantly reduce the active power output after filtering, thus greatly reducing the power factor.

Consider an SMPS without power factor correction (PFC), for example, the typical waveforms of its AC input voltage and input current from actual measurement are shown in Figure 5.1(a). Figure 5.1(b) shows the harmonic analysis of AC input current waveform with serious distortion, wherein the fundamental wave amplitude is defined as 100%, the amplitude of 3rd, 5th, …, 21st higher harmonics (all odd harmonics) is expressed as the percentage between it and the fundamental wave amplitude. As it is a symmetrical waveform

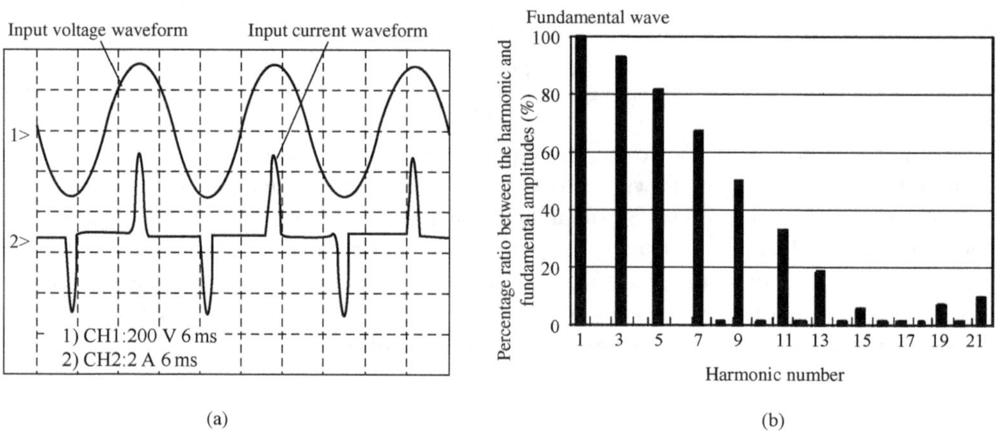

Figure 5.1 AC input waveform of SMPS without PFC. (a) Typical waveforms of input voltage and input current and (b) harmonic analysis chart

Table 5.1 Harmonic current limits of Class C equipment

Number of odd harmonics n	Percentage between the maximum harmonic current and the fundamental current (%)
3	30
5	10
7	7
9	5
$11 \leq n \leq 39$	3

of even harmonics, it can hardly be observed. The power factor of SMPS without adopting PFC circuit is only 0.5–0.6.

The power factor of ordinary SMPS is relatively low, which must be improved to reduce the reactive power of the grid and harmonic pollution, thus improving the quality of power supply.

Total harmonic distortion (THD) refers to the harmonic component difference (usually expressed as a percentage) between the output signal (harmonic and its double frequency component) and the input signal when signal source is used for input. The relationship between the power factor (λ) and the THD can be expressed as follows:

$$\lambda = \frac{1}{\sqrt{1 + (\mathrm{THD})^2}} \cos \phi \times 100\% \tag{5.6}$$

When AC input current and voltage are of the same phase, that is, $\cos \phi = 1$, the formula (5.6) can be simplified as

$$\lambda = \frac{1}{\sqrt{1 + (\mathrm{THD})^2}} \times 100\% \tag{5.7}$$

According to the certification standards for "ENERGY STAR" in the United States, for residential SSL lamps, the power factor $\lambda \geq 0.70$ ($P > 5$ W), and for commercial SSL lamps (including LED street lamps), $\lambda \geq 0.90$. According to the IEC61000-3-2 standard issued by IEC, the harmonic current is divided into four classes A, B, C, and D, and lighting (including dimming) equipment belongs to Class C, of which the harmonic current of the lighting equipment with input power greater than 25 W shall not exceed the limits specified in Table 5.1. For example, when $\lambda = 0.99$, the 3rd harmonic of LED lighting supply shall not exceed 29.7%, the 5th harmonic shall not exceed 10%, and so on.

5.1.2 Power Factor Correction Method

To avoid power factor reduction and harmonic pollution of the grid owing to the use of SMPS, PFC shall be considered in the design of the SMPS with large power. PFC (controller) is also known as power factor compensation (compensator). Customarily, "PFC" can represent not only PFC but also power factor controller, as the case may be.

PFC is used to keep the AC input current and the AC input voltage at the same phase and filter out current harmonics, increasing the power factor of equipment to a predetermined value close to 1. At present, it is increasingly important to increase the power factor, the fields requiring the use of PFC are gradually widening, and relevant international standards are constantly increased. It is required by IEC0500-3-2 mandatory standards that PFC shall be added between the bridge rectifier of power converter above 25 W and the electrolytic capacitor filter with large capacity. The European norm EN61000-3-2 specifies stringent requirements for harmonics of electric apparatus, including the maximum amplitude of the industrial frequency harmonics up to 39th. The national standard GB/T4549-93 implemented by China from March of 1994 also specifies stringent requirements for the quality of power energy and the harmonics of public grid.

The increase in power factor can bring tremendous social and economic benefits. For example, suppose that $\lambda = 0.6$, it indicates that the active power is only 60%, whereas if $\lambda = 0.95$, it indicates that the maximum active power of 95% can be achieved. Some systems must adopt PFC technology. Consider the aerospace field, for example, as the available power of electric generator in the spacecraft is limited, increasing the power factor means that the payload of a spacecraft can be increased without using the electric generator with larger power. Increasing the power factor of SMPS can also reduce the harmonic pollution of the grid and improve the quality of power supply.

In terms of working method, PFC can be divided into passive PFC (referred to as PPFC) and active PFC (referred to as APFC). PPFC circuit generally uses passive elements – inductors – for correction to reduce the phase difference between the fundamental current and the voltage of AC input to increase the power factor. APFC circuit adds a power conversion circuit between the input bridge rectifier and the output filter capacitor to correct the input current to the nondistortion sinusoidal wave of the same phase as the input voltage, making the power factor $\cos\phi$ close to 1. The circuit of the former is simple with low cost, but it is easy to generate noise, and its correction effect is worse than that of APFC. It needs to be pointed out that with the increasing popularity of PFC technology, usually "active PFC" is also referred to as PFC at present. Unless otherwise stated as "passive PFC" or "active PFC," "PFC" refers to active PFC (i.e., APFC).

For example, if SMPS is equipped with an APFC to make its power factor reach 0.98, a nearly perfect AC input waveform can be obtained as shown in Figure 5.2(a). Figure 5.2(b) is the calibrated harmonic analysis chart. Compared with Figure 5.1, the sinusoidal wave of AC input current is relatively ideal, and no higher harmonic components can be observed in the harmonic analysis chart.

5.2 Basic Principle of Passive PFC Circuit

5.2.1 Basic Circuit of Passive PFC

As mentioned earlier, the rectifier diode and the filter capacitor used in the input rectification filter circuit of SMPS belong to nonlinear elements. When sine wave voltage u is the input using AC, the waveform of AC input current i will be seriously distorted, becoming the spike pulse as shown in Figure 5.3. It is featured that the angle of flow of rectifier diode is significantly reduced. Although the mean square value of current is large, the average current is obviously reduced. For example, the conduction time t_C of ordinary silicon bridge rectifier is about 3 ms while the half cycle of 50 Hz AC is only 10 ms, so the angle of flow of the bridge rectifier is only

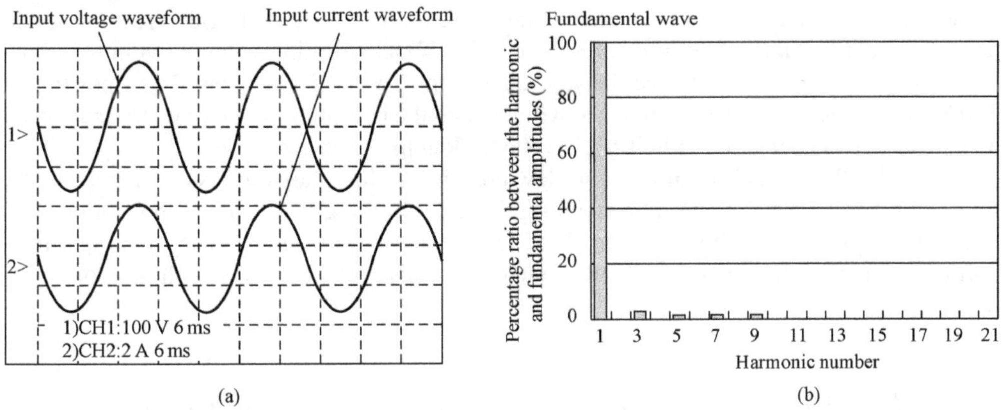

Figure 5.2 AC input waveform of SMPS flowing through active PFC. (a) Typical waveforms of input voltage and input current and (b) harmonic analysis chart

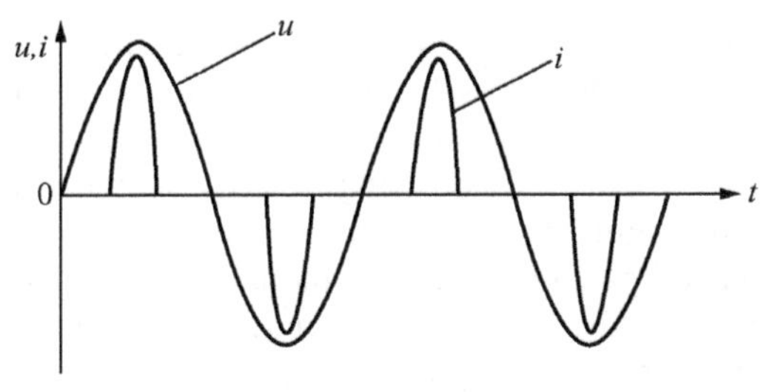

Figure 5.3 Waveform comparison between AC input voltage and AC input current

54°, and the corresponding conduction range is reduced from the ideal 0°–180° to 36°–90°. For this reason, PFC circuit can be added at the input stage of AC/DC converter.

See Figure 5.4 for a kind of PPFC input circuit used in 250 W PC power supply, wherein S refers to the selective switch of AC 110 and 220 V. When S is put in the position of 220 V, both

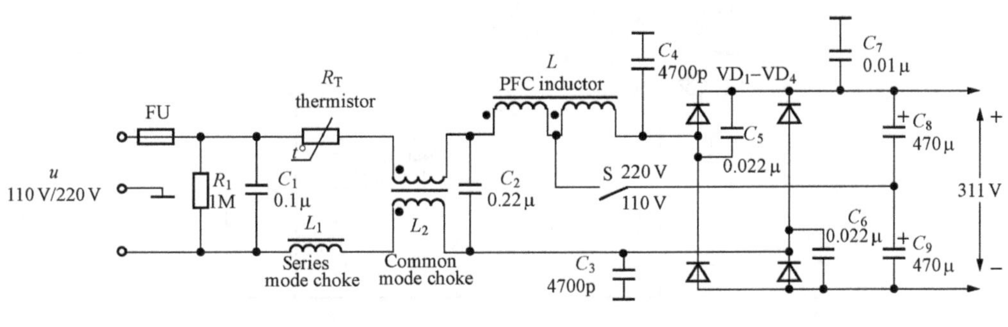

Figure 5.4 Passive PFC input circuit used in one kind of 250 W PC power supply

the left and right parts of PFC inductor winding are used with the full-bridge rectifier mode, and the output voltage of the bridge rectifier is 311 V DC pulsating voltage (without voltage regulation) and transmitted to the forward converter. When S is switched to the position of 110 V, only the left part of PFC inductor winding and the right part of bridge rectifier are used, and the circuit is changed into half-wave voltage-doubling rectifier mode.

Although PPFC circuit is simple with low cost, it has the following disadvantages: firstly, bulky PFC inductor has to be used, thus significantly limiting its practical application. Secondly, a changeover switch S has to be added for global use, which makes easy to bring serious harm to the power supply and the load resulted from misoperation (the switch in wrong position).

5.2.2 *PFC Working Principle Based on Passive Valley Fill Circuit*

"Valley fill circuit" is a new type of PPFC circuit. It is featured that the valley fill circuit behind the bridge rectifier is used to substantially increase the conduction time of rectifier and change the input current from spike pulse to the waveform close to sine waveform by filling valley points, thus increasing the power factor to about 0.9 or so. Compared with the traditional inductive PPFC circuit, it has such advantages as a simple circuit and extraordinarily effective in improving power factor as well as no need to use any bulky large inductor in the input circuit. It needs to be pointed out that as the valley fill circuit will increase power loss, it is only applicable to the low-cost LED driving power supply below 20 W. In addition, the valley fill circuit does have significant effect on increasing the power factor, but its THD is still comparatively large, failing to meet the harmonic requirements specified in European EN61000-3-2 and other standards. The harmonic frequency generated by it is far higher than 150 Hz, imposing no effect on the LED power supply, but it is easy to bring interference to other electronic devices.

See Figure 5.5 for the principle chart of two-stage valley fill circuit. The circuit uses two capacitors, so it is also known as two-capacitor valley fill circuit. The circuit all uses passive elements instead of any PFC inductor, and the bridge rectifier is constituted by VD_1–VD_4. The passive valley fill circuit only uses three diodes (VD_6–VD_8), two electrolytic capacitors (C_1 and C_2), and one resistor (R_1). VD_6–VD_8 apply 1N4007 silicon rectifying tubes. C_1 and

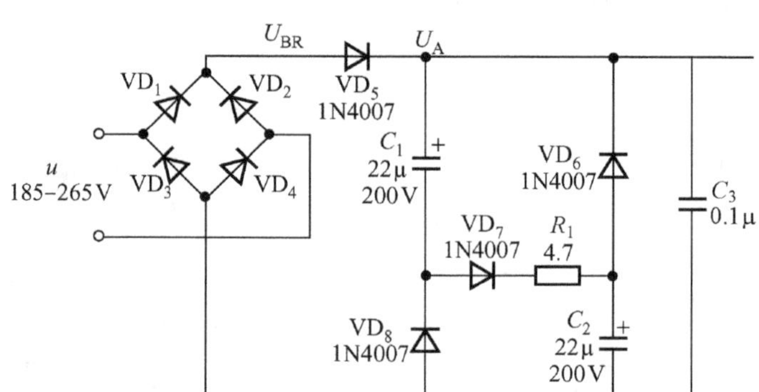

Figure 5.5 Principle chart of two-stage valley fill circuit

C_1 shall have equal capacity and use the electrolytic capacitor of $22\,\mu F/200\,V$. R_1 adopts the resistor of $4.7\,\Omega$ and $2\,W$, which can limit the impact current on C_1 and C_2 at power-on and suppress self-oscillation, but also consume some power. The valley fill circuit is featured that C_1 and C_2 are charged in series but discharged in parallel, and it can significantly increase the conduction time of rectifier by effectively extending the duration of AC input current. VD_5 refers to isolation diode, which can isolate the bridge rectifier from the valley fill circuit. C_3 is used to filter out high-frequency interference.

Assume that the effective value of AC input voltage is u, the peak voltage is U_P, the DC pulsating voltage of the bridge rectifier is U_{BR}, and the voltage at the right end of VD_5 is U_A (i.e., the total voltage of C_1 and C_2).

Stage 1: during the rising stage of AC positive half cycle, because of $U_{BR} > U_A$, VD_1, VD_4, VD_5, and VD_7 are conducted, so U_{BR} charges C_1 and C_2 along the series circuit of $C_1 \rightarrow VD_7 \rightarrow R_1 \rightarrow C_2$ and provides the load with current. Its charging time constant is very small and the charging is very fast.

Stage 2: when U_A reaches U_P, the total voltage on C_1 and C_2 is $U_A = U_P$. As C_1 and C_2 share equal capacity, their voltage drops are both $U_P/2$. At this time, VD_7 is conducted, while VD_6 and VD_8 are cut off owing to reverse bias.

Stage 3: when U_A begins to decline from U_P, VD_7 is cut off, stopping charging C_1 and C_2 immediately.

Stage 4: when U_A declines to $U_P/2$, VD_5 and VD_7 are cut off, and VD_6 and VD_8 begin to be conducted owing to forward bias, resulting in the charges in C_1 and C_2 being discharged by passing through the parallel circuit constituted by VD_6 and VD_8, respectively, to maintain the current in the load unchanged.

It is not difficult to find that power is supplied by the grid from stage 1 to stage 3, not only to the load but also charges C_1 and C_2 from stage 1 to stage 2. Power is provided to the load by the charges stored in C_1 and C_2 only in stage 4.

After entering the negative half cycle, C_1 and C_2 can still discharge to the load in parallel before VD_5 is conducted, so as to basically maintain the load current constant. For the situation after VD_2, VD_3, and VD_5 are conducted, readers can analyze by themselves by referring to the above.

In summary, two-stage valley fill circuit can be used to greatly extend the conduction time of the rectifier, thus expanding the conduction range in the positive half cycle to 30°–150° (30° exactly corresponds to $U_A = U_P \sin 30° = U_P/2$, and 150° corresponds to $U_A = U_P \sin 150° = U_P/2$). Similarly, the conduction range in the negative half cycle is expanded to 210°–330°. In this way, the waveform changes from narrow pulse to relatively close to sine wave. This equals to filling a large part of the valley point zone of the peak impulse current waveform, so it is called valley fill circuit. See Figure 5.6 for the timing waveform comparison among AC input voltage u, AC input current i, and point U_A.

The harmonic component of AC input current of SMPS is very large and the THD can be up to 100–150%. Assume the power factor as $\lambda = 0.55$ when valley fill circuit is not used, then the THD of AC input current THD = 151% as per formula (5.7). After the valley fill circuit is added, the power factor of SMPS can be increased to $\lambda = 0.92$–0.965 and correspondingly THD = 42.5–27.2%, which indicates that the THD is also improved to some extent.

Finally, a few points need to be pointed out:

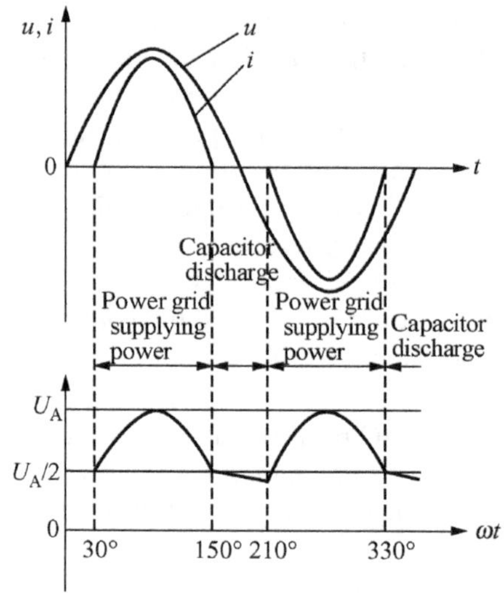

Figure 5.6 Timing waveforms of AC input voltage u, AC input current i, and point U_A

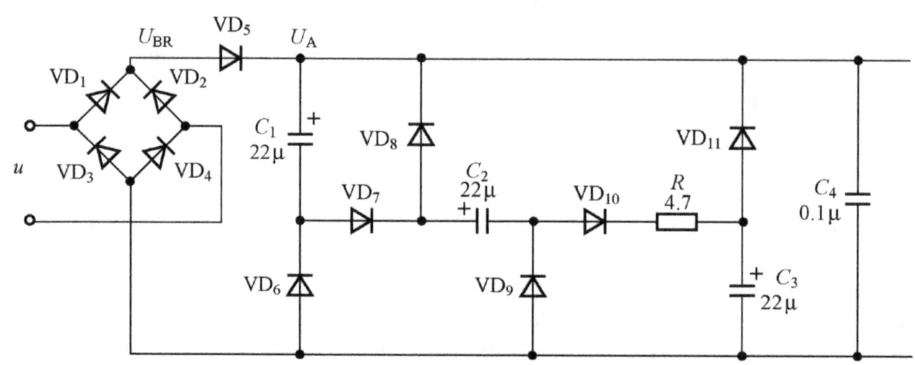

Figure 5.7 Three-stage valley fill circuit

1. Figure 5.5 shows two-stage valley fill circuit, which uses two capacitors. Three-stage valley fill circuit can also be used, which is also known as "three-capacitor valley fill circuit" and three capacitors are used. See Figure 5.7 for the three-stage valley fill circuit, wherein VD_5 refers to isolation diode. The circuit is featured that C_1, C_2, and C_3 are charged in series with a charging circuit of $U_{BR} \rightarrow VD_5 \rightarrow U_A \rightarrow C_1 \rightarrow VD_7 \rightarrow C_2 \rightarrow VD_{10} \rightarrow R \rightarrow C_3$. As C_1–C_3 share equal capacity, the voltage drops of all the three are $U_P/3$. Three parallel discharging circuits are, respectively, as follows: $C_1 \rightarrow$ the load of valley fill circuit $\rightarrow VD_6$; $C_2 \rightarrow VD_8 \rightarrow$ the load of valley fill circuit $\rightarrow VD_9$, and $C_3 \rightarrow VD_{11} \rightarrow$ the load of valley fill circuit. Three-stage valley fill circuit can significantly extend the conduction time of rectifier.
2. Applying two-stage valley fill circuit requires four diodes (including isolation diode, but excluding the four diodes in the bridge rectifier, the same below), and applying three-stage valley fill circuit totally requires seven diodes.

3. Despite the fact that the higher the order of valley fill circuit is, the more obvious the improvement effect on power factor will be, a more complicated circuit will necessarily require more elements. In addition, in the case of three-stage valley fill circuit, the parallel discharge voltage will be reduced to $U_P/3$, which shall be larger than the minimum mains voltage of single-chip SMPS, or the circuit cannot work normally.
4. The valley fill circuit can improve the utilization rate of the line current but will increase the ripple current in the load. However, in view that the brightness of LED light merely depends on the average current, the influence from the ripple current of LED driving power supply can generally be ignored.
5. The valley fill circuit does have significant effect on increasing the power factor, but its THD is still comparatively large, failing to meet the harmonic requirements specified by the international standard EN61000-3-2 for lighting equipment above 25 W. The harmonic frequency generated by it is far higher than 150 Hz, which will not affect the LED power supply, but it is easy to cause interference to other electronic devices.
6. As valley fill circuit will increase the loss of power supply, it is only applicable to the low-cost LED driving power supply below 20 W.

5.3 Design Examples of Passive PFC Circuit

See Figure 5.8 for the 9 W constant-current LED high-voltage driving power supply constituted by LNK306P product of LinkSwitch-TN series based on passive valley fill circuit. The circuit has the following main features:

1. The circuit is simple and low cost. Matching passive valley fill rectification filter circuit with LNK306P can realize PFC. When the AC input voltage range is 108–132 V, the power

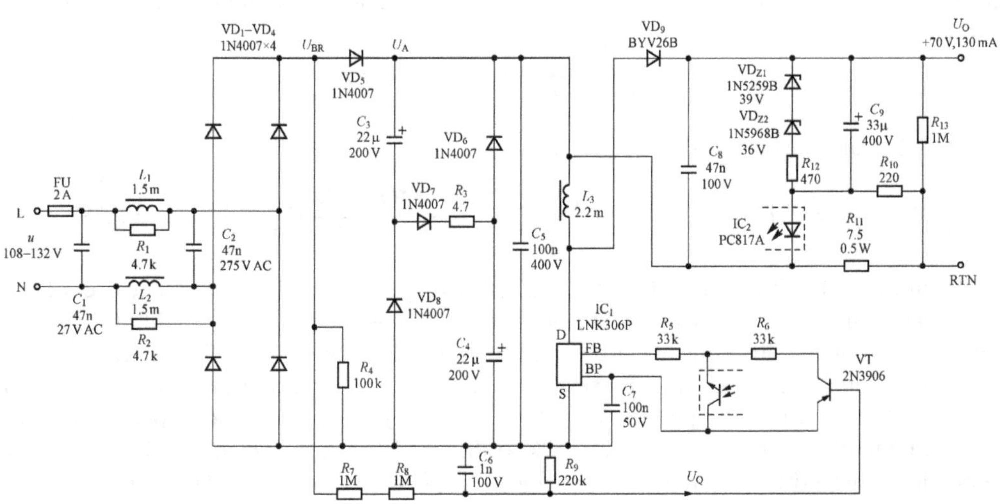

Figure 5.8 9 W constant-current LED high-voltage driving power supply based on passive valley fill circuit

factor can be increased to above 0.92, up to 0.965 at most (corresponding to the AC input voltage of 90 V). The power supply efficiency under full load is over 85%.
2. Double-type EMI (electromagnetic interference) filters (C_1, L_1, L_2, and C_2) are used at the AC input end, and R_1 and R_2 are bleeder resistors.
3. Transistor (VT, 2N3906) circuit is used to realize under-voltage protection (UVP).
4. The two-stage valley fill circuit is used as shown in Figure 5.5.

In Figure 5.8, L_3 refers to the power inductor of buck-boost converter. The output rectifier VD_9 uses 1 A/400 V super-fast recovery diode BYV26B, which is conducted during the shutdown of MOSFET in LNK306P, and transmits the energy stored in L_3 to the output filter capacitor C_8 and the load.

The output over-voltage protection (OVP) circuit is constituted by the voltage regulator tubes VD_{Z1} and VD_{Z2}, whose regulation values are 39 and 36 V (typical values), respectively. Once the load is disconnected, the output voltage can be clamped at about 75 V or so for protection, and the charges in C_8 can be discharged through R_{13}. C_9 refers to the noise reduction capacitor and is used to reduce output noise.

R_{11} refers to the current-detecting resistor, whose set rated output current is 130 mA. The voltage of R_{11} is applied to both ends of LED in optical coupler PC817A. The feedback signal can be connected to the feedback end FB of LNK306P through the phototransistor in PC817A and R_5. R_{12} refers to the set resistor of DC gain in feedback loop.

The UVP circuit is constituted by R_4, R_7–R_9, C_6, and the PNP type transistor VT. Whether the mains voltage is normal or not can be determined indirectly by sampling U_{BR}. Pulsating voltage can be obtained on R_9 after the voltage division of U_{BR} by R_7–R_9, and then the DC sampling voltage U_Q will be obtained after ripple is filtered in C_6, and it will be applied to the base electrode of VT. If the AC input voltage is too low, U_{BR} will be also reduced significantly. Once U_{BR} reaches the under-voltage threshold $U_{BR(UV)}$, U_Q will be reduced to below 5.1 V, making VT conducted and the collector current flow into the end FB to turn off MOSFET, thus realizing UVP. The under-voltage threshold is determined by the following formula:

$$\frac{R_9}{R_7 + R_8 + R_9} \cdot U_{BR\ (UV)} = 5.1\,\text{V} \tag{5.8}$$

It can be calculated from the formula that

$$U_{BR(UV)} = 5.1\,\text{V} \times \frac{R_7 + R_8 + R_9}{R_9} = 5.1\,\text{V} \times \frac{1\,\text{M}\Omega + 1\,\text{M}\Omega + 220\,\text{k}\Omega}{220\,\text{k}\Omega} = 51.5\,\text{V}$$

When MOSFET is conducted, VD_9 is cut off owing to reverse bias, and the current flows through the output filter capacitor C_4, the load, and the power inductor L_3, providing a constant current to the load with some electric energy stored in L_3. When MOSFET is turned off, the reverse potential generated in L_3 makes VD_9 conducted. The electric energy in L_3, through VD_9, continues to provide power to the load and charges C_8. LNK306P applies ON/OFF control method. When the current fed back to the end FB exceeds 49 μA, MOSFET will be prohibited from working within this switching cycle. The current at the end FB will be sampled again after entering the next switching cycle, and if the current is less than 49 μA, MOSFET will be allowed to work. The regulation for output voltage is realized by prohibiting (skipping over) or allowing a switching cycle to be finished.

Key points for design:

1. The reverse withstand voltage of the output rectifier VD_9 should be higher than the maximum DC output voltage with a 25% margin for the derating of VD_9.
2. In order to distribute voltage evenly, C_3 and C_4 shall share equal capacity and the total withstand voltage of them should be 400 V.
3. The power inductor L_3 uses TDK PC40EE EE19 ferrite core, winded 180 turns with ϕ0.29 mm enameled wire, and has an inductance of 2.2 mH (±12% error allowed).

5.4 Basic Principle of Active PFC Circuit

5.4.1 Basic Principle of Active PFC Boost Converter

See Figures 5.9 and 5.10, respectively, for the simplified circuit and the working principle of APFC boost converter. In the figures, BR is bridge rectifier, and DC pulsating voltage $U_I(t)$ will be obtained after AC sine wave voltage u is rectified.

The control loop is constituted by the PFC inductor (also known as boost inductor) L, the power switching tube V (MOSFET), the output rectifier VD, the input capacitor C_I (ceramic capacitor with small capacity), the output filter capacitor C_O (electrolytic capacitor with comparatively large capacity), PWM controller, DC sampling circuit, and AC sampling circuit. The PWM controller can simultaneously receive the signals from the two paths transmitted by DC sampling circuit and AC sampling circuit. Its output is PWM signal with fixed turn-on time (t_{ON}) and variable frequency, and the adjustment range of output duty ratio can be up to 0–100%.

Under the control of PWM signal, the converter forms two independent current loop when V is turned on, as shown in Figure 5.10(a). The first current loop is that $U_I(t)$ returns through L and V and stores electric energy in L, with voltage polarity being positive at left and negative at right, making VD turned off. The second current loop is that C_O discharges to the back-stage load to maintain the output voltage U_O unchanged. During the turn-on (t_{ON}) of V, inductive

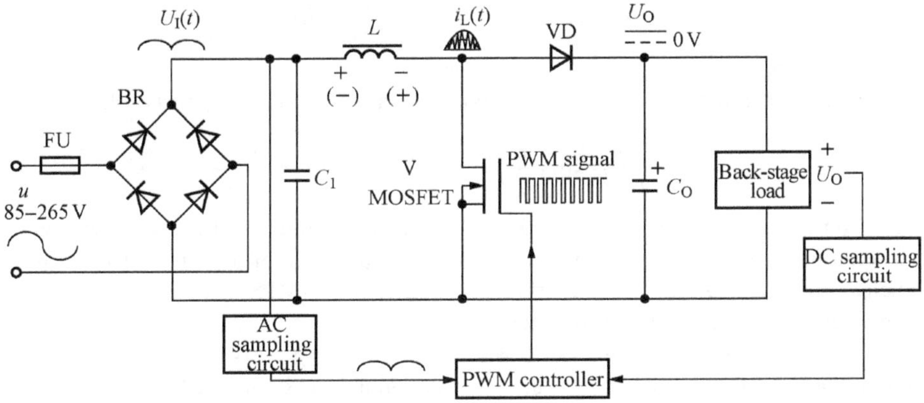

Figure 5.9 Simplified circuit of active PFC boost converter

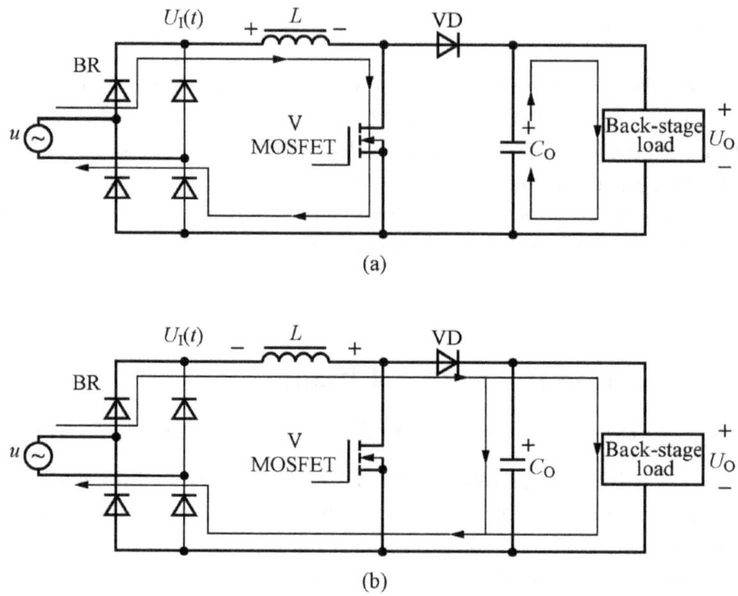

Figure 5.10 Working principle of active PFC boost converter. (a) During turn-on of V and (b) during turn-off of V

current $i_L(t)$ increases linearly from zero to the peak $i_{L(PK)}$. When V is turned off, the reverse potential generated in L is negative at left and positive at right, which is overlapped with $U_I(t)$ to realize the purpose of boosting voltage, thus making VD turned on as shown in Figure 5.10(b). The electric energy stored in L, after passing through VD, provides power to the load through one path and charges C_O through the other path. During the turn-off (t_{OFF}) of V, the current in L reduces from $i_{L(PK)}$ to zero.

Suppose that AC input voltage $u(t) = U_I \sin \omega t$. T is switching cycle, f switching frequency, and D the duty ratio. The formula of peak inductive current waveform can be expressed as follows according to Figure 5.10(a):

$$i_{L(PK)} = \frac{u(t)t_{ON}}{L} = \frac{U_I DT}{L} \cdot \sin \omega t = \frac{U_I D}{Lf} \cdot \sin \omega t \tag{5.9}$$

According to Formula (5.9), the maximum peak inductive current I_P is as follows:

$$I_P = \frac{U_I t_{ON}}{L} \cdot \sin\left(\frac{\pi}{2}\right) = \frac{U_I t_{ON}}{L} \cdot \sin 90° = \frac{U_I t_{ON}}{L} \tag{5.10}$$

According to Formulae (5.9) and (5.10), as long as the conduction time of power switching tube V remains fixed during one switching cycle, the waveform of peak inductive current will be the envelope line of $I_P \sin \omega t$, so that the input current of SMPS has the same phase as that of input voltage to increase the power factor.

See Figure 5.11 for the current waveform of APFC during work. It can be seen from the figure that the inductive current $i_{L(\omega t)}$ takes on triangular waves, strictly keeping synchronous with PWM signal. Each triangular wave includes the charging and discharging processes of L. Generally, the charging time of L is different from its discharging time. The figure shows

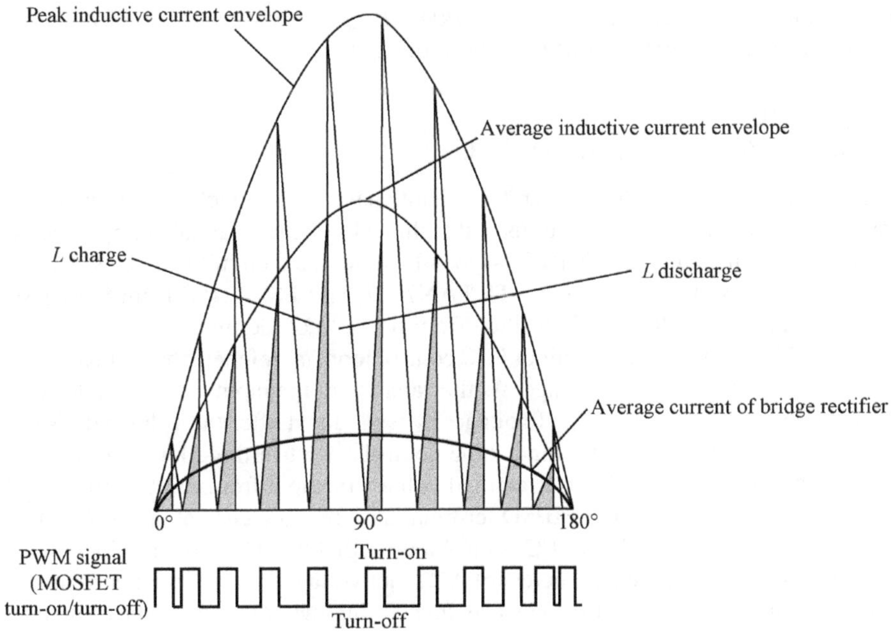

Figure 5.11 Current waveform of active PFC during work

the envelope curves of peak inductive current and average inductive current, respectively, at which time, the average of bridge rectifier is very close to a sine wave. When AC input voltage u changes in sine, the inductive current can automatically track the change of AC sine wave voltage and keep the same phase as that of it as long as the control circuit controls the turn-on and turn-off of V simply through PWM. As long as the switching frequency is high enough (above several tens of kilohertz), the AC input current can be very close to a sine wave. That is the basic principle of PFC boost converter. The circuit can simultaneously achieve the two functions of correcting the input power factor and increasing the output voltage.

It needs to be pointed out that:

1. The input capacitor C_I is dedicated to filtering out EMI with very small capacity, having no influence on the angle of flow of bridge rectifier. As VD has isolation effect, the current waveform of the bridge rectifier will not be affected by the back-stage filter capacitor C_O with comparatively large capacity.
2. The APFC circuit of critical conduction mode is simple with obvious effect on power factor compensation (the power factor can be greater than 0.95). The ripple of DC output voltage is very small, and it is not necessary to use the filter capacitor with very large capacity at the back stage.
3. Let power supply efficiency be η, the minimum AC input voltage be u_{min}, switching frequency be f, and the maximum output power of power supply be P_{OM}, and as $U_{I(min)} = \sqrt{2}\ u_{min}$, the formula for calculating PFC inductance is as follows (with a unit of H):

$$L \approx \frac{\eta U_{I(min)}^2}{4fP_{OM}} = \frac{\eta(\sqrt{2}u_{min})^2}{4fP_{OM}} = \frac{\eta u_{min}^2}{2fP_{OM}} \tag{5.11}$$

For example, when $u_{min} = 140\,V$, $f = 100\,kHz$, $P_{OM} = 150\,W$, and $\eta = 90\%$, $L \approx 588\,\mu H$ (in practice taken as $580\,\mu H$) according to Formula (5.11).

5.4.2 Basic Principle of Active PFC

In recent years, a variety of PFC integrated circuits have been launched into the market with a large quantity, whose control functions and technical indicators are also improved continuously. Typical products include NCP1650-type PFC integrated circuit from Onsemi, UC3852, UC3854, and UCC38050 produced by TI, FAN7527, and FSFR2100 from Fairchild of the United States, and TDA4862 and TDA4863 from Infineon of Germany.

See Figure 5.12 for the basic circuit of PFC control loop. u_L refers to the voltage after bridge rectification, also known as line voltage. As the capacity of the input filter capacitor C_1 is very small, u_L serves as full-wave rectified voltage. u_1 is obtained after the voltage division of u_L, which is applied to the AC IN terminal. The control loop has three kinds of input signals, which are, respectively, the full-wave rectified voltage u_1 input from AC IN terminal, the DC feedback voltage U_{FB} input from FB/SD terminal, and the line current signal i_{IN} input from I_S-terminal. The basic principle of PFC control loop is that the AC error amplifier controls the mains switch according to the parameters of AC input voltage and AC input current to change the input current into high-quality sine wave, thus letting the power factor be close to 1.

One input terminal of the reference multiplier is connected to u_1 and the other terminal is connected to the DC error voltage U_r. And then U_r is used to adjust u_1 so that the AC reference voltage (u_{REF}) output by the reference multiplier will be undistorted full-wave rectified waveform. The noninverting input terminal of AC error amplifier is connected to u_{REF}, and the high-frequency current signal i_2 output from the current-detecting amplifier is sent to the inverting input terminal according to the relationship expression $i_2 = ki_{IN}$. Meanwhile, u_1 outputs u_2 through average current compensation circuit, which is also added to the inverting input terminal of AC error amplifier. The AC error voltage output by the amplifier is u_r.

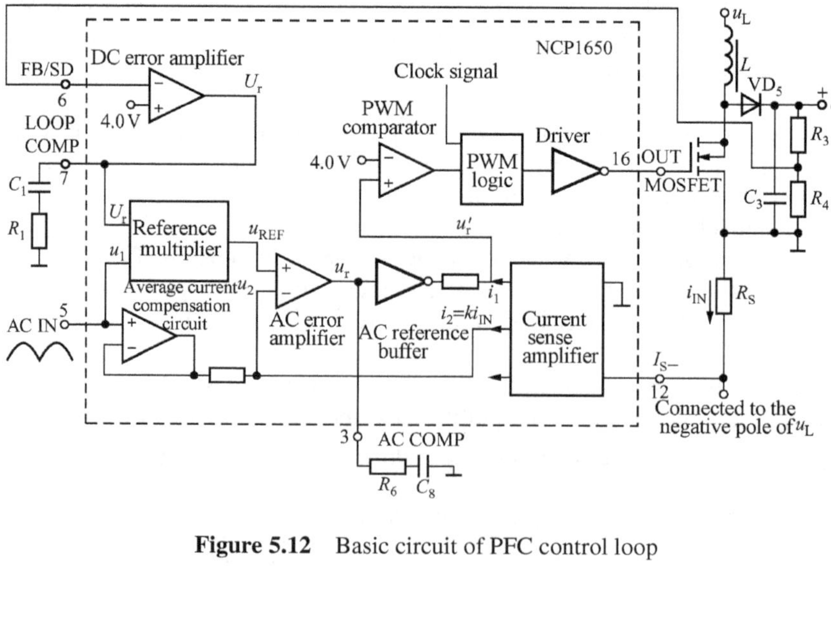

Figure 5.12 Basic circuit of PFC control loop

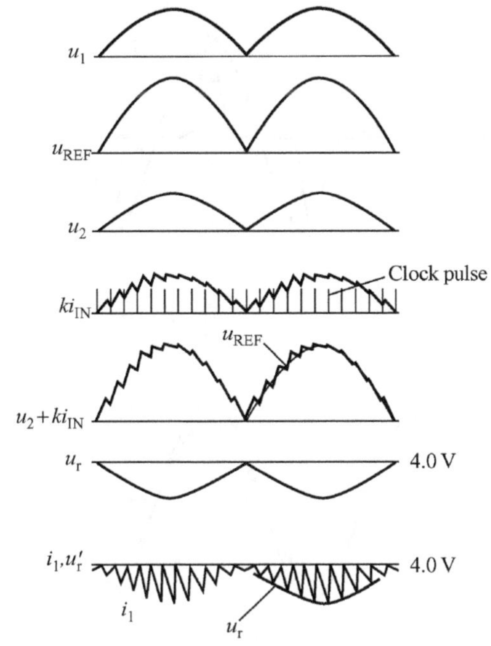

Figure 5.13 Work waveform in PFC circuit

See Figure 5.13 for the work waveform in PFC circuit, wherein 4.0 V is the internal reference voltage and u_r' the AC error voltage added with high-frequency current i_1, which is used as the input signal of PWM comparator. The changes of ki_{IN} in each clock cycle can be seen in the figure. In the waveform of u_r', the current signal i_1 can fully follow the change of u_r', thus realizing the purpose of PFC.

See Figure 5.14 for the comparison of the waveforms before and after PFC. Figure 5.14(a) shows the voltage and current waveforms of ordinary isolated power supply converter without PFC, whose current waveform is severely distorted. Figure 5.14(b) shows the u and i waveforms after PFC, whose current waveform is not distorted and has the same phase as that of the voltage waveform. The main reason for the current waveform distortion is that the AC current upon rectification fails to follow the change of voltage waveform. The function of PFC is to force line current to follow the change of line voltage waveform, and it can not only increase the power factor of AC power supply converter but also suppress harmonics to reduce the peak current and the RMS current and eliminate the phase drift of fundamental wave.

5.4.3 The Selection of Boost PFC Diode

See Figure 5.15 for the simplified circuit of boost PFC (the input rectifier bridge omitted). DC input voltage U_I is obtained after AC sine wave voltage is rectified. L is the PFC inductor, VD the PFC diode (also known as output rectifier), and C_O the output filter capacitor. The

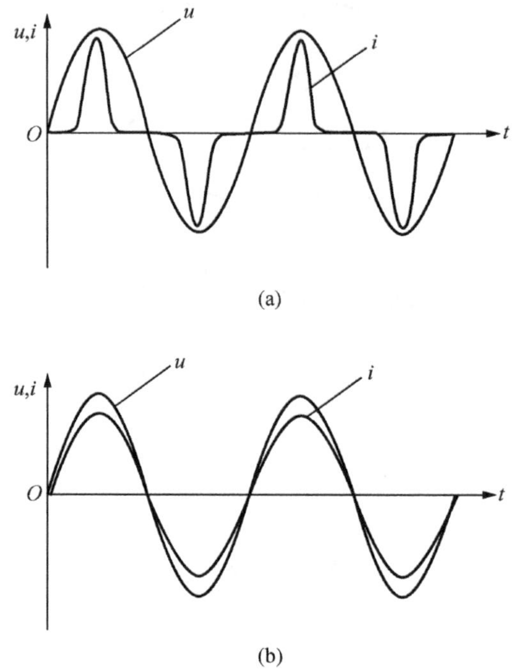

Figure 5.14 Comparison of waveforms before and after power factor correction. (a) Before correction and (b) after correction

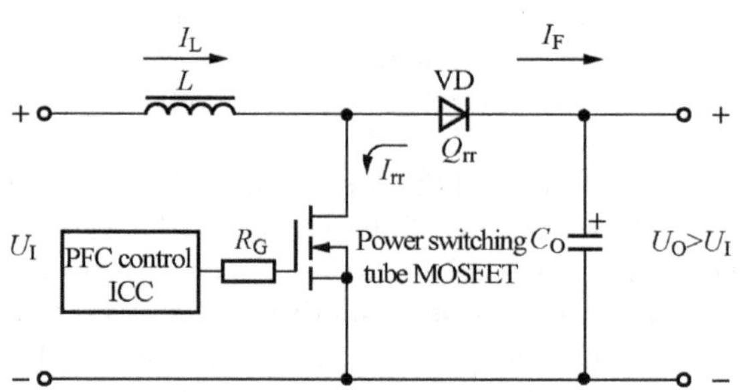

Figure 5.15 Simplified circuit of boost PFC

on/off state of power switching tube (MOSFET) is controlled by the PWM control IC. R_G is the gate limiting resistance. I_L and I_F are the current flowing through L and VD, respectively. Q_{rr} is the reverse recovery charge of VD, and I_{rr} is the reverse recovery current. At present, the PFC diodes in SMPS mostly use the super-fast recovery diode (SRD) that can withstand high voltage. However, the reverse recovery charge (Q_r) of SRD is comparatively large, which will not only form the reverse recovery current I_{rr} but also result in a nonideal reverse recovery waveform, thus inevitably reducing the conversion efficiency and causing EMI.

The following methods can be used to solve the above problems:

1. Use new SiC Schottky diodes, whose advantages include very fast switching without being affected by the chip junction temperature, especially the second- and third-generation SiC Schottky diodes, whose Q_{rr} is close to zero (typically 30 nC, wherein nC refers to nanocoulomb), the leakage current and the switching loss are very low, the forward current is 3–20 A, the forward conduction voltage drop is 1.7–2 V, and the reverse withstand voltage can be up to 600 V. The disadvantage of SiC Schottky diode is that its price is too high, difficult for mass popularization.
2. Use the high-quality and inexpensive Qspeed diode from Qspeed Semiconductor of the United States, whose current change rate (dI_F/dt) can be up to 1000 A/μs with very low reverse recovery charge and very flexible reverse recovery waveform, capable of improving the conversion efficiency of diode. As it does not generate high-frequency harmonics, which can not only simplify the design of EMI filter but also omit the buffer circuit, it is particularly suitable for the boost PFC circuit. Qspeed diode has the performance equivalent to SiC Schottky diode with a lower cost, so it can replace SiC Schottky diode. In addition, it can be used as the output rectifier in large-current high-voltage power supply in telecom and audio frequency to replace traditional Schottky diode. Qspeed diodes produced currently include three major series: X, Q, and H series. H series has the lowest switching loss and the highest efficiency. Among them, the working frequency ranges of X, Q, and H series are 50–80 kHz, 80–100 kHz, and 80–140 kHz, respectively. Reverse withstand voltage includes 300, 600 V, and other specifications.

See Figure 5.16 for the comparison of the reverse recovery current waveform between H series Qspeed diode and SRD. It can be seen from the figure that the reverse recovery time t_{rr1} of Qspeed diode is far less than that, t_{rr2}, of SRD, which indicates that Qspeed diode has "flexible" reverse recovery waveform compared with the "inflexible" reverse recovery waveform of SRD. See Table 5.2 for the main parameters of H series Qspeed diode, wherein $U_{RRM(MAX)}$ refers to the maximum reverse working voltage, $I_{F(AVG)}$ the average rectified current, and $U_{F(TYP)}$ the typical value of forward voltage drop.

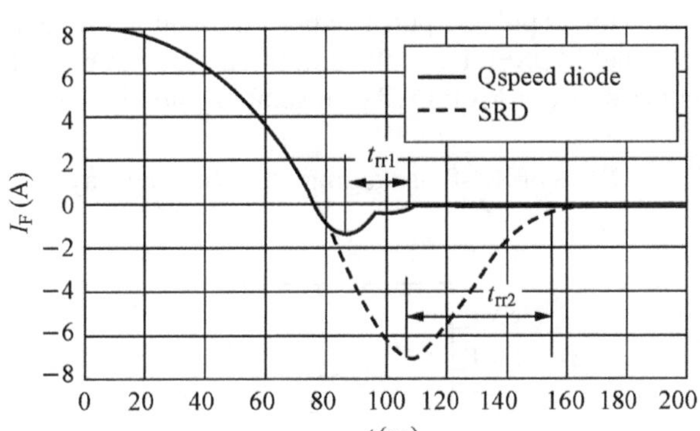

Figure 5.16 The comparison of the reverse recovery current waveform between H series Qspeed diode and SRD

Table 5.2 **Main** parameters of H series **Qspeed** diode

Model	$U_{RRM(MAX)}$ (V)	$I_{F(AVG)}$ (A) ($T_J = 150\,°C$)	$U_{F(TYP)}$ (V) ($T_J = 150\,°C$)	Q_{rr} (nC) ($T_J = 25\,°C$)	Q_{rr} (nC) ($T_J = 25\,°C$)
QH03TZ600	600	3	2.1	5.8	14.8
QH05TZ600	600	5	2.2	6.5	18.9
QH08TZ600	600	8	2.2	8.0	25.5
QH12TZ600	600	12	2.3	9.2	30

5.5 Design Examples of Active PFC Circuit

5.5.1 Working Principle of L6561 and L6562 Active PFC Converters

L6561 and L6562 are two types of PFC dedicated chip produced by ST with the same pin array and function. The only difference is that THD correction function is added at the back stage of internal multiplier of L6562. L6561 and L6562 are applicable to LED ballasts, SMPSs, and AC/DC power adapters with high power factor.

5.5.1.1 Main Features of L6561 and L6562

1. L6561 and L6562 can work in a wide input voltage range of AC 85–265 V, with the power factor and the power efficiency up to 0.99 and above 90%, respectively.
2. Start circuit and zero current-detecting circuit are configured internally to ensure that PFC converter works under the critical conduction mode.
3. They have over-voltage detection, under-voltage hysteresis lock out, current detection, disabling, and other functions. When the SMPS is not used temporarily, the output can be turned off externally to minimize the power consumption of the mains.
4. The precision of internal reference voltage is as high as 1%. The starting current is as low as 50 μA (typical value), and power supply current is only 4 mA (typical value).
5. Use high-performance totem pole output. The push–pull output stage is constituted by NPN transistor and N-channel MOSFET and the output current can be up to ±400 mA, capable of directly driving large-power MOSFET or insulated gate bipolar transistor (IGBT).

L6561 and L6562 use DIP-8 or SO-8 packaging with the pin array as shown in Figure 5.17.

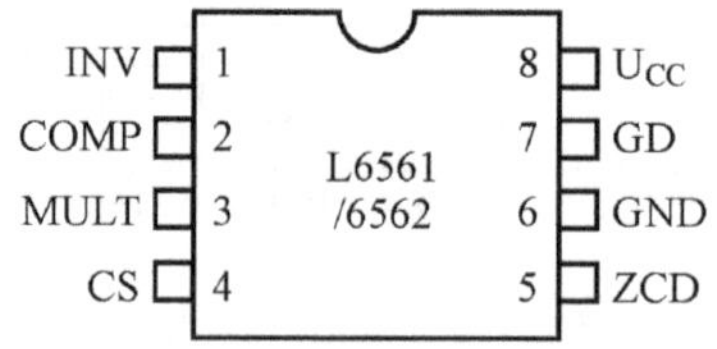

Figure 5.17 Pin array diagram of L6561 and L6562

The function of each pin is as follows: U_{CC} and GND terminals are connected with the mains voltage and common ground, respectively. INV refers to the inverting input terminal of error amplifier. The output voltage U_O is connected to INV terminal through resistance divider to provide feedback voltage. COMP refers to the output terminal of error amplifier connected with the external RC compensation network. MULT refers to the input terminal of internal multiplier. The output voltage of bridge rectifier is connected to terminal MULT through resistance divider to let the voltage signal of the terminal be proportional to the output voltage of bridge rectifier. CS refers to the peak current detection terminal of external power MOSFET. ZCD refers to the input terminal of zero current-detecting circuit. GD refers to the pin driving of the external MOSFET gate.

5.5.1.2 Working Principles of L6561 and L6562

See Figure 5.18 for the internal block diagrams of L6561 and L6562, which mainly include starting circuit, error amplifier, multiplier, voltage regulator (used to generate internal 7 V power supply), over-voltage-detecting circuit, reshaper, current comparator, the RS flip-flop, under-voltage comparator (UVLO), driving stage, the totem pole output stage (NPN transistor VT and N-channel MOSFET), zero current detection comparator, inhibit circuit, and gate circuit.

See Figure 5.19 for the basic working principles of L6561 and L6562. PFC control circuit uses double-loop feedback control method, on the one hand, controlling the input current

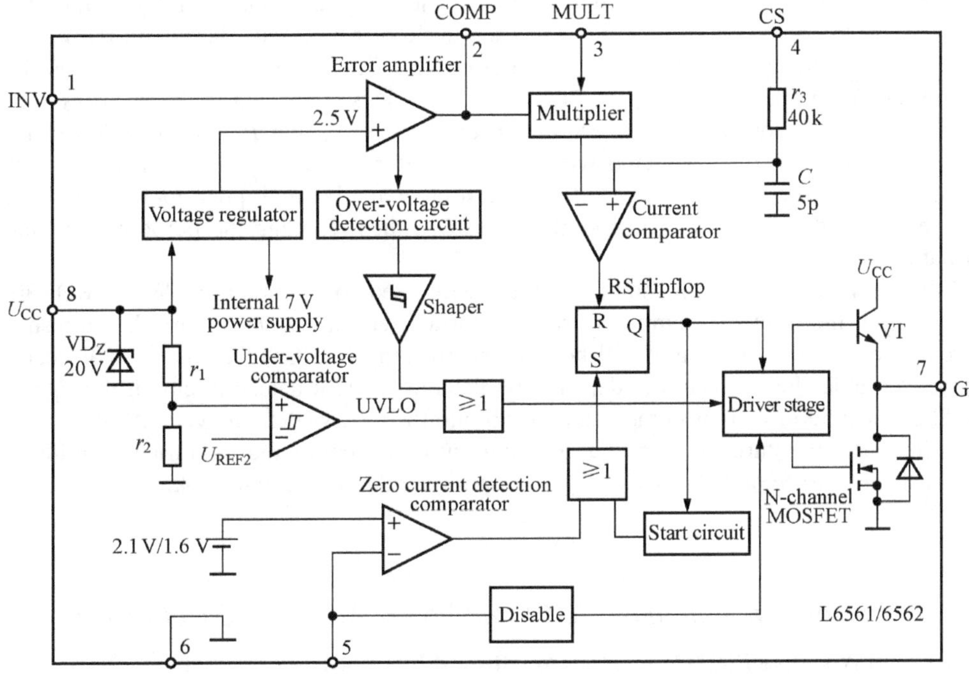

Figure 5.18 Internal block diagrams of L6561 and L6562

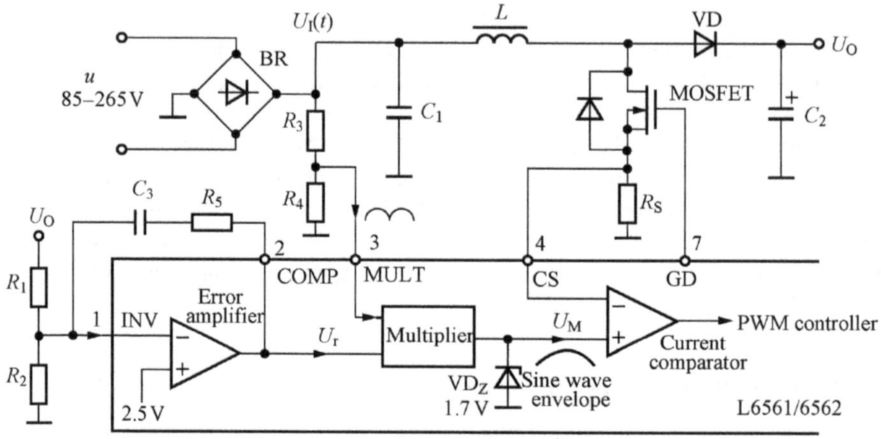

Figure 5.19 Basic working principles of L6561 and L6562

to be sine wave to obtain high power factor and, on the other hand, controlling the output voltage to be stable. The inner-loop feedback is to transmit the line voltage (half-cycle sine voltage) output from bridge rectification to the input terminal MULT of the multiplier after the voltage division of R_3 and R_4 as the sine voltage reference, letting the input current track the change of line voltage in real time. The output voltage U_M of multiplier, as the reference voltage of current comparator, is compared with the sampling voltage of MOSFET drain peak current to control the peak current in each cycle of MOSFET. The output terminal of current comparator is connected with PWM controller. The outer-loop feedback is used to control the DC output voltage U_O of PFC converter. U_O is connected to the inverting input terminal INV of error amplifier after voltage division of R_1 and R_2, and it will generate an error voltage U_r after compared with the 2.5 V reference voltage of noninverting input terminal. The error voltage then will be transmitted to the multiplier to let the output voltage U_M of the multiplier be proportional to U_O, thus realizing the purpose of voltage regulation.

Critical conduction mode is the PFC method with fixed conduction time. When MOSFET is turned on, the inductive current i_L rises along a fixed slope. Once it reaches threshold current, the current comparator will be inverted to turn off MOSFET, and i_L will decline along a variable slope. Once it is detected by the zero current-detecting circuit that $i_L = 0$, MOSFET will be turned on immediately to go into the next switching cycle. As there is no dead time after i_L returns to zero, the input current is still continuous and tracks the instantaneous change trace of AC input voltage u as per the rule of sine, thus making power factor approach 1.

5.5.2 Typical Application of L6561 and L6562 Active PFC Converters

5.5.2.1 80 W Active PFC Converter Constituted by L6561

See Figure 5.20 for the circuit of 80 W APFC converter constituted by L6561. The AC input voltage range of the power supply is 85–265 V, the output voltage $U_O = +400$ V, the output

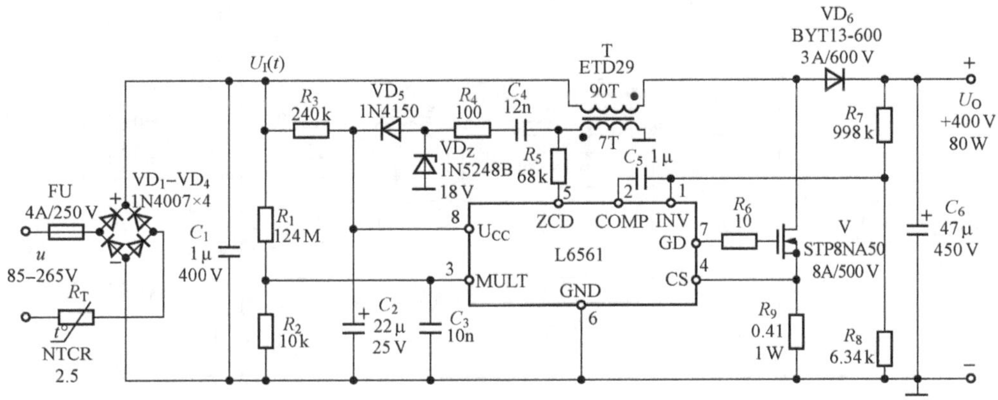

Figure 5.20 Circuit of 80 W active PFC converter constituted by L6561

current $I_O = 0.2\,\text{A}$, and the power factor can be up to 0.98 or more. At the beginning of turn-on, $U_I(t)$ provides a starting voltage to the chip through R_3. As the capacity of C_1 is very small, the influence on $U_I(t)$ waveform is negligible. R_T is the thermal resistor with negative temperature coefficient, which is used to limit current during power start-up.

The AC input voltage transmits a sine wave voltage signal to the terminal MULT of the internal multiplier of L6561 after bridge rectification and the voltage division of R_1 and R_2. The output voltage is transmitted to the inverting input terminal INV of the internal error amplifier after the voltage division of R_7 and R_8. The main winding of the transformer T is used as PFC inductor L. The output of the auxiliary winding is divided into two paths: one is connected to the input terminal ZCD of the zero current-detecting circuit through R_5 and the other provides mains voltage U_{CC} to L6561 through the DC blocking of C_4 and then the rectification and filtering by R_4, VD_5, and C_2. VD_5 uses 1N4150 small power switching diodes. VD_Z uses 1N5248B regulator tubes, capable of limiting U_{CC} under 18 V. C_3 refers to the bypass capacitor of terminal MULT, and C_5 refers to the compensation capacitor of error amplifier. The power switching tube uses STP8NA50 8A/500 V N-channel MOSFET. R_6 refers to the gate current-limiting resistance, R_9 refers to the current-detecting resistance, and the set peak drain current $I_{D(PK)} = 1.8\,\text{V}/R_9 = 2.2\,\text{A}$. The output rectifier VD_6 uses BYT13-600 3A/600 V FRD, with a reverse recovery time of 150 ns. The withstand voltage of output capacitor C_6 (47 μF) should be 450 V.

The transformer uses ETD29 ferrite core. The main winding has 90 turns of 10-strand ϕ0.20 mm enameled wires, and the auxiliary winding has 7 turns of ϕ0.15 mm enameled wires. The inductance of the main winding is 0.8 mH (±10% error allowed, similarly hereinafter).

5.5.2.2 80 W Active PFC Converter Constituted by L6562

See Figure 5.21 for the circuit of 80 W APFC converter constituted by L6562. The main differences between it and that of Figure 5.20 are as follows:

1. Four 1N4007 silicon rectifiers are replaced with DF06M bridge rectifier.
2. The compensation circuit of error amplifier is constituted by R_8 and C_6.

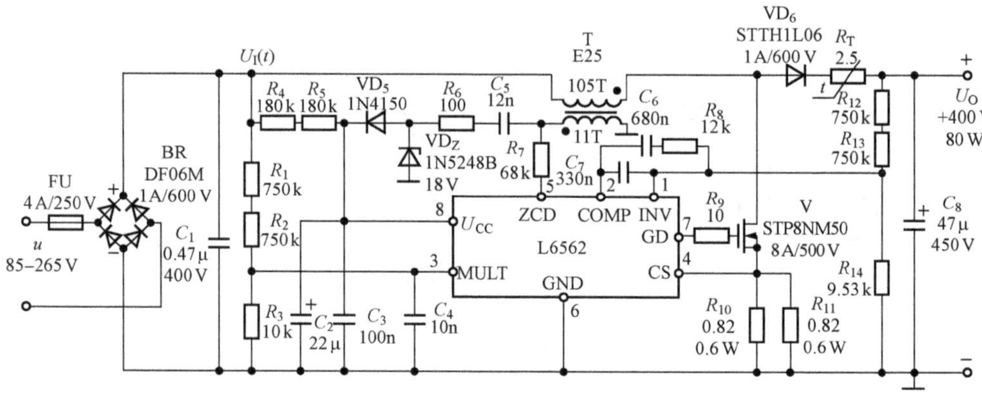

Figure 5.21 Circuit of 80 W active PFC converter constituted by L6562

3. The thermal resistor with R_T negative temperature coefficient is moved to the position behind the output rectifier VD_6.
4. To reduce the power consumption of each resistor, R_1 and R_2, R_4 and R_5, and R_{12} and R_{13} are, respectively, used to replace R_1, R_3, and R_7 in Figure 5.20.
5. The transformer uses E25 ferrite core. The main winding has 105 turns of 20-strand ϕ0.10 mm enameled wires, and the auxiliary winding has 11 turns of ϕ0.10 mm enameled wires. The inductance of the main winding is 0.7 mH.

5.6 Principle and Application of High-Power PFC

HiperPFS series is the boost single-stage high-power PFC controller IC with high-voltage power MOSFET newly launched by PI in November 2010, capable of constituting 85–1000 W SMPS with high power factor and high efficiency. HiperPFS is applicable to the preceding stage regulated power supply of distributed LED lighting with high power, PC power supply, high-power power supply adapter, large-screen LCD television, and industrial electrical equipment.

5.6.1 Performance Features of HiperPFS Series

1. HiperPFS series (hereinafter referred to as HiperPFS) integrates the PFC boost controller, the gate driver, and the high-voltage MOSFET of continuous conduction mode (CCM) in a ultra-thin chip to significantly simplify the design of the peripheral circuit of high-power PFC power supply and improve the efficiency at fractional loads. It uses the field effect tube (FET) inside the chip to detect the drain current of PFC converter, saving the external current-detecting resistor and reducing the power consumption of power supply.
2. HiperPFS includes 14 models. See Table 5.3 for the product classification and output power.
3. High power, high efficiency, and low power consumption. The AC input voltage range is 90–264 V. The maximum peak output power can be up to 1000 W. When the load is changed from 10% (fractional load) to 100% (full load), the power supply efficiency is

Table 5.3 Product classification and output power of HiperPFS

Model	90 V AC input			Model	180 V AC input	
	Minimum continuous output power (W)	Maximum continuous output power (W)	Peak output power (W)		Maximum continuous output power (W)	Peak output power (W)
PFS704EG	85	110	120	PFS723EG	255	280
PFS706EG	105	140	150	PFS724EG	315	350
PFS708EG	140	190	205	PFS725EG	435	480
PFS710EG	180	240	260	PFS726EG	540	600
PFS712EG	225	300	320	PFS727EG	675	750
PFS714EG	265	350	385	PFS728EG	810	900
PFS716EG	295	388	425	PFS729EG	900	1000

greater than 95%. When the AC input voltage is 230 V, the no-load power consumption is less than 130 mW, and the output voltage remains stable.

4. High power factor and minor harmonic distortion. The power factor can be up to 0.98.
5. HiperPFS is a boost PFC power supply with variable switching frequency. The variation range of switching frequency is 24–95 kHz (guaranteed value). It can adjust the frequency within the entire variation range of line voltage and cycle. It uses the spread spectrum technology to reduce the EMI related to the harmonics of fundamental switching frequency, not only simplifying the design of EMI filter but also reducing the inductance of boost PFC inductor.
6. Perfect protection functions, including UVP, OVP, over-temperature protection (OTP), voltage ramp/drop protection, over-current protection, and overload power-limiting function. The drain-source breakdown voltage of MOSFET is not less than 530 V.
7. High integration, capable of reducing the external dimension of power supply and increasing its power density.

5.6.2 *Working Principle of HiperPFS Series Products*

HiperPFS uses eSIP-7G packaging with the pin arrangement as shown in Figure 5.22(a) and (b). The function of each pin is as follows: V is the line voltage monitoring terminal connected with the peak detector internally. This terminal is connected to the output terminal of bridge rectifier through the high-resistance resistor *R*, which is used to detect the waveform of pulsating voltage (also known as line voltage) after bridge rectification. In order to filter out the switching noise of line voltage, a small-capacity ceramic capacitor *C* should be connected between the output terminal of rectifier and the signal ground to filter out the switching noise as a bypass. To reduce the power consumption of resistors, $R = 4\,\mathrm{M\Omega}$ and $C = 0.1\,\mu\mathrm{F}$ for PFS704–PFS708 and PFS710–PFS716, and $R = 9\,\mathrm{M\Omega}$ and $C = 0.047\,\mu\mathrm{F}$ for PFS723–PFS729. *R* should be the precision resistor with an error of 1%. This terminal has the input and output UVP. The terminal V shall be defined as the input under-voltage threshold when its input current is 27.50 μA and as the output under-voltage threshold when

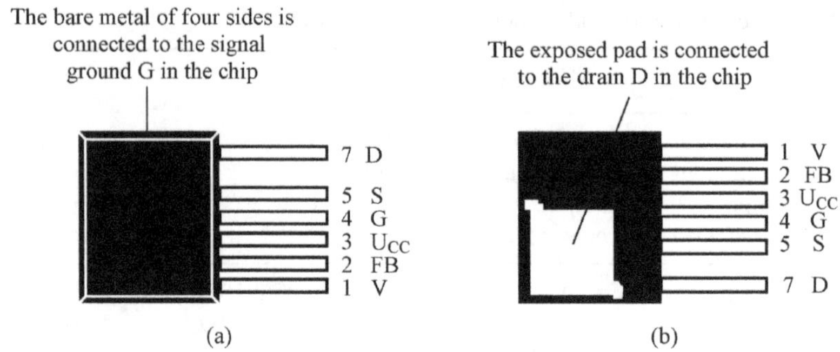

Figure 5.22 Pin array diagram of HiperPFS. (a) Front and (b) back

the input current is 24.50 μA. FB refers to the feedback terminal, connected to the feedback resistance network. Terminal FB can also be used to quickly detect the output over-voltage and under-voltage failures. It is required to connect a 10 nF bypass capacitor between terminal FB and the ground. U_{CC} refers to the bias power supply terminal, connected with the 12 V DC bias power supply (typical value, shall not exceed 13.4 V at most) and used to drive HiperPFS chip. S and D are connected to the source and the drain of internal MOSFET, respectively. G is the signal ground of feedback circuit and loop compensation circuit, which shall not be connected to the source.

See Figure 5.23 for the internal block diagram of HiperPFS, which mainly includes input line voltage interface, input voltage artificial circuit, the turn-off time setting circuit, transconductance error amplifier, fast over-voltage comparator, under-voltage comparator, low-pass filter, frequency smoothing circuit, comparator 1, comparator 2, the timer, the latch, the master

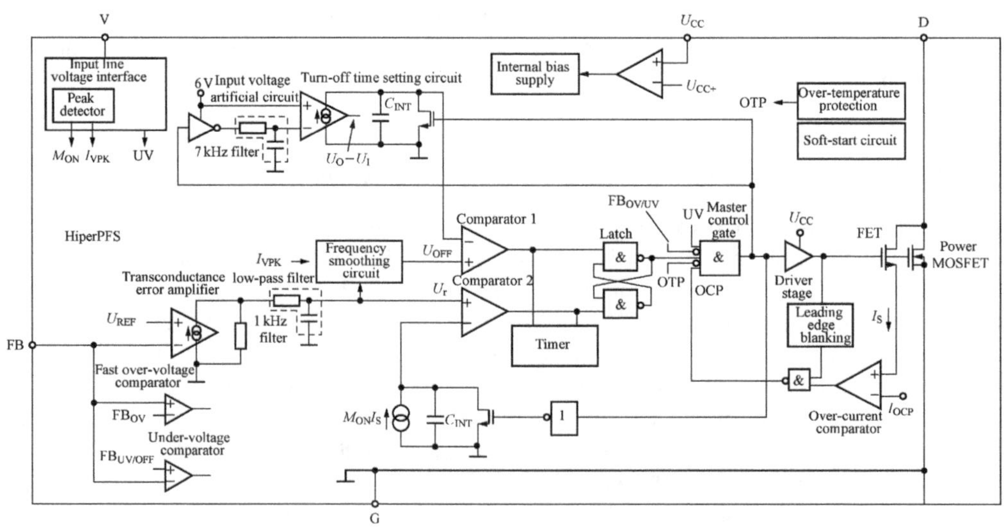

Figure 5.23 Internal block diagram of HiperPFS

control gate, the driving stage, the FET for detecting drain current, power MOSFET, internal bias supply, the OTP circuit, the over-current comparator, the leading edge blanking circuit, and the soft-start circuit.

The control algorithm employed by HiperPFS has the following features: its turn-off time T_{OFF} depends on the constant K_1 (with the unit of "volt second," i.e., V · s) and the turn-on time depends on the constant K_2 (with the unit of "ampere second," i.e. A · s). The algorithm allows the average input current to follow the change of input voltage within the time of T_{ON} and adjusts the output voltage and its waveform to reduce the harmonic component of input current, thus achieving the indicators of high power factor.

Let the input and output voltages of boost PFC be U_{I} and U_{O}, respectively, and the input current be I_{I}, then the turn-off time T_{OFF} and turn-on time T_{ON} are determined by the following formula:

$$T_{\mathrm{OFF}} = \frac{K_1}{U_{\mathrm{O}} - U_{\mathrm{I}}} \tag{5.12}$$

$$T_{\mathrm{ON}} = \frac{K_1}{U_{\mathrm{I}}} \tag{5.13}$$

$$I_1 = \frac{K_2}{T_{\mathrm{ON}}} \tag{5.14}$$

Put the formula (5.13) into the formula (5.14) to get

$$I_{\mathrm{I}} = U_{\mathrm{I}} \cdot \frac{K_2}{K_1} \tag{5.15}$$

By analyzing the formula (5.15), it is easy to see that the input current I_{I} and the input voltage U_{I} can be made directly proportional within a half cycle and follow the same waveform variation law just by controlling the constants K_1 and K_2, thus finally meeting the requirement of PFC.

The basic working principle of HiperPFS is that line voltage is first transmitted to the input line voltage interface through terminal V, and then outputs peak detection coefficient signal M_{ON}, peak current signal I_{VPK}, and input under-voltage signal U_{UV} through the peak detector. The output voltage of the input voltage artificial circuit and the turn-off time circuit is connected to the inverting input terminal of comparator 1. The error voltage U_{OFF} obtained after the frequency smoothing of peak current signal I_{VPK} is connected to the noninverting input terminal of comparator 1. The error voltage U_{r} used for adjusting output, which is obtained after the feedback voltage from the terminal FB passing through the transconductance error amplifier and the low-pass filter, is connected to the noninverting input terminal of comparator 2. The output from comparators 1 and 2 then passes through the latch and the driving stage to drive power MOSFET, so that the MOSFET can be turned on or off under the effect of feedback voltage to finally realize the purpose of voltage regulation.

HiperPFS is a boost PFC power supply with variable switching frequency. See Figure 5.24 for the relationship curve of the switching frequency and the conduction angle α of the line, corresponding to a certain AC input voltage. For example, when the AC input voltage u = 135 V and α changes from 0° to 90°, the switching frequency can be expanded from 23 to 94 kHz. Conversely, when α changes from 90° to 180°, the switching frequency is reduced from 94 to 23 kHz.

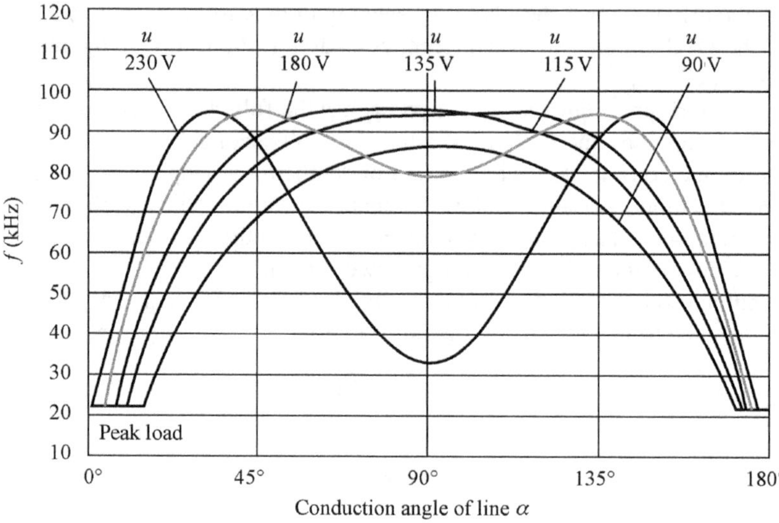

Figure 5.24 Relationship curve of the switching frequency and the conduction angle α of the line

5.6.3 347 W High-Efficiency Large-Power Boost PFC Power Supply Constituted by PFS714EG

See Figure 5.25 for the circuit of 347 W high-efficiency large-power boost PFC power supply constituted by PFS714EG, wherein the continuous output power is 347 W, the DC

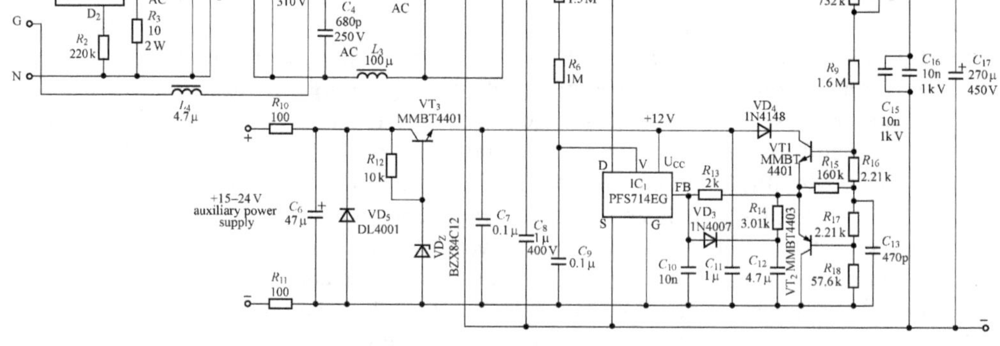

Figure 5.25 Circuit of 347 W high-efficiency large-power boost PFC power supply constituted by PFS714EG

output voltage is 380 V, and the rated output current is 0.913 A; moreover, the indicators of high power factor and high efficiency can be achieved from fractional load to full load.

The output terminal of PFC power supply is connected with buck LED driver (DC/DC converter), which can be the single-chip two-transistor forward converter and the flyback standby converter of HiperTFS series (TFS757–TFS764HG) newly launched by PI. The power supply uses two ICs: IC_1 (PFS714EG) and IC_2 (CAP006DG). The following introduces the working principles of each element circuit.

5.6.3.1 Input Protection Circuit and EMI Filter

Input protection circuit is constituted by the fuse (FU), the voltage-sensitive resistor (R_V), the X-capacitor zero loss discharger (CAP006DG), and negative temperature coefficient thermal resistor (R_T). When the power supply starts normal working, R_T is shorted out through the contact S of the relay to reduce the power loss to zero. EMI filter includes X capacitors C_1, C_2, and C_5 used to suppress the series-mode interference, Y capacitors C_3 and C_4 used to suppress the common-mode interference, the common-mode choke L_1, and the series-mode chokes L_2, L_3, and L_4, wherein L_4 uses 4.7 μH magnetic beads. When the power supply works normally, CAP006DG is kept disconnected to cut off the current in the bleeder resistors R_1 and R_2, reducing the power loss to close to zero. When AC power supply is turned off, the bleeder resistors are immediately turned on by CAP006DG to quickly discharge the charges stored in X capacitor to prevent the operator from suffering electric shock. BR refers to GBU806 8A/600 V bridge rectifier.

The output voltage of the bridge rectifier is transmitted to the line voltage monitoring terminal V of PFS714EG through R_4–R_6, which are all precision resistors with an error of $\pm 1\%$, and let $R_4 + R_4 + R_6 = 4\,\text{M}\Omega$. C_9 refers to the noise-canceling capacitor.

5.6.3.2 Boost PFC Converter

Boost PFC converter is constituted by the PFC inductor L_5, the rectifier VD_2, and PFS714EG. It is equivalent to a boost converter, on the one hand, correcting the power factor of input current from the power supply and, on the other hand, adjusting the DC output voltage. The rectifier VD_2 uses STTH8S06D 8A/600 V SRD, whose reverse recovery time is only 12 ns. VD_1 is used to make the inductor L_5 short out during power supply start-up to prevent the circuit from generating oscillation in the process of establishing output voltage. In addition, VD_1 charges the output capacitor C_{17}. VD_1 uses 1N5408 3 A/1000 V silicon rectifier. The thermal R_T resistor with negative temperature coefficient can limit the surge current during the start-up to prevent magnetic saturation of L_5. The capacitors C_{15} and C_{16} can be used to not only reduce the EMI but also prevent the MOSFET inside PFS714EG from generating voltage overshoot at the drain and the source at the turn-on and turn-off moments.

5.6.3.3 Auxiliary Power Supply

Auxiliary power supply is constituted by the resistors R_{10}–R_{12}, the capacitor C_6, the transistor VT_3, the diode VD_5, and 12-V voltage regulator VD_Z, which belongs to the series adjusting linear voltage regulator and is used to provide PFS714EG with +12 V mains voltage (with the maximum supply current of 3.5 mA) to keep it working normally. +12 V voltage can be obtained after +15–24V DC input voltage passing through auxiliary power supply. When output voltage exceeds +12 V, VD_Z will suffer reverse breakdown, thus performing clamping action. VD_5 is used to prevent auxiliary input voltage from being connected with reverse polarity to protect PFS714EG.

5.6.3.4 Feedback Circuit

During normal operation, both VT_1 and VT_2 are turned off. The output voltage U_O provides PFS714EG with feedback voltage after the voltage division of the resistance divider network R_7–R_9 and R_{16} (upper voltage-dividing resistor) as well as R_{17} and R_{18} (lower voltage-dividing resistors) and then passing through R_{15} and R_{13}. The feedback voltage in the case of rated output voltage is $U_{FB} = 6\text{V}$. R_7–R_9 and R_{16}–R_{18} are precision resistors with an error of ±1%. Complementary transistors VT_1 (NPN) and VT_2 (PNP) are biased by R_{16} and R_{17}, respectively, which are used for detecting the transient conditions of output voltage and providing the feedback terminal with information to enhance the rapid response capability of PFC and further improve the load transient response of SMPS. Use R_{13}–R_{15}, C_{12}, and C_{13} to correct the loop response of feedback network. R_{14} and C_{12} are the compensation components of feedback loop. VD_3 is used to prevent the feedback circuit from being loaded when U_{CC} is disconnected and it must be ordinary silicon rectifier (e.g., 1N4007) rather than SRD or FRD. C_{13} is the soft-start capacitor, which can reduce the overshoot of output voltage when power supply is started.

+12 V auxiliary voltage is connected to the emitter of VT_1 through VD_4. In the case of over-voltage output, VT_1 will be conducted immediately owing to the rise of base potential, thus making the feedback current I_{FB} increase, and then adjust the duty ratio D through PFS714EG, making $D \downarrow \rightarrow U_O \downarrow$ for protection. In the case of under-voltage failure, VT_2 will be conducted immediately, making the feedback current I_{FB} decrease; and by adjusting the duty ratio through PFS714EG, hence $D \downarrow \rightarrow U_O \downarrow$ for protection.

See Figure 5.26 for the relationship curve of the power supply efficiency η and output power P_O from actual measurement. See Figure 5.27 for the relationship curve of power factor λ and output power. It can be seen from the figures that when $u = 90\text{V}$ (AC) and $P_O \geq 150\text{W}$, $\lambda \approx 1.00$.

5.6.4 *Key Points of Circuit Design*

The following text introduces the key points of HiperPFS design with the circuit shown in Figure 5.25 as an example.

5.6.4.1 The Selection of HiperPFS Chip

1. Select the most appropriate chip from HiperPFS series products according to such factors as the required maximum output power, PFC efficiency, the total efficiency of power supply (including the poststage DC/DC converter), heat dissipation conditions, and cost targets.

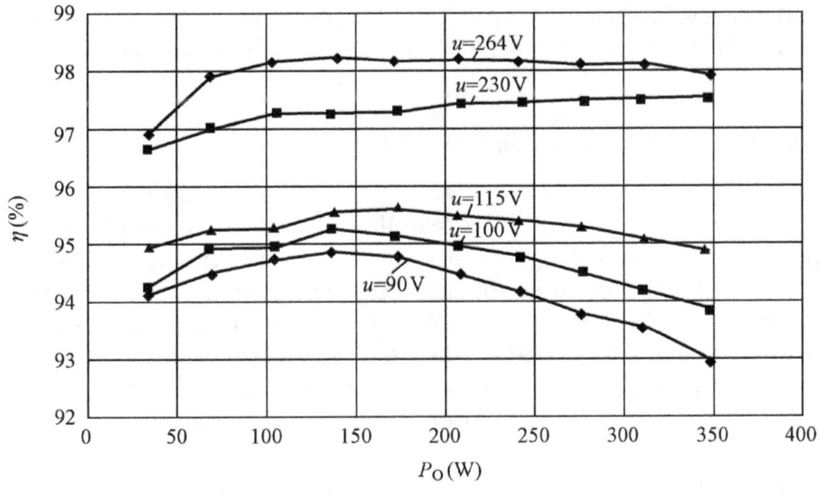

Figure 5.26 Relationship curve of efficiency and output power

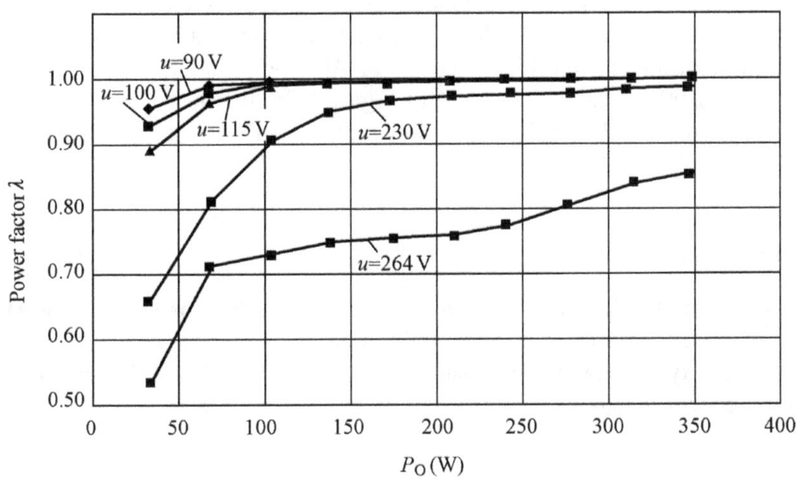

Figure 5.27 Relationship curve of power factor and output power

2. PFS704EG–PFS716EG can be selected for the AC input voltage with a wide range of 90–264 V. Under the minimum working voltage, the total efficiency should not be less than 93%. The PFC output rectifier should use SRD, and appropriate heat sink should be installed to let the junction temperature be no more than 100 °C. The rated output voltage is +380–385V.
3. PFS723EG–PFS729EG can be selected for AC 230 V (variation of ±15% allowed) fixed input voltage. Under the minimum working voltage, the total efficiency should not be less than 96%. Other technical requirements are the same as defined above.

4. The rated output voltage cannot exceed +390 V to prevent the source-drain voltage in MOSFET from being too high during the transient response of line and load.

5.6.4.2 Input Protection Circuit and EMI Filter

1. At the beginning of start-up, as the output filter capacitor is charged to the peak supply voltage, a large current will be generated at the AC input terminal, but the said current will be subjected to the impedance of the negative temperature coefficient thermal resistor R_T and the inductor in EMI filter as well as the restriction from the forward conduction voltage drop of input bridge rectifier. The rated current of fuse should be greater than the input current when PFC is turned off owing to a too low input voltage.
2. Usually, a 320-V voltage-sensitive resistor R_V will be used to suppress the surge voltage of the grid.
3. When the capacities of X capacitors C_1 (0.68 μF) and C_2 (1 μF) in EMI filter are close to or equal to 1 μF, a bleeder resistor needs to be connected in parallel. It is recommended to use CAPZero IC to eliminate the power loss caused by the bleeder resistor.
4. The output capacitor C_8 of input bridge rectifier should use the high-frequency filter capacitor with low equivalent series resistance (ESR). C_8 can reduce the ripple of input current and simplify the design of EMI filter. Capacity can usually be selected as per the proportional coefficient "0.33 μF/100 W." For AC 230 V input, the proportional coefficient "0.15 μF/100 W" can be selected. In Figure 5.25, the output power is 347 W, $C_8 = (0.33\ \mu F/100 W) \times 347 W = 1.145\,\mu F$, and actually 1 μF nominal capacity with a withstand voltage of 400 V should be taken.

5.6.4.3 PFC Inductor

PFC inductor should use 77324A7 magnetic core winded 108 turns with 125/40AWG multistrand wire, whose inductance is 1.38 mH (with a permissible error of ±8%). The maximum magnetic flux density of magnetic core should be less than 0.3 T and the peak magnetic flux density should be less than 0.42 T.

5.6.4.4 Output Rectifier

For the PFC power supply with 380 V rated output voltage, the output rectifier VD_2 should use the SRD with a reverse withstand voltage of 600 V. The rated forward conduction current I_F of output rectifier can be selected as per the proportional coefficient "(1.2–1.5 A)/100 W output power." VD_2 should actually use 8 A/600 V STTH8S06D SRD to allow a sufficient allowance.

5.6.4.5 The Selection of Output Filter Capacitor

In constituting the preceding-stage regulated power supply of DC/DC LED driver, the withstand voltage of output capacitor C_{17} must be no less than 450 V. The capacity of the capacitor depends on the required size of output ripple and the holding time of power supply. Let the rated output power be P_O, the supply holding time of C_{17} be t_{HOLD}, the rated and the minimum output voltage be U_O and $U_{O(min)}$, respectively, the maximum output current be $I_{O\ (max)}$, the switching frequency be f, the peak output ripple voltage be ΔU_O, and the power supply efficiency be η, and the formula for capacity fitting the supply holding time is as follows:

$$C_{17} = \frac{2P_O t_{HOLD}}{U_O^2 - U_{O(min)}^2} \tag{5.16}$$

The formula meeting the output ripple requirement is as follows:

$$C_{17} = \frac{I_{O\,(max)}}{2\pi f \Delta U_O \eta} \tag{5.17}$$

The requirements for supply holding time and output ripple can be met just by selecting a comparatively large one from the values calculated as per Formulas (5.16) and (5.17) as the capacity of C_{17}. In view of the fact that the capacity of electrolytic capacitor will be reduced with the increase of working time and the capacity has relatively large tolerance, actual capacity should be the value based on calculation that should be added with a proper capacity value.

5.7 Measures to Suppress PFC Electromagnetic Interference

5.7.1 Using EMI Filter to Suppress PFC Electromagnetic Interference

The interference signal of PFC is very complex, including the spark discharge interference generated by the electromagnetic relay, the self-oscillation (including ringing voltage), the spike interference, and the noise voltage (e.g., switching, capacitor, high-frequency transformer, and audio noise).

EMI filter is referred to as EMI filter, which plays an important role in suppressing the EMI of PFC. For example, in flyback PFC, pulsating voltage will be generated by frequent turn-on and turn-off of MOSFET, and each pulsating voltage has harmonics in its fundamental frequency. These harmonics will form spectral interference in a high-frequency range. In addition, most electronic components will radiate a certain type of EMI during work. Although some interference signal is very weak, it shall not be ignored when it starts to form interference toward the electronic equipment using the same grid or other electronic equipment in near areas. The said interference can be so weak that it causes hop count phenomenon on digital instruments or snow on TV screen and so severe that it seriously interferes with the flight control electronic equipment in the airplane.

EMI can be divided into two categories: conducted and radiated interferences. Conducted interference is transmitted through AC power supply. The frequency range of conducted interference is defined as 450 kHz–30 MHz by the European FCC standard. Radiated interference is the radio frequency interference transmitted through air with a frequency range of 30–960 MHz. It is also suggested by the FCC standard that the voltage level of EMI within the specified frequency range shall not exceed 48 $dB_{\mu V}$.

To facilitate comparing the role of EMI filter in PFC, the following only considers conduction noise, namely, the upper limiting frequency of EMI below 30 MHz. See Figure 5.28 for the principle of simple EMI filter to be used. L_1 and L_2 are common-mode chokes used to

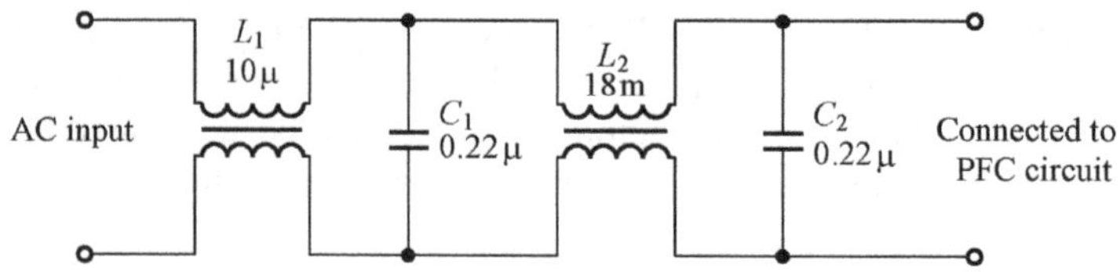

Figure 5.28 Principle of simple EMI filter

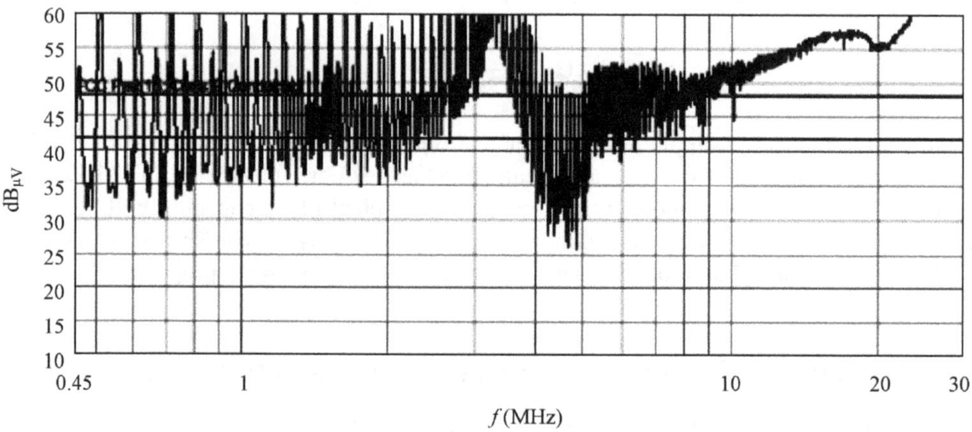

Figure 5.29 EMI waveform of PFC without EMI filter

suppress common-mode interference. C_1 and C_2 refer to wire-to-wire bypass capacitors used to filter out series-mode interference.

First, the actually measured EMI waveform of a PFC without EMI filter is shown in Figure 5.29. It can be seen from the figure that its EMI level is much higher than $60\,dB_{\mu V}$, far more than the upper limit of $48\,dB_{\mu V}$. Then, insert the EMI filter as shown in Figure 5.28 between the inlet terminal and PFC, and the actually measured EMI waveform is shown in Figure 5.30. Obviously, the EMI level is $45\,dB_{\mu V}$ only, lower than the upper limit of $48\,dB_{\mu V}$.

5.7.2 Other Measures to Reduce PFC EMI

To further reduce EMI, it is necessary to improve the circuit to meet the requirements for SMPS.

5.7.2.1 Selection of PFC Topology

Single-stage topology with high power factor is recommended to be used in the design of small- and medium-power PFC to achieve high power factor, high power supply efficiency, and a compact size. Traditional two-stage topology (PFC boost + DC/DC converter) is difficult to meet the requirements.

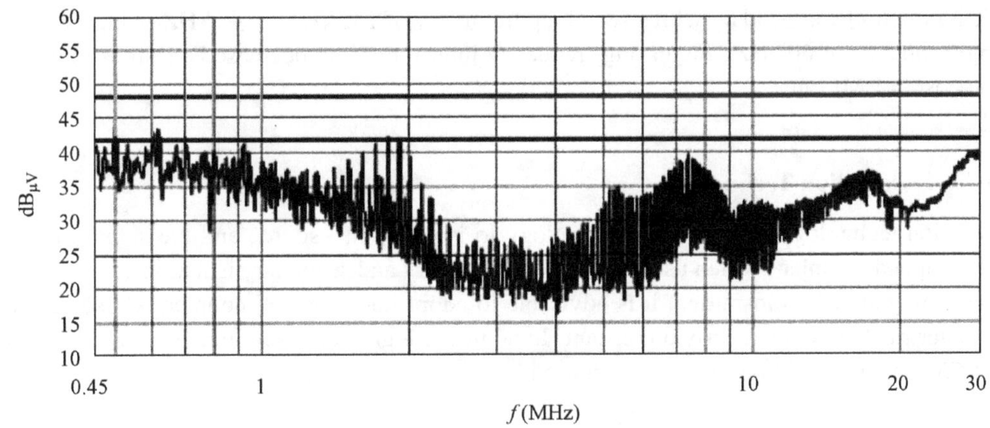

Figure 5.30 EMI waveform of PFC with EMI filter

Single-stage topology does not need PFC boost converter, reducing the quantity of components and overall system cost. Applying single-stage topology will have influence on the system, such as no primary high-voltage energy storage and short holding time of output voltage. In addition, the output ripple is relatively high, so more low-voltage output capacitors shall be used to make its response to dynamic load relatively slow.

5.7.2.2 Cutting off the Route of Interference Transmission

1. Increase the distance between interference sources (e.g., electric motors and relays) and PFC, isolating them with ground wire or adding a shield cover to PFC.
2. Reasonably divide the circuit board into several zones, properly arranging the zones of strong, weak, digital, and analog signal circuits.
3. The ground wire of power devices shall be grounded separately to reduce mutual interference. Power devices should be positioned as close as possible to the edge of printed circuit board.

5.7.2.3 PFC Frequency Jitter Technology

Frequency jitter refers to that switching frequency is controlled to change at a low frequency rate (e.g., 250 times/s) and by doing so, switching frequency f can be limited to jitter within a very narrow wave band. As switching frequency continuously changes in the vicinity of the rated value f, irrelevant to the higher harmonic interference of fixed frequency f, frequency jitter signal can be used to reduce the conduction noise of PFC.

In designing high-power PFC, the SMPS IC with a frequency jitter function can be selected, for example, for TOPSwitch-HX series single-chip SMPS IC with a maximum output power of 333 W, its frequency jitter range is ±5 kHz (with a switching frequency of 132 kHz) or ±2 kHZ (with a switching frequency of 66 kHz). By comparison, the frequency jitter range of

TOPSwitch-GX is ±4 kHz (with a switching frequency of 132 kHz) or ±2 kHZ (with a switching frequency of 66 kHz). Increasing frequency jitter range can decrease EMI, thus reducing the cost of EMI filter.

5.7.2.4 Isolation Technology

Isolation technology refers to the technology to isolate noise source and the signal line. In PFC, optical coupler is often used to realize the primary and secondary isolations. In isolating the analog signals transmitted, it is advisable to adopt linear optical coupler, whose current transfer ratio (CRT) is nearly a constant. In addition, high-frequency transformer, relay, and wiring isolation technologies are often used.

5.7.2.5 The Correct Choice for Ground Point

Power factor corrector has five types of ground wires: analog, digital, power, AC (connected to the ground G), and shield ground. In designing the circuitry, it should be determined whether to adopt the floating ground or the grounding connection, which kind of ground wire is necessary, and whether to use the single-point or multipoint groundings depending on technical conditions and practical situations.

5.7.2.6 Component Selection

The EMI filter with better performance can be used to further reduce EMI. To reduce the noise of components, metal film resistors and active devices with low noise should be used as much as possible. In addition, all components need to be subject to high- and low-temperature aging treatment to reduce temperature drift. Temperature compensation circuit can be added if necessary.

As for adding a single-stage LC filter behind the output capacitor, it is advisable to let $L = 47\,\mu\text{H}$ and $C = 0.1\,\mu\text{F}$.

5.8 PFC Configuration Scheme

5.8.1 Selection of PFC Type, Number of Stage, and Working Mode

5.8.1.1 Selection of PFC Type

SMPS can be divided into three types: SMPS without PFC, SMPS with PPFC, and SMPS with APFC. According to the certification standard for "ENERGY STAR" in the United States, PFC must be installed for the LED drive power supply above 5 W. In addition, the IEc0500-3-2 standard of international electro technical commission specifies a strict limit for the input harmonic of power supply, involving such Class D devices as personal computers, televisions, and monitors. PFC type should be selected according to the requirements for the power, power factor, and THD of SMPS.

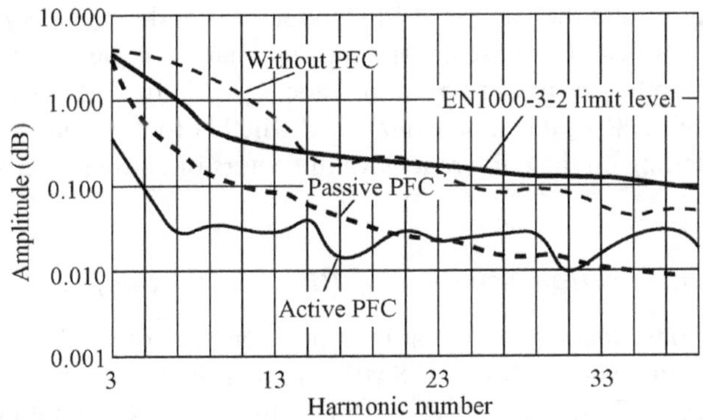

Figure 5.31 Comparison between the input harmonic amplitude of 250 W PC power supply and EN1000-3-2 international standard

See Figure 5.31 for the comparison between EN1000-3-2 international standard and the actually measured input harmonic amplitude of 250 W PC power supply under three conditions, that is, without PFC, with PPFC, and with APFC. It is not difficult to see that all the 3rd–15th harmonic amplitudes without PFC are higher than the limit level of EN1000-3-2. In using PPFC or APFC, harmonic amplitudes are lower than the limit level of EN1000-3-2. The 3rd–20th harmonic amplitudes using APFC are obviously lower than those using PPFC.

5.8.1.2 Selection of the Number of Stage of PFC Power Supply

1. Single-stage PFC power supply

   It combines the functions of PFC and DC/DC into one stage to form the single-chip IC of PFC + DC/DC converter, suitable for medium- and large-power SMPS of 40–100 W.
2. Two-stage PFC power supply

   Two-stage PFC is a commonly used PFC power supply, its first stage is PFC and the second stage is DC/DC converter. This type of power supply has relatively good harmonic suppression effect and can achieve a relatively high power factor. It can preset the DC input voltage of DC/DC stage owing to its independent PFC stage, so PFC has comparatively precise output voltage and strong load capacity, applicable to medium- and large-power SMPSs. The disadvantage is that it needs relatively more components and has low power density, thus increasing the cost and the loss.
3. Three-stage PFC power supply

   It contains the three stages of unit circuit of PFC, LLC harmonic converter, and DC/DC converter, suitable for high-power SMPSs above 100 W. If PFC and LLC are integrated into one chip (e.g., PFC + LLC controller PLC810PG), it will be a single-stage PFC power supply.

5.8.1.3 Selection of PFC Working Mode

PFC topology has two working modes. One is the boost converter working under CCM, which features that the current in PFC inductor is in continuous state, so it has large output power, suitable for SMPSs above 200 W. The other is the boost converter of critical conduction mode

(CRM), which features that the current of PFC inductor is at the boundary of continuous and discontinuous conductions, it can use cheap chips and has a simple circuit, is easy to design, and has no conduction loss of power switching tube, suitable for SMPSs below 100 W.

For SMPSs of 100–200 W, it can be determined whether to use continuous or critical conduction mode depending on the comprehensive indicators of entire power system.

5.8.2 *Four Configuration Schemes of PFC Power Supply*

The following text introduces several configuration schemes of PFC power supply through an example. The main design indexes of SMPS are as follows: AC input voltage range $u =$ 85–265 V (50/60 Hz), AC input power $P_I = 150$ W, the rated DC output power $P_O = 120$ W, switching frequency $f = 25$–476 kHz, the output voltage of PFC stage $U_{O1} = +400$ V (variation of ±8% allowed), the output voltage of DC/DC converter stage $U_{O2} = +12$ V, output current $I_O = 10$ A, and power supply efficiency $\eta = 80\%$.

By analysis, it can be known that when $P_O = 120$ W, PFC can use the boost converter of either continuous or critical conduction mode. There are four specific configuration schemes as follows.

1. Two-stage PFC power supply with fixed output-type boost converter of critical conduction mode

   See Figure 5.32 for the structural block diagram of two-stage PFC power supply with fixed output-type boost converter of critical conduction mode. The first stage uses the boost PFC chip MC33260 based on critical conduction mode. The second stage uses isolated DC/DC converter with a switching frequency of 200 kHz. In this configuration, +400 V fixed voltage is first output from PFC, and then reduced to +12 V by DC/DC converter for output.
2. Two-stage PFC power supply with adjustable output-type boost converter of critical conduction mode

   See Figure 5.33 for the structural block diagram of two-stage PFC power supply with adjustable output-type boost converter of critical conduction mode. The only difference between it and Figure 5.32 is that the output of PFC is changed to +200–400 V adjustable voltage.
3. Two-stage PFC power supply with fixed output-type boost converter of continuous conduction mode

   See Figure 5.34 for the structural block diagram of two-stage PFC power supply with fixed output-type boost converter of continuous conduction mode. It uses the boost PFC

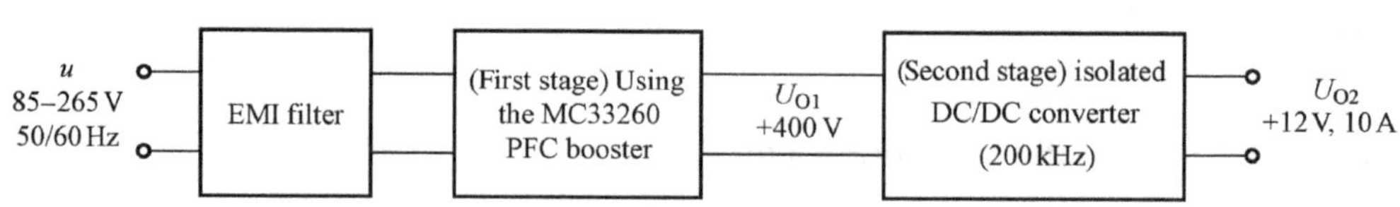

Figure 5.32 Structural block diagram of two-stage PFC power supply with fixed output-type boost converter of critical conduction mode

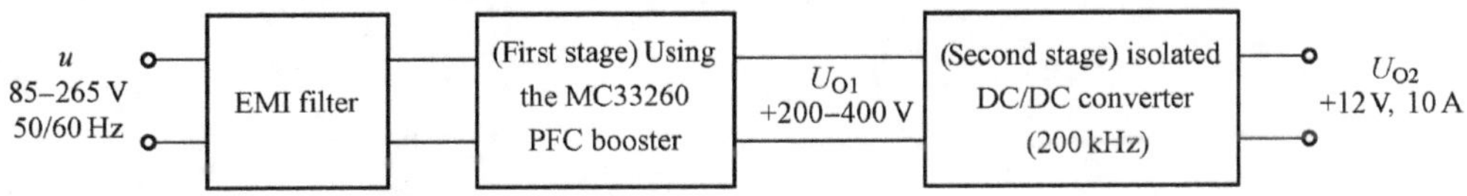

Figure 5.33 Structural block diagram of two-stage PFC power supply with adjustable output-type boost converter of critical conduction mode

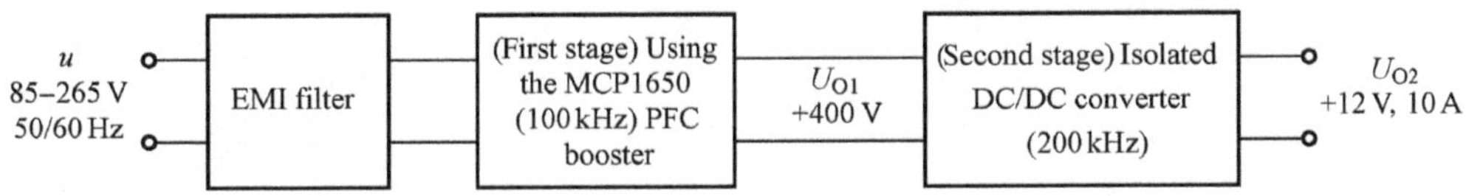

Figure 5.34 Structural block diagram of two-stage PFC power supply with fixed output-type boost converter of continuous conduction mode

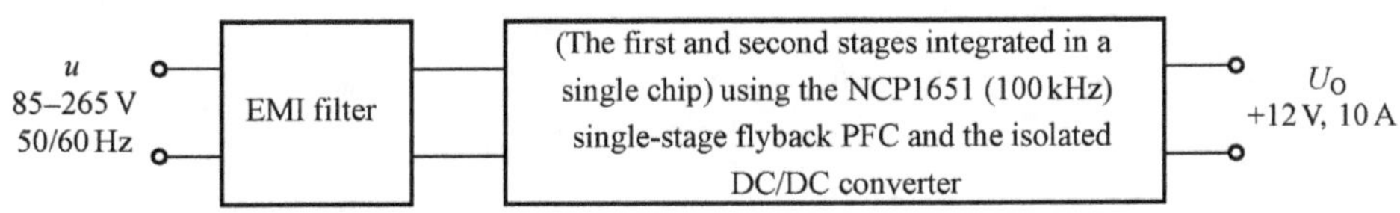

Figure 5.35 Structural block diagram of single-stage PFC power supply with flyback converter of critical conduction mode

chip NCP1650 based on continuous conduction mode with a switching frequency of 100 kHz. The other sections are the same as those in Figure 5.32.

4. Single-stage PFC power supply with flyback converter of critical conduction mode

   See Figure 5.35 for the structural block diagram of single-stage PFC power supply with flyback converter of critical conduction mode. NCP1651 is a single-stage PFC control chip suitable for flyback converters. It contains PFC and DC/DC converter, applicable to serve as medium- and large-power PFC power supplies with an output power of 50–200 W.

5. Boost high-power single-stage PFC power supply of continuous conduction mode

   The nonisolated high-power single-stage PFC chip (PFS723EG–PFS729EG) of continuous conduction mode of HiperPFS series newly launched by PI of the United States in November 2010 has a maximum continuous output power up to 255–900 W and a peak output power up to 280–1000 W.

6

Design of High-Frequency Transformer

6.1 Selection Method for Magnetic Cores by the Empirical Formula or Output Power Table

6.1.1 Selection Method for Magnetic Cores by the Empirical Formula

6.1.1.1 Classification of Commonly used Magnetic Cores

The high-frequency transformer of switching-mode power supply (SMPS) mostly uses EI or EE ferrite cores. See Figures 6.1 and 6.2, respectively, for the outline of EI and EE magnetic cores. See Tables 6.1 and 6.2, respectively, for the common dimension specifications of EI and EE magnetic cores, wherein L_e refers to the average length of magnetic path. It deserves pointing out that although the magnetic cores produced by different manufactures have the same outline, they are different in definitions of relevant dimensions and the actual dimensions, so the actual dimensions of magnetic cores should be determined by the information provided by the manufacturer or actual measurement.

RM ferrite core is also known as rectangular core, which is a kind of soft magnetic material between the pot-type and the EE magnetic cores. It is characterized by high output power and good heat dissipation. See Figure 6.3(a) and (b), respectively, for the structure and outline drawings of RM magnetic core. See Table 6.3 for the specifications of commonly used RM magnetic cores, wherein the unit of inductance coefficient A_L is nH/N^2.

Nano amorphous magnetic core is a new type of soft magnetic material, which is popular due to its advantages such as high permeability, high squareness ratio, and good high-temperature stability. The size of SMPS can be reduced by replacing the traditional ferrite core with nanocrystalline magnetic core.

Nanocrystalline magnetic core has the following features:

1. The initial permeability is very high ($\mu = 30,000–80,000$), and the permeability change is very small with the variation of magnetic flux density and temperature.
2. The magnetic core loss is very low, and it does not change with temperature within the range of -40 to $+120\,°C$.

Optimal Design of Switching Power Supply, First Edition. Zhanyou Sha, Xiaojun Wang, Yanpeng Wang, and Hongtao Ma.

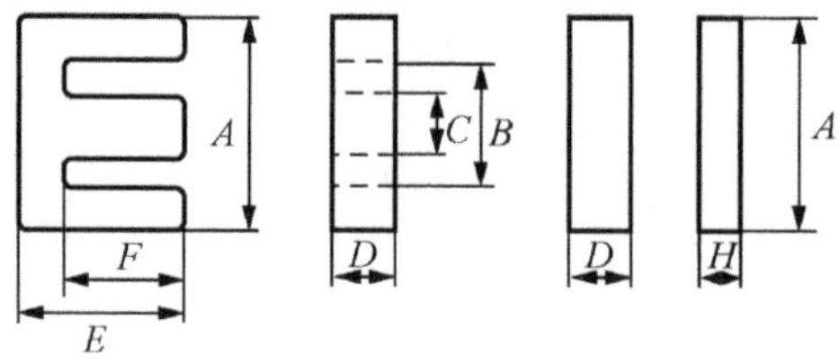

Figure 6.1 The outline of EI magnetic core

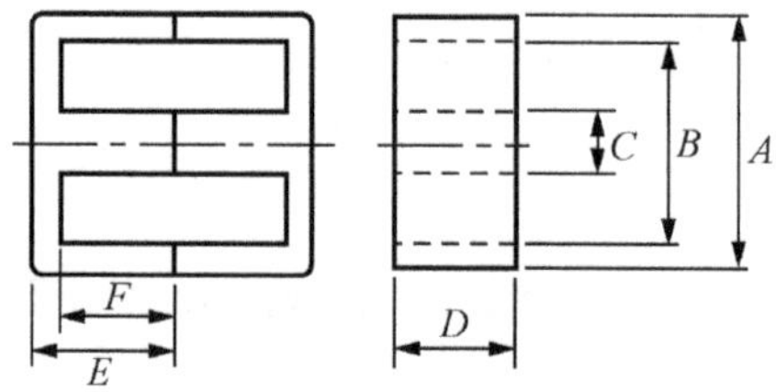

Figure 6.2 The outline of EE magnetic core

Table 6.1 Dimension specifications of commonly used EI magnetic cores

Model	*A* (mm)	*B* (mm)	*C* (mm)	*D* (mm)	*E* (mm)	*F* (mm)	*H* (mm)	L_e (cm)	Weight (g)	A_e (cm^2)	A_w (cm^2)	AP (cm^4)
EI16	16.0	11.8	4.0	4.8	12.0	10.87	2.0	3.59	3.6	0.19	0.42	0.08
EI19	19.0	14.2	4.85	4.85	13.6	11.3	2.4	3.96	4.5	0.23	0.53	0.12
EI22	22.0	13.0	5.75	5.75	14.55	10.55	4.5	3.96	10.0	0.41	0.38	0.16
EI25	25.4	19.0	6.35	6.35	15.8	12.5	3.2	4.8	10.0	0.40	0.79	0.32
EI28	28.0	18.7	7.2	10.6	16.75	12.25	3.5	4.86	23.5	0.83	0.70	0.58
EI30	30.25	20.1	10.65	10.65	21.3	16.3	5.5	5.86	33.5	1.09	0.77	0.91
EI33	33.0	23.6	9.7	12.7	23.75	19.25	5.0	6.75	40.6	1.18	1.34	1.58
EI40	40.5	26.8	11.7	11.7	27.3	21.3	6.5	7.75	59.0	1.43	1.61	2.30
EI50	50.0	34.5	15.0	15.0	33.0	24.5	9.0	9.5	112	2.27	2.39	5.43
EI60	60.0	44.5	15.8	15.8	35.9	27.5	8.5	11	138	2.44	3.95	9.64

3. The saturation flux density is very high ($B_S = 1.2\,T$), allowing to choose comparatively low switching frequency to reduce the cost of SMPS and electromagnetic interference (EMI) filter.
4. Magnetic core uses epoxy resin packaging with high mechanical strength, and it has no magnetostriction phenomenon and can withstand strong vibration.
5. It can replace the traditional ferrite core to reduce the size of SMPS and improve reliability.
6. Nanocrystalline magnetic core is also suitable for producing common-mode inductor (also known as common-mode choke) of EMI filter, which can get large inductance just with a few turns, thus reducing copper loss, saving wire material, and reducing the volume

Table 6.2 Dimension specifications of commonly used EE magnetic cores

Model	A (mm)	B (mm)	C (mm)	D (mm)	E (mm)	F (mm)	L_e (cm)	A_e (cm^2)	Weight (g)
EE10	10.3	7.9	2.45	4.65	5.7	4.45	2.73	0.11	1.5
EE13	13.3	10.0	2.7	6.15	6.2	4.65	3.08	0.18	2.8
EE16	16.1	11.8	4.0	4.8	7.4	5.3	3.71	0.19	3.5
EE19	19.0	14.3	4.6	4.8	8.2	5.7	4.02	0.22	4.6
EE22	22.0	12.8	5.75	5.75	9.4	5.4	3.98	0.41	9.2
EE25	25.0	17.5	7.2	7.2	12.5	8.9	5.77	0.52	16.0
EE30	30.0	19.5	6.95	7.05	15.0	10.0	6.56	0.6	22.0
EE33	33.2	23.5	9.7	12.7	14.0	9.65	6.7	1.17	39.8
EE41	41.3	28.0	12.7	12.7	16.8	10.4	7.75	1.61	64.6
EE50	50.0	34.6	14.6	14.6	21.3	12.75	9.59	2.28	113.5
EE65	65.0	45.0	20.0	27.1	32.5	22.5	14.7	5.32	402

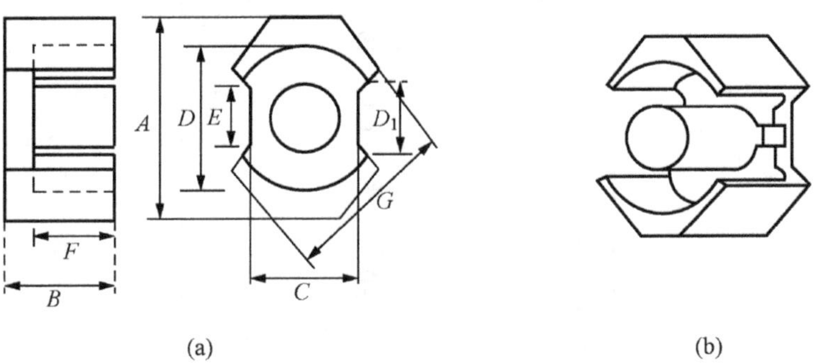

Figure 6.3 Structure and outline of RM magnetic core. (a) Structure drawing and (b) outline drawing

of common-mode inductor. The common-mode inductor made of nanocrystalline has very high common-mode insertion loss, which is capable of suppressing common-mode interference within a very wide frequency range. So it is not necessary to use any complex filter circuit. In addition, a spike suppressor can be made just by winding one or several turns of copper wire around the nanocrystalline magnetic core, which has excellent suppression effect on noise interference with a simple structure.

6.1.1.2 Selection Method for Magnetic Core by the Empirical Formula

The sectional area S_J of the magnetic core is the tongue width C (mm) multiplied by the depth D (mm) of the magnetic core. The effective sectional area of the magnetic core $A_e \approx S_J$. The formula is as follows:

$$A_e \approx S_J = CD \tag{6.1}$$

In view of the magnetic core loss and so on, the empirical formula between the maximum power handling P_M (W) of the high-frequency transformer and the effective sectional area

Table 6.3 The specifications of commonly used RM magnetic cores

Model	External dimensions (mm)								A_L (nH/N^2)	Weight (g/set)
	A	B	C	D	E	F	G	D_1		
RM4	10.8 ± 0.2	5.2 ± 0.2	4.5 ± 0.1	8.15 ± 0.2	3.8 ± 0.1	3.6 ± 0.1	9.6 ± 0.2	—	1150	1.7
RM5	14.3 ± 0.3	5.2 ± 0.1	6.6 ± 0.2	10.4 ± 0.2	4.8 ± 0.2	3.25 ± 0.2	12.05 ± 0.3	6.0	1800	3.0
RM6	17.6 ± 0.3	6.2 ± 0.1	8.0 ± 0.2	12.65 ± 0.3	6.3 ± 0.2	8.2 ± 0.2	14.4 ± 0.3	—	2400	5.3
RM8	22.75 ± 0.5	8.2 ± 0.2	10.8 ± 0.2	17.3 ± 0.3	8.4 ± 0.2	5.5 ± 0.1	19.3 ± 0.4	—	3150	12.5
RM10	27.85 ± 0.7	9.35 ± 0.2	13.25 ± 0.3	21.65 ± 0.5	10.7 ± 0.2	6.35 ± 0.2	24.70 ± 0.6	15.05 ± 0.7	4750	22.0
RM12	36.9 ± 0.7	12.25 ± 0.2	15.9 ± 0.3	25.5 ± 0.6	12.6 ± 0.2	8.55 ± 0.2	29.2 ± 0.6	13.4	5700	43.3
RM14	14.6 ± 0.6	10.25 ± 0.1	18.7 ± 0.3	29.5 ± 0.5	14.75 ± 0.3	5.7 ± 0.2	34.2 ± 0.5	17.0	10,250	54.2
RM14R	41.6 ± 0.6	14.4 ± 0.1	18.7 ± 0.3	29.5 ± 0.5	14.75 ± 0.3	10.7 ± 0.5	34.2 ± 0.5	17.0	13,900	68.4

Table 6.4 Corresponding relation between the commonly used magnetic core models and the output power

Model	External dimension (mm)			Output power (W)		Weight (kg)
	Length	Width	Depth	50 kHz	100 kHz	
EI10	11	10	9	3	6	0.008
EI13	13	12	10	4	8	0.01
EI16	17	16	14	5	9	0.011
EI19	20	19	16	8	13	0.012
EI22	23	21	18	14	20	0.016
EI25	26	22	19	20	30	0.021
EI28	29	22	22.5	42	58	0.035
EI30	31	29	26	61	95	0.054
EI35	37	33	28.5	100	150	0.078
EI40	42	38	29	160	250	0.11
EI45	46	41	33	260	391	0.15
EI50	52	44.5	38	430	650	0.195
EE8	9	8	7	2	4	0.008
EE10	11	10	10	3	6	0.009
EE13	13	11	11	4	8	0.011
EE16	17	15	15	5	9	0.017
EE19	21	19	22	8	13	0.021
EE30	31	24	29	61	95	0.044
EE35	37	31	54	100	150	0.051
EE40	42	37	60	160	250	0.119

A_e (cm^2) of the magnetic core is as follows:

$$A_e = 0.15\sqrt{P_M} \tag{6.2}$$

For example, when the rated output power of an SMPS is 55 W and the power supply efficiency $\eta = 70\%$, the rated input power of the high-frequency transformer $P_I = 55\,W \div 70\% = 78.6\,W$. Actually take $P_M = 80\,W$ and put it into Formula (6.2), then $A_e = 1.34\ cm^2$. EI40 magnetic core with an effective sectional area of magnetic core of $1.43\,cm^2$ (see Table 6.1) can be used.

6.1.2 *Selection Method for Magnetic Cores by Output Power*

See Table 6.4 for the corresponding relation between the commonly used magnetic core models and the output powers, which shows the reference output power at a switching frequency of 50 and 100 kHz as a reference for selecting magnetic cores.

See Table 6.5 for the comparison between the output power of small- and medium-power single-chip SMPSs recently recommended by PI and the magnetic core models as a reference for selecting the magnetic core of high-frequency transformer of single-chip SMPS.

Table 6.5 Comparison between the output power of small- and medium-power single-chip SMPSs and the magnetic core models

Output power range P_O (W)	Ferrite core model			
	Switching frequency of 66 kHz		Switching frequency of 132 kHz	
	Wound with enameled wire	Wound with three-layer insulated wire	Wound with enameled wire	Wound with three-layer insulated wire
0–10	EI22 EE19 EI22/19/6 EEL16 EF20 EI25 EEL19	EF12.6 EE13 EF16 EE16 EE19 EI22 EI22/19/6	EI22 EE19 EI22/19/6 EEL16	EF12.6 EE13 EF16 EE16
10–20	EI28 EEL22 EF25	EF20	EF20 EI25 EEL19	EE19 EI22 EI22/19/6 EF20
20–30	EI30 EPC30 EEL25	EF25	EI28	—
30–50	E30/15/7 EER28 ETD29 EI35 EI33/29/13-Z EER28L	EI28 EI30 E30/15/7 EER28	EEL22	EF25
50–70	EF32 ETD34	ETD29 EI35 EF32	EEL25 E30/15/7 EER28	EI28
70–100	EI40 E36/18/11 EER35	ETD34 E36/18/11 EI40	ETD29 EI35 EI33/29/13-Z EER28L EF32	EI30 E30/15/7 EER28 ETD29
100–150	ETD39 EER40 E42/21/15	ETD39 EER40	ETD34 EI40 E36/18/11 EER35	EI35 EF32 ETD34
>150	E42/21/20 E55/28/21	E42/21/15 E42/21/20 E55/28/21	ETD39 EER40 E42/21/15 E42/21/20 E55/28/21	E36/18/11 EI40 ETD39 EER40 E42/21/15 E42/21/20 E55/28/21

6.2 Waveform Parameters of the High-Frequency Transformer Circuit

The voltage and current waveforms of SMPS are relatively complex, including not only input sine wave and the half-wave or full-wave rectified wave but also the rectangular wave [PWM waveform (pulse width modulation)], saw-tooth wave (the primary current waveform of discontinuous current mode), trapezoidal wave (the primary current waveform of continuous current mode), and so on. There are three waveform parameters in high-frequency transformer circuit: waveform coefficient (K_f), form factor (k_f), and crest factor (k_P).

6.2.1 *Analysis of Waveform Coefficient and Waveform Factor*

6.2.1.1 Waveform Coefficient K_f

To facilitate analysis, AC sinusoidal current is applied to the input terminal of the high-frequency transformer without considering copper loss, and induced electromotive force e will be generated in the primary and secondary windings. According to Faraday's law of electromagnetic induction, $e = \mathrm{d}\Phi/\mathrm{d}t = \mathrm{d}(NAB\sin\omega t)/\mathrm{d}t = NAB\omega\cos\omega t$, wherein N refers to the number of turns of the winding, A the sectional area of transformer magnetic core, B the strength of magnetic induction generated by the alternating current, and the angular frequency $\omega = 2\pi f$. The effective value of sine wave voltage is as follows:

$$U = \frac{\sqrt{2}}{2} \times NAB \times 2\pi f = \sqrt{2}\pi NABf = 4.44NABf \tag{6.3}$$

The waveform coefficient of the sine wave defined in SMPS is $K_f = \sqrt{2}\pi = 4.44$. It is not difficult to calculate the waveform coefficient of the square wave according to the Fourier series $K_f = \frac{4\sqrt{2}}{2\pi} \times \frac{2\sqrt{2}\pi}{2} = 4$.

6.2.1.2 Waveform Factor k_f

The waveform factor defined in electronic measurement field is different from the waveform coefficient of SMPS, for the former refers to the ratio between the effective voltage (U_{RMS}) and the average voltage ($\overline{U}$), which is indicated as k_f to distinguish from K_f. The relevant formula is as follows:

$$k_f = U_{RMS}/\overline{U} \tag{6.4}$$

The crest factor (k_P) corresponding to it is defined as the ratio between the peak voltage (U_P) and the effective voltage with a formula as follows:

$$k_P = U_P/U_{RMS} \tag{6.5}$$

Take sine wave, for example, $k_f = \frac{\sqrt{2}U_P}{2} \div \frac{2U_P}{\pi} = \frac{\sqrt{2}\pi}{4} = 1.111$. It indicates that $K_f = 4k_f$, there is a four-time difference between them.

6.2.2 *Parameters of Six Kinds of Common Waveforms in SMPS*

See Table 6.6 for the parameters of six kinds of common waveforms in SMPS. As the average of square and trapezoidal waves is zero, it is replaced with the absolute average voltage $|\overline{U}|$. For rectangular wave, t_0 refers to the pulse width, T the cycle, and the duty ratio $D = t_0/T$.

6.3 Formula Derivation of Selecting High-Frequency Transformer Magnetic Core Based on AP Method

It should be pointed out that AP method is currently still recommended as an effective method for magnetic core selection. However, AP method was originally put forward for traditional industrial-frequency sine-wave iron-core transformer, so it is not appropriate to be directly used in the high-frequency transformers with complex waveforms, and the computing result could be inaccurate. It is necessary to derive and verify the calculating formula for AP method according to the concepts such as the waveform factor (k_f) defined in the electronic measurement field, the characteristic ripple factor (K_{RP}) of SMPS, the duty ratio (D), and the working waveforms of SMPS in continuous and discontinuous modes, so as to provide a scientific and practical approach to a correct selection of high-frequency transformer magnetic cores.

AP refers to the area product of the effective sectional area of magnetic core and the window area. The relevant formula is as follows:

$$\mathrm{AP} = A_w A_e, \tag{6.6}$$

wherein

AP magnetic core area product (cm^4),

A_w the window area (cm^2) of magnetic core suitable for being wound with wire. Its formula is as follows:

$$A_w = \frac{1}{2}(B - C)F \tag{6.7}$$

A_e effective sectional area of magnetic core (cm^2), $A_e \approx S_j = CD$,

D refers to the magnetic core thickness.

According to the calculated AP value, the required magnetic core model can be found by referring to the table. The following describes the principle of AP method and relevant points for attention.

Let the effective voltage of the primary winding be U_1, the number of turns of the primary winding be N_P, the AC magnetic flux density of the used magnetic core be B_{AC}, the magnetic flux be Φ, the switching cycle be T, the switching frequency be f, the waveform coefficient of the primary current be K_f, and the effective sectional area of magnetic core be A_e with a unit square centimeter ($1\,\mathrm{cm}^2 = 1\,\mathrm{m}^2 \times 10^{-4}$). According to the law of electromagnetic induction, it can be obtained that

$$U_1 = N_P \cdot \frac{d\Phi}{dt} = N_P \cdot \frac{B_{AC} A_e K_f}{T} \times 10^{-4} = N_P B_{AC} A_e K_f f \times 10^{-4} \tag{6.8}$$

Table 6.6 Parameters of six kinds of common waveforms in SMPS

Name	Waveform	Effective voltage U_{RMS}	Average voltage $\overline{U}$	Absolute average voltage $\lvert\overline{U}\rvert$	Waveform factor k_f	Crest factor k_p
Sine wave	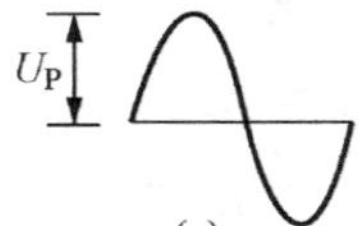 (a)	$0.707U_p\left(\frac{\sqrt{2}}{2}\cdot U_p\right)$	0	$0.637U_p\left(\frac{2}{\pi}\cdot U_p\right)$	1.111	1.414
Half-wave rectified wave	 (b)	$0.5U_p$	$0.318U_p\left(\frac{1}{\pi}\cdot U_p\right)$	$0.318U_p\left(\frac{1}{\pi}\cdot U_p\right)$	1.571	2
Full-wave rectified wave	 (c)	$0.707U_p\left(\frac{\sqrt{2}}{2}\cdot U_p\right)$	$0.637U_p\left(\frac{2}{\pi}\cdot U_p\right)$	$0.637U_p\left(\frac{2}{\pi}\cdot U_p\right)$	1.111	1.414
Square wave	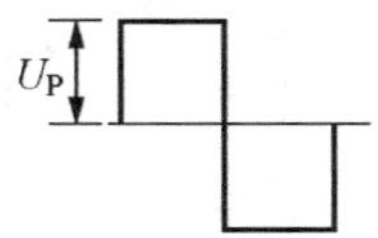 (d)	U_p	0	U_p	1	1
Rectangular wave	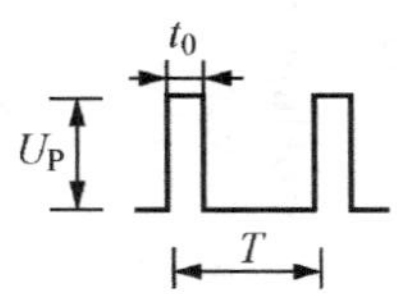 (e)	$\sqrt{\frac{t_0}{T}}\cdot U_p$	$\frac{t_0}{T}\cdot U_p$	$\frac{t_0}{T}\cdot U_p$	$\sqrt{\frac{T}{t_0}}$	$\sqrt{\frac{T}{t_0}}$
Saw-tooth wave	 (f)	$0.577U_p\left(\frac{\sqrt{3}}{3}\cdot U_p\right)$	0	$0.5U_p$	1.155	1.732

In view of the relational expression that $K_f = 4k_f$, it can be derived that

$$N_P = \frac{U_1 \times 10^4}{K_f B_{AC} A_e f} = \frac{U_1 \times 10^4}{4k_f B_{AC} A_e f} \tag{6.9}$$

Similarly, assume that the effective voltage of the secondary winding is U_S and the number of turns of the secondary winding is N_S, it can be obtained that

$$N_S = \frac{U_S \times 10^4}{4k_f B_{AC} A_e f} \tag{6.10}$$

Assume that the winding current density is J (unit : A/cm^2), and the sectional area of wire is

$$S_d = \frac{I}{J}\ (cm^2) \tag{6.11}$$

Let the window area utilization factor of high-frequency transformer be K_w, and the effective currents of the primary and secondary windings be I_1 and I_2, respectively. When the winding area is fully utilized,

$$K_w \cdot A_w = N_P \cdot \frac{I_1}{J} + N_S \cdot \frac{I_2}{J} \tag{6.12}$$

that is,

$$A_w = \frac{N_P}{K_W} \cdot \frac{I_1}{J} + \frac{N_S}{K_W} \cdot \frac{I_2}{J} \tag{6.13}$$

Put Formulas (6.9) and (6.10) into Formula (6.13), respectively,

$$\begin{aligned} A_w &= \frac{U_1 \times 10^4}{4K_W k_f B_{AC} A_e f} \cdot \frac{I_1}{J} + \frac{U_S \times 10^4}{4K_W k_f B_{AC} A_e f} \cdot \frac{I_2}{J} \\ &= \frac{U_1 I_1 + U_S I_2}{4K_W k_f J B_{AC} A_e f} \times 10^4\ (cm^2) \end{aligned} \tag{6.14}$$

And then it can be obtained that

$$AP = A_w A_e = \frac{U_1 I_1 + U_S I_2}{4K_W k_f J B_{AC} A_e f} \times 10^4 \times A_e = \frac{P_I + P_O}{4K_W k_f J B_{AC} f} \times 10^4 \tag{6.15}$$

The apparent power of high-frequency transformer refers to the total power withstood by the primary and secondary windings, that is, $S = P_I + P_O$. As $\eta = P_O / P_I$, $P_I + P_O = P_O/\eta + P_O = (1/\eta + 1)\ P_O = [(1 + \eta)\ /\ \eta]\ P_O$. Put it into Formula (6.15) to get

$$AP = A_w A_e = \frac{(1+\eta)\ P_O}{4\eta K_W k_f J B_{AC} f} \times 10^4\ (cm^4) \tag{6.16}$$

That is the basic formula to select magnetic cores by using AP method. The following sections carry out an in-depth analysis and proper simplification on Formula (6.16) from the perspective of engineering design, and the key is to make a further derivation of the parameters k_f and B_{AC} in the formula.

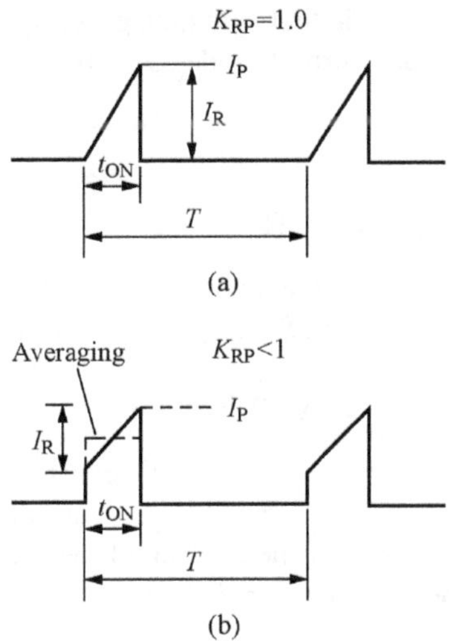

Figure 6.4 Primary current waveform. (a) Discontinuous current mode and (b) continuous current mode

6.3.1 *The Waveform Factor* k_f *of the Primary Current*

The primary voltage waveform can be approximately regarded as rectangular wave, that is, $k_f = \sqrt{T/t_0} = \sqrt{1/D} = 1/\sqrt{D}$; however, the primary current waveform is not rectangular wave, but saw-tooth wave (working under discontinuous current mode DCM) or trapezoidal wave (working under continuous current mode CCM). Take the discontinuous current mode, for example, the primary current waveform is the saw-tooth wave with periodic make-and-break as shown in Figure 6.4(a), which has primary saw-tooth wave current (high level) only when the power switching tube (MOSFET) is turned on and has zero primary current (low level) when the power switching tube is turned off. In the figure, K_{RP} refers to ripple factor, equal to the ratio between the primary pulsating current I_R and the peak current I_P, that is, $K_{RP} = I_R / I_P$. $K_{RP} < 1$ in continuous current mode, while $K_{RP} = 1$ in discontinuous current mode. Let conduction time be t_{ON} and switching period be T, so $D = t_{ON} / T$. For the saw-tooth wave with periodic make-and-break, the waveform factor of the primary current can be expressed as k'_f. The relevant formula is as follows:

$$k'_f = k_f \, t_{ON} / T = k_f D \tag{6.17}$$

As k_f of the periodical saw-tooth wave is 1.155, put it into Formula (6.17) to get

$$k'_f = 1.155D \tag{6.18}$$

In this case, k_f in Formula (6.16) should be replaced with 1.155 D.

Under continuous current mode, the primary current waveform is trapezoidal wave with periodic make-and-break, whose waveform factor is relatively complex. One way is to take the average of the waveform to turn it into rectangular wave, and then calculate according to the rectangular wave parameters as shown in Figure 6.4(b). The other way is to select magnetic core as per discontinuous current mode first, and then properly increase the magnetic core size to let SMPS work under continuous current mode by increasing the inductance of the primary winding.

6.3.2 *AC Magnetic Flux Density* B_{AC}

The AC magnetic flux density (B_{AC}) of magnetic core can be calculated according to the maximum magnetic flux density. For flyback SMPS, the relevant formula is as follows:

$$B_{AC} = B_M K_{RP} Z,$$

wherein Z refers to the loss distribution coefficient, representing the ratio between the secondary loss and the total loss. In extreme cases, $Z \to 0$ means that all loss occurs on the primary side and the load is disconnected. $Z \to 1$ means that all loss occurs on the secondary side and the load is short circuited. Generally take $Z = 0.5$, so

$$B_{AC} = 0.5 B_M K_{RP} \tag{6.19}$$

Put Formulas (6.18) and (6.19) together into Formula (6.16) to get

$$\mathrm{AP} = A_w A_e = \frac{0.433(1+\eta)\ P_O}{\eta K_W D J B_M K_{RP} f} \times 10^4\ (\mathrm{cm}^4) \tag{6.20}$$

That is the practical formula for selecting magnetic cores based on AP method. Formula (6.20) is derived according to the winding current and output power of unipolar transformer, applicable to the design of single-end forward or flyback high-frequency transformer. In the formula, the units of AP and P_O are cm^4 and W, respectively. The current density is generally $J = 200$–$600\,\mathrm{A/cm^2}$ (i.e., 2–6 $\mathrm{A/mm^2}$), with the maximum no more than $1000\,\mathrm{A/cm^2}$ (i.e., $10\,\mathrm{A/mm^2}$). The utilization factor of window area is generally $K_w = 0.3$–0.4. When high-frequency transformer has multiple windings, the sum of the product of the number of turns of each winding and the sectional area of wire should be calculated.

It can be known through further analysis that Formula (6.20) can be simplified as follows for discontinuous current mode ($K_{RP} = 1$):

$$\mathrm{AP} = A_w A_e = \frac{0.433(1+\eta)\ P_O}{\eta K_W D J B_M f} \times 10^4\ (\mathrm{cm}^4) \tag{6.21}$$

For continuous current mode ($0.4 < K_{RP} < 1$), assume that $K_{RP} = 0.7$, then Formula (6.20) can be simplified as

$$\mathrm{AP} = A_w A_e = \frac{0.62(1+\eta)\ P_O}{\eta K_W D J B_M f} \times 10^4\ (\mathrm{cm}^4) \tag{6.22}$$

For single-end forward high-frequency transformer, the maximum duty ratio $D_{max} < 0.5$. When the power supply efficiency $\eta = 80\%$, the actual window area utilization factor $K_W = 0.4$, the duty ratio $D = 0.4$, and $J = 400\,\mathrm{A/cm^2}$, Formula (6.20) can be simplified as

$$\mathrm{AP} = A_e A_w = \frac{152 P_O}{B_M K_{RP} f} \tag{6.23}$$

Formulas (6.21)–(6.23) are simplified according to different circuit structures and specified parameters, which will have errors in computing results when actual parameters are changed. A more accurate method is to use Formula (6.20). The following rules can be summarized by comparing Formulas (6.21)–(6.23):

1. Without taking into account factors such as magnetic core loss, magnetic core material difference, and the increase of magnetic core loss with the rise of switching frequency in the formulas above, AP calculated is the minimum, and the corresponding magnetic core size is also the minimum. In view of the above facts, the magnetic core of a larger size should be selected from the practical perspective.
2. For single-end flyback SMPS, its B_{AC} is relatively small ($B_{AC} = B_M K_{RP} Z$) and a little larger B_M can be taken, generally be 0.2–0.3 T. For push-pull, full-bridge, and half-bridge converters, $B_{AC} = 2B_M$. As B_M is relatively small, a little smaller B_{AC} should be taken to reduce magnetic core loss, generally 0.1–0.15 T. On the condition of the same output power, the volume of the high-frequency transformer required by full-bridge and half-bridge converters is the minimum.
3. On the condition of the same output power, the AP value under continuous current mode is larger than that under discontinuous current mode. This indicates that the volume of the high-frequency transformer required by continuous current mode is comparatively large, while that required by the discontinuous current mode is comparatively small.
4. Magnetic material manufacturers usually only provide A_e and A_w without giving the AP value directly. Some manufacturers do not provide A_w directly, in which case, the corresponding A_w and AP need to be calculated according to the relevant dimension parameters of magnetic cores for selecting proper magnetic core dimensions.

6.4 Design of Flyback High-Frequency Transformer

The high-frequency transformer of flyback SMPS is equivalent to a power inductor, and its amount of energy stored directly affects the output power of the SMPS. Therefore, the design of high-frequency transformer of flyback SMPS is actually the design of the power inductor. It includes several steps such as calculating the primary inductance L_P, selecting the magnetic core size, calculating the width of air gap δ, and calculating the number of turns of the primary winding N_P.

6.4.1 Design Method of Flyback High-Frequency Transformer

6.4.1.1 Calculating the Primary Inductance L_P

According to the energy storage formula for inductor

$$W = \frac{1}{2} I^2 L$$

The energy transmitted during each switching cycle is proportional to the squared value of pulsating current I_R directly. Assume that the switching frequency is f, the output power is P_O, the power supply efficiency is η, and the primary inductance is L_P, then the input power should be

$$P = \frac{P_O}{\eta} = \frac{1}{2} I_R^2 L_P f$$

After calculation,

$$L_P = \frac{2P_O}{\eta I_R^2 f}, \tag{6.24}$$

wherein the pulsating current $I_R = K_{RP} I_P$. The pulse coefficient K_{RP} is usually within the range of 0.4–1. For the same output power, when K_{RP} is comparatively large, the L_P required will be comparatively small, which is helpful to reduce the size of transformer. However, the copper loss of transformer will be increased. When the power supply efficiency is 80% and K_{RP} is 0.5, Formula (6.24) can be simplified as

$$L_P = \frac{2P_O}{0.8 \times (0.5 I_P)^2 f} = \frac{10 P_O}{I_P^2 f} \tag{6.25}$$

6.4.1.2 Selecting Magnetic Core Size

Magnetic core size of flyback SMPS high-frequency transformer can be selected by AP method. Proper magnetic core can be selected from a calculation according to Formula (6.22) or directly chosen from Tables 6.1 and 6.2 as per the computing result of Formula (6.2).

6.4.1.3 Calculating the Number of Winding Turns and the Diameter of Wire

Calculating the number of winding turns

After finishing selecting magnetic core, the number of winding turns of high-frequency transformer can be calculated according to the parameters of the magnetic core. As the number of the turns of the secondary winding can be calculated as per the transformer ratio, the key problem is to determine the number of the turns of the primary winding. For single-end flyback and forward converters, the maximum duty ratio (D_{max}) can be obtained when the minimum input voltage (U_{Imax}) is applied. Considering that the primary voltage waveform can be approximately regarded as rectangular wave, $K_f = 1/\sqrt{D}$ and $1/K_f = \sqrt{D}$, Formula (6.9) can also be expressed as

$$N_P = \frac{U_1 \sqrt{D} \times 10^4}{B_M K_{RP} f} \tag{6.26}$$

Before D_{max} is determined, it can be taken as 0.5 for calculating. It needs to be noted that the N_P obtained as per Formula (6.26) is only the minimum when the law of electromagnetic induction is met, so the actual number of turns should be slightly larger.

When selecting the number of the turns of the secondary winding, the maximum drain voltage that can be withstood by the induced voltage U_{OR} (also known as the secondary reflected voltage) and the power switching tube (MOSFET) needs to be considered. The maximum drain voltage is equal to the sum of the spike voltage generated by the input DC voltage, the induced voltage and the leakage inductance of high-frequency transformer, wherein the relation between U_{OR} and the number of the turns of the primary winding (N_P), that of the secondary winding (N_S), and the output voltage (U_O) is as follows:

$$U_{OR} = \frac{N_P}{N_S} \cdot (U_O + U_{F1}) \tag{6.27}$$

For flyback SMPS, U_{OR} is constant, usually between 85 and 165 V and typically 135 V. In the formula, U_{F1} refers to the forward voltage drop of output rectifier, which is usually 0.4 V for Schottky diode and 0.8 V for FRD. When U_O is relatively large, the forward voltage drop of output rectifier can be negligible.

After the number of the turns of the primary winding N_P is determined, the number of the turns of the secondary winding N_S can be calculated:

$$N_S = \frac{N_P}{U_{OR}} \cdot (U_O + U_{F1}) \tag{6.28}$$

When a high-frequency transformer has multiple secondary windings, the number of the turns of each winding can be calculated, respectively, according to different output voltages and the same U_{OR}.

Calculating the wire diameter

The selection of wire diameter relates to the root mean square (RMS) of the current flowing through the wire and the allowable current density. For the enameled wire with circular cross-section, the relation between the sectional area of the wire (S_d) and its diameter (d) is as follows:

$$S_d = \frac{\pi}{4} d^2$$

The relation between the RMS of the current I_{RMS} flowing through the wire and the sectional area of wire (S) and the current density (J) is as follows:

$$I_{RMS} = SJ$$

from which the formula for calculating wire diameter (d) can be obtained:

$$d = \sqrt{\frac{4I_{RMS}}{\pi J}} \tag{6.29}$$

For flyback SMPS, the current RMS of its high-frequency transformer winding relates to the maximum duty ratio (D_{max}) and the pulse coefficient (K_{RP}). The formula for calculating the primary current RMS (I_{RMS}) is as follows:

$$I_{RMS} = I_P \sqrt{D_{max} \left(\frac{K_{RP}^2}{3} - K_{RP} + 1 \right)}, \tag{6.30}$$

wherein I_P refers to the primary peak current.

The relation expression for the secondary peak current I_{SP}, the primary peak current I_P, and the numbers of the turns of the primary and secondary winding is as follows:

$$I_{SP} = I_P \cdot \frac{N_P}{N_S} \tag{6.31}$$

Table 6.7 The specifications of commonly used enameled wire

Nominal diameter (mm)	Maximum outside diameter (mm)	Sectional area of copper coil (mm^2)	DC resistance at 20 °C (Ω/m)	Wire size approximately as per British SWG	Current density (A/mm^2)		
					3	2.5	2
0.08	0.095	0.005027	3.487	44	0.0151	0.0126	0.0101
0.10	0.12	0.007854	2.237	42	0.0236	0.0196	0.0157
0.13	0.15	0.01327	1.322	39	0.0398	0.0332	0.0265
0.17	0.19	0.0227	0.773	37	0.0681	0.0568	0.0454
0.21	0.235	0.03464	0.506	35	0.104	0.0866	0.0693
0.25	0.275	0.04909	0.357	33	0.147	0.123	0.0982
0.29	0.33	0.06605	0.265	31	0.198	0.165	0.132
0.35	0.39	0.09621	0.182	29	0.289	0.241	0.192
0.41	0.45	0.132	0.133	27	0.396	0.33	0.264
0.51	0.56	0.2043	0.859	25	0.613	0.511	0.409
0.55	0.6	0.2376	0.737	24	0.713	0.591	0.475
0.62	0.67	0.3019	0.58	23	0.906	0.755	0.604
0.72	0.78	0.4072	0.43	22	1.22	1.02	0.814
0.80	0.86	0.5027	0.348	21	1.51	1.26	1.01
1.04	1.12	0.8495	0.206	19	2.55	2.12	1.7
1.20	1.28	1.131	0.155	18	3.39	2.83	2.26

The formula for the secondary current RMS (I_{SRMS}) is as follows:

$$I_{SRMS} = I_{SP}\sqrt{(1 - D_{max})\left(\frac{K_{RP}^2}{3} - K_{RP} + 1\right)} \tag{6.32}$$

The wire diameters of the primary and secondary windings can be calculated just by putting the current RMSs I_P and I_{SP}, respectively, into Formula (6.29).

The selection of wire diameter can be realized by referring to the table of the current RMSs of windings. See Table 6.7 for the common specifications of enameled wire, wherein the corresponding wire diameter can be found directly according to the required current. The current density of high-frequency transformer winding is typically 3–6 A/mm^2.

6.4.1.4 Calculating the Width of Air Gap δ

To prevent magnetic saturation of high-frequency transformer in flyback SMPS, air gap is usually added in the magnetic core. See Figure 6.5(b) for the magnetization curve of the magnetic core with air gap. Figure 6.5(a) shows the magnetization curve without air gap. It is easy to see that the maximum magnetic flux density B_M is not changed after air gap is added to the magnetic core, but the maximum magnetic field strength will be increased. This means that on the condition of the same B_M and winding number of turns, adding of air gap can raise the working current of winding and the magnetic saturation current of high-frequency transformer will be increased. In addition, after adding air gap, remanence B_r will be reduced and the variable

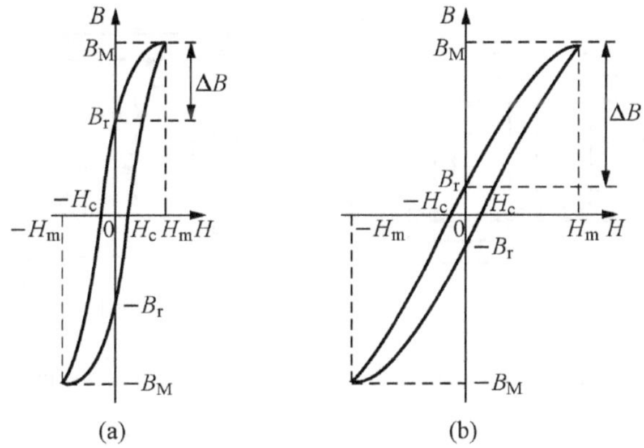

Figure 6.5 Magnetization curve of magnetic core. (a) Magnetization curve without air gap and (b) magnetization curve with air gap

quantity of magnetic flux density $\Delta B = B_M - B_r$ will be increased, thus improving the utilization of magnetization curve. Furthermore, adding air gap can linearize the magnetization curve, that is, the relative magnetic permeability variation reduces, which makes the winding inductance tend to be a constant value. These characteristic changes of high-frequency transformer after air gap is added are all helpful to improve the performance of flyback SMPS.

When the air gap width is comparatively small, the inductance of transformer winding, the winding number of turns, the sectional area of magnetic core, and the air gap width have a relation expression as

$$L \approx \frac{N^2 \mu_0 A_e}{\delta}$$

wherein μ_0 refers to the magnetic permeability in vacuum, equal to $4\pi \times 10^{-7}\,\mathrm{WB/(A \cdot m)}$. When the primary winding number of the turns of high-frequency transformer is N_P and its inductance is L_P, the formula for calculating air gap of transformer magnetic core will be as follows:

$$\delta \approx \frac{0.4\pi N_P^2 A_e}{L_P} \times 10^{-2}, \tag{6.33}$$

wherein the units of δ, A_e, and L_P are cm, cm^2, and μH, respectively.

It needs to be pointed out that the air gap width calculated here refers to the sum of the width of air gap in the magnetic path. As for EI and EE magnetic cores, electrical insulating paper (e.g., fish paper) with a certain thickness is usually used to generate air gap. As shown in Figure 6.6, as the air gap width is the sum of magnetic path gap, the magnetic core gap (the thickness of insulating paper) should be half of the air gap width, that is, $\delta/2$.

6.4.2 Design Examples of Flyback High-Frequency Transformer

Use the single-chip SMPS TOP226Y to design a 60-W flyback universal SMPS module with an AC input voltage of 85–265 V and an output of +12 V and 5 A. The design steps are as follows:

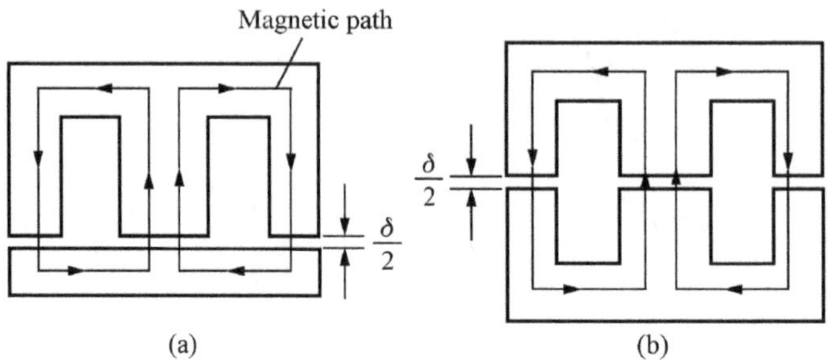

Figure 6.6 Air gap width and magnetic core gap. (a) EI magnetic core and (b) EE magnetic core

6.4.2.1 Calculating the Primary Inductance L_P.

The primary inductance can be calculated as per Formula (6.25). When the power supply efficiency is 80%, let the proportional coefficient K_{RP} between pulsating current (I_R) and peak current (I_P) be 0.7, the switching frequency of TOP226Y is 100 kHz and the drain current limit is $I_{LIMIT} = 2.25$ A, and when $I_P = 2.25$ A, I_R will be $I_R = K_{RP}I_P = 0.7 \times 2.25\,\text{A} = 1.58\,\text{A}$, and it can be obtained that

$$L_P = \frac{2P_O}{\eta I_R^2 f} = \frac{2 \times 60}{0.8 \times 1.58^2 \times 100\,\text{k}} = 600 \ \ (\mu\text{H})$$

When $K_{RP} = 1$, it can be calculated that $L_P = 296\,\mu$H. Therefore, L_P can be selected within the range of 296–600 μH. The medium value of 450 μH is selected for L_P in this example.

Note: There is another formula for calculating L_P:

$$L_P = \frac{(U_{Imin} - U_{DS(ON)})\ D_{max}}{I_R f} \approx \frac{U_{Imin} D_{max}}{I_R f} \tag{6.34}$$

wherein U_{Imin} is the minimum DC input voltage, $U_{DS\ (ON)}$ the conduction voltage drop of power switching tube, and D_{max} the maximum duty ratio. Typically, $U_{DS\ (ON)}$ is only several volts, which can be negligible. Assume that $U_{Imin} = 85\,\text{V} \times 1.2 = 102\,\text{V}$, $D_{max} = 0.6$, $I_R = 1.58$ A, and $f = 100$ kHz, the following can be obtained by putting them into Formula (6.34):

$$L_P \approx \frac{U_{Imin} D_{max}}{I_R f} = \frac{102 \times 0.6}{1.58 \times 100\text{k}} = 387 \ \ (\mu\text{H})$$

It is easy to find that 387 μH obtained through calculation is comparatively close to $L_P =$ 450 μH selected in this example.

It needs to be pointed out that L_P in Formula (6.24) is calculated according to the input power P_I, for $P_O/\eta = P_I$. L_P in Formula (6.34) is calculated according to the minimum DC input voltage U_{Imin}, which is certain to meet the requirements when U_I is comparatively large and D is comparatively small as long as it meets the requirements when SMPS is under the most disadvantageous input condition (U_I is the minimum U_{Imin} and D is the maximum duty ratio D_{max}). That is the main difference between Formula (6.34) and (6.24). It is normal that there is a certain deviation between the results calculated as per the two formulas. Readers can determine which formula to be chosen according to actual situations and design experience.

6.4.2.2 Selecting Magnetic Core

When AP method is used to select magnetic core, either Formula (6.20) or (6.2) can be used for estimation. When Formula (6.20) is used and $\eta = 80\%$, $P_O = 60\,\text{W}$, $K_W = 0.35$, and $D = 0.5$, B_M should be within the range of 0.2–0.3 T for flyback SMPS. Now take $B_M = 0.25\,\text{T}$, $K_{RP} = 0.7$, and $f = 100\,\text{kHz}$, and put them into Formula (6.20) to get

$$\begin{aligned}\text{AP} = A_w A_e &= \frac{0.433(1+\eta)\ P_O}{\eta K_W D J B_M K_{RP} f} \times 10^4 \\ &= \frac{0.433 \times (1+0.8) \times 60}{0.8 \times 0.35 \times 0.5 \times 400 \times 0.25 \times 0.7 \times 100\text{k}} \times 10^4 \\ &= 0.48\ (\text{cm}^4)\end{aligned}$$

According to $\text{AP} = 0.48\,\text{cm}^4$, the minimum magnetic core specification proximate to the AP is EI28 found from Table 6.1, whose $\text{AP} = 0.58\,\text{cm}^4$. In view of various factors such as the magnetic core loss, at least EI30 magnetic core whose $\text{AP} = 0.91\,\text{cm}^4$ and $A_e = 1.09\,\text{cm}^2$ should be selected.

When Formula (6.2) is used for estimation, $A_e = 1.16\,\text{cm}^2$, whose proximate value is $A_e = 1.18\,\text{cm}^2$ of EI33 magnetic core according to Table 6.1. It can be known that the results obtained by using the two methods are basically consistent. To meet the requirements for output power within a wide voltage range, EI33 magnetic core is selected practically in this example.

6.4.2.3 Calculating the Primary Number of Turns N_P

The primary number of turns can be calculated directly according to Formula (6.26). In this example, $U_{Imin} = 102\,\text{V}$, $D_{max} = 0.5$, $B_M = 0.25$, $K_{RP} = 0.7$, and $f = 100\,\text{kHz}$, so it can be obtained that

$$N_P = \frac{U_I \sqrt{D_{max}} \times 10^4}{B_M K_{RP} f} = \frac{102 \times \sqrt{0.5} \times 10^4}{0.25 \times 0.7 \times 100\text{k}} = 41.2\ (\text{turns})$$

In practice, $N_P = 41$ turns is taken.

Putting $I_P = 2.25\,\text{A}$, $D_{max} = 0.5$, and $K_{RP} = 0.7$ into Formula (6.30), it can be obtained that the maximum primary current RMS I_{RMS} is 1.17 A. Take $6\,\text{A/mm}^2$ as the current density, so the six-strand enameled wire of $\phi 0.51\,\text{mm}$ can be chosen to winding in parallel according to Table 6.7.

6.4.2.4 Calculating the Secondary Winding Number of Turns N_S

The secondary winding number of turns N_S can be calculated according to Formula (6.28). When U_O is 12 V, U_{OR} is 130 V, and U_{F1} is 0.5 V,

$$N_s = \frac{N_P}{U_{OR}}(U_O + U_{F1}) = \frac{41}{130} \times (12 + 0.5) = 3.9(\text{turns})$$

In view of the resistance loss of copper wire, in practice $N_S = 4.5$ turns is taken.

Putting $I_P = 2.25\,A$, $N_P = 41$ turns, and $N_S = 4.5$ turns into Formula (6.31), it can be obtained that the secondary peak current $I_{SP} = 20.5\,A$. Then, putting $I_{SP} = 20.5\,A$, $D_{max} = 0.5$, and $K_{RP} = 0.7$ into Formula (6.32), it can be calculated that the secondary current RMS, $I_{SRMS} = 7.64\,A$. Take $6\,A/mm^2$ as the current density, so the enameled wire of $\phi 1.2$ mm should be selected, but in practice the eight-strand enameled wire of $\phi 0.45$ mm is chosen for winding in parallel. The current of feedback winding N_F is comparatively small, so it is advisable that feedback voltage should be slightly higher than 12 V, and in practice, the enameled wire of $\phi 0.3$ mm is chosen for winding four turns..

6.4.3 Calculating Air Gap Width

In flyback SMPS, the air gap size of high-frequency transformer magnetic core has a relatively large impact on the performance of power supply. The air gap width can be calculated according to Formula (6.33). In the example, $N_P = 41$ turns, $L_P = 450\,\mu H$, and $A_e = 1.17\,cm^2$, it can be calculated that

$$\delta \approx \frac{0.4\pi N_P^2 A_e}{L_P} \times 10^{-2} = \frac{0.4\pi \times 41^2 \times 1.17}{450} \times 10^{-2} = 0.055(\text{cm}) = 0.55\ (\text{mm})$$

Inserting the fish paper with a thickness of 0.275 mm between EI magnetic cores, the effective air gap width will be about 0.55 mm (0.275 mm × 2).

6.4.3.1 Verifying the Maximum Magnetic Flux Density B_M

Let $I_P = I_{LIMIT} = 2.25\,A$, and put L_P, N_P, and A_e into the following formula, then

$$B_M = \frac{I_P L_P}{N_P A_e} \times 10^{-2} = \frac{2.25 \times 450}{41 \times 1.17} \times 10^{-2} = 0.21\,(\text{T})$$

The value of B_M calculated from the formula is within the range of 0.2–0.3 T, which can meet the design requirements.

6.4.3.2 Verifying the Magnetic Saturation Current

The purpose of verifying the maximum magnetic flux density B_M is to prevent magnetic saturation from occurring in high-frequency transformer during working. Owing to the deviation in magnetic core parameters and other reasons, the computed value of B_M is only a theoretical one. Measuring magnetic saturation current directly is the best way to verify whether magnetic saturation will occur during the working of high-frequency transformer. According to the method introduced in Section 9.5, the magnetic saturation current of high-frequency transformer is measured by oscilloscope, and the actual value of it is 4.0 A, about 1.7 times of the actual peak current (2.25 A), which can ensure no magnetic saturation during the working of high-frequency transformer.

6.5 Design of Forward High-Frequency Transformer

6.5.1 Steps in Designing Forward High-Frequency Transformer

1. Design steps. Calculating the total output power → using the area product method (AP method) to select magnetic core → calculating the number of the turns of the primary winding → calculating the number of the turns of the secondary winding → calculating such parameters as the wire diameter.
2. Main formulas. The minimum number of the turns of the primary winding is

$$N_{\mathrm{P(min)}} = \frac{U_{\mathrm{I\ (min)}} D_{\max}}{\Delta B A_{\mathrm{e}} f} \tag{6.35}$$

wherein $N_{\mathrm{P(min)}}$ is the minimum number of the turns of the primary winding, $U_{\mathrm{I\ (min)}}$ the minimum DC input voltage, and $D_{\max}$ the maximum duty ratio. ΔB is the variation of magnetic flux density, and the ΔB of single-end forward type is $\Delta B = B_{\mathrm{m}} - B_{\mathrm{r}} = B_{\mathrm{AC}}$. A_{e} is the effective sectional area of magnetic core (cm^2) and f is the switching frequency.
3. Calculating the turn ratio n

$$n = \frac{N_{\mathrm{P}}}{N_{\mathrm{S}}} = \frac{U_{\mathrm{I\ (min)}} D_{\max}}{U_{\mathrm{O}} + U_{F1}} \tag{6.36}$$

4. Calculating the number of the turns of the secondary winding N_{S}

$$N_{\mathrm{S}} = n N_{\mathrm{P}} \tag{6.37}$$

5. Calculating the diameter of the primary winding wire D_{Pm}
   a. the average input current I_{AVG}

$$I_{\mathrm{AVG}} = \frac{P_{\mathrm{O}}}{\eta U_{\mathrm{Imin}}} \tag{6.38}$$

   b. the primary peak current I_{P}

$$I_{\mathrm{P}} = \frac{I_{\mathrm{AVG}}}{(1 - 0.5K_{\mathrm{RP}})\ D_{\max}} \tag{6.39}$$

   c. the primary RMS current I_{RMS}

$$I_{\mathrm{RMS}} = I_{\mathrm{P}} \sqrt{D_{\max}\left(\frac{K_{\mathrm{RP}}^2}{3} - K_{\mathrm{RP}} + 1\right)} \tag{6.40}$$

   d. Select an appropriate current density and then calculate the wire diameter.The current density of the primary winding wire can be selected as 4–6 A/mm^2. According to the value of J, the diameter of the primary winding wire can be calculated as follows:

$$d_{\mathrm{P}} = \sqrt{\frac{4I_{\mathrm{RMS}}}{\pi J}} \tag{6.41}$$

6. Calculating the diameter of the secondary winding wire D_{Pm}
   a. the secondary peak current I_{SP} (A)

$$I_{SP} = I_P \cdot \frac{N_P}{N_S} \tag{6.42}$$

   b. the secondary RMS current I_{SRMS} (A)

$$I_{SRMS} = I_{SP}\sqrt{(1 - D_{max}) \cdot \left(\frac{K_{RP}^2}{3} - K_{RP} + 1\right)} \tag{6.43}$$

   c. the ripple current of output filter capacitor I_{RI} (A)

$$I_{RI} = \sqrt{I_{SRMS}^2 - I_O^2} \tag{6.44}$$

   d. the minimum diameter (bare wire) of secondary winding wire D_{Sm} (mm)

$$D_{Sm} = 1.13\sqrt{\frac{I_{SRMS}}{J}} \tag{6.45}$$

The following items should be noted in designing forward SMPS:

1. Synchronous rectification technology can be selected for forward SMPS with low voltage and large current.
2. Magnetic reset of single-end forward SMPS. The disadvantage of single-end forward DC/DC converter is that the high-frequency transformer must be reset during the turn-off of power tube to prevent the saturation of the transformer magnetic core, so it is generally necessary to add a magnetic reset circuit (also known as the transformer reset circuit).

It is not required to take into account magnetic reset in designing half-bridge/full-bridge forward converter. The reason is that the exciting currents of the positive and negative half cycle in the primary winding are equal with opposite directions, making the magnetic flux variation of transformer magnetic core move up and down symmetrically, and the maximum variation range of magnetic flux density B is $\Delta B = 2B_m$, so the DC component of the magnetic core can be balanced out. Similarly, it is also unnecessary to take into account magnetic reset in designing push-pull converter.

6.5.2 *Three Kinds of Commonly used Magnetic Reset Circuit*

See Figure 6.7 for the three kinds of commonly used magnetic reset circuit. Figure 6.7(a) refers to the reset circuit constituted by auxiliary winding, while Figure 6.7(b) refers to the clamping circuit constituted by R, C, and VD_Z, and Figure 6.7(c) the active clamping circuit constituted by MOSFET V_4. The three kinds of magnetic reset methods have their advantages and disadvantages, respectively. Specifically speaking, the auxiliary winding reset method will make the structure of transformer complex, while the R, C, and VD_Z clamping method is passive clamp, whose advantage is that the magnetic reset circuit is simple and capable of absorbing the spike voltage generated by the leakage inductance of high-frequency transformer. However, the clamping circuit itself also consumes magnetic field energy. Active clamping method is the most efficient among the above three methods, but it will also increase the circuit cost.

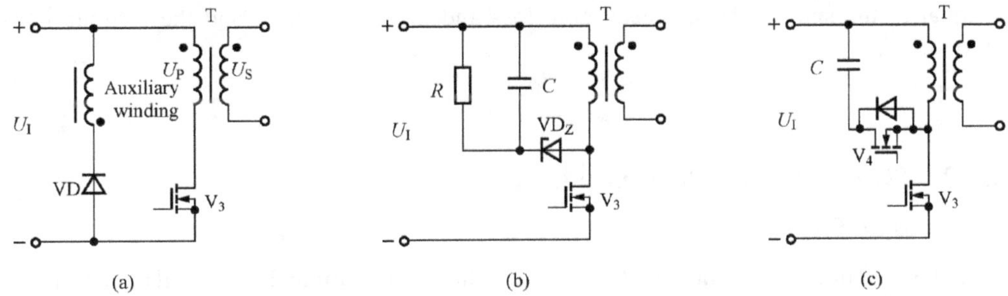

Figure 6.7 Three kinds of magnetic reset circuit commonly used in single-end buck synchronous rectifier. (a) Reset circuit constituted by auxiliary winding, (b) clamping circuit constituted by *R*, *C*, and VD_Z, and (c) active clamping circuit

6.6 Loss of High-Frequency Transformer

6.6.1 Loss of High-Frequency Transformer

High-frequency transformer loss is the main reason for the temperature rise of high-frequency transformer. High-frequency transformer loss mainly includes copper loss and magnetic core loss. Let P be the loss of high-frequency transformer, the symbol of copper loss is P_{Cu}, and the symbol of magnetic core loss can be P_{CORE}, and when the magnetic flux density increases, the magnetic hysteresis loss increases but the copper loss is reduced. When the magnetic core loss is approximately equal to the copper loss, the selected magnetic flux density is the best. See Figure 6.8 for the relation curve of the total loss of high-frequency transformer and the magnetic flux density. The total loss of high-frequency transformer

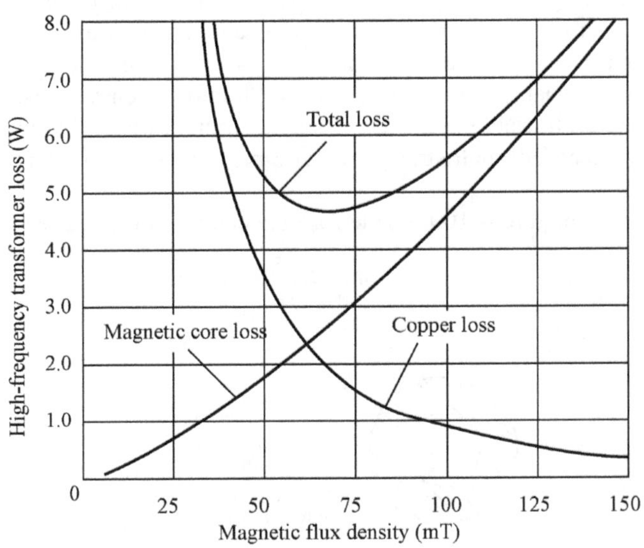

Figure 6.8 Relation curve of the total loss of high-frequency transformer and the magnetic flux density

is equal to the sum of the magnetic core loss and the copper loss, and the formula is as follows:

$$P = P_{Cu} + P_{CORE} \tag{6.46}$$

6.6.2 Skin Effect and Proximity Effect

6.6.2.1 Skin Effect

When high-frequency current flows through a conductor, the current flowing will be centralized on the conductor surface, which is called skin effect. Current is limited to a part of cross-section of a conductor due to the skin effect, which not only reduces the effective utilization area of wire but also increases equivalent resistance. Skin effect can be expressed as "skin depth" d under the frequency. See Figure 6.9 for the diagram for skin effect. Current tends to flow through the shadow area of the wire. The remaining part of the wire is the unused area. The skin depth is determined by the following formula:

$$d = \sqrt{\frac{1}{\pi f \mu \sigma}}, \tag{6.47}$$

wherein f is the switching frequency, μ the magnetic permeability of wire, and σ the electrical conductivity of wire. The unit of d is centimeter.

For copper wire, at an ambient temperature of 20 °C, Formula (6.47) can be simplified as

$$d = \frac{6.61}{\sqrt{f}} \tag{6.48}$$

Obviously, when switching frequency is fixed, skin depth is a constant. Therefore, the unused area of copper wire can be reduced by winding multistrand wire in parallel.

For example, when the switching frequency $f = 100\,\text{kHz}$, it can be obtained that skin depth $d = 0.0209\,\text{cm}$ by putting it into Formula (6.48). The minimum inner diameter of wire selected $D_{min} = 2d = 0.0418\,\text{cm} = 0.418\,\text{mm}$ and the enameled wire with an inner diameter of $D = 0.45\,\text{mm}$ is selected in practice, corresponding to 25# enameled wire in the American Wire Gage (AWG). This indicates that in designing the high-frequency transformer of 100 kHz SMPS, the maximum allowable diameter of single-strand wire of the winding is 0.45 mm. Once this value is exceeded, multistrand fine wire must be wound in parallel to reduce skin effect.

When ambient temperature is 100 °C, the formula can be simplified as

$$d = \frac{7.65}{\sqrt{f}} \tag{6.49}$$

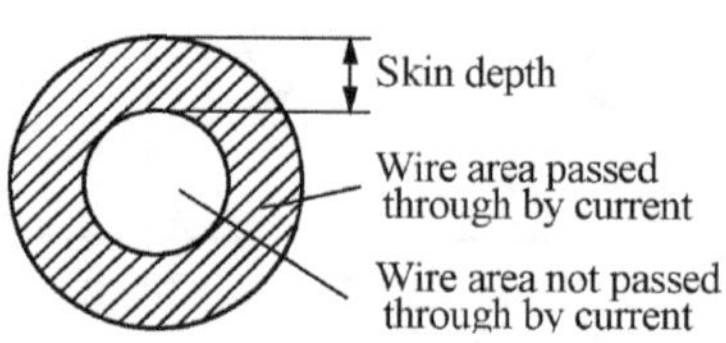

Figure 6.9 Diagram for skin effect

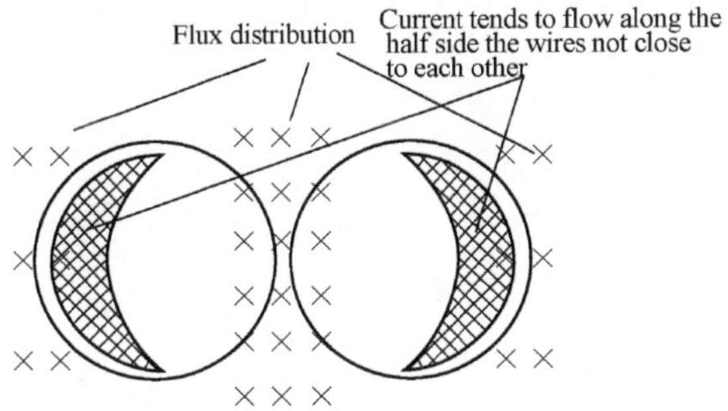

Figure 6.10 Diagram for proximity effect

6.6.2.2 Proximity Effect

Proximity effect relates to the magnetic field between two wires with a very short distance. When the switching currents in two adjacent wires of high-frequency transformer are in the same direction, the currents tend to flow along the half wire side not close to each other as shown in Figure 6.10, wherein × refers to the direction of the magnetic field. Similarly, if the directions of switching current are opposite, the current will tend to flow along the half wire side close to each other. Both of the above situations will reduce the effective wire area. When a multilayer structure is used by high-frequency transformer, the influence of proximity effect is greater than that of skin effect. Try to use fewer multilayer structures in designing.

7

Examples of SMPS Optimization Design

7.1 Multioutput SMPS Design

Many electronic products (e.g., electronic instruments, microcomputers, color TVs, set-top boxes, and video recorders) require multiple regulated power supplies. The following section describes in detail the optimization design of multioutput SMPS through a typical example.

7.1.1 Circuit Design Scheme of Multioutput Single-Chip SMPS

7.1.1.1 Determining the Technical Indicators of Multioutput

Assume that the SMPS to be designed has three outputs: the main output U_{O1} (5 V, 2 A, and 10 W), the auxiliary output U_{O2} (12 V, 1.2 A, and 14.4 W), and U_{O3} (30 V, 20 mA, and 0.6 W). The gross output is 25 W. See Table 7.1 for the detailed technical indicators. The voltage regulation performance of each output is crucial for the design of circuit structure and high-frequency transformer. Typically, the stability of main output is higher than that of auxiliary output. Now take +5 V as the main output, specifically for CMOS and TTL digital circuits with the load regulation $S_I \leq 1\%$. The load regulation of the other two auxiliary outputs $S_I \leq \pm 5\%$.

7.1.1.2 Determining the Feedback Circuit

The multioutput feedback circuit has four types: the basic feedback circuit, the improved basic feedback circuit, the optocoupler feedback circuit with regulator tube, and the optocoupler feedback circuit with TL431. The fourth has the best voltage regulation performance. See Table 7.2 for the four optional types of feedback circuit for multioutput. It shall be noted that the load regulation of multioutput S_I is greater than that of single output, and the indicators of main output are superior to those of auxiliary output.

1. The basic feedback circuit uses the feedback winding to indirectly acquire the variation signal of output voltage, so it is not necessary to adopt any optical coupler. The circuit of

Optimal Design of Switching Power Supply, First Edition. Zhanyou Sha, Xiaojun Wang, Yanpeng Wang, and Hongtao Ma.

Table 7.1 Technical indicators of multioutput single-chip SMPS

Main output			Auxiliary output						Gross output P_O (W)
Output 1			Output 2			Output 3			
U_{O1} (V)	I_{O1} (A)	P_{O1} (W)	U_{O2} (V)	I_{O2} (A)	P_{O2} (W)	U_{O3} (V)	I_{O3} (mA)	P_{O3} (W)	
+5 (±5%)	0.4–2.0	10	+12 (±10%)	0.12–1.20	14.4	+30 (±10%)	10–20	0.6	25

Table 7.2 Four optional types of feedback circuits for multioutput

Feedback circuit type	Load regulation S_I		Circuit description
	Main output (%)	Auxiliary output (%)	
Basic feedback circuit	±5	> ±10	A regulator tube is connected in parallel to the output terminal to improve the load regulation at light load
Improved basic feedback circuit	±2.5	> ±10	A regulator tube and a capacitor are added in the feedback circuit
Optocoupler feedback circuit with regulator tube	±2	> ±5	The reference voltage is provided by the regulator tube
Optocoupler feedback circuit with TL431	±1	≤ ±5	The reference voltage with high stability is provided by TL431. In addition to the feedback signals provided by the main output, other outputs also provide feedback in certain proportion

the scheme is the simplest one, but the stability of SMPS is not high, and it is difficult to reduce the load regulation S_I to below ±5%. If it is only intended to improve the load regulation at light load, a suitable regulator tube can be connected in parallel to the output terminal to let the regulated voltage $U_Z = U_{O1}$, at which time $S_I < \pm 5\%$ at light load.

2. Improved (also known as enhanced) basic feedback circuit. It is featured that a 22-V regulator tube is connected in series and then a 0.1-μF capacitor is connected in parallel in the feedback circuit.
3. Optocoupler feedback circuit with regulator tube. Regulated voltage of a regulator tube is applied as the secondary reference voltage. The output voltage is determined by the sum of the following three values: the regulated voltage (U_Z) of regulator tube, the forward voltage drop (U_F) of LED in optical coupler, and the voltage drop (U_{R1}) of series resistor R_1 used to control loop gain. When the deviation of U_Z is less than 2%, the load regulation of main output can be controlled within ±2%. The disadvantage of the circuit is that the stability of reference voltage is not high and feedback is provided to the main output only, not to the other auxiliary outputs, so the voltage stability of auxiliary output is poor.

4. Multioutput optocoupler feedback circuit with TL431. It has the following features: (i) TL431 adjustable precision shunt regulator is introduced to constitute the secondary error current amplifier and then precisely adjust the main output through the optical coupler. (ii) In addition to the main feedback signals provided by the main output, other auxiliary outputs also provide feedback to the 2.50-V reference terminal of TL431 in certain proportion, which is important to comprehensively improve the voltage regulation performance of multioutput SMPS. (iii) The load regulation of main output is up to ±1%.

7.1.1.3 Designing the SMPS Circuit

See Figure 7.1 for the 25-W multioutput SMPS designed according to the abovementioned principles.

In the circuit, a TOP223Y three-terminal single-chip SMPS is applied with the AC input voltage range of 85–265 V. The secondary side of high-frequency transformer has three separate windings, and the optocoupler feedback circuit with TL431 is designed only at the main output terminal (+5 V).

Multioutput SMPS also has two working modes: (i) the discontinuous mode (DCM), which is featured that the magnetic core with small size can be used for the high-frequency transformer under the same output power, and (ii) the continuous mode (CCM), which is featured that it can improve the utilization of TOPSwitch. Multioutput SMPS generally uses CCM, as the focus is no longer the size of high-frequency transformer but rather how to achieve the best match of multiple secondary windings and the printed circuit.

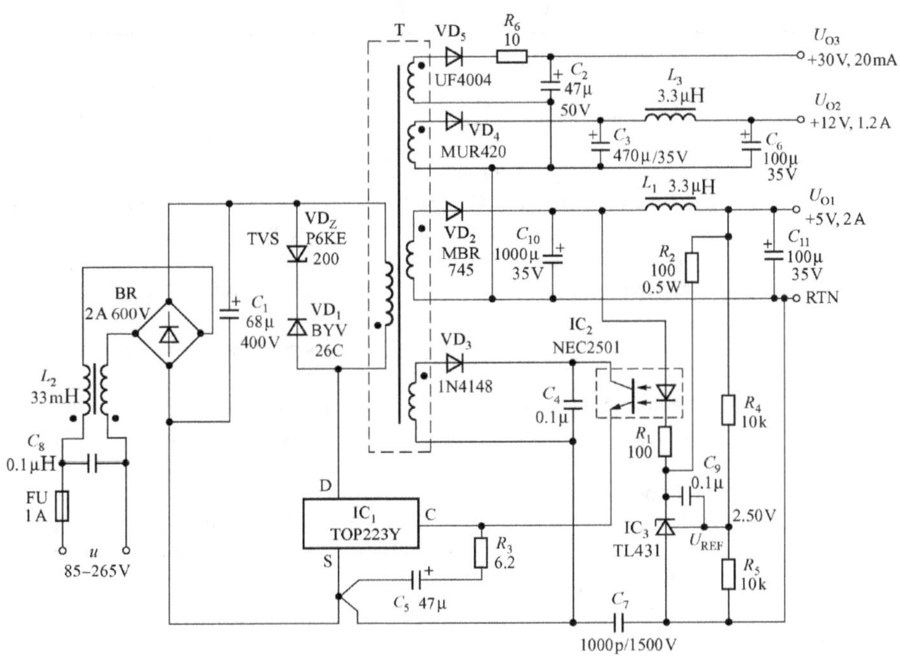

Figure 7.1 25 W multioutput SMPS circuit

7.1.2 Design of Multioutput High-Frequency Transformer

EE29 ferrite core is employed for high-frequency transformer with the effective magnetic flux area $S_J = 0.76\,cm^2$. The air gap width $\delta = 0.38\,mm$. The effective frame width is 26 mm. 77 turns of $\phi 0.3\,mm$ enameled wire are adopted for primary winding and 9 turns of $\phi 0.3\,mm$ enameled wire for bias winding. The secondary winding includes two types: the separate and superimposed windings. See Table 7.3 for the comparison between two types of secondary windings, whose structures are shown in Figure 7.2(a) and (b), respectively. The superimposed winding can improve the cross-regulation characteristic of auxiliary output. Each output winding of the separate winding type uses an individual wire and transmits only the current relating to its specific load, which is flexible in determining the sequence of the windings. Designers can design two secondary windings separately and provide electrical isolation between them. When using the superimposed winding, the high-voltage winding is superimposed on the low-voltage winding. The same wire is adopted for these superimposed windings so that the low-voltage winding can share the load current of high-voltage winding. The initial terminal of superimposed winding refers to the common point (typically the ground point) of all superimposed windings. After the frame is wound with necessary turns of winding, the winding of each superimposed winding is ended at the frame pin of high-frequency transformer. The next superimposed winding is wound from the end pin of the previous winding. Therefore, each winding with higher voltage is "superimposed" on the next winding with lower voltage but shall be provided with electrical isolation with the primary and feedback windings. In addition, the separate and superimposed windings can be combined for use.

Table 7.3 Comparison between two types of secondary windings

Winding method	Advantages	Disadvantages
Separate winding	Flexible sequence The output with higher output current can be positioned close to the primary side to minimize the energy loss caused by leakage inductance	It will generate the peak charging effect in the output filter capacitor owing to its large leakage inductance, thus worsening the load regulation at light load High manufacturing cost There are more pins on the frame (6 in total)
Superimposed winding	It can enhance the magnetic coupling It can improve the voltage regulation performance at light load There are fewer pins on the frame (only 4) Low manufacturing cost	The winding with the minimum (or maximum) voltage shall be close to the primary side Not flexible enough to reduce the leakage inductance under high current

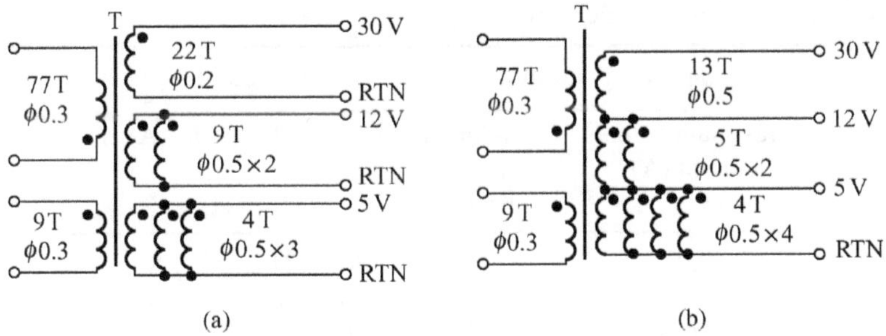

Figure 7.2 Two types of secondary windings. (a) Separate winding and (b) superimposed winding

In Figure 7.2(b), the +5 V winding provides the +12 V winding with partial turns and the ground terminal, and the +30 V winding contains the +5 and +12 V windings and newly added turns. The wire diameter of each winding shall meet the requirement of the sum of current flowing through it from itself and other outputs. The superimposed winding is a kind of advanced technology, which can not only save the wire to reduce winding size and costs but also increase the mutual inductance between windings to enhance the degree of coupling. For example, when the +5 V output is at full load and the +12 and +30 V outputs are at light load, the leakage inductance of these windings can be reduced as +5 V is concurrently applied as part of +12 and +30 V, so that the filter capacitor in +12 and +30 V output circuits can be prevented from being charged to the peak by the spike voltage (also known as peak charging effect) owing to the leakage inductance, thus avoiding unstable output voltage. The disadvantage of superimposed winding is that it is not flexible enough to determine which secondary winding shall be close to the primary winding. Now the +5 V winding is adopted as the initial terminal of secondary winding.

If the separate winding is used, one winding selected from +5 V (2 A) and +12 V (1.2 A) can be close to the primary side in view that most power are from them. The best sequence is that the +5 V winding is wound first, and then the +12 and +30 V in turn, to realize best coupling between the secondary windings with minimum leakage inductance. Conversely, if the +30 V winding is close to the primary winding, the power supply efficiency will be reduced and the interference will be increased owing to the large leakage inductance from +5 and +12 V windings.

When winding, the multistrand wire is recommended to be connected in parallel and then wound on the frame in parallel to ensure good coverage, enhancing the degree of coupling between the primary and secondary sides.

When calculating the number of turns of each secondary winding, the same turn ratio per volt (N_V) can be applied. Assume that the secondary winding number of turns of the main output is N_S, the output voltage is U_{O1}, and the forward conduction voltage drop of output rectifier is U_{F1}, then N_V is determined by the following formula:

$$N_V = \frac{N_S}{U_{O1} + U_{F1}} \tag{7.1}$$

Put $N_S = 4$ turns, $U_{O1} = 5\,V$, and $U_{F1} = 0.4\,V$ (the voltage drop of Schottky rectifier) into Formula (7.1), it can be obtained that $N_V = 0.74$ turns/V. The number of turns of other windings can also be calculated.

Table 7.4 Selection of output rectifier tubes

Output voltage U_O (V)	Specified indicator		Rectifier tube model and parameter		
	Maximum output current (A)	Minimum withstand voltage (V)	Model	I_F (A)	U_{RM} (V)
5	2.0	30	MBR745	7.8	45
12	1.2	70	MUR420	4.0	200
30	20 mA	170	UF4004	1.0	400

For the +12 V output, it is known that $U_{O2} = 12\,\text{V}$ and $U_{F2} = 0.7\,\text{V}$ (the voltage drop of FRD), so $N_{12} = 0.74\ \text{turns/V} \times (12\,\text{V} + 0.7\,\text{V}) = 9.4$ turns, actually 9 turns.

For the +30 V output, it is known that $U_{O3} = 30\,\text{V}$ and $U_{F3} = 0.7\,\text{V}$ (the voltage drop of silicon rectifier at low current output), so $N_{30} = 0.74\ \text{turns/V} \times (30\,\text{V} + 0.7\,\text{V}) = 22.7$ turns, actually 22 turns in practice.

When selecting the parameters of output rectifier, the following principles shall be obeyed: the rated working current (I_F) of rectifier shall be at least three times of the maximum output current of the output path, and the maximum reverse working voltage (U_{RM}) of rectifier shall exceed the required minimum withstand voltage (U_R). See Table 7.4 for the models and parameters of the output rectifier selected as per the above principles. It is easy to find that all the technical indicators of the selected rectifier tubes have some margin.

7.2 Methods to Improve the Cross-Load Regulation of Multioutput SMPS

Cross-load regulation refers to the change rate of each output voltage when one load of multi-output SMPS is changed. It is a key parameter to measure the voltage regulation performance of multioutput SMPS.

In the multioutput SMPS circuit shown in Figure 7.1, feedback signals are from the +5 V main output only and other outputs have no feedback circuit. Thus, when the load current of +5 V output changes, the stability of +12 V output will be affected. The solution is to connect a feedback circuit to the +12 V output as shown in Figure 7.3. Connect the resistor R_6 in parallel between the +12 V output terminal and the reference terminal of TL431 and increase the resistance of R_4 from 10 to 20 kΩ. The +12 V output also provides part of feedback signal, so the stability of the output path can be improved. Before improvement, when the load current of +5 V main output changes from 0.5 to 2.0 A (i.e., from 25% to 100% of the full-load current), the load regulation S_I of +12 V output is ±2%. After improvement, $S_I = \pm 1.5\%$. See Figure 7.4 for the comparison of load characteristic curves before and after improvement. The following text describes the method to design the feedback circuit of +12 V output.

The amount of feedback from +12 V output is determined by the resistance of R_6. Assume that the feedback amount of 12 and 5 V outputs shall be both equal to half of the total feedback amount, that is, the feedback proportion factor $K = 50\%$. In this case, the current flowing through R_6 and R_4 shall be equal, that is, $I_{R_6} = I_{R_4}$. The voltage U_{REF} at the reference terminal

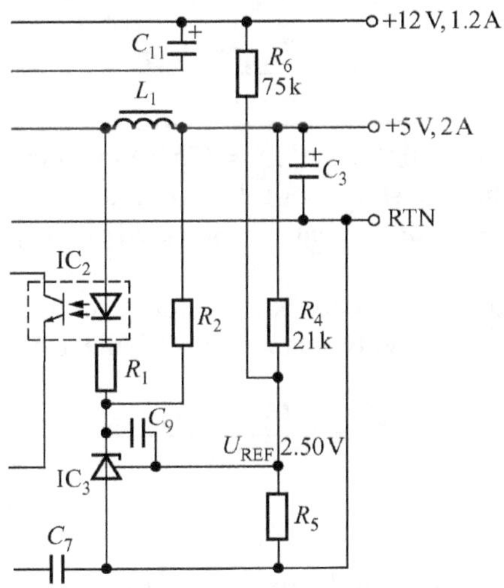

Figure 7.3 Circuit with feedback simultaneously provided by 5-V and 12-V outputs

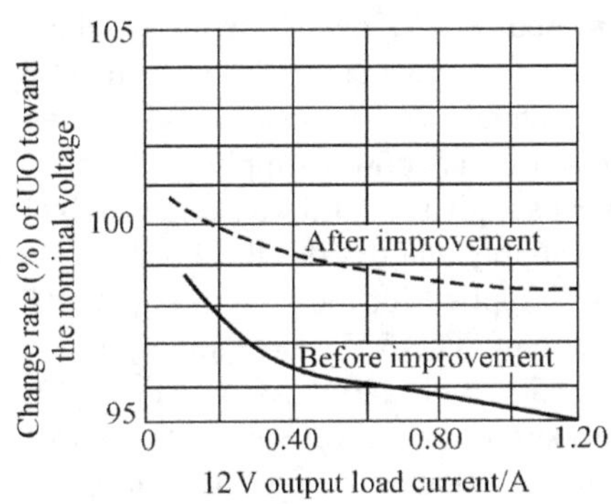

Figure 7.4 Comparison of load characteristic curves before and after improvement

of TL431 is 2.50 V. Before improvement, all feedback current flows through R_4, so

$$I_{R_4} = \frac{U_{O1} - U_{REF}}{R_4} = \frac{5\,V - 2.50\,V}{10\,k\Omega} = 250\,\mu A$$

After improvement, 50% current flows through R_6, that is, $I_{R_6} = 250\,\mu A/2 = 125\,\mu A$. The resistance of R_6 is determined by the following formula:

$$R_6 = \frac{U_{O2} - U_{REF}}{I_{R_6}} \tag{7.2}$$

Substituting $U_{O2} = 12\,V$, $U_{REF} = 2.50\,V$, and $I_{R_6} = 125\,\mu A$ into Formula (7.2), it can be obtained that $R_6 = 76\,k\Omega$, nominal resistance of $75\,k\Omega$ is desirable. As I_{R_4} has been reduced from $250\,\mu A$ to $I'_{R_4} = 125\,\mu A$, it is necessary to adjust the resistance of R_4 according to the following formula:

$$R_4 = \frac{U_{O1} - U_{REF}}{I'_{R4}} \tag{7.3}$$

Putting $U_{O1} = 5\,V$, $U_{REF} = 2.50\,V$, and $I'_{R_4} = 125\,\mu A$ into Formula (7.3), it can be obtained that $R_4 = 20\,k\Omega$. In view that the stability of +5 V output will be slightly reduced after the connection to R_6, the resistance of R_4 shall be increased slightly for compensation, actually taken $R_4 = 21\,k\Omega$.

Two points shall be noted: firstly, feedback circuit can also be added to the +30 V output by referring to the above method. Secondly, when $K \neq 50\%$, the resistance of R_6 can be calculated as per the following formula:

$$R_6 = \frac{U_{O2} - U_{REF}}{K \times 250 \times 10^{-6}} \tag{7.4}$$

7.3 Design of PC SMPS with Magnetic Amplifier

AT power supply is almost used by all SMPSs in early PC (e.g., from 286 to 586). The output power of AT power supply is generally 150–250 W with four outputs (+5, −5, +12, and −12 V) in total. In addition, a power good (PG) signal is provided to the mainboard. The disadvantage of AT power supply is that the shutdown can only be realized by turning off the AC power supply, incapable of software shutdown. At present, AT power supply has been out of the market with the popularization of ATX power supply.

The popular ATX2.01 power supply standard was launched by Intel in 1997. Compared with AT power supply, ATX power supply mainly adds the 3.3 V output voltage and a PS-ON signal, wherein 3.3 V power supply is applied for the low-voltage CPU, greatly reducing the power consumption of mainboard circuit. The 5 V power supply is also known as auxiliary power supply. The output voltage is 5 V with 220 V AC power supply. PS-ON signal is the level signal provided by the mainboard to the power supply to control other voltage outputs of the power supply. Functions such as software startup/shutdown and network-based remote wakeup can be realized just through the 5 V power supply and PS-ON signal. The power supply will be turned on when the PS-ON signal sent by the mainboard to the power supply is at low level and turned off when the PS-ON signal is at high level. The main technical indicators of ATX power supply include the output power, the safety standards (e.g., China's CCEE certification), electromagnetic interference (EMI) characteristics, the delay time of "power fail" (PF), "power good" signal, and so on.

The power of PC SMPS must meet the requirement of entire machine and reserve some margin. Currently, as PC is developed in the "green" energy-saving and environment-friendly direction, high source power is not always better. The power of PC power supply specified in the Micro-ATX standard newly launched by Intel is only 145 W, and can even be reduced to 90 W. ATX power supply has become the mainstream product of PC power supply.

7.3.1 Main Circuit Design of 145 W Multioutput PC SMPS

See Figure 7.5 for the main circuit of 145 W multioutput PC SMPS constituted by TOP247Y.

The AC input voltage range is 90–130 V (typically 110 V) or 180–265 V (typically 220 V).

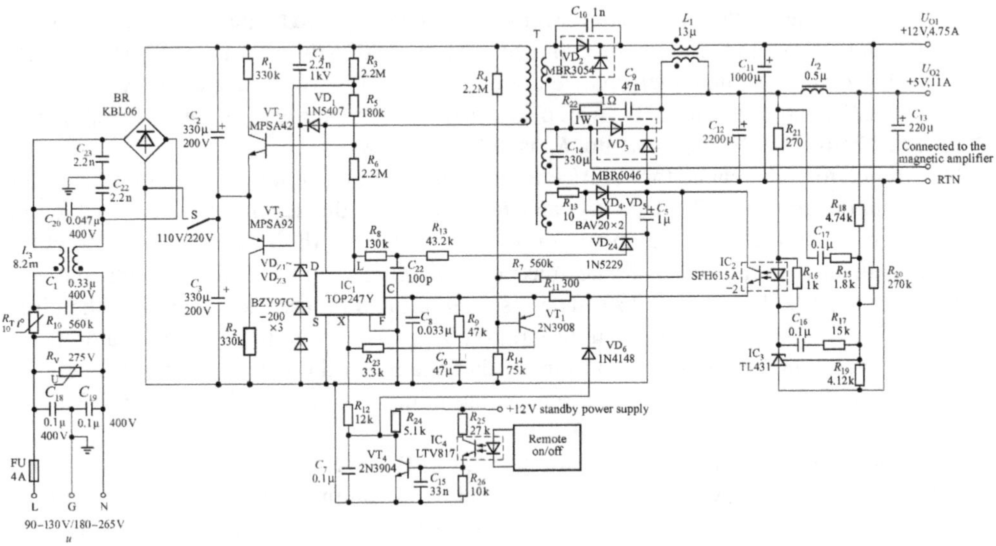

Figure 7.5 Main circuit of 145-W multioutput PC SMPS constituted by TOP247Y

The three outputs are U_{O1} (+12 V, 4.75 A), U_{O2} (+5 V, 11 A), and U_{O3} (+3.3 V, 10 A), respectively. In order to be compatible with AT power supply, high-frequency transformer has no special +3.3 V winding but adopts the 5 V winding voltage to obtain +3.3 V output through the external magnetic amplifier circuit, simplifying the design of high-frequency transformer. The voltage regulation performance can be further improved through the magnetic amplifier. The gross power is 145 W and the peak output power is up to 160 W. The remote on/off circuit is added to remotely control the turn-on and turn-off of SMPS. The power supply efficiency $\eta \geq 71\%$. When the input power is only 0.91 W, the output power is up to 0.5 W with the power consumption of only 0.41 W, meeting the requirement that the power consumption of power supply shall not exceed 1 W under such condition. S refers to the 110 V/220 V AC input voltage selector switch. When S is closed, 110-V voltage doubling rectifying circuit is selected. Here, the equalization resistor is replaced with transistors VT_2 and VT_3 and resistors R_1, R_2, R_3, R_5, and R_6 to constitute the equalizing circuit of filter capacitors C_2 and C_3. This circuit can reduce the resistance loss. During the circuit design, MPSA42 high-voltage NPN transistor and MPSA92 high-voltage PNP transistor are adopted for VT_2 and VT_3, respectively, which are complementary geminate transistors with the following main parameters: $U_{(BR)CEO} = 300\,V$, $I_C = 0.5\,A$, $P_D = 0.625\,W$, and $h_{FE} = 25$ times. When S is disconnected, 220 V AC is selected. In this case, C_2 and C_3 are connected in series and the total capacitance becomes 165 μF.

R_V is the voltage-sensitive resistor. When the surge voltage of the grid exceeds 275 V, R_V is broken down rapidly for clamping protection. R_T is the negative temperature coefficient resistor used for current limit protection when the power supply is turned on. The EMI filter at the AC input terminal is constituted by C_{18}, C_{19}, and C_1, the common-mode chokes L_3, C_{20}, C_{22}, and C_{23}, and R_{10}, wherein, C_1, C_{22}, and C_{23} are safety capacitors (X capacitor). R_{10} is the bleeder resistor, which can release the charge accumulated in the capacitor when the power supply is turned off.

The under-voltage at the start-up of power supply is determined by the total series resistance of R_3, R_5, and R_6. When the AC mains voltage is below 180 V, the start-up of SMPS is prohibited. In addition, a separate under-voltage protection circuit is constituted at pin X by resistors R_4, R_{14}, and R_{23} and transistor VT_1, which allows the power supply to continue working at the DC voltage below 140 V after start-up. R_7 is delay resistor.

The "regulator tube/capacitor reset/clamping" protection circuit is constituted by diode VD_1, regulator tubes VD_{Z1}–VD_{Z3}, C_4, and R_{22} and C_9 in the secondary circuit, which can provide a reset voltage and clamp the drain voltage within the safe range (below 600 V) regardless of any circumstances. The maximum magnetic flux density of high-frequency transformer shall be less than 0.25 T. The reset circuit can be used together with the circuit automatically reducing the maximum duty ratio (D_{max}) to prevent the magnetic saturation of high-frequency transformer and avoid circuit damage when the load is in short circuit. The circuit that can automatically reduce the maximum duty ratio is constituted by R_8, R_{13}, C_{22}, VD_{Z4}, and VD_5.

The remote on/off circuit is constituted by R_{12}, C_7, R_{24}, VT_4, C_{15}, R_{25}, R_{26}, optical coupler IC_4, and VD_6. When turned on, VT_4 is conducted by the output signal of IC_4, so the pin X is connected to the control terminal C through resistors R_{12}, VD_6, and R_{11}. When turned off, IC_4 and VT_4 are in the off state, so the pin X is connected to the external +12 V standby power supply through R_{12} and R_{24} to let TOP247 enter into the off state. The +12 V standby power supply provides the control terminal of TOP247 with current through R_{24} and VD_6 to reduce the power consumption of SMPS to 2 mW. R_{11} is the bias resistor.

Precision optocoupler feedback circuit is constituted by the optical coupler IC_2 (SFH615A) and the adjustable precision shunt regulator IC_3 (TL431). 5 V is applied for SMPS as the main output and 12 V as the auxiliary output. 3.3 V is obtained after the 5 V winding voltage passes through the external magnetic amplifier circuit.

7.3.2 *Circuit Design of 3.3-V Magnetic Amplifier*

The output 3 of PC is U_{O3} (+3.3 V, 10 A). However, the output current of traditional linear regulator is generally only below several amperes. In the case of high output current, the radiator with relatively large size shall be configured owing to its low power supply efficiency and high caloric value, which further increases the size of SMPS and the cost. The output of +3.3 V and 10 A can be obtained through multioutput SMPS, but the circuit will be more complex and it will be easy to generate noise interference. In recent years, the magnetic amplifier regulator circuit constituted by high-performance amorphous alloy magnetic ring is widely used in multioutput PC SMPS, especially suitable for high output current from 1 A to tens of amperes with good regulation performance, high efficiency, small size, and low cost, worthy of popularization. The magnetic amplifier regulation circuit changes the delay time of magnetic reset through the controllable magnetic saturation inductor and realizes precise regulation by finely adjusting the pulse width.

See Figure 2.13 for the basic principle of magnetic amplifier regulator circuit in flyback SMPS. See Figure 7.6 for the 3.3-V magnetic amplifier regulator circuit in PC SMPS. The magnetic amplifier is constituted by the sampling circuit (R_{24} and R_{26}), the adjustable precision shunt regulator (TL431), the magnetic reset control circuit (3 A/40 V PNP power tube TIP32), the controllable magnetic saturation inductor (L_4), and so on. The sampling voltage U_Q, obtained after the voltage division of the 3.3 V voltage by R_{24} and R_{26}, is connected to the

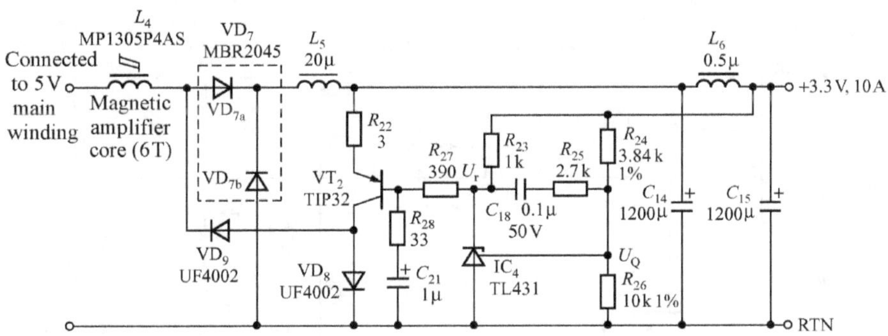

Figure 7.6 Voltage regulator circuit of 3.3-V magnetic amplifier in PC SMPS

output voltage setting terminal (U_{REF}) of TL431. After comparing U_Q with the 2.5-V band-gap reference voltage, the error voltage U_r will be obtained, which is applied to the base electrode of VT_2 through R_{27}. The collector current of VT_2 flows through SRD VD_9 (UF4002) to the right terminal of L_4. An MBR2045 20 A/45 V Schottky geminate transistor VD_7 produced by Samsung is shared by the output rectifier and the freewheeling diode, which contains the rectifier VD_{7a} and the freewheeling diode VD_{7b}. C_{14} is the output filter capacitor. L_6 and C_{15} constitute the postfilter.

The working principle of magnetic amplifier is analyzed as follows: when the MOSFET in TOP247Y is conducted, the output rectifier VD_{7a} is turned off, VD_{7b} is turned on, and the electrical energy stored in C_{14} and C_{15} provides the load with current. At this time, L_4 has high impedance for high-frequency switching current. When MOSFET is turned off, VD_{7a} is turned on not immediately, but after a delay time. Owing to the presence of the magnetic reset current, the forward current in the secondary winding must first cancel out the magnetic reset current and then flow through L_2 to make it enter into magnetic saturation and have low impedance, thus letting VD_{7a} be conducted. The duration of magnetic reset is the delay time blocking the output. At that time, the output is turned on to not only provide current to the load but also store some energy in the output filter capacitors C_{14} and C_{15}, so that the output voltage can be remained unchanged during the turn-off of VD_{7a}.

For example, when the output voltage U_{O1} (3.3 V) is increased owing to the sudden reduction of load, the sampling voltage U_Q will rise to increase the error voltage U_r. After U_r passes through VT_2 and VD_9, the magnetic reset current rises to prolong the magnetic reset time and reduce the output pulse width, so that U_{O1} is reduced to 3.3 V again, and vice versa. Therefore, the magnetic amplifier can be equivalent to a pulse width modulator, which realizes the precise regulation by finely adjusting the pulse width. That is the regulation principle of magnetic amplifier.

7.4 Design of Synchronous Rectification DC/DC Converter

See Figure 7.7 for the circuit of 3.3-V and 5-A (16.5 W) DC/DC converter using synchronous rectification technology. A DPA424R is adopted with the DC input voltage range 36–75 V and the power supply efficiency of up to 87%. C_1, L_1, and C_2 constitute the EMI filter at the input terminal to filter out the EMI introduced by the grid. R_1 is applied to set the under-voltage

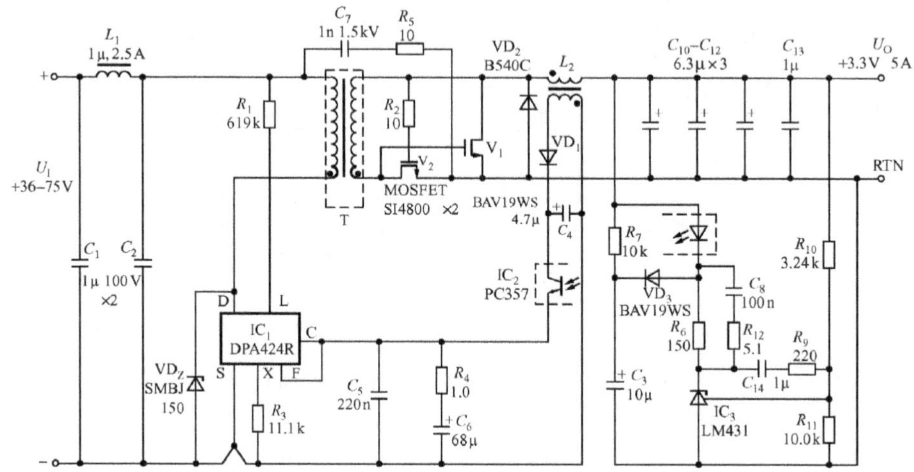

Figure 7.7 Circuit of 3.3-V and 5-A (16.5 W) DC/DC converter using synchronous rectification technology

(U_{UV}) and the over-voltage (U_{OV}). When $R_1 = 619\,\text{k}\Omega$, $U_{UV} = 619\,\text{k}\Omega \times 50\,\mu\text{A} + 2.35\,\text{V} = 33.3\,\text{V}$, and $U_{OV} = 619\,\text{k}\Omega \times 135\,\mu\text{A} + 2.5\,\text{V} = 86.0\,\text{V}$. When the input voltage is too high, R_1 can decrease linearly the maximum duty ratio to prevent magnetic saturation. R_3 is the limit current setting resistor. When $R_3 = 11.1\,\text{k}\Omega$, the drain limit current $I'_{LIMIT} = 0.6I_{LIMIT} = 0.6 \times 2.50\,\text{A} = 1.5\,\text{A}$. The rectifier VD_{Z1} (SMBJ150) in circuit is used to clamp the drain voltage, which is capable of ensuring the reset of high-frequency transformer.

Synchronous rectification technology is applied for the power supply and the MOSFET SI4800 with low drain-source on-state resistance as the rectifier to greatly reduce the rectifier loss, thus improving the efficiency of DC/DC converter. SI4800 is the MOSFET produced by Philips through TrenchMOS^{TM} technology, whose on and off state can be controlled by the logic level with the drain-source on-state resistance of $0.0155\,\Omega$ only. The maximum drain-source voltage $U_{DS\ (max)}$ of SI4800 is 30 V, the maximum gate-source voltage $U_{GS\ (max)}$ is ±20 V, and the maximum drain current is 9 A (25 °C) or 7 A (70 °C). The peak drain current is up to 40 A. The maximum power consumption is 2.5 W (25 °C) or 1.6 W (70 °C). The turn-on time t_{ON} of SI4800 is 13 ns (including the turn-on delay time $t_{d(ON)} = 6$ ns and the rise time $t_R = 7$ ns), the turn-off time t_{OFF} is 34 ns (including the turn-off delay time $t_{d\ (OFF)} = 23$ ns and the fall time $t_F = 11$ ns), and the transconductance $g_{FS} = 19$ S. The working temperature range is −55 to + 150°C. A freewheeling diode VD is available in SI4800, connected in parallel between the drain source with reverse polarity (the negative pole is connected to D and the positive pole to S) to protect the MOSFET. The reverse recovery time t_{rr} of VD is 25 ns.

The synchronous rectifier V_2 is driven by the secondary voltage and R_2 is the gate load of V_2. The synchronous rectifier V_1 is driven by the reset voltage of high-frequency transformer and works only when V_2 is turned off. When the Schottky diode VD_2 is turned off, some energy is stored in the common-mode choke (also known as power inductor) L_2. When the high-frequency transformer is reset, VD_2 is turned on, so the electric energy in L_2 continues to provide current to the load through the loop constituted by VD_2. Therefore, it is called VD_2 freewheeling diode. The output of auxiliary winding provides bias voltage to the receiver tube of optical coupler through the rectification and filtering of VD_1 and C_4. C_5 is the bypass capacitor at control terminal. The time of power-on start-up and automatic restart is determined by C_6.

After the voltage division by R_{10} and R_{11}, the output voltage is compared with the 2.50 V reference voltage in adjustable precision shunt regulator LM431 to generate the error voltage, which controls the duty ratio of DPA424R through the optical coupler to regulate the output voltage. R_7, VD_3, and C_3 constitute the soft-start circuit to prevent the output voltage from overshooting at the time of turning on the power supply. At the time of power-on, LM431 does not work as the voltage drop across C_3 cannot change suddenly. The voltage of C_3 gradually rises with the increase of output voltage from the rectifier filter and the charging of C_3 by R_7, so that LM431 enters into normal work. In the soft-start process, the output voltage rises gradually, and eventually reaches the stable value of 3.3 V.

Key design points:

1. 20 turns of ϕ0.40 mm enameled wires divided into two layers are applied for the primary winding of high-frequency transformer, and three turns of four-strand ϕ0.40 mm enameled wires for the secondary winding. The inductance of primary winding $L_P = 600\,\mu H$ (±25% error allowed at 400 kHz), and the maximum leakage inductance $L_{P0} = 1\,\mu H$. The resonant frequency of high-frequency transformer is not less than 3.8 MHz.
2. When designing the high-frequency transformer reset, the influence of the gate load R_2 of V_1 on the reset waveform of transformer must be considered. The gate capacitance C_{GS1} of synchronous freewheeling diode V_1 is used as the load of transformer reset, and the selected V_1 shall be reliably reset at the minimum ($U_{I(min)}$) and maximum ($U_{I(max)}$) input voltage. The drain-source on-state resistance $R_{DS(ON)}$ and the total gate charge Q_G of synchronous rectifier and freewheeling diode shall be very small.
3. To improve the power supply efficiency, the Schottky diode with low voltage drop shall be adopted for VD_2.
4. In the case of low or medium voltage input, DPA-Switch chip with relatively high power can improve the power supply efficiency.

7.5 Design of SMPS for Peak-Power-Output Audio Power Amplifier

See Figure 7.8 for the SMPS circuit of 75 W (with the peak power of 126 W) audio power amplifier constituted with PKS607Y of PeakSwitch series capable of peak power output. The power supply has the following features:

1. It belongs to the multioutput flyback SMPS, not only having excellent regulation characteristics but also enjoying good transient response and cross-load regulation. The five outputs include two main outputs U_{O1} (+26 V, 1.45 A) and U_{O2} (−26 V, 1.45 A) and three auxiliary outputs U_{O3} (+15 V, 150 mA), U_{O4} (+5 V, 150 mA), and U_{O5} (−15 V, 150 mA). The continuous output power is 75 W and the peak output power is up to 126 W. The two outputs of +26 and −26 V can provide 1.45 A continuous output current (or 2.42 A peak output current), which can be used as the power supply of OCL audio power amplifier. ±26 V dual power supply can be applied for this kind of power amplifier circuit to eliminate the output capacitor, improving the low-frequency response of audio power amplifier.

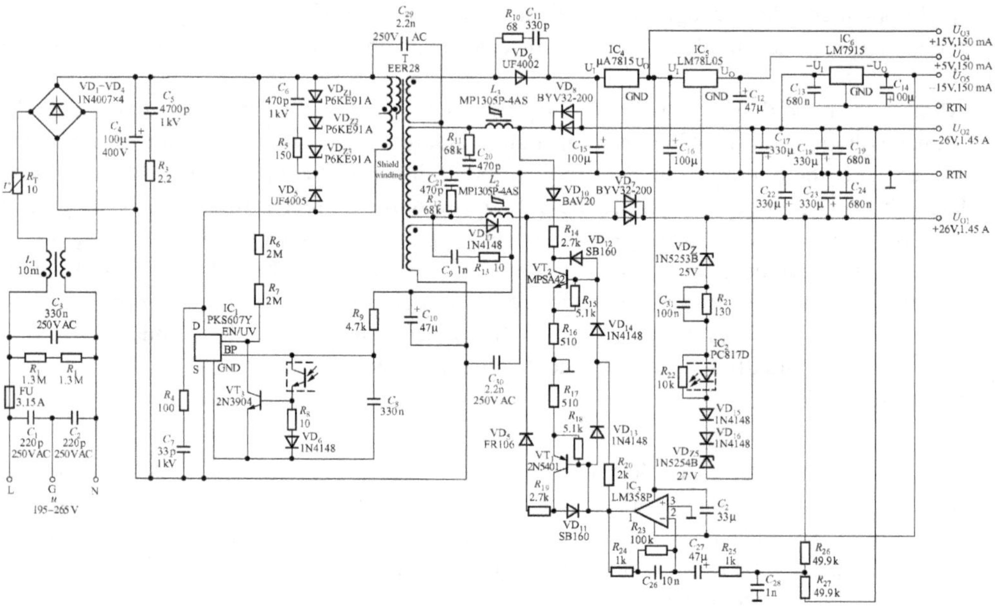

Figure 7.8 SMPS circuit of 75-W (with the peak power of 126 W) audio power amplifier

2. The two outputs of OCL circuit work alternately and have the maximum output power at low frequency. In the extreme case that one output is at full load and the other at zero load, almost all of the energy is provided to the full load output. This requires that the two outputs of +26 and −26 V have good cross-load regulation, so that when the transient variance of one load is up to 25–100%, the two outputs can still be balanced with the load regulation remained within the range of ±5%. In view of this, a magnetic amplifier is used in the circuit of +26 and −26 V main outputs to constitute the regulation circuit of high-frequency magnetic amplifier. The magnetic amplifier can precisely control the output of SMPS, not only enhancing the regulation precision but also significantly improving the cross-load regulation.
3. The three auxiliary outputs of +15, +5, and −15 V are provided by the standard linear regulators μA7815, LM78L05, and LM7915, respectively.
4. An 18 V winding is specially designed for the high-frequency transformer to reduce the input–output voltage drop of μA7815 and LM78L05 and the power consumption of regulator.
5. The power supply efficiency at full load is greater than 82%, and the no-load power consumption at AC 265 V input is less than 800 mW.

Magnetic amplifier is constituted by the sampling circuit, the error amplifier, the magnetic reset circuit, and the controllable magnetic saturation inductor. R_{26} and R_{27} are the sampling resistors of total output of +26 and −26 V. An LM358 dual op-amp (half used currently) is used to constitute the error amplifier. Inductors L_2 and L_3 are connected through the transistors VT_1 (PNP 2N5401) and VT_2 (NPN MPSA42), respectively, in the magnetic reset circuit. When +26 and −26 V outputs are balanced, the sampling voltage is zero.

Consider +26 V output for example, when MOSFET is turned on, the energy is stored in the high-frequency transformer, at which time L_2 shows high impedance for the high-frequency switching current and VD_7 is turned off owing to reverse bias. When the MOSFET in PKS607Y is turned off, the energy stored in the high-frequency transformer is transmitted to the secondary side. However, L_2 is reset as the output current of magnetic reset control circuit flows through L_2 (from right to left, considered as the forward current), so the current of secondary winding flowing through L_2 is changed from negative value to positive value after some delay time, and then rise rapidly to let L_2 enter into magnetic saturation and show low impedance, thus making the +26 V output rectifier VD_7 (16 A/200 V superfast recovery geminate transistors BYV32-200) turned on. The above delay time is the duration of magnetic reset, that is, the time blocking the output. After that, the output is turned on to not only supply current to the load but also store some energy in the output filter capacitor, thus maintaining the output voltage constant when VD_7 is turned off.

For example, when U_{O2} becomes more negative owing to the transient load decrease of −26 V output, the sampling voltage U_Q passing through R_{26} and R_{27} is negative. U_Q is applied to the reverse input terminal of error amplifier, so the error voltage U_r of LM358 is positive. At this point, VD_{13} is turned off and VD_{14} is turned on. U_r is connected to the base of VT_2 through VD_{14}, which becomes negative through the phase inversion of VD_{14} and adjusts the delay time of magnetic reset through R_{14} and VD_{10} to prolong the output blocking time of L_2, letting U_{O2} rise to −26 V again and keep the balanced output with +26 V. That is the regulation principle of magnetic amplifier.

In summary, the controllable magnetic saturation inductor in the regulator circuit is equivalent to a controllable switching, which can finely adjust the pulse width just by changing the delay time of magnetic reset, greatly improving the cross-regulation characteristics.

L_2 and L_3 are made by winding MP1305P-4AS amorphous alloy magnetic ring. MP1305P-4AS is the amorphous alloy magnetic ring produced by Allied-Sicnal Inc. with the outer and inner diameters and the height of 14.4, 7.9, and 6.6 mm, respectively. It has the advantages of high magnetic permeability, rectangular *B-H* hysteresis loop with high aspect ratio, low magnetic core loss, and good high temperature stability. In addition, the external current is adopted to change the magnetic flux, and it is easy to realize magnetic saturation or magnetic reset.

The voltage feedback circuit of ±26 V output is constituted by the high-gain optical coupler IC_2 (PC817D), the voltage regulator tube VD_{Z4} (1N5253B) and VD_{Z5} (1N5254B), the temperature compensation diodes VD_{15} and VD_{16} (two diodes of 1N4148), and so on. The output voltage U_O is set by the stable voltage U_{Z4} of regulator tube VD_{Z4}, the stable voltage U_{Z5} of VD_{Z5}, the voltage drop U_{R21} of R_{21}, the turn-on voltage drops U_{VD15} and U_{VD16} of VD_{15} and VD_{16}, and the turn-on voltage drop U_{LED} of LED in the optical coupler PC817D. Relevant formula is as follows:

$$U_O = U_{Z4} + U_{Z5} + U_{R21} + U_{VD15} + U_{VD16} + U_{LED} \tag{7.5}$$

wherein, $U_{Z4} = 25\,V$, $U_{Z5} = 27\,V$, $U_{R21} + U_{VD15} + U_{VD16} \approx 2V$, and $U_{LED} \approx 1$ V, so $U_O = 52\,V$.

The rang of AC input voltage is 195–265 V. The EMI filter is constituted by C_1–C_3 and L_1. C_1, C_2, and L_1 can filter out the common-mode interference introduced from the power line, and L_1 is the common-mode choke. C_3 is used to filter out the series-mode interference. R_1 and R_2 are bleeder resistors, which can release the charges accumulated in the capacitor when the power supply is turned off. R_T is the negative temperature coefficient resistor (10Ω at room temperature) and can limit the surge current at the moment of power-on. DC input voltage can be obtained after the rectification and filtering of the AC input voltage by VD_1–VD_4 and C_4, which is applied to one terminal of the primary winding of high-frequency transformer (T). The other terminal of primary winding is connected to the drain D of MOSFET in PKS607Y. Drain clamp protection circuit is constituted by the transient voltage suppressor VD_{Z1}–VD_{Z3}, the SRD VD_5, and the resistor–capacitor components C_6 and R_5, which are capable of clamping the drain voltage within the safe range. Input under-voltage detection circuit is constituted by R_6 and R_7. When the input voltage is too low, PKS607Y turns off the output as long as the current flowing into terminal EN/UV is less than 25 μA, thus realizing under-voltage protection.

The skip-cycle method is applied for the on/off controller in PKS607Y to realize the regulation purpose. When the load requires the peak power of power supply, the on/off controller skips only a small number of switching cycles to achieve 126 W peak power output within a short time and maintain the output voltage constant. In the case of continuous power output, it skips more clock cycles to reduce the output power. That is the regulation principle of PKS607Y.

Key design points:

EER28 ferrite core is adopted for high-frequency transformer. The primary winding is divided into two parts, wound with 16 and 17 turns of ϕ0.25 mm two-strand enameled wire, respectively. The bias winding is wound with 5 turns of ϕ0.30 mm two-strand enameled wire. +26 and −26 V windings are wound with 8 turns of ϕ0.60 mm two-strand enameled wire. The 18 V winding is wound with 6 turns of ϕ0.33 mm two-strand enameled wire. A shielding layer is added, respectively, among the primary, secondary, and bias windings. The inductance of primary winding $L_P = 151\,\mu H$ (±20% deviation allowed) and the maximum leakage inductance $L_{P0} = 5\,\mu H$. The resonant frequency of high-frequency transformer is more than 2 MHz.

7.6 Design of Industrial Control Power Supply Based on Voltage-Doubling Rectifier

See Figure 7.9 for the circuit of 1.25 W low-voltage input industrial control SMPS constituted by the TinySwitch-III series product TNY274P. The range of AC voltage input is 18–30 V, the output voltage is +5 V, and the output current is 250 mA. The power supply efficiency is up to 65%.

When the AC input voltage is 30 V, the no-load power consumption is less than 100 mW. This power supply can be adopted as the auxiliary power supply for industrial control devices.

VD_1, VD_2, C_1, and C_2 constitute the voltage-doubling rectifier circuit. R_1 and R_2 are voltage-sharing resistors used to balance the voltage drops of C_1 and C_2. 100 μF/110 V electrolytic capacitors with relatively large capacity shall be adopted for C_1 and C_2 to ensure that the DC voltage after rectification is no less than +50 V. The on/off controller in TNY274P receives the feedback voltage of secondary winding through the optical coupler PC817A and maintains the output voltage stable by enabling or disabling the on/off state of MOSFET. The current flowing through the LED in PC817A is proportional to the output voltage U_O, so the current I_{EN} pulled out from terminal EN/UV by the phototransistor in PC817A is also proportional to U_O. Once $I_{EN} > 115\,\mu A$, TNY274P skips the switching cycle. If $I_{EN} < 115\,\mu A$, the switching will be re-enabled.

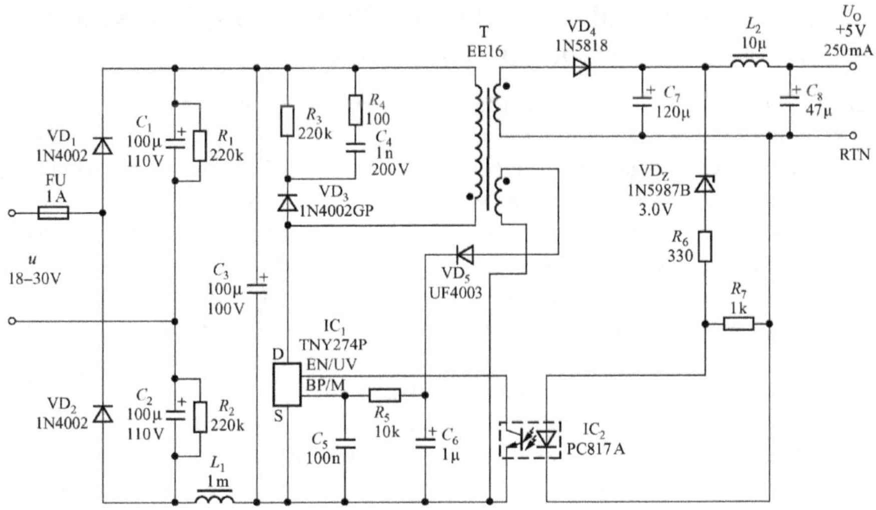

Figure 7.9 Circuit of 1.25-W low-voltage input industrial control SMPS constituted by TNY274P

The drain clamp circuit is constituted by the blocking diode VD_3, R_3, R_4, and C_4, which is capable of limiting the drain voltage 700 V or less. R_3, R_4, and C_4 constitute the absorption circuit. The output voltage of bias winding provides TNY274P with bias voltage through the rectification and filtering of VD_5 and C_6. 1N5818 Schottky rectifier is applied for output rectifier VD_4 with the rated rectified current $I_d = 1\,A$ and the maximum reverse working voltage $U_{RM} = 30V$. Electrolytic capacitor with low equivalent series resistance (ESR) is adopted for output filter capacitor C_7 to reduce the output voltage ripple. The postfilter constituted by L_2 and C_8 can attenuate the high-frequency switching noise.

The output voltage is set by the sum of voltage drops of rectifier VD_Z, resistor R_6, and LED in PC817A. 3 V rectifier 1N5987B is applied for VD_Z and the voltage drop sum of R_6 and LED is about 2 V, so the set $U_O = 3\,V + 2\,V = 5\,V$. Proper resistance adjustment of R_6 and R_7 can realize the fine adjustment of output voltage.

Key design points:

1. EE16 ferrite core is introduced for high-frequency transformer. The primary winding is wound by 64 turns of $\phi0.25$ mm enameled wire. The secondary winding is wound by 10 turns of $\phi0.28$ mm enameled wire. The bias winding is wound by 32 turns of $\phi0.25$ mm two-strand enameled wire. The inductance of primary winding $L_P = 780\,\mu H$ ($\pm12\%$ error allowed) and the maximum leakage inductance $L_{P0} = 80\,\mu H$. The minimum resonant frequency of primary winding is 1 MHz.
2. To fully utilize the clamping energy to improve the no-load efficiency, the reverse recovery time of blocking diode VD_3 shall not exceed 2 μs, and VD_3 shall be either glass passivated rectifier (e.g., 1N4002GP) or FR107 fast diode. The peak drain voltage can also be reduced to below 650 V by properly increasing the resistance of R_3.
3. For further improvement of the output accuracy, adjustable precision shunt regulator TL431 (or PC817 and LM431) can be used to replace the ordinary regulator 1N5987B.

7.7 Design of Industrial Control Power Supply Based on Suspension High-Voltage Constant Current Source

See Figure 7.10 for the circuit of industrial control power supply with 3 W ultra-wide input range constituted by TNY280P of TinySwitch-III series. The notable features of the power supply are that the AC voltage input range is very wide (18–265 V), the output voltage is +5 V, and the output current is 600 mA. The power supply efficiency is up to 65% with the no-load power consumption of below 200 mW at the AC input voltage of 230 V. The application field of the power supply includes the auxiliary power supply used for industrial control.

The minimum drain voltage for normal start-up and work of TinySwitch-III series product is 50 V. Typically, when the AC input voltage $u > 85$ V, the chip can provide the self-bias voltage. However, when $18\text{ V} < u < 75\text{ V}$, the chip cannot provide enough bias voltage to maintain normal work, greatly limiting the application of TinySwitch-III series product under low voltage. To solve the above problem and let TinySwitch-III work normally at ultra-low AC input voltage, a suspension (also known as floating) high-voltage constant current source shall be added outside TNY280P to supply power to the bypass terminal BP/M at low voltage. See Figure 7.11(a) for the circuit of suspension constant current source. It includes a 7.5-V rectifier VD_{Z1} (1N5236B), a PNP transistor VT_1 (ZTX558), a NPN transistor VT_2 (ZTX458), diodes VD_2 and VD_4, and resistors R_4–R_6, wherein VD_2 is the half-wave rectifier and VD_4 is the isolation diode, capable of isolating the constant current source from other circuits.

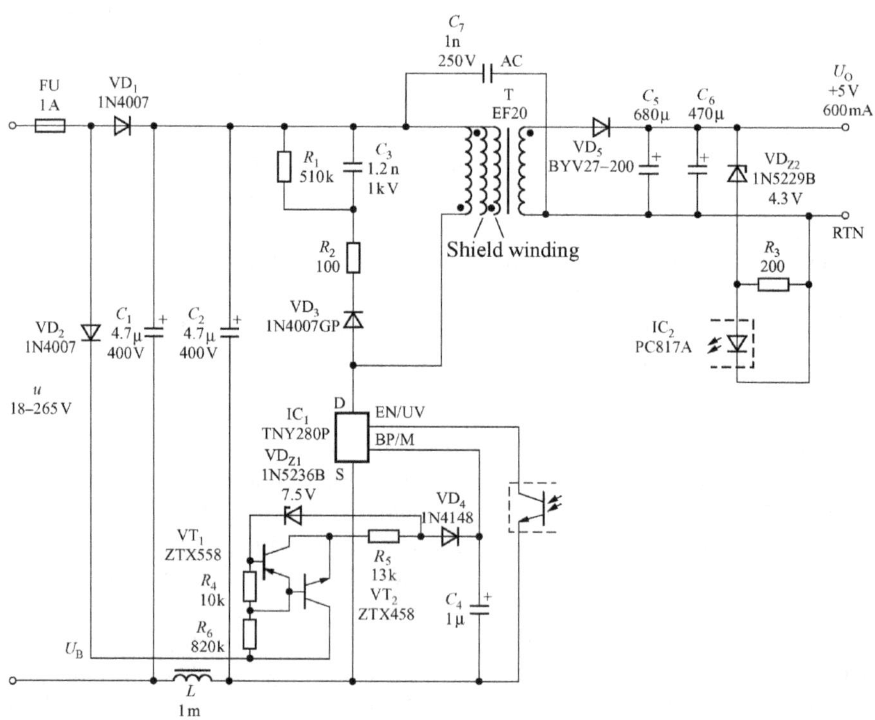

Figure 7.10 Circuit of industrial control power supply with 3 W ultra-wide input range constituted by TNY280P

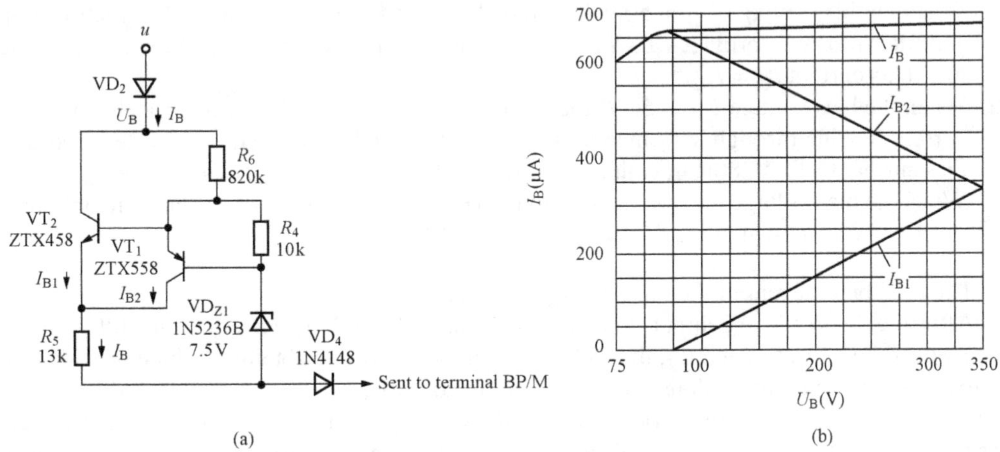

Figure 7.11 Circuit of suspension constant current source. (a) Circuit and (b) relation curve of bias voltage and total bias current

ZTX558 and ZTX458 are high-reverse-voltage transistors produced by Zetex Semiconductors. ZTX558 belongs to PNP high-reverse-voltage transistor with the following main parameters: the collector–emitter reverse breakdown voltage $U_{(BR)CEO} = -400\,\text{V}$ when the base is disconnected, the collector–base reverse breakdown voltage (i.e., the collector reverse breakdown voltage) $U_{(BR)CBO} = -400\,\text{V}$ when the emitter is disconnected, the maximum collector current $I_{CM} = -200\,\text{mA}$, the common-emitter current amplification factor $h_{FE} \leq 300$, and the maximum power consumption $P_{CM} = 1\,\text{W}$. ZTX458 belongs to NPN high-reverse-voltage transistor with the following main parameters: $U_{(BR)CEO} = 400\,\text{V}$, $U_{(BR)CBO} = 400\,\text{V}$, the maximum collector current $I_{CM} = 300\,\text{mA}$, $h_{FE} \leq 300$, and $P_{CM} = 1\,\text{W}$.

The pulsating direct current bias voltage U_B will be obtained after the half-wave rectification of 18–265 V AC voltage by VD_2, which is applied to the input terminal of suspension constant current source. The constant current source can provide the terminal BP/M of TNY280P with about 600 μA constant current within the entire input voltage range.

First, assume that transistor VT_2 is adopted for the circuit only, it can be regarded that the regulator tube VD_{Z1} provides the base of VT_2 with a reference potential U_{B2}. The sum of the emitter junction voltage (U_{BE2}) of VT_2 and the voltage drop (U_{R5}) of R_5 is equal to the regulated voltage U_Z of regulator tube, and U_{BE2} is approximately constant when the ambient temperature does not change, so the variation range of bias current provided by the regulator tube is very large, which will cause the deviation of the set constant current. To overcome the above problem, another constant bias current shall be provided by the PNP transistor VT_1 and R_4. Letting the emitter junction voltage of VT_1 equal U_{BE1}, the constant bias current set through R_4 is $I_{B1} = U_{BE1}/R_4$. Obviously, I_{B1} is not affected by the change of input voltage.

See Figure 7.11(b) for the relation curve of the bias voltage (U_B) and the total bias current (I_B) obtained through circuit simulation. It can be seen from the figure that VT_2 provides a constant bias current I_{B2} at low input voltage and VT_1 provides a constant bias current I_{B1} at high input voltage. Specifically, it can be divided into the following three conditions:

1. When the bias voltage $U_B \approx 50\,V$ (i.e., the DC input voltage $U_I \approx 50\,V$ after rectification and filtering), VT_2 provides TNY280P with a constant bias current I_{B2}, in which case the total bias current $I_B = I_{B2}$.
2. When the bias voltage $U_B > 50\,V$, the current flowing through VT_2 decreases linearly, the current flowing through VT_1 increases linearly, VT_1 and VT_2 together provide a constant bias current to TNY280P, in which case the total bias current $I_B = I_{B1} + I_{B2}$ and $I_{B2} > I_{B1}$.
3. When the bias voltage reaches the maximum ($U_B = 375\,V$), the bias current is mainly provided by VT_1, in which case $I_B = I_{B1} + I_{B2}$ and $I_{B1} > I_{B2}$.

The total bias current of the circuit is $I_B \approx 600\,\mu A$.

After the 18–265 V AC input voltage passes through the half-wave rectifier filter circuit constituted by VD_1, C_1, and C_2, the DC input voltage U_I will be obtained, which provides the flyback SMPS with high-voltage direct current. In addition, a pi filter is constituted by C_1, C_2 and inductor L to reduce the series-mode EMI. A Y capacitor C_7 is introduced between the primary and secondary windings of high-frequency transformer to filter out common-mode interference. The drain clamping circuit is constituted by VD_3 (1N4007GP), R_1, R_2, and C_3. The rectifier VD_5 is BYV27-200 2 A/200 V SRD with the reverse recovery time $t_{rr} < 25\,ns$. The output voltage is determined by the sum of voltage drop of the regulator tube VD_{Z2} and the LED in optical coupler PC817A. VD_{Z2} is a 4.3-V regulator tube 1N5229B, the forward voltage drop of LED is approximately 1 V, and the set no-load output voltage is 5.3 V.

The on/off control mode is applied for TNY280P, which receives the feedback voltage of secondary winding through the optical coupler and keeps the output voltage constant by enabling and disabling the on/off of internal MOSFET. Once the current from terminal EN/UV exceeds the turn-off threshold current (115 μA), the switching cycle will be skipped. When the current from terminal EN/UV is less than the turn-off threshold current, the switching cycle will be re-enabled.

Key design points:

1. EF20 ferrite core is employed for high-frequency transformer. The primary winding is wound by 32 turns of $\phi 0.33\,mm$ enameled wire. The secondary winding is wound by 8 turns of $\phi 0.40\,mm$ enameled wire. The inductance of primary winding $L_P = 278\,\mu H$ ($\pm 12\%$ error allowed) and the maximum leakage inductance $L_{P0} = 12\,\mu H$. The minimum resonant frequency of primary winding is 1 MHz.
2. As the input voltage range of the power supply is very wide, the inductance L_P of primary winding must be small enough (actually 278 μH) to let TNY280P work at the edge of continuous mode. However, the current rise rate di/dt will be increased when L_P is small, so the TinySwitch-III series chip with relatively high power can be applied if necessary.
3. The resistance of R_1 in the clamping circuit shall not be too small, or the no-load power consumption will rise.

7.8 Design of StackFET™ Technology-Based Micro-SMPS

See Figure 7.12 for the circuit of micro-SMPS constituted by LNK364P with 12 V and 250 mA ultra-wide input voltage range.

StackFET™ (superimposed FET) proprietary technology and flyback topology are introduced, which can provide the rated power within a very wide input range. The AC input voltage

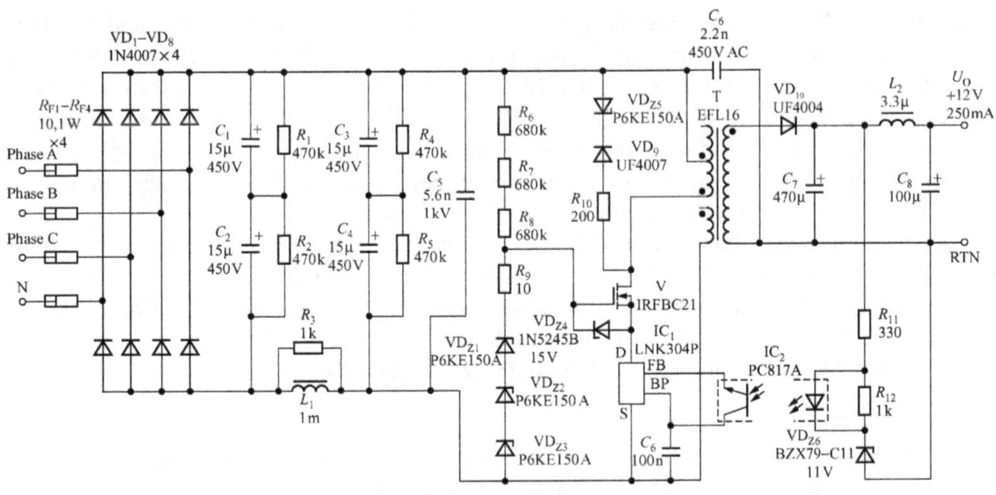

Figure 7.12 Circuit of micro-SMPS constituted by LNK364P with 12 V and 250 mA ultra-wide input voltage range

can be either 57–580 V single-phase AC or three-phase four-wire system (phases A, B, and C and neutral N) AC, which can work normally even if any phase is in power shortage or the neutral is not connected. The output is +12 V and 250 mA. It has the functions such as automatic restart, open-loop protection, over-load protection, and short circuit protection. The circuit is suitable for industrial instrumentation fields such as three-phase energy meter.

A DC high voltage will be obtained after the rectification and filtering of three-phase alternating current by VD_1–VD_8, C_1, and C_2. R_{F1}–R_{F4} are all fusible resistors. To increase the withstand voltage of capacitor, two 450-V electrolytic capacitors C_1 and C_2 are connected in series, capable of withstanding up to 900 V high input voltage. R_1 and R_2 are equalizing resistors, capable of equalizing the voltage drops of C_1 and C_2, preventing any capacitor from being broken down resulted from too high voltage drop. In addition, the two resistors provide the capacitors with bleed-off circuit after power-off. L_1 and C_5 constitute the EMI filter, and R_3 is the bleeder resistor.

A MOSFET with 700 V drain-source breakdown voltage is integrated in LNK364P. When the maximum AC input voltage $U_{I(max)} = 580$ V, the maximum primary voltage of high-frequency transformer is close to 1050 V (including the primary induced voltage, also known as secondary reflected voltage), much higher than 700 V. To prevent the internal MOSFET from being damaged, another high-voltage MOSFET V must be superimposed on the drain as external MOSFET. Now, the IRFBC20 N-channel MOSFET with the following main parameters is used: the drain-source breakdown voltage $U_{(BR)DS} = 600$ V, the drain-source on-state resistance $R_{DS\ (ON)} = 4.4\,\Omega$, the maximum drain current $I_{D(max)} = 2.2$ A, and the maximum drain power consumption $P_{D(max)} = 50$ W. The working principle is that the source of V is driven by the drain of LNK364P first, and then the primary winding of high-frequency transformer is driven by the drain of V. As the drain voltage of LNK364P is limited to 450 V by the regulator tubes VD_{Z1}–VD_{Z3}, the total maximum peak drain voltage is increased to 450 V + 600 V = 1050 V after the external MOSFET is superimposed, capable of meeting the needs of the circuit. That is the working principle of StackFET circuit.

R_6–R_8 are adopted to provide the gate of IRFBC20 with start-up voltage, and R_9 (10 Ω) is the damping resistor, preventing high-frequency self-oscillation. The regulator VD_{Z4} can protect the gate-source voltage of IRFBC20. The drain clamping circuit is constituted by the transient voltage suppressor VD_{Z5} (P6KE150A), the SRD VD_9 (UF4007), and the resistor R_{10}, which can prevent the total maximum peak drain voltage from reaching or exceeding 1050 V owing to the superimposition of the spike voltage generated by the leakage inductance of high-frequency transformer on the drain of external MOSFET.

It shall be noted that the work of LNK364P is not affected by the circuit structure of Stack-FET. When the internal MOSFET is turned on, the external MOSFET is also turned on to apply the input voltage to the primary winding. Once the primary winding current reaches the limit current threshold of LNK364P, MOSFET will be turned off. The output voltage can be maintained stable just through the on/off controller in LNK364P.

Key design points:

1. EEL16 ferrite core is adopted for high-frequency transformer. The primary winding is wound by 184 turns of ϕ0.13 mm enameled wire. The secondary winding is wound by 30 turns of ϕ0.29 mm enameled wire. A shielding layer is added between the primary and secondary windings. The inductance of primary winding $L_P = 3.5\,\text{mH}$ (±10% error allowed) and the maximum leakage inductance $L_{P0} = 160\,\mu\text{H}$. The resonant frequency of high-frequency transformer is above 500 kHz.
2. When the neutral N is grounded, C_3, C_4, R_4, and R_5 can be eliminated, but shall be remained for decoupling of the input circuit.
3. Regulator tubes VD_{Z1}–VD_{Z3} can be replaced with a P6KE540 transient voltage suppressor. If the minimum AC input voltage is 100 V, the capacity of C_1–C_4 can be reduced to 10 μF.
4. R_1, R_2, R_4, and R_5 shall be 0.5-W resistors.
5. In order to reduce the switching loss, several layers of insulating tape can be added to the interlayer of primary winding to reduce the distributed capacitance of high-frequency transformer.

7.9 Design of Power Supply for the Digital TV Set-Top Box

Set-top box is the key technology of interactive television (ITV), which can help broadband multimedia services with powerful functions such as digital broadcast television, video and music on demand, Kara-Ok, three-dimensional games, high-speed Internet access, online shopping, and voice prompt. The set-top box has strict requirements for power supply, generally requiring the multioutput SMPS of high efficiency, small size, and light weight, which shall also have good electromagnetic compatibility.

See Figure 7.13 for the power supply circuit of 35 W digital TV set-top box. The five outputs are U_{O1} (+30 V, 100 mA), U_{O2} (+18 V, 550 mA), U_{O3} (+5 V, 2.5 A), U_{O4} (+3.3 V, 3 A), and U_{O5} (−5 V,100 mA), respectively, wherein +5 and +3.3 V outputs are the main output and others are the auxiliary output. When the AC input voltage $u = 220\,\text{V} \pm 15\%$, the total output power is up to 38.5 W. If the wide-range voltage input (u = 85–265 V) is applied, the total output power will be reduced to 25 W. It can be used as the SMPS of set-top box, video cassette recorder (VCR), camera and video cassette recorder (CVCR), and DVD. The set-top box SMPS has the efficiency of above 77% and the functions of under-voltage and over-voltage protection.

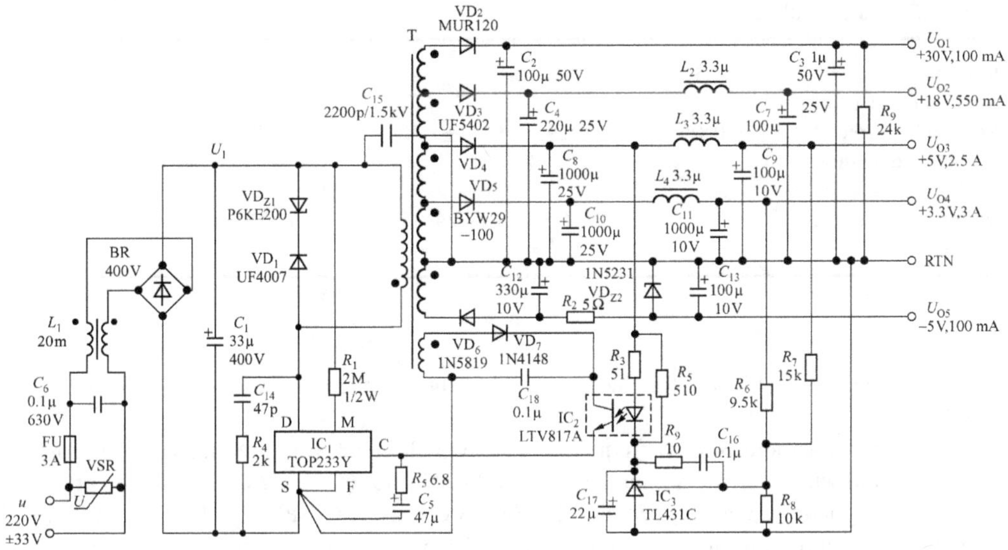

Figure 7.13 Power supply circuit of 35 W digital TV set-top box

Three ICs are applied for the power supply: TOP233Y (IC_1), optical coupler LTV817A (IC_2), and adjustable precision shunt regulator TL431C (IC_3). In order to reduce the size of high-frequency transformer and improve the coupling degree of magnetic field, the secondary winding adopts the superimposed winding. The absorption circuit constituted by R_4 and C_{14} can reduce the inference generated by radio frequency noise toward video equipment such as TV. If necessary, the control terminal (C) of switching frequency can be connected instead of the selecting terminal (F) to select the half-frequency mode, further reducing the sensitivity of TV toward video noise.

In order to withstand the lightning voltage that may enter from the grid line, a voltage-sensitive resistor VSR with the nominal voltage $U_{1\,\mathrm{mA}} = 275\,\mathrm{V}$ is connected in parallel at the AC input terminal. $U_{1\,\mathrm{mA}}$ refers to the voltage across the voltage-sensitive resistor when 1 mA DC current flows through the element. R_1 is the set resistor. The set under-voltage and over-voltage are 100 and 450 V (DC), respectively, when the 2 MΩ, 1/2 W resistor is used.

R_6, R_7, and R_8 are proportional feedback resistors applied to let the 5 and 3.3 V power supplies give feedback according to certain proportion. The load regulations of the outputs of 5 and 3.3 V are all up to ±5%. R_9 and C_{16} constitute the corresponding correction network of TL431C. C_{17} is the soft-start capacitor. When $C_{17} = 22\,\mu\mathrm{F}$, the soft-start time can be increased by 4 ms, so the total soft-start time is 14 ms after adding the 10 ms already owned. Other outputs have no feedback, whose output voltages are determined by the turn ratio of high-frequency transformer. As the output power of −5 V power supply is very low, the resistor R_2 and the regulator tube VD_{Z2} can be used for voltage regulation. R_9 is the dummy load of +30 V output, which can reduce the no-load and light-load voltage of the circuit. In view of the high output power of 5-V, 3.3-V, and 18-V power supplies, three power supplies are added to the poststage LC filter (L_3 and C_9, L_4 and C_{11}, and L_2 and C_7) to reduce the output

Table 7.5 Output voltage deviation

Nominal voltage of five outputs (V)	3.3	5	18	30	−5
Allowable output voltage deviation (V)	±0.16	±0.15	±1.35	±2.25	±0.375
Output voltage range (V)	3.15–3.36	4.91–5.34	17.24–19.01	28.71–30.76	−4.93 to − 5.23

ripple voltage. See Table 7.5 for the measured output voltage deviation range of the power supply.

TOP233Y is characterized by frequency jitter, which is helpful to reduce EMI. As long as the component values of safety capacitor C_{15} and EMI filter (C_6 and L_1) are selected properly, the electromagnetic radiation generated by SMPS can meet CISPR22 (FCCB) international standards. The common-mode interference of TOP233Y can be reduced to the minimum by connecting one terminal of C_{15} to the positive pole of U_I. It shall be noted that C_{15} and C_6 are both called safety capacitor. The only difference is that C_{15} is connected between the high voltage and the ground to filter out the common-mode interference generated by the coupling capacitance of primary and secondary windings, which is called "Y capacitor" according to IEC950 international standards. C_6 is connected to the inlet terminal of AC power supply, which is specially applied to filter out the series-mode interference between the grid lines and is called "X capacitor."

In actual production, EEL25 ferrite core can be adopted for the high-frequency transformer. The inductance of primary winding is 1.1 mH ± 10%, the leakage inductance is less than 12 μH, and the resonant frequency of primary winding shall be no less than 1.5 MHz.

7.10 Design of Mobile Phone Charger with USB Interface

Mobil phone charger has many types and different specifications, which not only brought a lot of inconvenience to users but also caused great waste owing to the absence of uniform standard. According to statistics of the Ministry of Information Industry, more than 100 million mobile phone chargers are sold or replaced every year, costing more than one billion yuan. Therefore, it is specified by China in 2007 that the USB interface is the standard interface of mobile phone chargers. The voltage of USB interface is +5 V, capable of supplying current up to 500 mA. After the adoption of the standard, mobile phones can be charged directly through the USB interface of computers. This has not only made mobile phone charging more convenient but also realized the data exchange between mobile phone and computer. The following text describes a kind of mobile phone charger meeting the USB interface voltage standard, which can realize constant voltage/current charging of mobile phones with USB interface when computer is absent.

See Figure 7.14 for the circuit of 5 V and 550 mA (2.75 W) USB charger constituted by LinkSwitch-II series LNK613D, wherein the AC input voltage range is 85–265 V, the output is +5 V and 550 mA, and the maximum output power is 2.75 W.

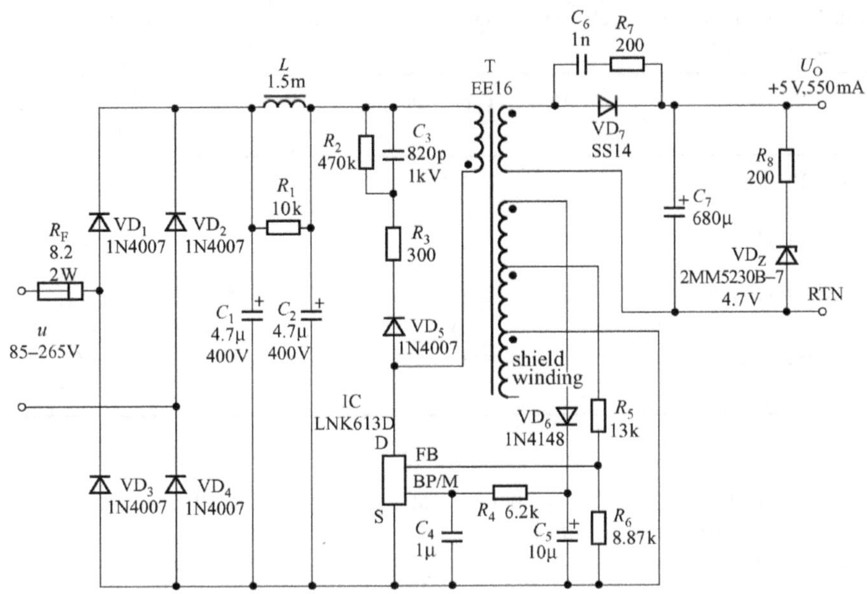

Figure 7.14 Circuit of 5 V and 550 mA (2.75 W) USB charger constituted with LNK613D

Primary control method is applied, eliminating the secondary control circuit and optical coupler. The constant voltage and current accuracies are ±5% and ±10%, respectively. Within the entire load range, the average power supply efficiency is up to 74% and the no-load power consumption is less than 40 mW.

The fusible resistor with cold resistance of 8.2Ω is adopted for R_F, which not only can replace the fuse but also have the function of current limiting protection during charging. DC high voltage will be obtained after the AC input voltage is rectified by VD_1–VD_4 first and then filtered by C_1 and C_2. The pi EMI filter constituted by C_1, C_2, and L can attenuate series-mode interference. R_1 is the damping resistor (optional). RCD clamp protection circuit constituted with VD_5, R_2, R_3, and C_3 is applied to limit the spike voltage caused by the leakage inductance.

The output rectifier VD_7 is SS14 1 A/40 V Schottky diode. C_7 is the output filter capacitor. 4.7 V regulator tube VD_Z and resistor R_8 constitute the dummy load to prevent the rise of output voltage at no load. Feedback resistors R_5 and R_6 are introduced to set the maximum working frequency and the output voltage of the constant voltage stage. In order to improve the regulation accuracy of output voltage and output current, precision resistors are applied for R_5 and R_6 with 1% error.

The working principle of the charger is that the output voltage is kept stable with the skip-switching-cycle method under the regulation of on/off controller during the constant voltage stage. In the case of light load (adopting trickling charge for the mobile phone battery), it can reduce the limit current to decrease the magnetic flux density of high-frequency transformer, thus reducing audio noise and switching loss. The limit current rises with the increase of load current, so the switching cycles skipped are gradually decreased. When no switching cycle is skipped, it means that the peak power point is reached. In this case, the on/off controller of LNK613D automatically switches to the constant current mode, the

Table 7.6 Corresponding relation between the bypass capacitor and the output voltage compensation factor

Model	Bypass capacitor (μF)	Output voltage compensation factor k
LNK613P/G/D	1	1.035
	10	1.055
LNK614 P/G/D	1	1.045
	10	1.065
LNK615 P/G/D	1	1.050
	10	1.070
LNK616 P/G/D	1	1.060
	10	1.090

output voltage decreases with the further rise of load current, and the output current is kept constant to realize constant current output.

Key design points:

1. EE16 magnetic core is adopted for high-frequency transformer. The primary winding is wound by 128 turns of ϕ0.13 mm enameled wire. The secondary winding is wound by 7 turns of triple insulated wire with the specification of 22. The bias winding is wound by 6 turns of ϕ0.25 mm enameled wire. The shield winding is wound by 23 turns of ϕ0.29 mm enameled wire. The inductance of primary winding $L_P = 2.58$ mH ($\pm 10\%$ error allowed) and the maximum leakage inductance $L_{P0} = 130\,\mu$H. The minimum resonant frequency is 500 kHz.
2. LNK613D obtains self-bias voltage through the bypass capacitor C_4. For LNK613D, the capacity of C_4 also determines the compensation value of output lead voltage drop. Typically, $C_4 = 1\,\mu$F, in which case the output voltage compensating factor is 1.035. For other models of LinkSwitch-II series, appropriate bypass capacitors can be selected according to Table 7.6 to realize optimum compensation for the voltage drop output lead. For example, if the output lead resistance of mobile phone charger is 300 mΩ, 0.15 V voltage drop will be formed at 500 mA output current, thus reducing the output voltage by 0.15 V. The nominal output voltage is 5.0 V, but the actual output voltage is 4.85 V, reduced by 3%. In this case, $C_4 = 1\,\mu$F and the output voltage compensation factor $k = 1.035$ shall be selected to increase the output voltage to $4.85\,\text{V} \times 1.035 = 5.02\,\text{V} \approx 5.0\,\text{V}$ after compensation, fully consistent with the nominal output voltage of 5.0 V.

8

Key Design Points of SMPS

8.1 SMPS Design Requirements

8.1.1 SMPS Design Requirements

In terms of switching mode power supply (SMPS) design, firstly, select the SMPS topology type. Secondly, select the switching components [e.g., integrated circuits such as SMPS, switching regulator, or pulse width modulation (PWM) modulator] and main external components according to the electrical properties of SMPS to design the circuit of SMPS. Thirdly, make the prototype (or computer simulation) according to the indicators such as mechanical properties (size, weight, structure, installation direction, and printed circuit board (PCB) layout), environmental working conditions (ambient temperature, relative humidity, and ventilation conditions), electromagnetic compatibility (EMC), service life, reliability, and price factor. Finally, complete the SMPS optimization design according to the test data (or simulation result) of prototype. The components shall be tested and filtered before making prototype.

SMPS design includes the following: the input voltage type (AC or DC), the AC voltage range and the grid frequency, the rectifier filter mode, the power supply type (universal or special SMPS and its topology), whether it is necessary to isolate the primary and secondary sides, the low or high voltage input, the adjustment range of output voltage and output current, the voltage or current regulation accuracy, the power supply efficiency, the output ripple voltage, the load characteristics (using the constant voltage output, constant current output, or constant voltage/current output, and whether the load is accumulator, motor, or LED display), the control characteristics (using the voltage or current control mode, and whether the functions of external turn-off, remote control, and digital control shall be configured), the single or multiple output, the protection circuit (e.g., soft start-up, input under-voltage/over-voltage protection, output over-current protection, short circuit protection, over-heat protection, and the protection circuit preventing reverse polarity of input voltage), the transient response requirements, whether it is necessary to add the digital display and fault alarm functions, the EMC requirements, the environmental working conditions, the size, the weight, and so on. The above shall be determined by designers according to actual situations.

Optimal Design of Switching Power Supply, First Edition. Zhanyou Sha, Xiaojun Wang, Yanpeng Wang, and Hongtao Ma.

8.1.1.1 Electrical Performance Indicators of SMPS

The electrical performance of SMPS mainly includes the following:

1. Input characteristics
   The input voltage type and voltage range, the grid frequency, and the harmonic distortion.
2. Output characteristics
   The output voltage, the output current, the regulation characteristics (voltage and load regulation), the transient response, the output ripple voltage and ripple current, and the output noise voltage.
3. Control mode and functions
   Voltage-mode control, current-mode control, and external turn-off, remote control, and digital control function.
4. Protection function
   If necessary, the followings can be added: input and output voltage and current monitors, protective relays, alarm, automatic/manual reset circuit, and so on.

If conditions permit, electrical insulation and EMC tests shall be carried out on the prototype.

8.1.1.2 Mechanical Property Indicators

Size, weight, and so on.

8.1.1.3 Environmental Working Conditions

Ambient temperature, relative humidity, heat dissipation conditions (natural and fan cooling), and so on.

8.1.1.4 Reliability Indicator

The reliability indicator usually refers to mean time between failures (MTBF). MTBF shall generally be greater than 100,000 h.

8.1.1.5 Cost Indicator

Under the premise of guaranteed performance indicators, the cost of SMPS shall be minimized to raise the price/performance ratio, creating conditions for commercialization.

8.1.2 Precautions for SMPS Design

8.1.2.1 Influence of the Parasitic Effect of Capacitor on Regulation Performance

The capacitor can be regarded as an ideal pure capacitor at low-frequency stage. However, the distribution parameters of capacitor shall not be ignored at high-frequency stage owing to the parasitic effect of capacitor. See Figure 4.2(c) for the equivalent circuit of actual capacitor at high frequency. For the output filter capacitor, its equivalent series resistance (ESR) R_{ESR} may cause the following influence: first, affecting the value of output ripple voltage. The higher the ESR is, the poorer will be the filtering effect and the higher the ripple voltage. When the ripple voltage exceeds the maximum rated value, the filter capacitor may be damaged. Second, when the ripple current flows into or out of the filter capacitor, power loss will be caused in ESR to make the filter capacitor emit heat. Third, too high ESR may make the switching regulator unstable. ESR relates to temperature. ESR grows rapidly when the temperature is below 10 °C.

Equivalent series inductance (ESL) can cause spike voltage at the output terminal, limiting the high-frequency characteristic of output filter capacitor and affecting its service life. If necessary, a film or ceramic capacitor can be connected in parallel to the output filter capacitor to improve its high-frequency performance.

8.1.2.2 Magnetic Core Type and Radiated Noise of High-Frequency Transformer/Inductor

The magnetic core selection of high-frequency transformer or inductor will affect the cost, size, and radiated noise of SMPS. Magnetic flux is an important indictor to assess high-frequency transformer or inductor noise. See Figure 8.1(a) and (b), respectively, for the flux paths of strip and toroidal cores. The flux direction of strip core is from one end to the other end, passing through the air between the two ends with the maximum radiated noise. The flux path of toroidal core is round with relatively small radiated noise.

Some air gap shall be remained in the magnetic core of high-frequency transformer to prevent core saturation. The air gap width is generally 0.02–0.1 mm. See Figure 8.2(a), (b), and (c), respectively, for the flux paths of EI, EE, and POT cores commonly used in SMPS. The magnetic flux density near the air gap is very high and it is easy to generate magnetic flux noise, so the flux shall not reach other core positions across the air gap. EI and EE cores are widely used in SMPSs owing to their low price and easy manufacture. POT ferrite core has the minimum radiated noise and is easier to manufacture than the toroidal core.

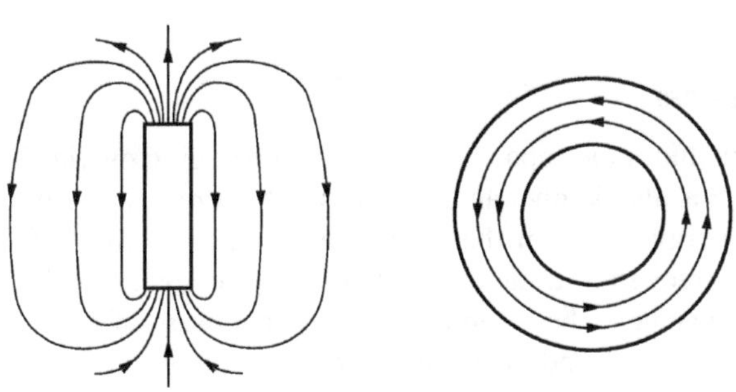

(a) (b)

Figure 8.1 Comparison between flux paths of (a) strip and (b) toroidal cores

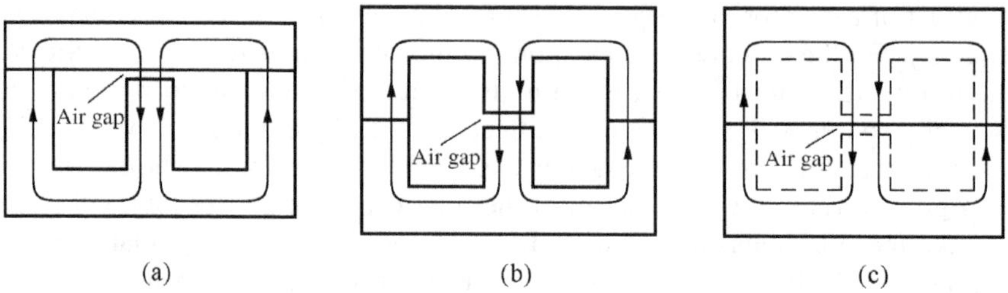

Figure 8.2 Flux paths of (a) EI, (b) EE, and (c) POT cores

8.2 Design of High-Efficiency SMPS

There are many factors affecting the efficiency of SMPS, such as component mode and package selection, radiator design, design and manufacturing process of high-frequency transformer, and conduction voltage drop of rectifier. The following section firstly describes the reasons for SMPS power loss, then the principles of high-efficiency SMPS, and finally the methods to improve efficiency.

8.2.1 Power Loss of SMPS

The power loss of SMPS includes three parts: transmission loss, switching loss, and other losses.

8.2.1.1 Transmission Loss

The transmission loss is composed of two parts. The first part is the transmission loss caused by the on-state resistance $R_{DS\ (ON)}$ of MOSFET. For example, the $R_{DS\ (ON)}$ of early SMPS chip TOP227Y is 2.6 Ω (typically, the same below) and that of the new product TOP262 is ≤ 0.90 Ω. The smaller the on-state resistance is, the lower will be the transmission loss. The second part is the loss on the current sensing resistor R_S.

8.2.1.2 Switching Loss

The switching loss includes the capacitance loss and switching overlap loss of MOSFET. The capacitance loss here is also known as CU^2f loss, which refers to the loss resulted from the release of electric energy stored in the output capacitance of MOSFET and the distributed capacitance of high-frequency transformer at the beginning of each switching cycle. The smaller the Miller capacitance between the gate and the drain of MOSFET is, the faster will be the switching of MOSFET and the smaller the switching loss.

Overlap loss is caused by the switching time of MOSFET. During the turn-on and turn-off of MOSFET, MOSFET is subject to the simultaneous effect of effective voltage and current, which results in long switching overlap time of MOSFET, and causes the loss.

8.2.1.3 Other Losses

1. Loss of start circuit. There is a start circuit in SMPS, which will be turned off automatically after the chip is activated, so there will be no loss after entering into normal working. PWM controller shall be connected to an external start circuit, which will cause certain power loss.
2. Loss of PWM controller. The loss of PWM includes the loss caused by the control circuit itself and the power loss of MOSFET drive control.
3. The loss of input bridge rectifier (BR) accounts for about 2% of the total power loss of SMPS. In case of low voltage and high current output, the loss of output rectifier can account for more than 10% of the total power loss of SMPS.
4. The loss of drain clamp protection circuit is about 1–1.2 W.
5. The magnetic core loss of high-frequency transformer (grows with the rise of switching frequency) and the skin effect loss of winding wire.
6. Loss of other circuits. It includes the sum of loss of the current limiting resistor (negative temperature coefficient resistor, NTCR) in electromagnetic interference (EMI) filter, the bleeder resistor in X capacitor, the input BR, the input filter, the output filter, and the feedback circuit.

8.2.2 Design Principles of High-Efficiency SMPS

8.2.2.1 General Principles

1. The designed SMPS shall try to work under the maximum duty ratio D_{max}.
2. For the wide AC input range 85–265 V, the induced voltage of primary winding U_{OR} shall be as high as possible and selectable within the range 100–135 V to ensure that the high-frequency transformer can transmit large enough energy at 85 V AC input voltage.

8.2.2.2 Specific Principles

1. Primary side circuit
    a. By increasing the primary winding inductance (L_P), the high-frequency transformer can work under continuous mode. In this case, the power losses of power switching tube and high-frequency transformer are relatively low.
    b. Connect an NTCR in series at the input terminal, which can limit current at the beginning of power-on and work under hot state (i.e., low resistance state) after power supply start-up, capable of reducing the power loss of current limiting resistor. The indicators of BR shall have enough margin. In order to reduce energy consumption, the nominal rectified current of BR must be greater than the rated current.
    c. Estimate the capacitance of input filter capacitor correctly. The output power per watt shall correspond to 3 μF at AC 85–265V input, that is, the proportional factor is 3 μF/W. When the AC 230 V input is fixed, the proportional factor is 1 μF/W.
    d. The most suitable PWM modulator or SMPS IC must be selected to improve the power supply efficiency and reduce costs.
2. High-frequency transformer
    a. Select low-loss core material, suitable shape, and correct winding method to minimize the leakage inductance. If the installation space allows, magnetic core with relatively large size is helpful to reduce core loss.

b. It is recommended to use multistrand wire for the secondary winding to reduce the loss resulted from the skin effect of winding wire. The copper loss of wire can be reduced by appropriately increasing the wire diameter of secondary winding.

3. Secondary side circuit
   a. The nominal current of the selected output rectifier shall be at least three times of typical continuous output current. In the case of high current output, it is recommended to use the Schottky diode with low forward voltage and extremely short reverse recovery time (t_{rr}).
   b. The ESR of output aluminum-electrolytic filter capacitor shall be as low as possible.

8.2.3 Methods to Improve SMPS Efficiency

8.2.3.1 Appropriately Increasing the Inductance of Primary Winding L_P

The power supply efficiency can be improved by appropriately increasing the primary inductance of high-frequency transformer to let the SMPS work under continuous current mode. This is because after L_P is increased, the peak current I_P and RMS current I_{RMS} of the primary side can be reduced, thus reducing the loss of output rectifier and filter capacitor. In addition, the energy stored in the leakage inductance L_{P0} of high-frequency transformer can be reduced, which is proportional to $I_P{}^2$ and consumed within each switching cycle of primary clamp circuit.

8.2.3.2 Selecting Appropriate Parameters of D_{max} and U_{OR}

Considering TOPSwitch-GX series single-chip SMPS, for example, the maximum duty ratios (D_{max}) obtained by the load circuit directly affect the power loss allocation between the primary and secondary sides when the DC input voltage is the minimum (U_{Imin}). However, the D_{max} here is not the upper limit of duty ratio owned by TOPSwitch-GX itself, but a limit to be set externally. It not only relates to U_{Imin}, U_O, and the forward voltage drop U_{F1} of output rectifier but also depends on the turn ratio between the primary and secondary windings ($n = N_P/N_S$). Relevant formula is as follows:

$$D_{max} = \frac{U_O + U_{F1}}{U_{Imin}/n + U_O + U_{F1}} \tag{8.1}$$

It is easy to see that the maximum duty ratio can be increased by raising the turn ratio. Formula (8.1) can also be changed to the expression of turn ratio:

$$n = \frac{N_P}{N_S} = \frac{D_{max} U_{Imin}}{(1 - D_{max})(U_O + U_{F1})} \tag{8.2}$$

The turn ratio can also determine the primary induced voltage U_{OR} (also known as the secondary reflected voltage) during the turn-off of MOSFET. Relevant formula is as follows:

$$U_{OR} = n(U_O + U_{F1}) \tag{8.3}$$

In this case, the drain voltage is equal to the sum of the primary DC voltage U_I, the induced voltage U_{OR}, and the spike voltage caused by leakage inductance. As the total voltage limits the transformer ratio of high-frequency transformer, it limits the maximum duty ratio of SMPS.

This limit effect can be reflected in the maximum recommended value of U_{OR}. U_{OR} is generally allowed to be 80–140 V, and the maximum recommended value can be 135 V. The design steps are $n \rightarrow U_{OR} \rightarrow D_{max}$. Setting the turn ratio of high-frequency transformer can let U_{OR} reach the recommended value, thus automatically setting D_{max} to the maximum recommended value.

8.2.3.3 Reducing the Loss of Input Bridge Rectifier

For the BR constituted with diode, its nominal current shall be greater than the I_{RMS} when $u = u_{min}$. The formula of I_{RMS} is as follows:

$$I_{RMS} = \frac{P_O}{\eta u \cos \varphi} \tag{8.4}$$

wherein, the power factor of input circuit $\cos \varphi = 0.6$–0.8. Choose a BR with relatively large capacity and let it work under small current to reduce the voltage drop and power loss of BR.

8.2.3.4 Reducing the Loss of Power MOSFET

The basic requirements for power MOSFET are as follows: the drain current shall be large enough, the drain-source breakdown voltage shall be high enough, and the drain-source on-state resistance ($R_{DS\ (ON)}$) and output capacitance (C_{OSS}, i.e., the total drain-source distributed capacitance) shall be as small as possible. The smaller the $R_{DS\ (ON)}$ is, the lower will be the transmission loss. The smaller the C_{OSS} is, the lower will be the switching loss. However, it shall be noted that it is not always good to simply pursue small on-state resistance. Some MOSFETs has very small $R_{DS\ (ON)}$, helpful to reduce the transmission loss, but the switching loss is increased owing to relatively high C_{OSS}. Therefore, comprehensive evaluation shall be implemented for the above two technical indicators.

For AC 220 V input voltage, the withstand voltage of discrete MOSFET shall be 1000 V. In order to prevent breakdown, the MOSFET with withstand voltage of 600 V shall not be used.

8.2.3.5 Reducing the Loss of High-Frequency Transformer

High-frequency transformer is an important component for energy storage and transmission and has relatively large influence on power supply efficiency.

1. DC loss. The DC loss of high-frequency transformer is caused by the copper loss of windings. To improve efficiency, the current density of wire is generally $J = 4$–$6\,\text{A/mm}^2$ and up to $10\,\text{A/mm}^2$ under good heat dissipation conditions.
2. AC loss. The AC loss of high-frequency transformer is caused by the skin effect and the core loss. The skin effect will cause the AC equivalent impedance of wire to be much higher than copper resistance. The penetration capacity of high-frequency current toward conductor is proportional to the square of switching frequency, so the wire radius shall not exceed twice the depth reachable by the high-frequency current to reduce AC copper loss. See Figure 8.3

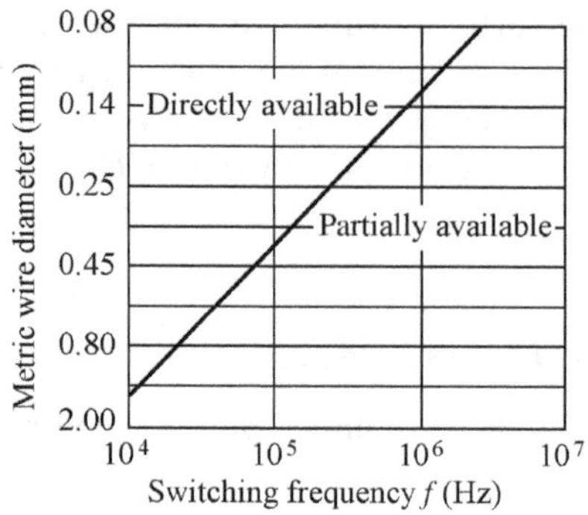

Figure 8.3 Relation curve of wire diameters and switching frequencies

for the relation curve of available metric wire diameters and switching frequencies. For example, when $f = 132\,\text{kHz}$, the wire diameter theoretically can be $\varphi 0.4\,\text{mm}$. However, in order to reduce skin effect, multistrand wires with diameter less than $\varphi 0.4\,\text{mm}$ are usually wound in parallel rather than a piece of thick wire.

The core loss of high-frequency transformer also reduces the power supply efficiency. Its AC flux density can be evaluated as the following formula:

$$B_{AC} = \frac{0.4\pi N_P I_P K_{RP}}{2\delta} \tag{8.5}$$

wherein, N_P is the turns of primary winding, I_P the primary peak current, K_{RP} the ratio between primary ripple current and peak current, and δ the air gap width of magnetic core with the unit of mm. "Ampere-turns" is used as the unit of magnetic flux in Formula (8.5). The loss of ferrite core at 132 kHz shall be less than $50\,\text{mW/cm}^3$.

3. Leakage inductance. When designing the low-loss high-frequency transformer, the leakage inductance must be minimized. This is because the spike voltage amplitude is increased with the rise of leakage inductance, thus increasing the loss of primary clamp circuit, which will inevitably reduce the power supply efficiency. The measures to reduce leakage inductance are as follows: (i) reducing the turns of primary winding, (ii) increasing the winding width, (iii) increasing the ratio between the height and width of winding (referred to as aspect ratio), (iv) reducing the insulation layer between windings, and (v) increasing the coupling degree between windings. The optimization design of high-frequency transformer uses ordinary high-strength enameled wire for the primary and bias windings and triple insulated wire for the secondary winding, which can significantly reduce the leakage inductance.
4. Winding arrangement. To reduce the leakage inductance, the windings shall be arranged in concentric as shown in Figure 8.4. Figure 8.4(a) uses triple insulated wire for the secondary side. Figure 8.4(b) uses enameled wire only, wherein, safety margin shall be remained and reinforced insulation layer shall be added between the secondary and bias windings. For multioutput power supply, the secondary side with the maximum output power shall be close to the primary side to reduce magnetic field leakage. When the secondary turns are very small, multistrand wires shall be wound in parallel and dispersedly on the entire skeleton to increase the coverage area.

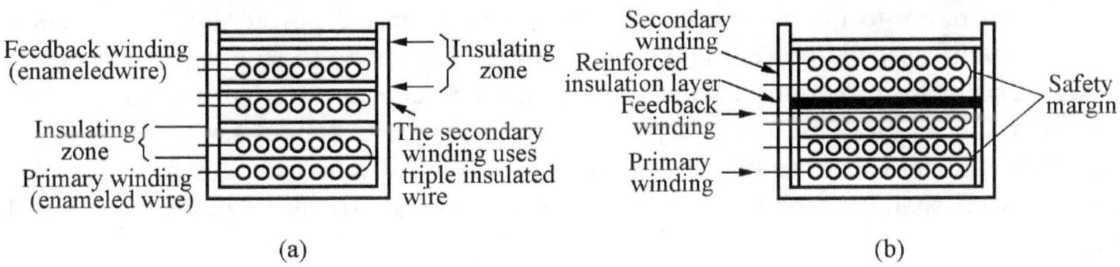

Figure 8.4 Winding arrangement. (a) Using triple insulated wire for the secondary side and (b) using enameled wire only

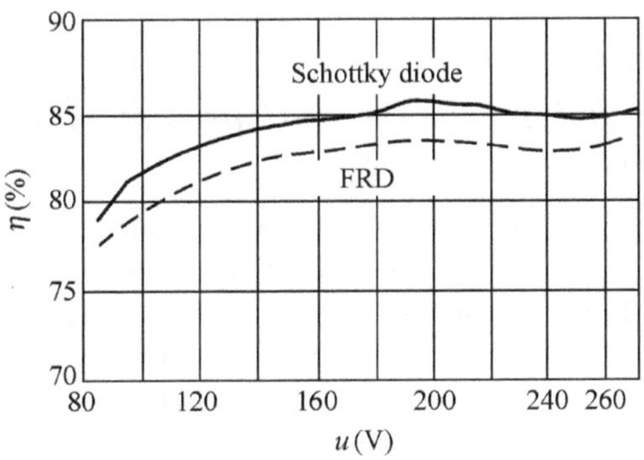

Figure 8.5 Relationship between efficiency and AC input voltage of two kinds of rectifier

8.2.3.6 Reducing the Loss of Output Rectifier

The output rectifier is an important reason of the reduction of power supply efficiency, the loss of which accounts for about 1/4–1/5 of the total loss. The transmission loss of rectifier mainly depends on its conduction voltage drop U_F. Assuming that the average rectified current is I_d and the duty ratio of current waveform in rectifier is D, the transmission loss of output rectifier can be calculated as per the following formula for flyback SMPSs:

$$P_D = I_d U_F \ (1 - D) \tag{8.6}$$

Therefore, Schottky rectifier shall be selected for low-voltage and high-current rectifying and FRD for high-voltage rectifying. See Figure 8.5 for the relation curve of efficiency and AC input voltage of two rectifier types. It is easy to see that when the AC voltage $u = 220\,V$, their efficiencies are, respectively, 85% and 83.4%.

8.3 Methods of Reducing No-Load and Standby Power Consumption of SMPS

The no-load power consumption of SMPS refers to the electric energy consumed when the power supply load is turned off with no function to be implemented. The standby power consumption refers to the electric energy consumed when the power supply is in standby state, in which case, the power supply enters into sleep mode to stop supplying power to the load, except for providing a little quantity of electricity to the small load (e.g., CPU in TV) under special circumstances only. According to the IEC 16301 standard (Section 4.5) of the International Electrotechnical Commission, the no-load and standby power consumptions of SMPS shall both be less than 5 mW, which brings stricter requirements for SMPS design.

The energy efficiency of SMPS is equal to the ratio between the rated output power P_O and the no-load power consumption P_K:

$$K = \frac{P_O}{P_K} \times 100\% \tag{8.7}$$

8.3.1 Methods to Eliminate the Power Consumption of Bleeder Resistor after Start-Up

The series-mode capacitor in EMI filter is also known as X capacitor, which is located between the two input terminals of power supply, and used to filter out power supply noise. X capacitor continues high-voltage charge storing after the AC power supply is turned off, so it is dangerous once the user touches the power plug inadvertently. The solution is connecting a bleeder resistor in parallel to X capacitor for discharging to meet the security requirements. However, the disadvantage of the solution is that when the SMPS works normally, the bleeder resistor will cause constant power consumption to reduce the power supply efficiency. The resistance range of bleeder resistor R is typically 150 kΩ–1.5 MΩ, corresponding to 220 V AC. The power consumption of R is up to 48–323 mW, much higher than the upper limit of 5 mW. Therefore, effective measures must be taken during the design of EMI filter to significantly reduce the power consumption of bleeder resistor.

8.3.1.1 Working Principle of CAPZero Series

A new type of self-powered two-terminal CAPZero series – X capacitor zero-loss arrester IC – was launched by PI in April 2010. It does not need any external bias circuit and has strong capability of suppressing common and series-mode interference without grounding. X capacitor zero-loss arrester can be equivalent to a smart high-voltage switch S, which is turned on when the SMPS works normally, letting the power consumption of bleeder resistor be close to zero by switching off the current of bleeder resistor. Although the turn-off resistance of MOSFET is not infinitely great, the power consumption of bleeder resistor can be reduced to below 5 mW, which has been close to zero loss, fully meeting the IEC 16301 international standard. When the AC power supply is turned off, the device can quickly turn on the bleeder resistor to automatically discharge the X capacity safely and allows using the X capacitor with larger capacity without increasing standby power consumption.

X capacitor zero-loss arrester is suitable for the AC/DC converter with X capacitor (greater than 100 nF), requiring very low standby power consumption. It integrated two MOSFETs with the withstand voltage of 825 V (or 1000 V) internally, capable of meeting various SMPS

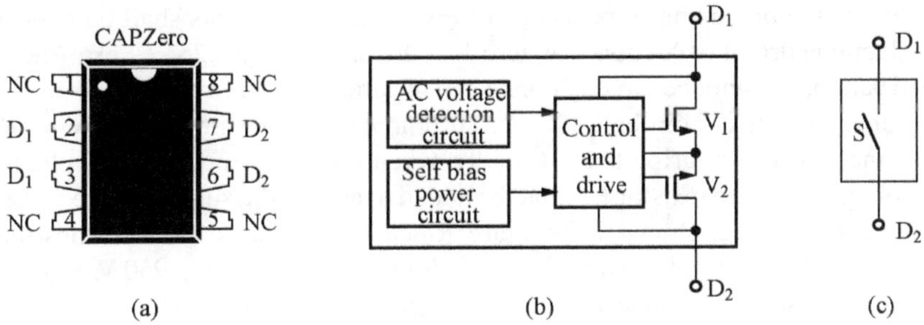

Figure 8.6 Pin array and internal block diagram of X capacitor zero-loss arrester. (a) Pin array, (b) internal block diagram, and (c) equivalent circuit

design requirements and suitable for the SMPSs of PC, server, TV, printer, and laptop as well as power adapters requiring very low no-load power consumption.

X capacitor zero-loss arrester uses SO-8 package with the pin array and internal block diagram, respectively, as shown in Figure 8.6(a) and (b). D_1 and D_2 are two outlet terminals, and others are none connection pins (NC). The chip mainly contains an AC voltage detection circuit, a self-bias power circuit, a control and drive circuit, and two complementary field effect transistors (FETs) (MOSFET) V_1 and V_2. The chip can automatically detect the AC input voltage and properly control the turn-on and turn-off of V_1 and V_2, equivalent to the smart switch S as shown in Figure 8.6(c). The supply current is only 21.7 μA during its work, and the maximum ambient temperature is up to +105 °C.

8.3.1.2 Typical Application of CAPZero Series

See Figure 8.7 for the typical application of X capacitor zero-loss discharger in EMI filter, wherein, components such as the common-mode capacitor (also known as Y capacitor, used to suppress common-mode interference) are omitted. R_1 and R_2 are external bleeder resistors with the total resistance of R. C_1 and C_2 are X capacitors. X capacitor zero-loss arrester can be

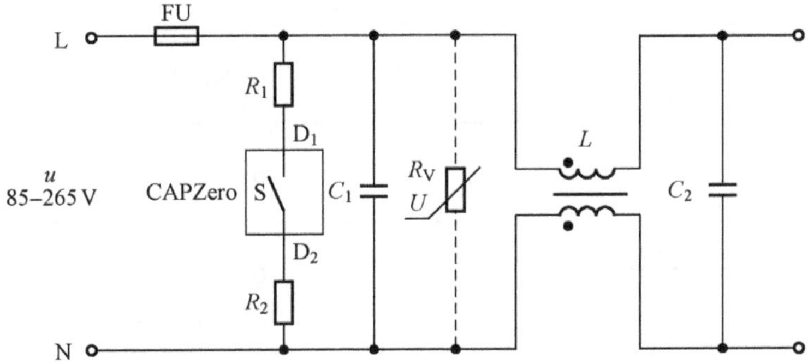

Figure 8.7 Typical application of X capacitor zero-loss arrester in EMI filter

located either at the preceding or poststage of EMI filter. Two points shall be noticed: firstly, as the conduction current of X capacitor zero-loss arrester is only 0.25–2.5 mA (the minimum guaranteed value), it cannot be directly connected in parallel to the two terminals of X capacitor to avoid damage resulted from high bleeder current and current limiting resistors R_1 and R_2 must be connected series during use. Secondly, when the series-mode surge voltage exceeds 1 kV, the voltage-sensitive resistor R_V shall be added to absorb the surge voltage at the AC inlet terminal as the dotted lines shown in the figure. R_V can be omitted when the series-mode surge voltage is below 1 kV. Thirdly, when the peak drain voltage exceeds 950 V, a 33 pF/1000 V ceramic capacitor shall be connected in parallel between terminals D_1–D_2 to attenuate the peak drain voltage, in which case, the increased power load will not exceed 0.5 mW with AC input of 230 V and 50Hz.

See Table 8.1 for the selection of CAPZero series. Take Figure 8.7 for example, the total X capacitance $C = C_1 + C_2$, and the total bleeder resistance $R = R_1 + R_2$. Typically, $R_1 = R_2$. During the design, the time constant determined by R and C is $\tau = RC = 0.75\,\text{s}$, whose maximum is $\tau = 1\,\text{s}$.

Table 8.1 Selection of **CAPZero** series

Model	Forward and reverse withstand voltages between pins D_1–D_2 (V)	Total X capacitance C (μF)	Total bleeder resistance R (MΩ)	Time constant τ (s)
CAP002DG	825	0.50	1.500	0.75
CAP012DG	1000			
CAP003DG	825	0.75	1.000	0.75
CAP013DG	1000			
CAP004DG	825	1.0	0.750	0.75
CAP014DG	1000			
CAP005DG	825	1.5	0.480	0.72
CAP015DG	1000			
CAP06DG	825	2	0.360	0.72
CAP016DG	1000			
CAP007DG	825	2.5	0.300	0.75
CAP017DG	1000			
CAP008DG	825	3.5	0.200	0.70
CAP018DG	1000			
CAP009DG	825	5.0	0.150	0.75
CAP019DG	1000			

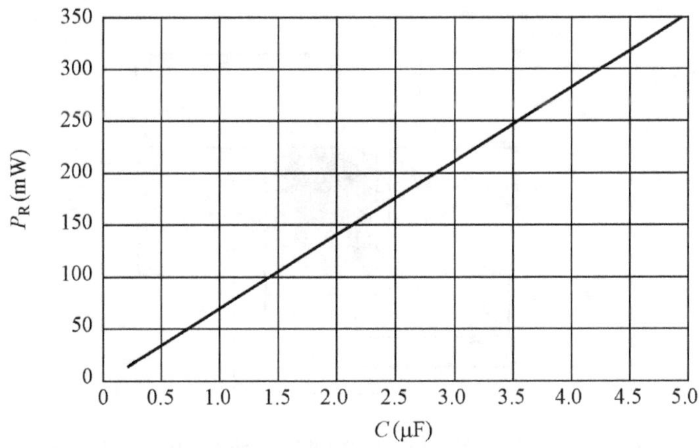

Figure 8.8 Relation curve of X capacitor and bleeder resistor power consumption ($u = 230\,\text{V}$ and $\tau = 0.75\,\text{s}$)

See Figure 8.8 for the relation curve of X capacitor and bleeder resistor power consumption drawn according to the data in Table 8.1 when the AC input voltage $u = 230\,\text{V}$ and $\tau = 0.75\,\text{s}$. It can be seen from Table 8.1 and Figure 8.8 that when $C = 0.5\,\mu\text{F}$ (corresponding to $R = 1.5\,\text{M}\Omega$), the constant power consumption of R is $P_R = 35\,\text{mW}$, and when $C = 5\,\mu\text{F}$ (corresponding to $R = 150\,\text{k}\Omega$), $P_R = 350\,\text{mW}$, which is much greater than 5 mW.

When no NTCR is used, P_R can be regarded as the power consumption of EMI filter P_{EMI}. For 50 Hz AC, the relation curve of the power consumption of EMI filter and the AC input voltage u is shown in Figure 8.9. It can be seen that when $u = 220\,\text{V}$, $P_{EMI} = 3.8\,\text{mW}$, and $P_{EMI} < 5\,\text{mW}$ within the variation range of $u = 80\text{–}265\,\text{V}$.

See Figure 8.10 for the layout of X capacitor zero-loss arrester in the PCB, wherein, the dashed box refers to X capacitor. R_1 and R_2 are surface-mount resistors.

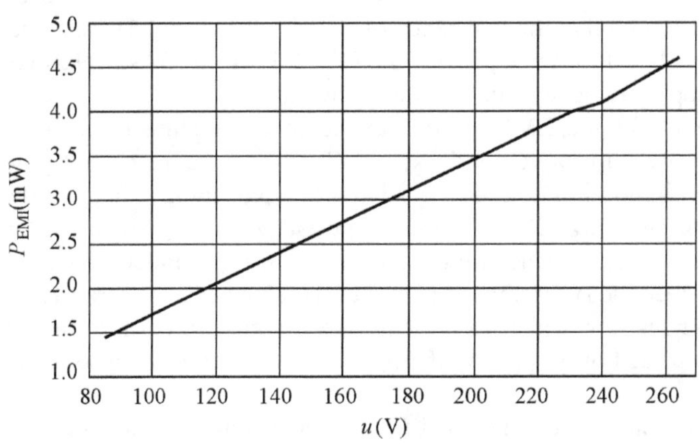

Figure 8.9 Relation curve of the power consumption of EMI filter and the AC input voltage

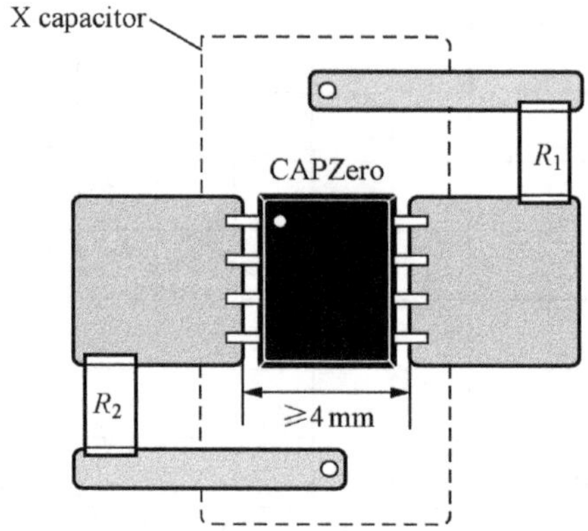

Figure 8.10 Layout of X capacitor zero-loss arrester in the printed circuit board

8.3.2 Methods to Eliminate the Power Consumption of Thermal Resistor after Turn-On

The power thermal resistor with negative temperature coefficient can be used for instantaneous current limiting protection when the SMPS is switched on. At the beginning of power-on, the voltage drop in filter capacitor cannot change suddenly and the capacitive reactance tends to zero, so the instantaneous charging current is very high and easy to damage the high-voltage electrolytic capacitor. To solve this problem, the power thermal resistor R_T is usually used to replace the ordinary current limiting resistor and the current limiting value can be slightly higher. Its work is featured that the resistance of R_T is relatively high at the beginning of power-on, delivering good instantaneous current limiting effect. R_T constantly emits heat with the current flowing through, so the power consumption is significantly reduced owing to its rapid decrease of resistance. However, the thermal resistor will reduce the no-load power consumption of SMPS. For example, when the SMPS has 220 V AC input, the maximum output power of 150 W, and the efficiency at full load of 92%, the current flowing through R_T is $(150\,W \div 92\%) \div 220\,V = 0.74\,A$ when a $5\,\Omega$ power thermal resistor is used at atmospheric temperature. Assuming that the resistance of R_T is reduced to $2.5\,\Omega$ after heating, the power consumption will be $(0.74\,A)^2 \times 2.5\,\Omega = 1.37\,W$. $1.37\,W \div (150\,W \div 92\%) \times 100\% = 0.84\%$, so the power supply efficiency can be reduced by 0.84%.

See Figure 8.11 for the EMI circuit capable of eliminating the power consumption of thermal resistor after turn-on. C_1 and C_2, and C_5 and C_6 are, respectively, used to constitute the two-stage common-mode capacitor, with the preceding stage used to suppress the common-mode noise above 30 MHz and the poststage used to suppress the resonance peak within the intermediate frequency range. The common-mode inductance L is used to suppress the low and intermediate interference below 1 MHz. C_3 and C_4 are series EMI filter capacitors. R_1 is the bleeder resistor, which can release the charges stored in C_3 and C_4 when the AC power supply is turned off to prevent the operator from being shocked owing to the touch of power plug.

The thermal resistor R_T can be used for instantaneous current limiting protection at the beginning of power-on. When the power supply enters into normal working, the output voltage U_{JD} passing through the on/off control line of relay is 0 V. Connect the voltage $(U_{CC} - 0\,V)$ to

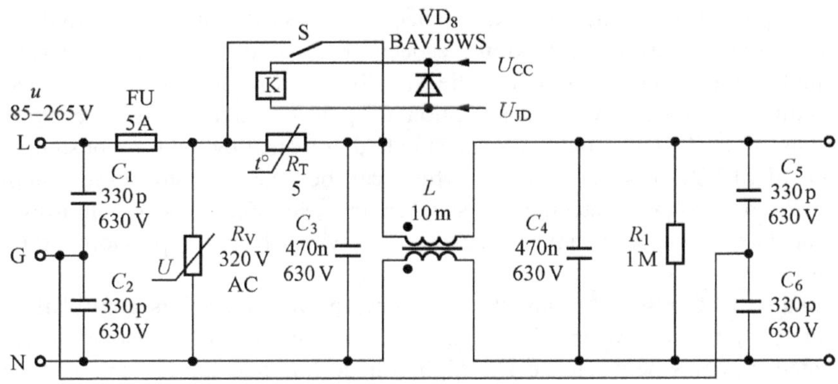

Figure 8.11 EMI circuit capable of eliminating the power consumption of thermal resistor after turn-on

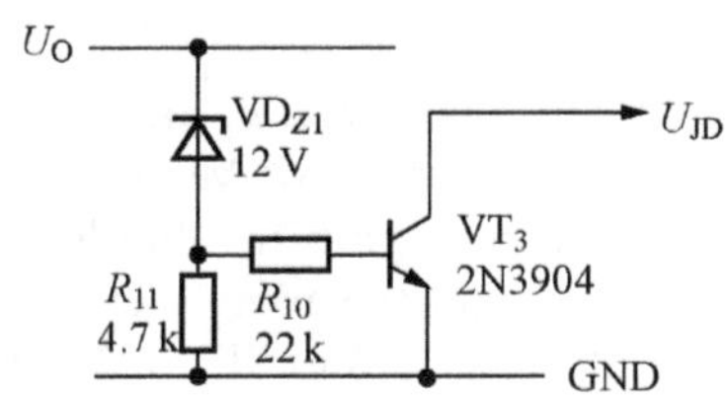

Figure 8.12 On/off control circuit of relay

the relay winding to let the contact S be absorbed, so R_T is in short circuit to reduce the power consumption of R_T to zero. Only when the power supply is turned on, the on/off control line is disconnected to release the relay winding owing to power-off, thus disconnecting the contact S. Connecting a freewheeling diode VD_8 in parallel across both ends of the relay winding, a bleed-off circuit can be provided to the back electromotive force for protection.

See Figure 8.12 for the on/off control circuit of relay. Once U_O reaches the specified value ($U_O \geq 15$ V), the regulator tube will suffer reverse breakdown to enable the transistor VT_3 of open collector with the collector output of 0 V, that is, $U_{JD} = 0$ V. In this case, current flows through the relay winding to let the contact S absorbed, so R_T is in short circuit to reduce the power consumption of R_T to zero. Only when the power supply is turned on, V_{T3} is turned off and the on/off control line is disconnected to release the relay winding owing to power-off, thus disconnecting the contact S to stop supplying power to the relay winding. When U_O is less than 15 V, the regulator with relatively low regulation value shall be used. This design scheme can improve the power supply efficiency by about 0.5–1.5%.

8.3.3 Methods to Eliminate the Power Consumption of Sense Resistor in Standby Mode

8.3.3.1 Working Principle of SENZero Series

In order to minimize the power consumption of SMPS in standby mode (or at zero load), the SENZero series – zero-loss high-voltage sense-signal on/off IC – was newly launched by PI in August 2010, which can disconnect unnecessary unit circuit under standby, zero-load, or remote turn-off conditions, especially disconnecting the sense resistor and the DC high-voltage line to eliminate the standby power consumption of sensor resistor and reduce the total power consumption of power system, thus meeting the strict requirements of IEC 16301 international standard for standby power consumption. The application of SENZero is very simple, which can be added to any power supply using a resistor for bus voltage detection to save energy. SENZero has an ultra-low leakage current, capable of not only maximizing energy savings but also providing short circuit protection.

Traditionally, LLC resonant converter and other power controllers control the work of power supply by monitoring the DC input high-voltage U_I, so a resistance divider must be connected in series between the DC high-voltage line and the low-voltage monitoring terminal. When the SMPS is in the standby mode, the power consumption of resistance divider is up to a few hundred milliwatts, significantly increasing the standby power consumption of SMPS. SENZero chip can be used to satisfactorily resolve the above problem.

Currently, the SENZero series has two models: SEN012 (dual channel) and SEN013 (triple channel). The maximum leakage current of MOSFET integrated inside is only 1 μA at the high voltage of 375 V (DC), belonging to ultra-low leakage current. The power consumption of each channel is less than 0.5 mW. When the AC input voltage is 230 V, the standby power consumptions of SEN012 and SEN013 with the MOSFET disconnected are, respectively, less than 1 and 1.5 mW. The measured total standby power consumption of SEN013 is as low as 0.79 mW < 1.5 mW, which is completely negligible, and provides the best solution to thoroughly eliminate the standby power consumption of sense resistor. SENZero can be configured with the AC/DC converter with high-voltage impedance signal path, suitable for the power supply having strict requirements for standby or no-load power consumption as well as laser printers, household appliances, servers, and network equipment.

SENZero uses the small SO-8 package and its pin array and internal block diagram are, respectively, shown in Figure 8.13(a), (b), and (c). The pin functions are as follows: S_1–S_3 are source terminals, D_1–D_3 are drain terminals, GND is the common ground, and NC is the nonconnection pin. Terminal U_{CC} shall be connected with the mains voltage U_{CC}, and its input current is less than 0.5 mA. When $6\,V < U_{CC} < 16\,V$, the internal MOSFET is fully turned on. When $U_{CC} = 0\,V$ (ground potential), the internal MOSFET is turned off. SENZero mainly contains a bias circuit, a drive controller, and 2–3 (depending on the product model) MOSFETs with withstand voltage of 650 V. In addition, there is an internal pull-down resistor (not shown in Figure 8.13) at the gate of each MOSFET, which can ensure no device will be damaged in case pin U_{CC} is disconnected.

8.3.3.2 Typical Application of SENZero Series

See Figure 8.14 for the typical application circuit of SEN013 in two-stage PFC power supply. This kind of power supply is featured that the first stage is PFC (with the output of DC high-voltage U_{O1}) and the second stage is DC/DC converter (with the output of DC low-voltage U_{O2}) and is usually used as high-power SMPS. The power end of SEN013 is connected to system U_{CC} through the switch S. Only when the standby signal from the system is received, S is

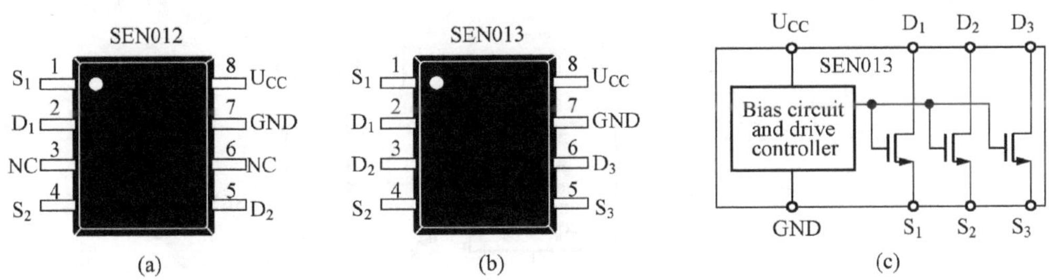

Figure 8.13 Pin array and internal block diagram of SENZero. (a) SEN012 (dual channel), (b) SEN013 (triple channel), and (c) internal block diagram of SEN013

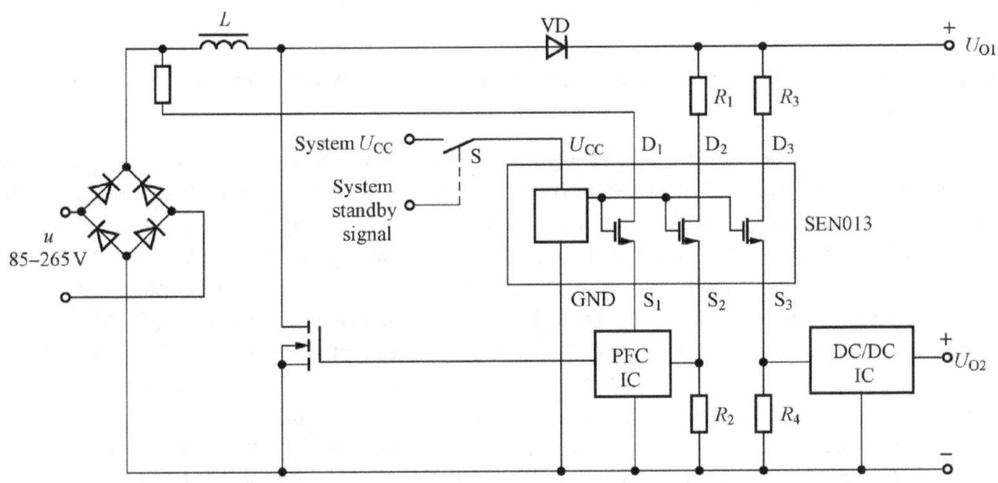

Figure 8.14 Typical application circuit of SEN013 in two-stage PFC power supply

disconnected to turn off the internal MOSFET. The three MOSFETs in SEN013 are, respectively, connected in series in the circuits of PFC IC, sense resistors (R_1 and R_2) of PFC, and sense resistors (R_3 and R_4) of DC/DC IC. The control toward PFC IC and DC/DC IC is realized by U_{CC} supplying power to SEN013. Once U_{CC} is disconnected, PFC IC and DC/DC IC will be turned off and the two detection circuits are simultaneously disconnected. When U_{CC} is added again during start-up, PFC IC, DC/DC IC, and the detection circuit will recover normal work immediately. It shall be noted that the maximum allowable voltages between pins U_{CC}-GND and S-GND of SENZero are, respectively, 16 and 6.5 V, and the on-state resistance of internal MOSFET is about 500 Ω at room temperature, so SENZero can be connected in series to high-resistance circuits only. Thus, the on-state resistance accounts for a small part of the total series resistance, not affecting the normal operation of the original circuit. It is not necessary to connect a bypass capacitor to pin U_{CC} of SEN013. In addition, SENZero can be used in the triple-stage PFC power supply, which typically includes the following three stages: PFC, LLC resonant converter, and DC/DC converter.

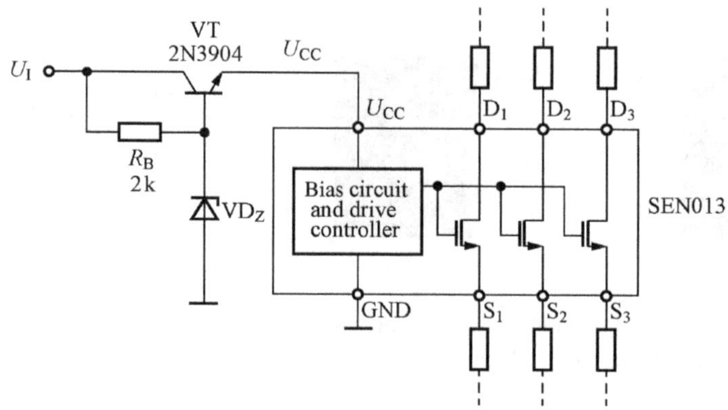

Figure 8.15 A power configuration mode of SENZero

8.3.3.3 Power Configuration Modes and Remote Shutdown Function of SENZero Series

See Figure 8.15 for a power configuration mode of SENZero. The DC voltage U_I without regulation will be obtained after the voltage in the bias winding of SMPS passing through the rectifier filter, which will be first sent to the simple linear regulator constituted with the transistor VT (2N3904), the base bias resistor (R_B), and the regulator tube (VD_Z) to obtain the regulated output U_{CC}, and then supply power to SENZero. This configuration mode can realize the control of the voltage U_{CC} when $U_I > 16$ V. When the power is turned off, SENZero will be turned off immediately as long as U_I is less than the regulated value U_Z of VD_Z. When the bias winding voltage is regulated or the voltage at pin U_{CC} can be maintained within the range of $6\,V < U_{CC} < 16\,V$, the linear regulators (VT, R_B, and VD_Z) can be omitted.

See Figure 8.16 for the power configuration capable of realizing remote shutdown function, wherein, U_I supplies power to SENZero through the NPN transistor VT. VT not only has the preregulation function but also can realize on/off control. When the control signal U_K is at high level, VT is conducted to allow SENZero work. When U_K is at low level, VT is turned off to disable SENZero.

8.4 Stability Design of Optocoupler Feedback Control Loop

8.4.1 Basic Requirements for Optocoupler Feedback Control Loop

The optocoupler feedback loop (hereinafter referred to as feedback loop) of SMPS is the key circuit of SMPS, which belongs to the negative feedback system. It is used to control the output voltage to be stable in all cases. The feedback loop design is also directly related to the technical indicators of SMPS such as the voltage regulation, the load regulation, and the transient response.

A bode diagram can be generated using the amplitude-frequency and phase-frequency characteristic curves, wherein, the gain and frequency coordinates use the logarithmic scale and the phase adopts the dominant scale. See Figure 8.17 for the bode diagram example of SMPS, which can provide convenience for calculating the loop gain and phase as well as stability

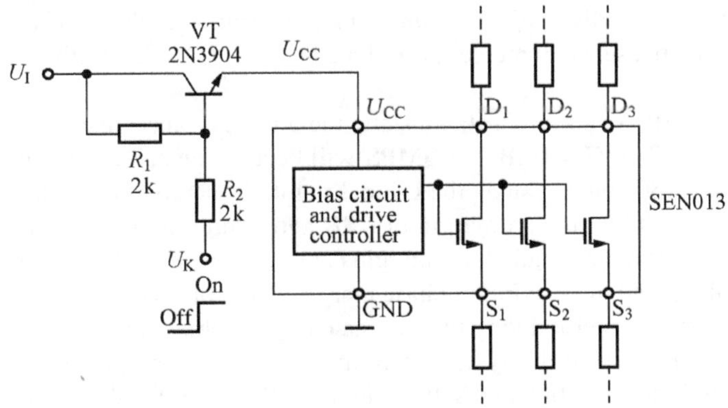

Figure 8.16 Power configuration capable of realizing remote shutdown function

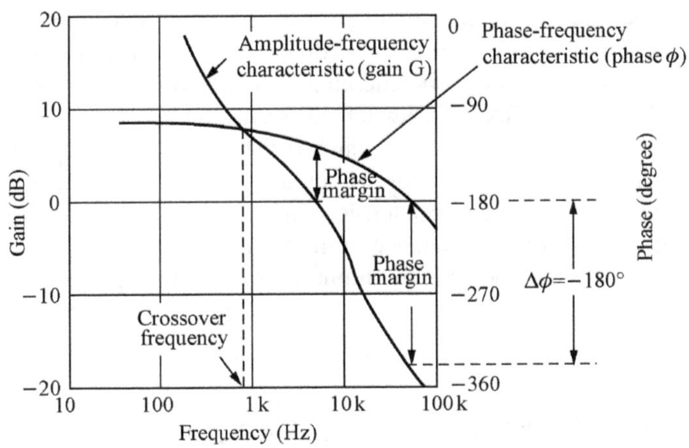

Figure 8.17 Bode diagram example of SMPS

analysis. The gain curve in bode diagram reflects the total amplification capability of the two main circuits of SMPS main loop (also known as power stage) and optocoupler feedback control loop (also known as feedback stage) toward different frequency signals. The crossover frequency in bode diagram indicates the frequency point of the amplitude-frequency characteristic curve (also known as gain curve) crossing over line 0 dB. When the denominator of transfer function is zero, a pole will be generated, which corresponds to the pole generated when the gain in bode diagram decreases at the slope of 20 dB/decade. In the frequency range, a zero point will be generated when the molecule of transfer function is equal to zero, which corresponds to the point the gain in bode diagram increases at the slope of 20 dB/decade and has 90° phase advance.

The indicators measuring the stability of flyback SMPS are gain and phase margins, that is, the allowable gain and phase ranges before the power supply become unstable. Specifically, the gain margin refers to the gain (actually attenuation) when the phase $\varphi = -180°$, that is, the value from 0 dB to $-G$ (dB). $\Delta\varphi$ in Figure 8.17 represents the phase difference. The phase margin refers to the phase when the gain is reduced to 0 dB ($G = 0$ dB), that is, the value from φ (degree) to $-180°$.

In order to let SMPS enter into a stable state, the gain and phase margins of feedback loop must be designed. When $G = 0$ dB, the SMPS will become unstable if the phase lag is more than $-180°$ with $-180°$ phase lag as the critical point. Typically, the feedback loop shall be designed to have a minimum gain margin of 6 dB and a minimum phase margin of 45°. The recommended minimum gain and phase margins are, respectively, 10 dB (absolute value) and 60° (absolute value), and the maximum phase margin shall not exceed $-180°$. Thus, the SMPS can be ensured to be stable for a long time in the case of ambient temperature or the transient load change, so as to prevent the system stability from being disrupted owing to the failures such as error amplifier saturation and self-excited oscillation of feedback loop.

8.4.2 *Stability Design of Optocoupler Feedback Control Loop*

This section describes the specific method of stability design of optocoupler feedback control loop (hereinafter referred to as optocoupler feedback loop) by an instance. See Figure 8.18 for the basic structure of feedback loop, wherein, the TOPSwitch single-chip SMPS includes TOPSwitch-GX, TOPSwitch-HX, TOPSwitch-JX series, and so on. C_O is the output filter capacitor. R_{F5} and C_{F3} are, respectively, the shunt resistor and bypass capacitor at the control terminal. R_{F1} and R_{F2} are sampling resistors of output voltage. C_{F1} is the compensation capacitor. R_{F3} is the loop gain adjustment resistor. R_{F4} and C_{F2} constitute the phase boost network (optional). The primary winding inductance of the SMPS $L_P = 827\,\mu$H, the secondary equivalent inductance $L_E = 41\ \mu$H, the ESR of output filter capacitor $R_{ESR} = 33\,\text{m}\Omega$, the working

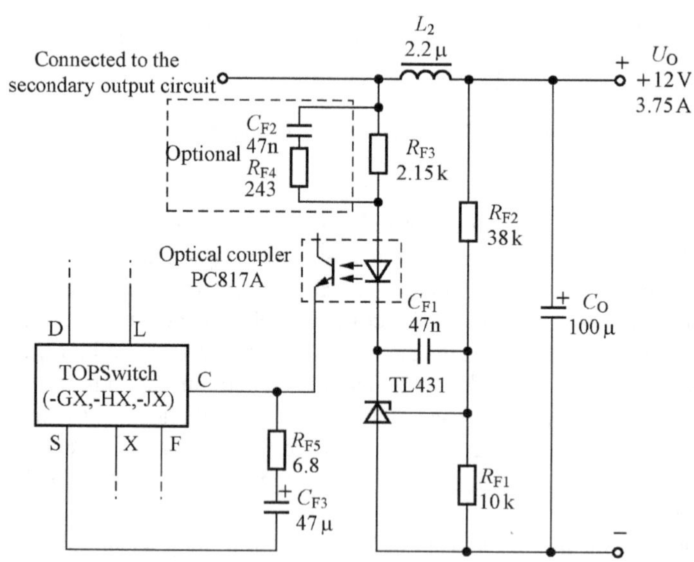

Figure 8.18 Basic structure of feedback loop

duty ratio $D = 0.55$, and the output load impedance $R_O = 3.2\,\Omega$. It is not necessary to take into consideration the influence of resonant frequency from post-filter on the loop response in the design, because the resonant frequency of post-filter is required to be greater than 10 kHz only just through reasonable design.

8.4.2.1 Performance Parameters Setup of Feedback Loop

1. Calculating the expected right-half-plane zero (referred to as RHP Zero) frequency f_{RHP}.

   When the flyback SMPS works in continuous conduction mode (CCM), its transfer function contains an inherent RHP zero. When the load current increases, it is necessary to increase the primary winding current of high-frequency transformer. Flyback SMPS must increase the duty ratio to achieve this goal. The output current flows to the load only when the MOSFET is turned off and the output rectifier is conducted. However, the switching frequency is fixed, so increasing the duty ratio will definitely increase the conduction time of MOSFET, thus reducing the conduction time of output rectifier. The consequence is that the load current is actually reduced, causing the output voltage to be reduced at the initial time. From then on, the primary winding current rises continuously to finally let the average current of output rectifier reach the rated value. The above phenomenon actually requiring reducing the current of output rectifier before increasing it is called RHP Zero. Obviously, the Zero will generate a phase delay. RHP Zero occurs in the CCM only.

   The calculation formula of RHP Zero frequency f_{RHP} is as follows:

$$f_{RHP} = \frac{R_O}{2\pi L_E D} \tag{8.8}$$

   wherein, R_O is the output load impedance, L_E the secondary equivalent inductance, and D the duty ratio of power switching tube. Put $R_O = 3.2\,\Omega$, $L_E = 41\,\mu\text{H}$, and $D = 0.55$ into Formula (8.8) to get $f_{RHP} \approx 23\,\text{kHz}$.
2. Selecting the crossover frequency $f_{CROSSOVER}$. For TOPSwitch-JX series single-chip SMPS, the switching frequency is 132 kHz (full-frequency mode) or 66 kHz (half-frequency mode). The crossover frequency shall be less than one-tenth of the switching frequency (i.e., $f_{CROSSOVER} \le 0.1f$) and less than one-fifth of the RHP Zero frequency (i.e., $f_{CROSSOVER} \le 0.2f_{RHP}$). The reasonable crossover frequency $f_{CROSSOVER}$ is 500 Hz–3 kHz, which obviously meets the requirement of $f_{CROSSOVER} < 0.1f = 6.6\,\text{kHz}$ (half-frequency mode) and $f_{CROSSOVER} < 0.2f_{RHP} = 4.6\,\text{kHz}$. $f_{CROSSOVER} = 1\,\text{kHz}$ is a good starting point for more than 80% designs. The higher the $f_{CROSSOVER}$ is, the wider will be the frequency band, but the noise sensitivity will be increased.
3. Selecting the required open-loop phase margin with at least 45° phase margin. The allowable phase margin range is 45°–75°, and the recommended phase margin is 60°.

8.4.2.2 Selecting the output filter capacitance C_O

When the switching frequency $f = 66\,\text{kHz}$, the output ripple voltage $\Delta U_O = 100\,\text{mV} = 0.1\,\text{V}$, the ripple current of capacitor $\Delta I_O = 0.4I_O = 0.4 \times 3.75\,\text{A} = 1.5\,\text{A}$, and the ESR of the selected electrolytic capacitor $R_{ESR} = 33\,\text{m}\Omega = 0.033\,\Omega$, the minimum C_O can be calculated according to Formula (4-2-4):

$$C_{O(min)} = \frac{1}{8f\left(\frac{\Delta U_O}{\Delta I_O} - R_{ESR}\right)} = \frac{1}{8 \times 66\,\text{kHz}\left(\frac{0.1\,\text{V}}{1.5\,\text{A}} - 0.033\,\Omega\right)} = 56.3\,\mu\text{F}$$

Actually taking the nominal capacity of $C_O = 100\,\mu F$.

To avoid the resonance of output circuit, the resonant frequency of LC circuit constituted by the secondary equivalent inductor L_E and the output filter capacitor C_O shall be greater than 500 Hz. The secondary equivalent inductance L_E is determined by the following formula:

$$L_E = \frac{L_P\left(\frac{N_S}{N_P}\right)^2}{(1-D)^2} \tag{8.9}$$

wherein, L_P is the primary winding inductance of high-frequency transformer, N_P the turns of primary winding, N_S the turns of secondary winding, and D the duty ratio of power switching tube. The secondary equivalent inductance of the power supply is $L_E = 41\,\mu H$.

8.4.2.3 Selecting the bypass capacitor C_{F3} and the shunt resistor R_{F5} at the control terminal

C_{F3} is powered by the internal circuit of TOPSwitch and determines the automatic restart time. The range of C_{F3} is 10–100 μF and the recommended value is 47 μF.

R_{F5} is a small resistor connected in series with C_{F3}. After the resistance of R_{F5} is connected in series with the ESR R_{ESR} of C_{F3}, their total impedance can be remained relatively stable owing to $R_{F5} \gg R_{ESR}$. The allowable range of R_{F5} is 0–22 Ω and the recommended value is 6.8 Ω.

8.4.2.4 Selecting the Optical Coupler

The current transfer ratio CTR of optical coupler is equal to $(I_E/I_{LED}) \times 100\%$, wherein, I_E is the emitter current of infrared receiver tube and I_{LED} is the current flowing through LED. CRT has significant impact on the entire loop gain. The recommended CTR range is 80–160% and the nominal value is 100%. PC817A optical coupler can be used, of which CTR = 80–160% and $I_{LED} = 1\,mA$ (typically). The steady-state operating current decreases with the rise of CTR, which can reduce the no-load power consumption. When LTV817D optical coupler is used, CTR = 300–600% and $I_{LED} = 100\,\mu A$, in which case the resistance of gain adjustment resistor R_{F3} must be increased.

8.4.2.5 Selecting the Adjustable Reference Voltage Source

The single-chip SMPS contains an error amplifier internally, so it is required to configure an external adjustable reference voltage source only. When $U_O > 3.3\,V$, TL431 adjustable precision shunt regulator (the internal reference voltage $U_{REF} = 2.5\,V$) can be used. When $U_O \leq 3.3\,V$, it is recommended to use LMV431 adjustable precision shunt regulator ($U_{REF} =$ 1.24 V, can be approximately regarded as 1.25 V).

8.4.2.6 Selecting the Sampling Resistors R_{F1} and R_{F2}

Sampling resistors R_{F1} and R_{F2} are used to get a proper sampling voltage for TL431, not affecting the loop gain and phase. The value of sampling voltage is set by the resistance dividers R_{F2} and R_{F1}. The resistance range of R_{F1} is 2–50 kΩ, and the recommended value is 10 kΩ. Too high resistance of R_{F1} will cause the bias current of error amplifier to be too small.

Once R_{F1} is selected, the resistance of R_{F2} can be calculated according to the values of output voltages U_O and U_{REF}. The formula is as follows:

$$R_{F2} = R_{F1} \cdot \frac{U_O - U_{REF}}{U_{REF}} \tag{8.10}$$

When $R_{F1} = 10\,\text{k}\Omega$, $U_O = 12\,\text{V}$, and $U_{REF} = 2.5\,\text{V}$ are known, it can be calculated that $R_{F2} = 38\,\text{k}\Omega$ according to Formula (8.10). The accuracies of R_{F1} and R_{F2} are 1%.

8.4.2.7 Selecting the Compensation Capacitor C_{F1}

Crossover frequency is determined by the zero frequency f_{ZERO} of RC network constituted by R_{F2} and C_{F1} and the pole frequency f_{POLE} (7 kHz) of TOPSwitch control terminal. Generally set $f_{ZERO} \approx 100\,\text{Hz}$. The pole frequency f_{POLE} of control terminal C of TOPSwitch is 7 kHz and the crossover frequency $f_{CROSSOVER} = 1\,\text{kHz}$. The formula of f_{ZERO} is as follows:

$$f_{ZERO} = \frac{f_{CROSSOVER}^2}{f_{POLE}} \tag{8.11}$$

It is easy to find out that $f_{ZERO} = 142.9\,\text{Hz}$, actually taken 120 Hz.

The capacitance of compensation capacitor C_{F1} is determined as follows:

$$C_{F1} = \frac{1}{2\pi f_{ZERO} R_{F2}} \tag{8.12}$$

Put $f_{ZERO} = 120\,\text{Hz}$ and $R_{F2} = 38\,\text{k}\Omega$ into Formula (8.12) to get $C_{F1} = 34.9\,\text{nF}$, actually taking $C_{F1} = 47\,\text{nF}$.

8.4.2.8 Selecting the Gain Adjustment Resistor R_{F3}

The total gain (G) of feedback loop is equal to the gain (K_{TOP}) of TOPSwitch control circuit multiplied with the TL431 open-loop voltage gain (K_{TL431}), that is, $G = K_{TOP}K_{TL431}$, wherein, $K_{TL431} \approx 53\,\text{dB}$ (the frequency range is 1–10 kHz). However, G is neither the higher the better nor the smaller the better. When G is too high, the output voltage will track back and forth around the mean value to let the output voltage fluctuate severely and even occur oscillation in serious cases. Conversely, too low G will cause unstable output voltage, so that the dynamic response becomes poor owing to the adjustment lag resulted from improper voltage track. The feedback loop gain is set by R_{F3}. The smaller the R_{F3} is, the higher will be the current I_{LED} flowing through LED, so the control action becomes stronger to increase the gain. However, the power consumption of feedback loop will be increased. Conversely, $R_{F3} \uparrow \rightarrow I_{LED} \downarrow \rightarrow G \downarrow$. This indicates that the gain curve will move up or down when adjusting the resistance of R_{F3}, but the phase response of loop is not affected by R_{F3}.

Table 8.2 Allowable resistance range of R_{F3} (the **CTR** of the used optical coupler is equal to **100%**)

Output voltage U_O/V	Allowable resistance range of TOPSwitch-JX series	Allowable resistance range of other TOPSwitch series
5	200 (initial value)–470 Ω (maximum)	100 (initial value)–233 Ω (maximum)
12	910Ω–2.7 kΩ	470Ω–1.3 kΩ
15	1.3–3.6 kΩ	680–1.8k Ω
24	2–6.8 kΩ	1–3.3 kΩ
48	3.9–14.7 kΩ	2–7.3 kΩ

There are two methods to select R_{F3}. One is look-up table method. When the CTR of the used optical coupler is equal to 100%, see Table 8.2 for the allowable resistance range of R_{F3} for single-chip SMPSs of TOPSwitch-JX series and other TOPSwitch series. When $U_O = 12\,V$, actually take $R_{F3} = 2.15\,k\Omega$.

The other method is calculating R_{F3} according to the following formula when the gain margin of crossover frequency is known to be G (with the unit of dB):

$$R_{F3} = 10^{\frac{G(dB)}{20}}\ (\Omega) \tag{8.13}$$

For example, assuming that G (dB) = 66.6 dB, put it into Formula (8.13) to get $R_{F3} = 2138\,\Omega = 2.138\,k\Omega$.

When CTR > 100%, the resistance of R_{F3} can be increased according to corresponding proportion. For example, when CTR = 400% and the output voltage is constant (i.e., the feedback loop gain is constant), the initial and maximum values of R_{F3} shall be enlarged to 4 times, respectively. However, too large R_{F3} will limit the current I_{LED} of optical coupler, so R_{F3} shall not exceed the maximum value. During commissioning, R_{F3} can be gradually increased from its initial value until the regulation performance and the transient response of load are both the best.

8.4.2.9 Selecting the R_{F4} and C_{F2} Network Increasing the Phase Margin (Optional)

Resistor R_{F4} and C_{F2} are optional, which are dedicated to improve the phase margin and particularly suitable for multioutput SMPSs.

$$R_{F4} = \frac{R_{F3}}{9} \tag{8.14}$$

$$C_{F2} = \frac{1}{10 \times 2\pi R_{F4} f_{CROSSOVER}} = \frac{9}{10 \times 2\pi R_{F3} f_{CROSSOVER}} \tag{8.15}$$

When $R_{F3} = 2.15\,k\Omega$ and $f_{CROSSOVER} = 1\,kHz$ are, respectively, substituted into Formulas (8.14) and (8.15), it can be obtained that $R_{F4} = 239\,\Omega$ and $C_{F2} = 67\,nF$. The nominal value 243 Ω of E196 series is taken as R_{F4} and the nominal capacity 47 nF as C_{F2}. Additional phase margin of about 30° can be provided using R_{F4} and C_{F2}.

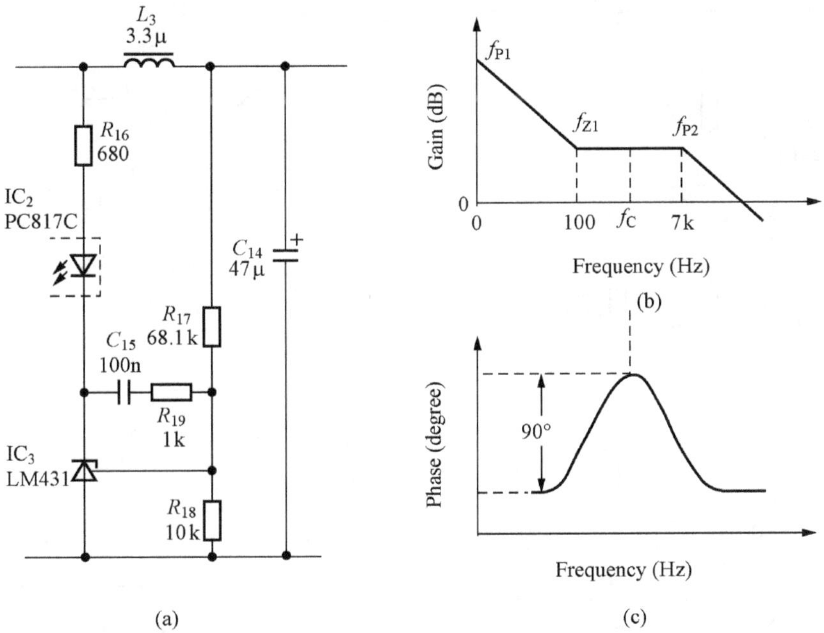

Figure 8.19 (a) Feedback loop before change and its (b) amplitude-frequency and (c) phase-frequency characteristic curves

8.4.3 Design Examples of Increasing Phase Margin

For the SMPS constituted by TOP269EG of TOPSwitch-JX series, the AC input voltage $u = 85$–$265\,V$ and the output is $U_O = 19\,V$ and $I_O = 3.42\,A$. See Figure 8.36 for relevant circuit. In the power supply, the phase compensation network of R_{22} and C_{14} is connected in parallel to the gain resistor R_{21} of feedback loop. See Figure 8.19(a), (b), and (c), respectively, for the feedback loop before change and its amplitude-frequency and phase-frequency characteristic curves. f_{P1}–f_{P3} in the amplitude-frequency characteristic curve refer to pole frequencies. There are two poles (f_{P1} and f_{P2}, referred to as dual pole frequency) in Figure 8.19, wherein, the f_{P1} set by C_{15} is equal to 0 Hz, the f_{Z1} set by C_{15}, R_{19}, and R_{17} is equal to 100 Hz, and the f_{P2} set by the resistor capacitor units in TOP269EG is equal to 7 kHz.

See Figure 8.20(a), (b), and (c) for the feedback loop after change and its amplitude-frequency and phase-frequency characteristic curves. There are three poles (f_{P1}, f_{P2} and f_{P3}, referred to as three-pole frequency) in Figure 8.20(b). f_{Z1} and f_{Z2} refer to zero frequency, which can also be one or several. The f_{P1} set by C_{15} is equal to 0 Hz, the f_{Z1} set by C_{15}, R_{24}, and R_{25} is equal to 100 Hz, and the f_{Z2} set by R_{21}, R_{22}, and C_{14} is equal to f_C, which refers to the crossover frequency, representing the frequency corresponding to the total open-loop gain of 1 (i.e., 0 dB). The crossover frequency must be less than 1/10 of the switching frequency. The f_{P2} set by TOP269EG is equal to 7 kHz. The f_{P3} set by R_{21}, R_{22}, and C_{14} is equal to $10f_C$. After R_{22} and C_{14} are added, the phase difference of phase-frequency characteristic curve is increased from 90° to 135°, that is, increased by 45°. The total phase margin of bode diagram can be

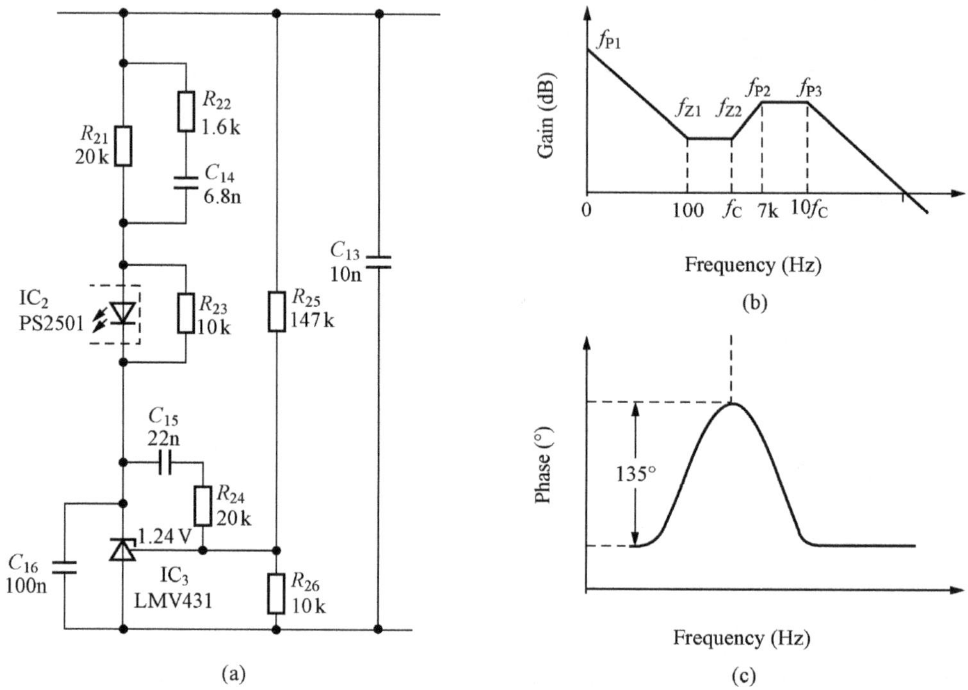

Figure 8.20 (a) Feedback loop after change and its (b) amplitude-frequency and (c) phase-frequency characteristic curves

increased by 30° through the above change, improving the stability of feedback loop. The bode diagram of SMPS can be measured with frequency response analyzer.

It shall be noted that R_{21} will not be affected at low-frequency stage after the R_{22} and C_{14} network and the original gain resistor R_{21} are connected in parallel. However, when f_C exceeds, the capacitive reactance decreases continuously to have more obvious parallel connection effect toward R_{21}, so the total resistance is reduced to increase the feedback loop gain. This is also reflected in Figure 8.20(b), the amplitude-frequency characteristic curve at the frequency stage f_C-$10f_C$ raised to a certain extent.

8.5 SMPS Layout and Wiring

8.5.1 General Principles of SMPS Layout and Wiring

8.5.1.1 SMPS Layout Principles

1. During the planning of component layout, the size and shape of PCB shall be determined first. The isolated SMPS is divided into the input and output sides and requires electrical isolation between the two sides, so the PCB is usually designed to be in rectangular shape. First, arrange all component packages of PCB evenly in a rectangular, with the primary components on the left, the secondary components on the right, and setting aside

four mounting holes. After that, draw a rectangle on the keep-out layer to surround all components with some safety margin left. The size of PCB is basically determined.

2. The layout is generally started from the high-frequency transformer. Arrange the high-frequency transformer in the middle of PCB with the primary side at left and the secondary side at right. The input filter capacitor, the primary winding, and the power switching tube constitute a relatively large impulse current loop. The secondary winding, the rectifier (or freewheeling) diode, and the output filter capacitor constitute another relatively large impulse current loop. The two loops shall have compact layout with short lead to reduce the leakage inductance, thereby reducing the loss of absorption circuit to improve the power supply efficiency. Appropriately increase the space between the primary components with high voltage and properly implement necessary fine adjustment for the size of PCB to finalize the entire layout of PCB.
3. The size of PCB shall be moderate. When the size is too large, the printed line will be lengthened to increase the impedance, thereby not only decreasing the noise immunity but also increasing the cost. However, too small size will cause poor heat emission and make the printed line susceptible to the interference of adjacent wires.

8.5.1.2 SMPS Wiring Principles

1. SMPS wiring mainly considers the wire width selection and isolation space issues. The wiring of ground wire and the selection of sampling points shall be particularly noted, which will directly affect the performance indexes of power supply. Single-clad board shall be given priority for selection during layout to reduce the production cost of PCB, but it will increase the difficulty of wiring. If necessary, some crossover lines shall be designed to complete the circuit connection.
2. During PCB wiring, when two thin parallel wires are too close, the signal waveform will be delayed to generate reflection noise at the transmission line terminals.
3. Select reasonable wire width. The pulse interference generated in the printed wire by transient current is mainly caused by the distributed inductance of printed wire, so the distributed inductance of printed wire shall be minimized. In view that the distributed inductance of printed wire is proportional to the length of printed wire and inversely proportional to the width of printed wire, the short and wide wire is beneficial.
4. Although paralleling wiring can reduce the distributed inductance of wire, it will not only increase the mutual inductance and distributed capacitance between wires but also easily cause series-mode interference. When designing more complex power systems, well-shaped network wiring structure can be applied. The specific method is as follows: use horizontal wiring on one side of PCB and vertical wiring on the other side and then connect with plated-through-hole in the cross-hole.

8.5.2 Notes for SMPS Layout and Wiring

8.5.2.1 Layout of SMPS Key Components

Reasonable layout of key components is also very important. For example, the ground terminal of bypass capacitor at the output terminal must be low-noise reference ground and

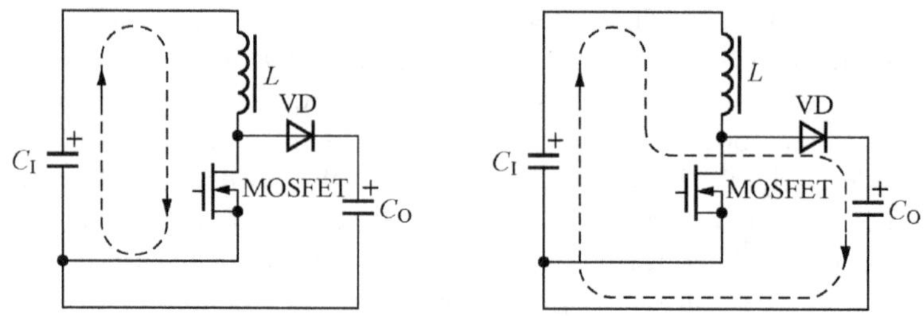

Figure 8.21 Two current paths of power circuit of the controller. (a) Input circuit and (b) output circuit

connected to the analog ground (AGND), allowing small signals only and the ground terminal of sampling resistance divider, being away from the power ground (PGND) with high current passing through as far as possible. This is crucial to realize the isolation between the low-noise reference ground and the high-noise PGND. This layout can prevent relatively large switching current from entering into the battery or power supply through the AGND loop to cause interference.

See Figure 8.21(a) and (b) for the two current paths of power circuit of the controller. The two current paths are, respectively, the input and output circuits. Figure 8.21(a) shows the path of the current flowing through the input circuit when MOSFET is turned on. Figure 8.21(b) illustrates the path of the current flowing through the output circuit when MOSFET is turned off. Shortening the layout distance between the components of the two circuits, high current can be limited in the power circuit of the controller, far away from the circuit of low-noise components. Positions of C_I, L, MOSFET, VD, and C_O shall be as close as possible. Short and wide printed wire can be applied for layout to improve efficiency, reduce the ringing voltage, and avoid causing interference to the low-noise circuit.

Compromise proposals are often adopted in the layout of the above two current loops. When some components in the loop are required to be installed nearby, it shall be determined whether incontinuous current flows through them. Installing components nearby can minimize the distributed inductance and the positions of incontinuous-current component are very important to reduce the distributed inductance.

8.5.2.2 SMPS Grounding

The "ground" in circuits is assumed to be the reference potential (usually zero potential). However, the resistance of ground wire is not equal to zero, so voltage drop will be generated when the ground current flows through the ground wire. In DC or low-frequency circuits, the ground wire is relatively simple, wherein, the voltage drop generated when current flows through the ground wire is formed by the DC resistance of wire. However, in high-frequency SMPSs, "ringing" (referring to the decaying oscillation) and spike voltages will be generated when the pulse current flows through the ground wire owing to the distributed inductance of ground wire.

See Figure 8.22 for an example of unreasonable ground wire layout, wherein, the multipoint grounding method is introduced and L_{01}–L_{03} are distributed inductance.

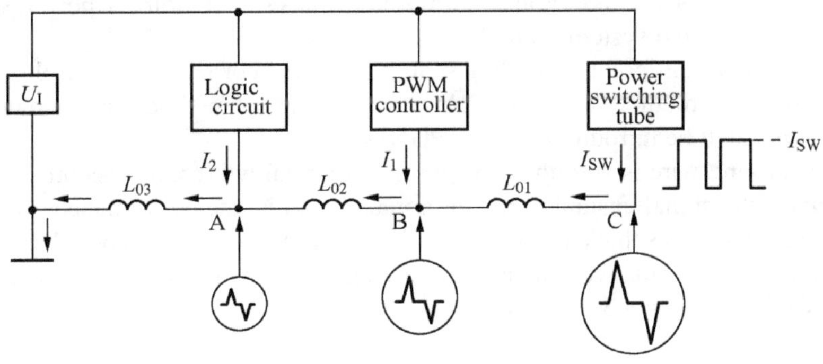

Figure 8.22 Multipoint grounding method

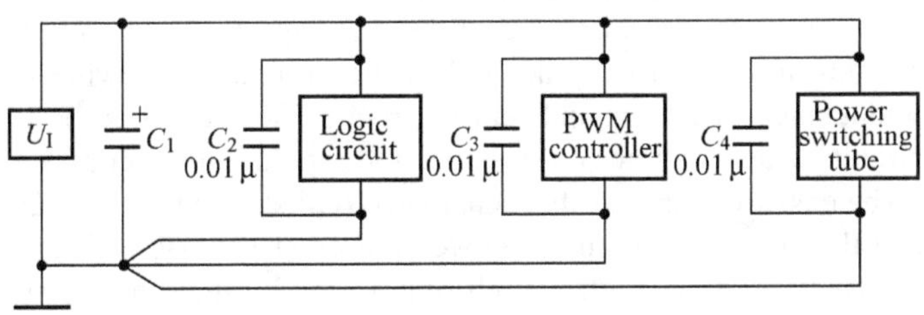

Figure 8.23 Single-point grounding method

The switching current I_{SW} of power switching tube, the return current I_1 of PWM controller, and the return current I_2 of logic circuit flow through the ground wire to form three return circuits. I_{SW} flows through L_{01}–L_{03} in turn. Corresponding to the rise and descent processes of I_{SW}, the three points of A, B, and C will generate positive and negative spike voltages $U_{spike\ (A)}$–$U_{spike\ (C)}$, of which the formula is $U_{spike\ (A–C)} = L_{01–03}\ (di/dt)$. The spike voltage amplitudes of points A, B, and C are different with the maximum amplitude near the power switching tube. It shall be noted that the spike voltages of point B and A are, respectively, applied to the PWM chip and the logic circuit, which will cause false action to result in failures such as over-voltage. In addition, when the load current rises, the ground current will be increased, and it will be easy to cause self-oscillation.

When designing the ground wire, the single-point grounding method as shown in Figure 8.23 is recommended to eliminate the influence of distributed inductance in the ground wire. In addition, a good-quality high-frequency electrolytic capacitor (e.g., tantalum capacitor) C_1 shall be connected in parallel between the input voltage terminal and the ground. A decoupling capacitor C_2–C_4 shall be respectively connected in parallel between the high-voltage terminal of each unit circuit and the ground, 0.01 μF ceramic or film capacitor usable. When there is high current pulse in the circuit, more capacitors can be connected in parallel.

Summary:

1. For electronic equipment, grounding is an effective method to control interference. Grounding and shielding can often be applied to get twice the result with half the effort, solving most of the interference problems. The ground wire in power systems can be approximately

divided into primary, secondary, analog (also known as signal ground), power, shield (connected to cabinet) and system ground.

2. Single-point and multipoint grounding shall be selected correctly. AGND allows the multipoint grounding method, and the single-point grounding method (also known as Kelvin connections) shall be introduced for the PGND.
3. When the ground wire is very thin, the ground potential will change according to the current to make the signal unstable and deteriorate the antinoise performance. Therefore, the ground wire shall be as thick as possible, the wire width of which can be selected as per the margin three times of the working current flowing through. The width of PGND wire shall generally be greater than 3 mm.

8.5.2.3 Reducing Noise Interference

The noise interference is mainly generated in three ways: the first is the switching noise. When the grounding return current of power circuit flows through the ground return circuit of controller (IC), switching noise will be generated in the ground wire owing to the distributed resistance and inductance in the ground return circuit. The second is the inherent noise of ground return circuit, which will not only reduce the regulation (or constant current) output accuracy but also be easy to interfere other sensitive circuits on the same PCB. The third is the switching noise at the positive terminal of power supply or battery, which can be coupled to other components (including the controller chip) powered by the same power supply to fluctuate the reference voltage. If it is found that the voltage across the terminals of input bypass capacitor is unstable, one-stage RC filter can be added in front of the power supply pin of controller, which helps stabilize the supply voltage. In addition, the larger the loop area is, the AC current flowing through will generate stronger magnetic field and significantly increase the possibility of generating interference. Locate the bypass capacitor at input terminal close to the power circuit to reduce the area surrounded by the input current loop, avoiding interference generation.

8.5.2.4 Bypass Capacitor Connection

1. If a node is found to be particularly sensitive to noise, the bypass capacitor can be adopted to reduce the sensitivity of the node to series-mode interference. Typically, the bypassing effect can be achieved just by connecting a small capacitor between the node and the ground or between the node and the input high-voltage line. When selecting bypass capacitors, make sure it has the low enough impedance within the frequency range possible to cause problems. The ESR and ESL will increase the high-frequency impedance, so ceramic capacitors with low ESR and ESL are particularly suitable to be applied as high-frequency bypass capacitor.
2. The installation position of bypass capacitor is also very important. To suppress high-frequency interference, bypass capacitor can be directly connected in parallel during the wiring of bypass signal wire. See Figure 8.24(a) and (b), respectively, for the two wiring methods of surface-mount bypass capacitor. Figure 8.24(a) is incorrect wiring, because the two wires connected in series with the bypass capacitor will increase the ESR and ESL of bypass capacitor to raise the high-frequency impedance, thereby deteriorating

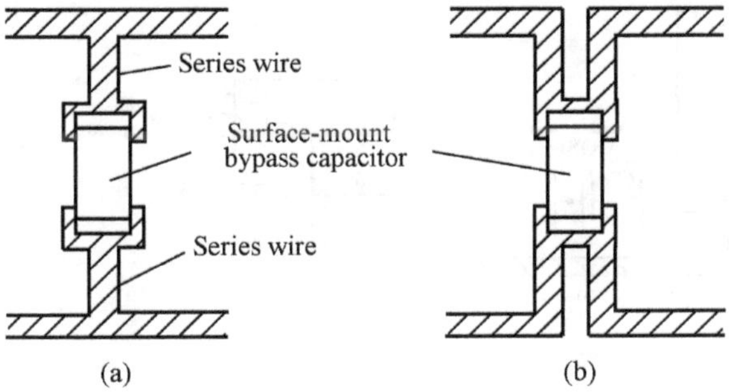

Figure 8.24 Two wiring methods of surface-mount bypass capacitor. (a) Incorrect wiring and (b) correct wiring

the high-frequency bypassing effect. Figure 8.24(b) is correct wiring, wherein, the distribution parameters of wire can help the bypass capacitor better filter out high-frequency interference.

3. Regardless of the kind of power supply, it is impossible that the source impedance is zero. This means that when the controller draws rapidly changing current from the power supply, the main voltage will change. To reduce this transient change, the input bypass capacitor can be installed at the input terminal. Sometimes, ceramic and electrolytic capacitors are adopted in parallel to prevent high current from entering into the power circuit, thereby avoiding interfering the low-noise circuit.
4. Some nodes do not allow the bypassing measure, because this will change its frequency characteristic. For example, the resistance divider applied for feedback does not allow connecting any bypass capacitor, or phase distortion will be caused to destroy the stability of feedback loop.

8.5.3 Notes for PCB Design

The following points shall be noted when designing PCB:

1. AGND is the common ground terminal of bypass capacitor (except the input bypass capacitor, C_I) for the AGND pin, the ground terminal of sampling resistance divider, and any specific pin of controller. AGND can use relatively wide and long lead, of which the current is very small and relatively stable, so it is not necessary to consider factors such as the lead resistance and the distributed inductance. PGND includes the ground terminals of input capacitor C_I and output capacitor C_O and the source of MOSFET. PGND wire must be short and wide lead to reduce lead impedance and improve power supply efficiency. See Figure 8.25(a), (b), and (c) for the three connection methods of analog and PGND, which can ensure that no switching current flows through the AGND wire.
2. When designing multilayer PCB, one intermediate layer can be adopted as shielding layer. Place power components in the top layer of PCB and small power components in the bottom layer. This layout can reduce interference.

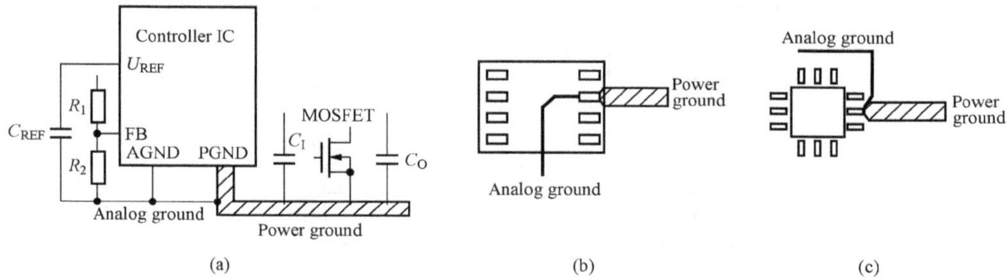

Figure 8.25 Three connection methods of analog and power ground. (a) Method one, (b) method two, and (c) method three

3. When designing the PCB of SMPS, the source shall adopt the single-point connection method. Connect the negative terminal of input filter capacitor and the return terminal of bias supply to the source pin S together. The bypass capacitor must be close to pin BP and as close as possible to pin S. It is easy to generate conducted and radiated noise when high current of MOSFET flows through long leads. The output rectifier must be as close as possible to the output filter capacitor. The output terminal of high-frequency transformer shall be connected to the negative pole of aluminum-electrolytic capacitor with short lead.

8.6 Design of Constant Voltage/Current SMPS

Constant voltage/current output SMPS can be referred to as the constant voltage/current source, which is featured in two control loops, one is the voltage control loop and the other is the current control loop. When the output current is small, the voltage control loop acts with regulation characteristics, which is equivalent to a constant voltage source. When the output current is close to or reaches the rated value, the current control loop maintains I_O constant, which becomes a constant current source now. This power supply is particularly suitable for battery chargers and special motor drivers. The following section describes a kind of low-cost constant voltage/current output SMPS, of which the current control loop is constituted by the transistor. This power supply is featured in simple circuit and low cost, and it is easy to produce.

8.6.1 Working Principle of Constant Voltage/Current Output SMPS

See Figure 8.26 for the circuit of 7.5 V and 1 A constant voltage/current output SMPS. A TOP200Y SMPS (IC_1) is adopted and is configured with PC817A linear optical coupler (IC_2). About 82–375 V DC high voltage U_I will be obtained after the 85–256 V AC input voltage u passes through the EMI filter (L_2 and C_6), the BR, and the input filter capacitor (C_1), which is then connected to the drain of TOP200Y after passing through the primary winding. The drain clamp protection circuit constituted by VD_{Z1} and VD_1 can limit the spike voltage generated by the leakage inductance of high-frequency transformer within a safe range. BZY97-C200

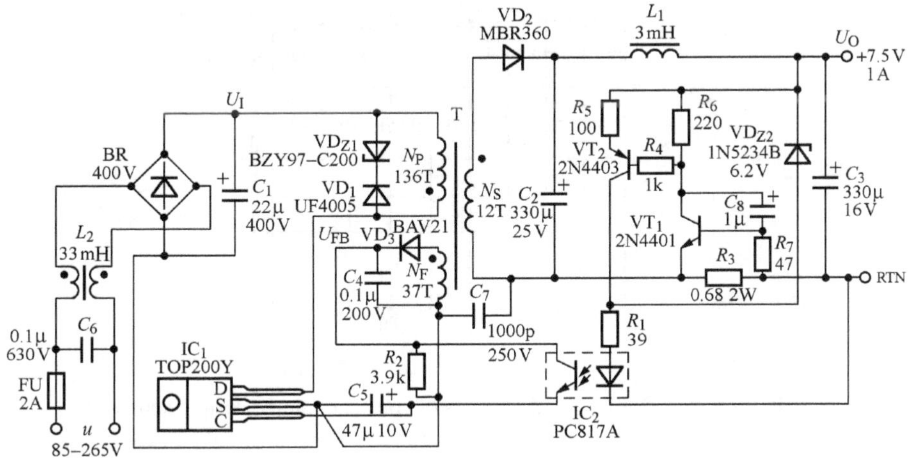

Figure 8.26 Circuit of 7.5 V and 1 A constant voltage/current output SMPS

transient voltage suppressor with the clamp voltage $U_B = 200\,\text{V}$ is applied for VD_{Z1}. UF4005 SRD is adopted for VD_1. +7.5 V output will be obtained after the secondary winding voltage is rectified and filtered by VD_2 and C_2 and then filtered by L_1 and C_3. 3A/70 V Schottky diode is adopted for VD_2. The output voltage of bias winding will generate the feedback voltage $U_{FB} = 26\,\text{V}$ after the rectification and filer of VD_3 and C_4, which provides the phototriode with bias voltage. C_5 is the bypass capacitor, which is also employed as frequency compensation capacitor and determines the automatic restart frequency. R_2 is the dummy load of bias winding, which can prevent the feedback voltage U_{FB} from increasing at no load.

The power supply has two control loops. The voltage control loop is constituted by 1N5234B 6.2 V regulator (VD_{Z2}) and optical coupler PC817A (IC_2), which is applied to let the SMPS work in constant voltage output mode when the output current is small. In this case, current flows through VD_{Z2} and the output voltage is determined by the regulation value (U_{Z2}) of VD_{Z2} and the forward voltage drop (U_F) of LED in optical coupler. The current control loop is constituted by the transistors VT_1 and VT_2, the current sensing resistor R_3, the optical coupler IC_2, resistors R_4–R_7, and the capacitor C_8, wherein, R_3 is dedicated to detecting the output current. VT_1 is the 2N4401 NPN silicon transistor, which can be replaced with the domestic model 3DK4C. VT_2 is 2N4403 PNP silicon transistor, which can be replaced with the domestic model 3DK9C. R_6 and R_5 are, respectively, adopted to set the collector currents I_{C1} and I_{C2} of VT_1 and VT_2. R_5 also determines the DC gain of current control loop. C_8 is the phase correction capacitor applied to prevent the loop from generating self-oscillation. At the beginning of power-on or automatic restart, the transient peak voltage will make VT_1 conducted, so now R_7 is applied to limit its emitter junction current. R_4 is employed to bypass the conduction current of VT_1 through VT_2, preventing the current from flowing through R_1. The start-up process of current control loop is as follows: when I_O rises to about 1 A, $U_{R3} \uparrow \rightarrow VT_1$ conducted, $\rightarrow U_{R6} \uparrow \rightarrow VT_2$ conducted, and the collector of VT_2 supplies current to the optical coupler to let $U_O \downarrow$. Reverse breakdown of VD_{Z2} does not occur and no current flows through VD_{Z2} owing to the reduction of U_O, so the voltage control loop is in open circuit and the SMPS

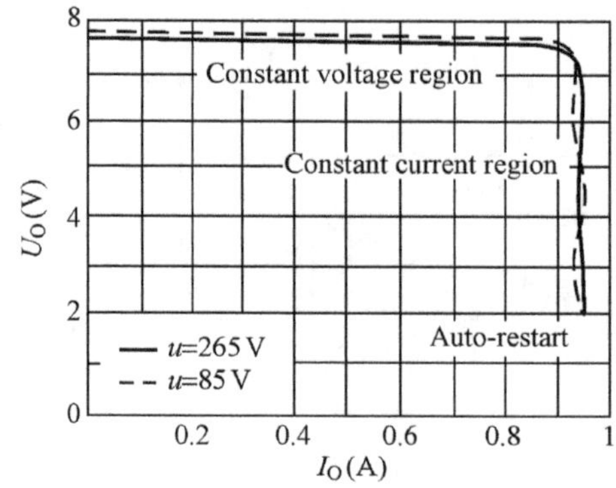

Figure 8.27 Output characteristics of constant voltage/current source

automatically enters into the constant current mode. C_7 is the safety capacitor, capable of filtering out the common-mode interference caused by the primary and secondary coupling capacitances.

The power supply can work not only in 7.5 V regulated output state but also under the 1 A controlled current. When the ambient temperature range is 0–50 °C, the accuracy of constant current output is about ±8%.

See Figure 8.27 for the output voltage–output current (U_O–I_O) characteristics of the power supply. It can be seen from the figure that the power supply has the following obvious features: (i) When $u = 85$ V (AC) or 265 V (AC), the change of characteristic curve is very small, which indicates that the output characteristic is basically not influenced by the variation of AC input voltage. (ii) It is in the constant voltage region when $I_O < 0.90$ A and in the constant current region when $I_O \approx 0.98$ A. U_O decreases rapidly with the slight increase of I_O. (iii) When $U_O \leq 2$V, VT_1 and VT_2 cannot continue to supply adequate working current to the optical coupler. In this case, the current control loop does not work, but the primary current is still limited by the maximum limit current $I_{LIMIT\,(max)}$ of TOP200Y. At this time, $U_{R6} \uparrow$ and the working current of optical coupler is rapidly reduced by VT_1 and VT_2 to force TOP200Y enter into auto-restart. This indicates that once the current control loop is out of control, the constant current mode will be converted to the auto-restart state to reduce I_O, thus protecting the chip.

8.6.2 Circuit Design of Constant Voltage/Current Output SMPS

See Figure 8.28 for the unit circuits of voltage and current control loops.

8.6.2.1 Design of Voltage Control Loop

The output voltage of constant voltage source is determined by the following formula:

$$U_O = U_{Z2} + U_F + U_{R1} = U_{Z2} + U_F + I_{R1}R_1 \tag{8.16}$$

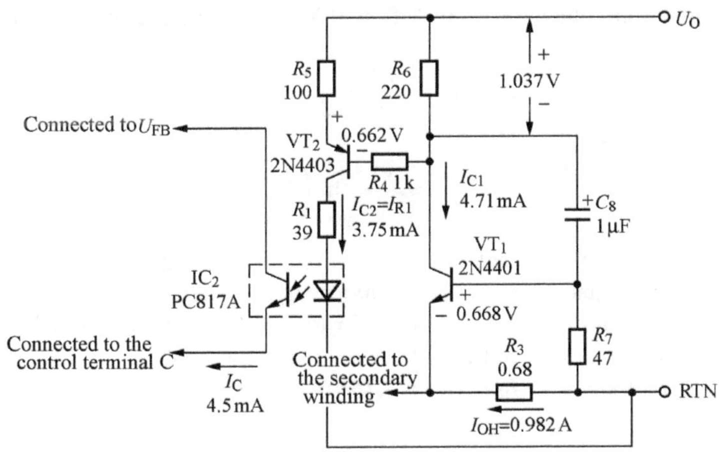

Figure 8.28 Unit circuits of voltage and current control loops

wherein, $U_{Z2} = 6.2\,\text{V}$ (i.e., the regulated voltage of regulator VD_{Z2}, see Figure 8.26), $U_F = 1.2\,\text{V}$ (typically), and only the voltage drop U_{R1} of R_1 needs to be determined. Let the current of R_1 be I_{R1}, the collector current of VT_2 be I_{C2}, and the input current of optical coupler (i.e., the working current of LED) be I_F. Obviously, $I_{R1} = I_{C2} = I_F$, and they vary with u, I_O, and the current transfer ratio CTR of optical coupler. The variation range of the control terminal current I_C of TOP200Y is 2.5 mA (corresponding to the maximum duty ratio D_{max}) –6.5 mA (corresponding to the minimum duty ratio D_{min}). Now take the intermediate value $I_C = 4.5\,\text{mA}$. I_C flows into the control terminal from the emitter of phototriode, so

$$I_{R1} = \frac{I_C}{\text{CTR}} \tag{8.17}$$

After the values of I_C and CTR are determined, it is easy to calculate I_{R1}. The linear optical coupler shall be adopted for the single-chip SMPS and require CTR = 80–160%. The intermediate value of 120% can be taken. Put $I_C = 4.5\,\text{mA}$ and CTR = 120% into Formula (8.17) to get $I_{R1} = 3.75\,\text{mA}$. When $R_1 = 39\,\Omega$, $U_{R1} = 0.146\,\text{V}$. Finally, put it into Formula (8.16) to get

$$U_O = U_{Z2} + U_F + U_{R1} = 6.2\text{V} + 1.2\text{V} + 0.146\text{V} = 7.546\text{V} \approx 7.5\text{V}$$

8.6.2.2 Design of Current Control Loop

Current control loop is constituted by VT_1, VT_2, R_1, R_3–R_7, C_8, and PC817A. The following text is intended to finally calculate the expected value of constant output current I_{OH}. In Figure 8.28, VT_1 is the base bias resistance of R_7. The base current is very low and the current of R_3 is very high, so it can be regarded that the emitter junction voltage drop U_{BE1} of VT_1 is completely dropped on R_3. Relevant formula is as follows:

$$I_{OH} = \frac{U_{BE1}}{R_3} \tag{8.18}$$

The emitter junction voltage drops of VT_1 and VT_2 can be calculated as per the following two formulas:

$$U_{BE1} = \frac{kT}{q} \cdot \ln\left(\frac{I_{C1}}{I_S}\right) \tag{8.19}$$

$$U_{BE2} = \frac{kT}{q} \cdot \ln\left(\frac{I_{C2}}{I_S}\right) \tag{8.20}$$

wherein, k is Boltzmann constant, T is the ambient temperature (expressed in thermodynamic temperature), and q is the electronic charge. When $T_A = 25\,°C$, $T = 298\,K$ and $kT/q = 0.0262\,V$. I_{C1} and I_{C2}, respectively, refer to the collector current of VT_1 and VT_2. I_S is the reverse saturation current of transistor. For small power tube, $I_S = 4 \times 10^{-14}\,A$.

It has been known that $I_{R1} = I_F = I_{C2} = 3.75\,mA$, so

$$U_{BE2} = \frac{kT}{q} \cdot \ln\left(\frac{I_{C2}}{I_S}\right) = 0.0262 \ln\left(\frac{3.75\,mA}{4 \times 10^{-14}\,A}\right) = 0.662\,V$$

$I_{E2} \approx I_{C2}$, so $U_{R5} = I_{C2}R_5 = 3.75\,mA \times 100\,\Omega = 0.375\,V$, which leads to $U_{R6} = U_{R5} + U_{BE2} = 0.375\,V + 0.662\,V = 1.037\,V$. When $R_6 = 220\,\Omega$, $I_{R6} = I_{C1} = U_{R6}/R_6 = 4.71\,mA$. The following formula uses this value to calculate U_{BE1} to determine the resistance of current sensing resistor R_3:

$$U_{BE1} = 0.0262 \ln\left(\frac{4.71\,mA}{4 \times 10^{-14}\,A}\right) = 0.668\,V$$

$$R_3 = \frac{U_{BE1}}{I_{OH}} = \frac{0.668\,V}{1.0\,A} = 0.668\,\Omega$$

The closest nominal resistance is $0.68\,\Omega$. Put it into Formula (8.18) to get

$$I_{OH} = \frac{0.668\,V}{0.68\,\Omega} = 0.982\,A$$

Considering the temperature coefficient of the emitter junction voltage U_{BE1} of VT_1 is $\alpha_T \approx -21\,mV/°C$, so when the ambient temperature rises to 25 °C, I_{OH} is reduced to

$$I'_{OH} = \frac{U_{BE1} - \alpha_T \Delta T}{R_3} = \frac{0.668\,V - (2.1\,mV/°C) \times 25\,°C}{0.68\,\Omega} = 0.905\,A$$

The constant current accuracy is

$$\gamma = \frac{I'_{OH} - I_{OH}}{I_{OH}} \times 100\% = \frac{0.905 - 0.982}{0.982} \times 100\% = 7.8\% \approx -8\%$$

It indicates that the calculated result is consistent with the design indexes.

8.7 Design of Precision Constant Voltage/Current SMPS

The following section describes a precision constant voltage/current output SMPS, which uses TOP214Y single-chip SMPS and is configured with low-power-consumption dual operational amplifier and adjustable precision shunt regulator to constitute the voltage and current control loops. Compared with the control loop constituted by transistors, it enjoys the advantages such as high constant voltage and current accuracy, simple peripheral circuit, small resistance of current sensing resistor, low power consumption, and the ability to improve the power supply efficiency. In addition, its power supply efficiency is up to 80%. This power supply is suitable to be applied as the fast battery charger of laptops and video recorders.

8.7.1 Working Principle of Precision Constant Voltage/Current Output SMPS

See Figure 8.29 for the circuit of 15 V and 2 A precision constant voltage/current output SMPS.

Four ICs are used for the circuit: TOP214Y single-chip SMPS (IC_1), PC816A linear optical coupler (IC_2), adjustable precision shunt regulator TL431C (IC_3), and low-power-consumption dual operational amplifier LM358 (IC_4, containing two operational amplifiers IC_{4a} and IC_{4b}). The circuit has the following features: (i) IC_{4b}, sampling resistors R_3 and R_4, and IC_3 are applied to constitute the voltage control loop. IC_{4a} is adopted to constitute the current control loop. (ii) Voltage and current control loops work as per the "logic or gate" principle, that is, at any time, the loop with high level output plays the control role. (iii) The secondary bias winding N_{SB} is added to supply power to the control loop. The secondary bias voltage U_{SB} can automatically follow the change of DC input voltage U_I to keep the constant current characteristic when the output voltage U_O of power supply decreases significantly and enter into auto-restart state only when $U_O \leq 0.8$ V.

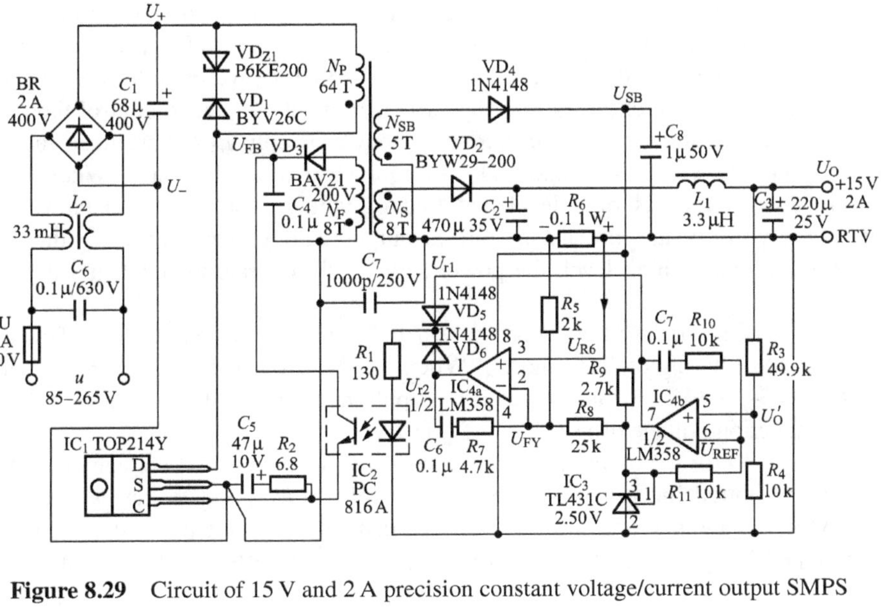

Figure 8.29 Circuit of 15 V and 2 A precision constant voltage/current output SMPS

(iv) A TOP214Y single-chip SMPS is adopted to let the rated output power reach 30 W. In the case of wide-range voltage ($u = 85$–265 V, AC) input, the maximum output power of TOP214Y is $P_{OM} = 42$ W. (v) When the current control loop constituted by operational amplifier, the resistance of current sensing resistor R_6 can be reduced to 0.1 Ω, the rated voltage drop $U_{R6} = 0.1\,\Omega \times 2\,A = 0.2\,V$, the power consumption is reduced to 0.4 W, and the ratio between the power consumption and the output power is only $(0.4\,W/30\,W) \times 100\% = 1.3\%$, improving the power supply efficiency. (vi) VD_{Z1} uses P6KE200 transient voltage suppressor. VD_1 uses BYV26C 2.3A/600 V SRD. VD_2 uses BYW29-200 8A/200 V SRD. In view of the very small working current of VD_4, 1N4148 FRD can be applied. After the maximum value of feedback voltage U_{FB} is raised to 46 V, the working voltage of optical coupler also rises to 40 V, so PC816A optical coupler with $U_{(BR)\ CEO} = 70\,V > 40\,V$ is employed here.

In Figure 8.29, the secondary winding voltage will generate +15 V output after the rectification and filtering of VD_2, C_2, L_1, and C_3. R_3 and R_4 are sampling resistors. U_O will generate the sampling voltage U_O' after the voltage division of R_3 and R_4, which will be connected to the noninverting input terminal of IC_{4b}. The reference voltage U_{REF} generated by TL431C is 2.50 V (with the exact value of 2.495 V), connected to the inverting input terminal of IC_{4b}. After the comparison of U_O' and U_{REF} by IC_4, the output of error signal U_{r1} will be generated. The said error signal is then converted into the current signal through VD_5 and R_1, which flows into the LED of optical coupler to control the duty ratio of TOP214Y, maintaining U_O constant in the constant voltage region. The phase correction network of voltage control loop is constituted by C_7, R_{10}, and R_{11}. After short-circuiting the cathode (pin 3) of TL431C to the output voltage setting terminal (pin 1), the output voltage $U_{REF} = 2.50\,V$. R_9 is the current limiting resistor, which can limit the working current of TL431C within the range 1–10 mA.

IC_{4a} is the voltage comparator of current control loop, of which the noninverting input terminal is connected with current sensing signal U_{R6} and the inverting input terminal is connected with the voltage U_{FY} of voltage divider. The voltage divider is constituted by R_5, R_8, and TL431C. After the comparison of U_{R6} and U_{FY} by IC_{4a}, the output of error signal U_{r2} will be generated. The said error signal is then converted into the current signal through VD_6 and R_1, which flows into the LED of optical coupler to control the duty ratio of TOP214Y, maintaining the output current of power supply I_{OH} constant in the constant current region. Obviously, VD_5 and VD_6 are equivalent to an "or gate." When the output of current control loop is high level and that of voltage control loop is low level, the power supply will operate in constant current output state. Conversely, when the output of voltage control loop is high level, the power supply will operate in constant voltage output state.

The voltage of secondary bias winding N_{SB} will generate the bias voltage U_{SB} after the rectification and filtering of VD_4 and C_8. When the AC input voltage u changes from 85 to 265 V, $U_{SB} = 5$–28.3 V. U_{SB} is dedicated to supplying power to LM358 and TL431C. The forward conduction voltage drop U_{F4} of VD_4 is 1 V.

See Figure 8.30 for the output characteristics of precision constant voltage/current source. The solid and dotted lines in the figure, respectively, correspond to the two cases of $u = u_{min} = 85$ V (AC) and $u = u_{max} = 265$ V (AC). It can be seen in the figure that the two curves are overlapped in the constant voltage region and slightly different in the constant current region.

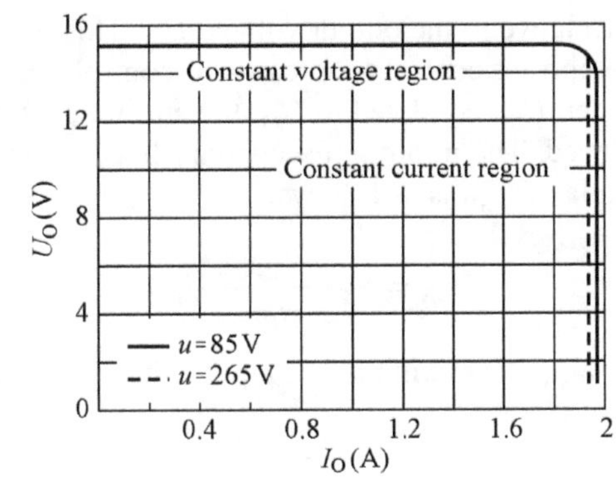

Figure 8.30 Output characteristics of precision constant voltage/current source

8.7.2 Circuit Design of Precision Constant Voltage/Current Output SMPS

8.7.2.1 Design of Voltage Control Loop

The output voltage of the power supply in the constant voltage region is determined by the following formula:

$$U_O = U_{REF} \cdot \frac{R_3 + R_4}{R_4} = 2.50\,V \times \left(1 + \frac{R_3}{R_4}\right) \tag{8.21}$$

The total series resistance of R_3 and R_4 shall be appropriate. Too large resistance is easy to cause noise interference and too small resistance will increase the circuit loss. Generally, $R_4 = 10.0\,k\Omega$ is desirable. Put this resistance into Formula (8.21) to get $R_3 = 50.1\,k\Omega$. The standard resistance of E196 series closest to the above result is $49.9\,k\Omega$.

8.7.2.2 Design of Current Control Loop

The expected value I_{OH} of constant current output of the power supply is determined by the following formula:

$$I_{OH} = \frac{U_{REF} R_5}{R_6 R_8} \tag{8.22}$$

When selecting the resistance of R_5, the influence of load on TL431C and the error caused by the LM358 input bias current shall be considered. Generally, take $R_5 = 2\,k\Omega$. When $R_6 = 0.1\,\Omega$ and $I_{OH} = 2\,A$, the current sensing signal $U_{R6} = 0.2\,V$. Put $U_{REF} = 2.50\,V$, R_5, and R_6 into Formula (8.22) to get $R_8 = 25\,k\Omega$.

8.7.2.3 Design of Secondary Bias Supply

It can be seen in Figure 8.29 that the secondary bias winding N_{SB} and the primary winding N_P have the same voltage polarity with corresponding positions of identical terminals, so VD_4 and TOP214Y can be conducted simultaneously. This means that U_{SB} can follow the change of DC input voltage U_I, irrelative to the output voltage U_O. This is critical. Only in this way, it can be ensured that the current control loop can still control the output current when U_O is very low. Otherwise, when U_{SB} is related to U_O, the current control loop may fail to work normally when U_O is reduced. The minimum secondary bias voltage of the power supply is $U_{SB\,(min)} = 5\,V$. The formula of U_{SB} is as follows:

$$U_{SB} = U_I \cdot \frac{N_{SB}}{N_P} - U_{F4} \tag{8.23}$$

To calculate the turns of N_{SB}, the minimum DC input voltage U_{Imin} shall be determined first. Relevant formula is as follows:

$$U_{Imin} = \sqrt{2u_{min}^2 - \frac{2P_O\left(\frac{1}{2f_L} - t_C\right)}{\eta C_{IN}}} \tag{8.24}$$

wherein, P_O is the rated output power, f_L the grid frequency, t_C the response time of BR (typically 3 ms), η the power supply efficiency, and C_{IN} the input filter capacitor. Put $u_{min} = 85\,V$, $P_O = 30\,W$, $f_L = 50\,Hz$, $t_C = 3\,ms$, $\eta = 80\%$, and $C_{IN} = C_1 = 68\,\mu F$ into Formula (8.24) to get $U_{Imin} = 82\,V$. Then, put $U_I = U_{Imin} = 82\,V$, $U_{SB} = U_{SB\,(min)} = 5\,V$, and $U_{F4} = 1\,V$ into Formula (8.23) to get $N_{SB} = 4.7$ turns, actually taken as 5 turns.

The maximum DC input voltage $U_{Imax} = \sqrt{2}\ u_{max} = \sqrt{2} \times 265\,V = 375\,V$, at which time, U_{SB} reaches the maximum $U_{SB\,(max)} = 28.3\,V$, not exceeding the maximum mains voltage (32 V) of LM358.

8.7.2.4 Design of Bias Winding

To maintain I_{OH} unchanged when U_O is reduced, the polarity of bias winding N_F is the same as that of the primary winding. The feedback voltage U_{FB} will be obtained after the output voltage is rectified and filtered by VD_3 and C_4, which is required to be no less than 9 V. The formula to calculate the turns of N_F is as follows:

$$N_F = \frac{N_P(U_{FB} + U_{F3})}{U_{Imin}} \tag{8.25}$$

Put $N_P = 64$ turns, $U_{FB} = U_{FB\,(min)} = 9\,V$, $U_{F3} = 1\,V$, and $U_{Imin} = 82\,V$ into Formula (8.25) to get $N_F = 7.8$ turns, actually taken as 8 turns. When U_{Imin} in the formula is replaced with $U_{Imax} = 375\,V$, the maximum value of U_{FB} can be obtained: $U_{FB\,(man)} = 45.9\,V$. The minimum voltage at the control terminal of TOP214Y is 5.5 V and the actual working voltage of optical coupler is 45.9 – 5.5 V = 40.4 V.

8.7.2.5 Resistance of the series resistor R_1 of optical coupler

R_1 is not only the current limiting resistor of LED but also determines the gain of control loop. Its formula is as follows:

$$R_1 = \frac{(U_{SAT} - U_{F6} - U_F) \cdot \mathrm{CTR}_{min}}{I_{Cmax}} \tag{8.26}$$

wherein, the forward saturation voltage of LM358 is $U_{SAT} = 3.5\,V$, the forward voltage drop of VD_6 is $U_{F6} = 0.65\,V$, the forward voltage drop of LED in optical coupler is $U_F = 1.2\,V$, the minimum current transfer ratio of PC816A is $CTR_{min} = 80\%$, and the maximum current at the control terminal of TOP214Y is $I_{Cmax} = 10\,mA$. It is not difficult to know that $R_1 = 132\,\Omega$, actually taken as $130\,\Omega$. When the resistance of R_1 is too high, the control sensitivity will be reduced. Too small resistance will cause the control loop to work unstably, and even self-oscillation.

8.8 Design of Remote Turn-Off Circuit for SMPS

8.8.1 Remote Turn-Off Circuit of TOPSwitch-GX

The multifunctional terminal (M) of TOPSwitch-GX series single-chip SMPS has multiple functions, which can realize control functions such as the remote control after configured with peripheral circuits.

8.8.1.1 Circuit Controlling the On/Off of SMPS with the Active-ON Method

See Figure 8.31 for the circuit controlling the on/off of SMPS with the Active-ON method.

The on/off control signal (ON/OFF) is connected to Terminal M through NPN transistor VT, which can be replaced with an optical coupler or manual switch. When $ON/\overline{OFF} = 1$ (high level), VT is conducted and Terminal M is at low level. In this case, $U_C > U_M$ and current flows through the internal circuit of control terminal into Terminal M without any under-voltage or over-voltage state (i.e., $I_{UV} < I_M < I_{OV}$), so the SMPS can be started to work. When $ON/\overline{OFF} = 0$ (low level), VT is turned off and Terminal M is disconnected to let $I_M = 0$, so the output of SMPS is turned off immediately. This method is equivalent to positive logic control, which is called Active-ON method because the SMPS can be activated when ON is high level. SMPS is in micropower consumption state after turned off. When Terminal M is disconnected, the drain current of TOPSwitch-GX is only 0.6 μA (typically).

8.8.1.2 Circuit Controlling the On/Off of SMPS with Active-OFF Method

See Figure 8.32 for the circuit controlling the on/off of SMPS with the Active-OFF method. Now connect the transistor VT between Terminal M and C and connect the resistor R_{MC}

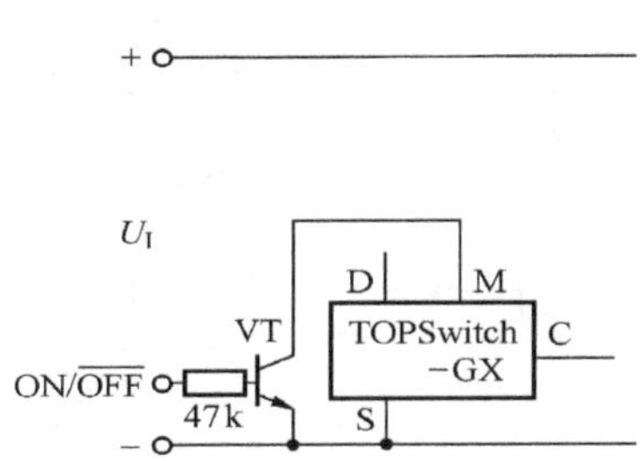

Figure 8.31 Circuit controlling the on/off of SMPS with the Active-ON method

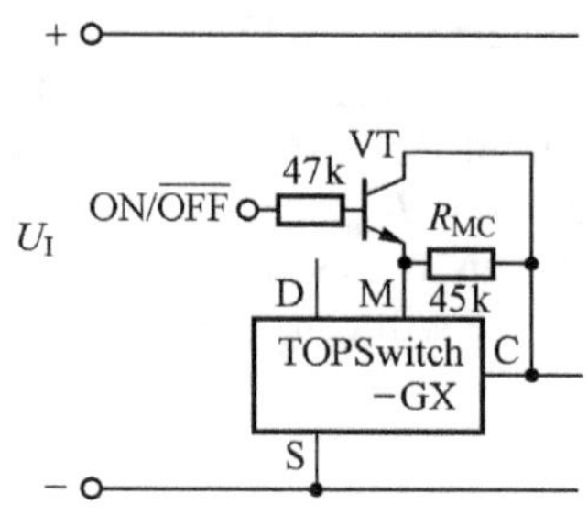

Figure 8.32 Circuit controlling the on/off of SMPS with the Active-OFF method

between the emitter and the collector of VT. Only when the control signal $ON/\overline{OFF} = 0$, VT is turned off and the current of control terminal flows through the external 45 kΩ resistor into Terminal M to form I_M, which will activate the SMPS immediately. When $ON/\overline{OFF} = 1$, VT is conducted to let Terminal M and C in short circuit. As the two terminals have equal potential, $I_M = 0$ to turn off the SMPS immediately. In this case, the drain supply current of chip is reduced to 1.0 μA (typically). This method is equivalent to negative logic control, which is called Active-OFF method because the SMPS is activated only when OFF is at low level.

8.8.2 Remote Turn-Off Circuit of PC SMPS

See Figure 8.33 for the remote on/off control interface circuit of multioutput PC main power supply (see Figure 7.5 for the main circuit of PC SMPS). The following text analyzes the working principle of the interface circuit.

When the turn-on wire ($\overline{ON}$) is grounded, VT_3 is conducted. VT_4 is conducted through the optical coupler IC_4 (LTV817) and then the Terminal X of TOP247Y is at low level through R_{12} to let the MOSFET in TOP247Y enter into on/off state. Before the discharge of C_{16}, the +5 V main output of PC has reached the rated value, so the +5 V main output maintains VT_3 in conduction state through R_{34} and lets TOP247Y still in the on/off state. Once the wiring voltage exceeds the under-voltage threshold, TOP247Y will enter into auto-restart mode until C_{16} discharges. At this point, the turn-on wire must be grounded again for restart. In the case

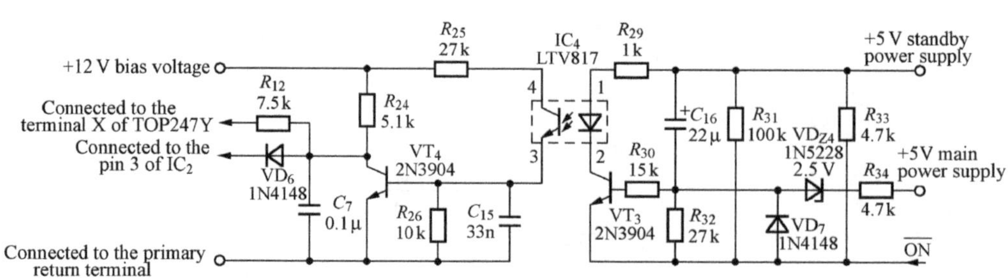

Figure 8.33 Remote on/off control interface circuit of PC SMPS

of over-voltage output, TOP247Y will stop on/off conversion. PC SMPS can be restarted only by disconnecting the turn-on wire from the ground first and then connecting the turn-on wire to the ground.

When the turn-on wire $\overline{\text{ON}}$ is disconnected from the ground, the voltage of Terminal $\overline{\text{ON}}$ is raised to +5 V through the pull-up resistor R_{33}. In this case, TOP247Y is prohibited for work.

Notes:

1. When the remote turn-on wire is grounded (main power supply enabled), TOP247Y will automatically start up when an AC voltage is applied to PC SMPS. However, if the main power supply is failed to be activated owing to too slow rise of AC voltage (e.g., using the autotransformer for AC voltage regulation), the turn-on wire ($\overline{\text{ON}}$) shall be grounded again to resend a turn-on signal.
2. Connect an ON/OFF control switch to the turn-on wire to turn on or off the PC SMPS.

8.9 Typical Application and Printed Circuit Design of New Single-Chip SMPS

The new single-chip SMPS has notable features in terms of circuit design and production process. The following section focuses on its printed circuit design for the reference of technical personnel participating in the research or development of new SMPS products.

8.9.1 Typical Application and Printed Circuit Design of TOPSwitch-HX Series Single-Chip SMPS

8.9.1.1 Typical Application of TOPSwitch-HX Series Single-Chip SMPS

See Figure 8.34 for the circuit of 65 W (19 V, 3.42 A) laptop power adaptor constituted by TOP259Y of TOPSwitch-HX series. The power adaptor is designed to be very compact with a few components. TOP259Y adopts the standard TO-220 package. The AC input voltage range is 90–265 V. The average power supply efficiency is above 87.5%. The no-load power consumption at AC230V is less than 250 mW, in line with the requirements of international energy saving standard "Energy Star." TOP259Y adjusts the duty ratio by changing the current at control terminal to keep the output voltage stable. It has the over-load and over-heat protection functions and multiple control modes. Now under the full-frequency (132 kHz) PWM control mode, the high-frequency transformer core with small size can be applied. The laptop power adaptor can keep long-term work in the sealed enclosure at 40 °C ambient temperature. TOPSwitch-HX series products can provide constant efficiency within the entire specified load range.

The AC voltage passes through the EMI filter (C_1, R_1, R_2, and L_1–L_3), then the 3 A/800 V bridge rectifier (BR, 3KBP08M) and the filter capacitor (C_2) to get a DC high voltage. Common-mode chokes L_1 and L_2, respectively, filter the low-frequency and high-frequency common-mode interference. L_3 and C_1 are adopted to filter out series-mode interference. The capacitance of C_1 is greater than 0.1 μF, so C_1 shall be discharged through bleeder resistors R_1 and R_2 after power-off. C_5 is the primary decoupling capacitor, capable of reducing series-mode interference. C_{11} is the safety capacitor (Y capacitor), applied to filter out the

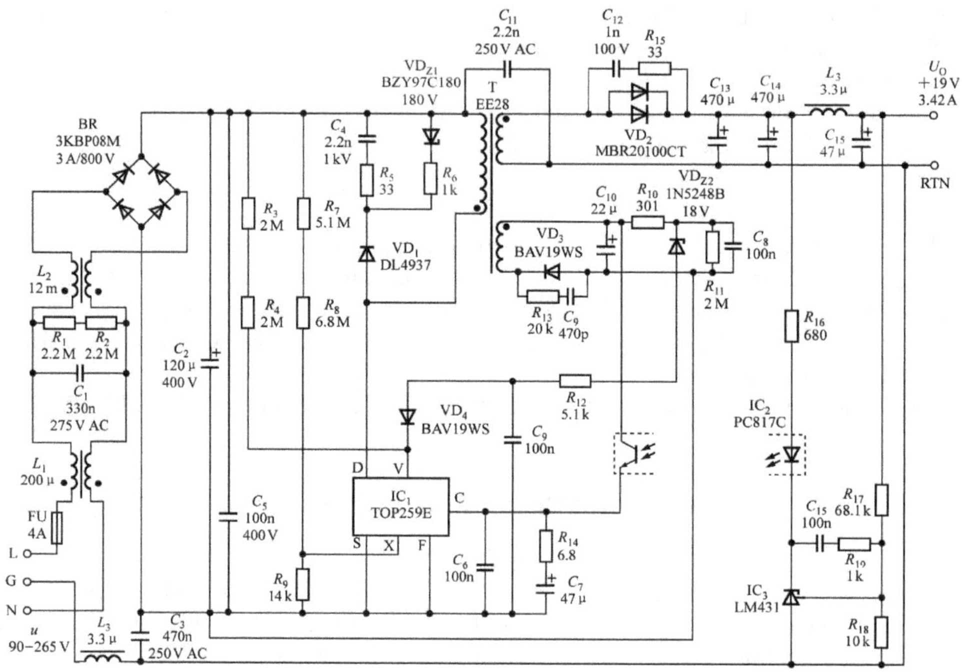

Figure 8.34 Circuit of 65 W laptop power adaptor constituted by TOP259Y

common-mode interference caused by the coupling capacitance primary and secondary windings. The drain clamp circuit is constituted by the 180 V regulator VD_{Z1} (BZY97C180), the FRD (DL4937), C_4, R_5, R_6, and so on. When $R_3 + R_4 = 4\,\text{M}\Omega$, the set input under-voltage threshold $U_{UV} = 97\,\text{V}$ and the input over-voltage threshold $U_{OV} = 445\,\text{V}$.

The on-state resistance of MOSFET is adopted for the peak drain current limit circuit in TOP259Y as current sampling resistance. The default limit current I_{LIMIT} has been preset internally. However, the value of I_{LIMIT} may be too large when designing power supplies. For example, in the case of TOP259Y derating, the value of I_{LIMIT} is expected to be reduced to provide reliable protection. Actually, the limit current can be externally controlled within (30%–100%) I_{LIMIT} just by setting resistor R_9. When $R_9 = 14\,\text{k}\Omega$, the set external self-protection limit current (I'_{LIMIT}) is only 0.52 times of the internal self-protection limit current (I_{LIMIT}) of TOP259Y, that is, $I'_{LIMIT} = 0.52 I_{LIMIT} = 0.52 \times 5.15\,\text{A} = 2.68\,\text{A}$. The capacitor C_{10}, the regulator VD_{Z2} (1N5248B), the resistor R_{12}, and VD_4 at both ends of bias winding are adopted to constitute the output over-voltage protection circuit. Under over-load cases, the voltage of C_{10} will rise. Once the voltage of C_{10} exceeds 18 V, VD_{Z2} will suffer reverse breakdown to trigger the voltage monitoring terminal (V) to turn off the output, thus realizing the protection purpose. The delay circuit constituted by R_{11} and C_8 can prevent false triggering caused by noise interference. VD_4 is the isolation diode, which can isolate the input voltage detection circuit and the output over-voltage protection circuit to prevent the output over-voltage protection circuit from becoming the load of pin V.

+19 V and 3.42 A regulated output can be obtained after the rectification and filtering of secondary winding voltage by VD_2, C_{13}, C_{14}, L_3, and C_{15}. 20 A/100 V Schottky geminate transistor MBR20100CT is applied for the output rectifier VD_2. Connect the RC buffer circuit (C_{12} and R_{15}) in parallel to VD_2 to attenuate leakage inductance oscillation and reduce EMI.

Precision optocoupler feedback circuit is constituted by PC817C and LM431. After sampled by R_{17} and R_{18}, the output voltage is compared with the 2.5 V reference voltage in LM431 to generate an error voltage and then change the duty ratio of TOP259Y through PC817C. C_{15} and R_{19} are the phase correction network. R_{16} is applied to set the DC gain of feedback loop.

Key design points:

1. The maximum drain voltage of TOP259Y shall not exceed 650 V. The values of C_4 and R_5 can be adjusted if necessary.
2. VD_{Z2} shall have proper regulated voltage U_Z to prevent false triggering. The selection principle is as follows: U_Z shall be greater than the bias voltage measured at low input voltage or full load.
3. When the output over-voltage protection circuit needs the latch function, the resistance of R_{12} shall be 20 Ω.
4. The EE28 ferrite core is adopted for the high-frequency transformer. The primary winding adopts 17 × 2 turns of two-strand φ0.40 mm enameled wire divided into two layers, and the secondary winding adopts six turns of four-strand φ0.40 mm enameled wire located between the above two layers. The bias winding adopts five turns of three-Standard φ0.25 mm enameled wire. The inductance of primary winding is 343 μH (±5% error allowed), and the maximum leakage inductance is 4 μH. The resonant frequency is greater than 1.1 MHz. Add a shielding layer, respectively, among the primary, secondary, and bias windings.

8.9.1.2 Printed Circuit Design of TOPSwitch-HX Series Single-Chip SMPS

See Figure 8.35 (some components different from those shown in the circuit of Figure 8.34) for the printed circuit of 65 W laptop power adapter designed by referring to Figure 8.34. The following points shall be noted when designing the printed circuit:

1. The negative pole of input filter capacitor at the source pin of TOPSwitch-HX is connected to the bias winding loop using single-point (Kelvin) method, so that the surge current directly returns to the input filter capacitor from the bias winding to enhance the resistance to surge current. The bypass capacitor at the control terminal shall be as close as possible to the source and the control pin.
2. All components with the source as the reference potential and connected to the voltage detection terminal (V) and the external limit current setting terminal (X) shall also be as close as possible to the source and corresponding pin. Do not let the switching current of MOSFET flow through the source connecting line. The switching current must return to the negative pole of input filter capacitor through a separate path, rather than using the same path as that of any peripheral components of control terminal (C) and terminal M, V, or X.
3. One end of the voltage monitoring resistor shall be as close as possible to terminal V and the other end shall be connected to the positive pole of input filter capacitor. In addition, the above connection shall be far away from the power switching circuit.

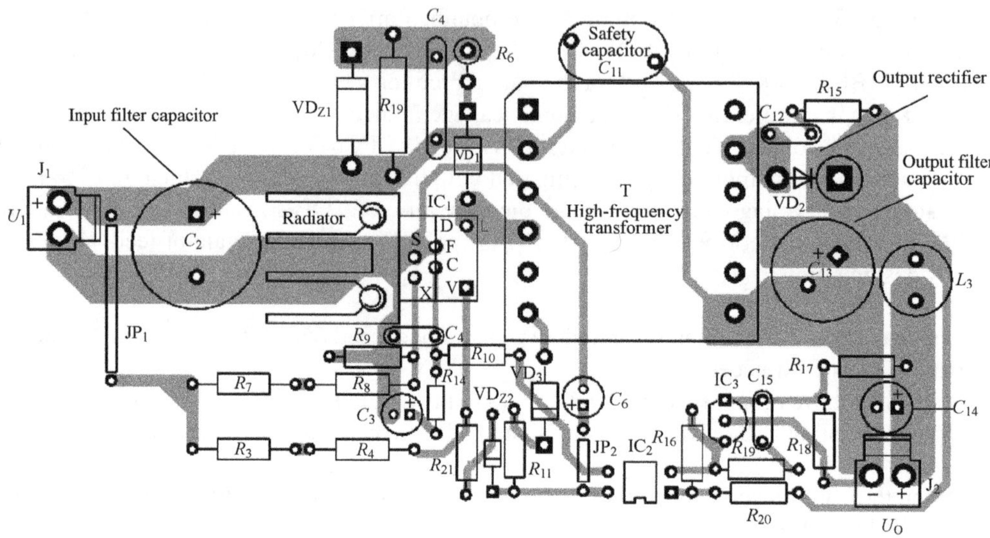

Figure 8.35 Printed circuit of 65 W laptop power adapter

4. Connect a high-frequency bypass capacitor C_6 (100 nF) in parallel to the control terminal capacitor C_7 (47 μF) to filter out high-frequency noise.
5. The safety capacitor (i.e., Y capacitor) C_{11} shall be connected in parallel between the primary DC high-voltage input terminal and the secondary return terminal.
6. The small radiating fin of Y package (TO-220) or E package (eSIP-7C) is connected to the source in the chip. The external radiator can be directly fixed on the small radiating fin. A layer of heat conducting silicone grease shall be coated on the contact surface between the external radiator and the small radiating fin.
7. When using components with P (DIP-8), G (SMD-8), or M (DIP-10) package, the copper sheet area close to the source pin under the component has the function of heat radiation. When designing double-sided PCB, the via hole connecting the top and bottom layers can be applied to improve the heat dissipation effect.
8. The copper foil area at the positive and negative pins of output rectifier shall be large enough to facilitate the heat dissipation of rectifier.

8.9.2 Typical Application and Printed Circuit Design of TOPSwitch-JX Series Single-Chip SMPS

8.9.2.1 Typical Application of TOPSwitch-JX Series Single-Chip SMPS

TOPSwitch-JX series is the sixth generation of single-chip SMPS IC. See Figure 8.36 for the circuit of 65 W (19 V, 3.42 A) power adapter with low no-load power consumption and high efficiency constituted by the product TOP269EG of this series. The circuit of the power supply is basically the same as that of 65 W (19 V, 3.42 A) laptop power adapter shown in Figure 8.34. The main differences between them are as follows:

1. TOP269EG of TOPSwitch-JX series with more advanced performance is adopted to replace TOP259Y of TOPSwitch-HX series to reduce the no-load power consumption (less than 100 mW) and improve the protection function.
2. The GBU8J 8A/600 V BR manufactured with glass passivated process is applied to replace ordinary 3KBP08M BR to let the reverse recovery time close to the indicator of FRD and accelerate the response speed, not only improving the power factor, but also reducing conduction noise.
3. The total resistance of line voltage detection resistor ($R_3 + R_4$) is increased from 4 to 10.2 MΩ to reduce the no-load power consumption from 26 to 10 mW. The input under-voltage threshold set by R_3 and R_4 is 95 V (DC) and the over-voltage threshold is 980 V (DC), which is equivalent to disabling the OVP function. In the design, the input filter capacitor C_2 can withstand the series-mode surge voltage of above 2 kV, so the drain voltage of TOP269EG will not exceed the specified value (generally limited to 680 V).
4. The limit current of TOP259Y is set through R_5, R_6, and R_{19}, and the total resistance of ($R_5 + R_6$) is also increased from 11.9 to 20 MΩ.
5. The single resistor in the clamp circuit is replaced with multiple resistors (R_8–R_{10} and R_{11}–R_{12}), respectively, connected **in parallel** to prevent resistors from heating. R_{11} and R_{12} are adopted to attenuate high-frequency ringing and reduce EMI.
6. The output over-voltage protection circuit constituted by transistor VT1, regulator VDZ2, R15, R16, and C7 is added. Once the control circuit has open-loop failure, the output voltage UO and the bias winding voltage UB will rise rapidly. When U_B exceeds the sum of the breakdown voltage ($U_Z = 14$ V) of VD_{Z2} and the emitter junction voltage ($U_{BE} \approx$ 0.7 V) of VT_1, VT_1 will be conducted immediately to let the current flowing into pin V exceed the latch shutdown threshold current (typically $I_{OV\ (LS)} = 336\,\mu A$) to quickly turn off the output, thus realizing output over-voltage protection.
7. To reduce the loss of bias winding at no load, the capacity of C_5 is increased to 56 μF to let the minimum voltage of C_5 to be about 9 V, which is the minimum bias voltage required by optocoupler.
8. The low-voltage adjustable precision shunt regulator LMV431 with 1.24 V reference voltage is applied to replace the traditional 2.5 V adjustable precision shunt regulator TL431 to reduce the LED forward current I_{LED} of infrared emission tube in optical coupler PS2501 from 1 mA to 100 μA.
9. Transistor VT_2 is added in the optocoupler feedback circuit, which constitutes the Darlington tube with very high-current amplification factor together with the infrared receiving tube in optical coupler PS2501, capable of reducing the bias current of infrared receiving tube side to 1 mA.
10. The phase compensation network constituted by R_{22} and C_{14} is connected in parallel to the gain resistor R_{21} in control loop to increase the phase margin of control loop from 90° to 135°, thus improving the stability of control loop. I_{LED} is reduced to 100 μA when using LMV431, only 1/10 of the original value, so the resistance of R_{21} shall also be increased by 10 times, that is, from 680 Ω to 6.8 kΩ. However, in view that proper increase of the resistance of R_{21} can further raise the loop gain, R_{21} is actually taken as 20 kΩ. The increase of R_{23} is intended to provide LMV431 with appropriate bias current to ensure its proper work.

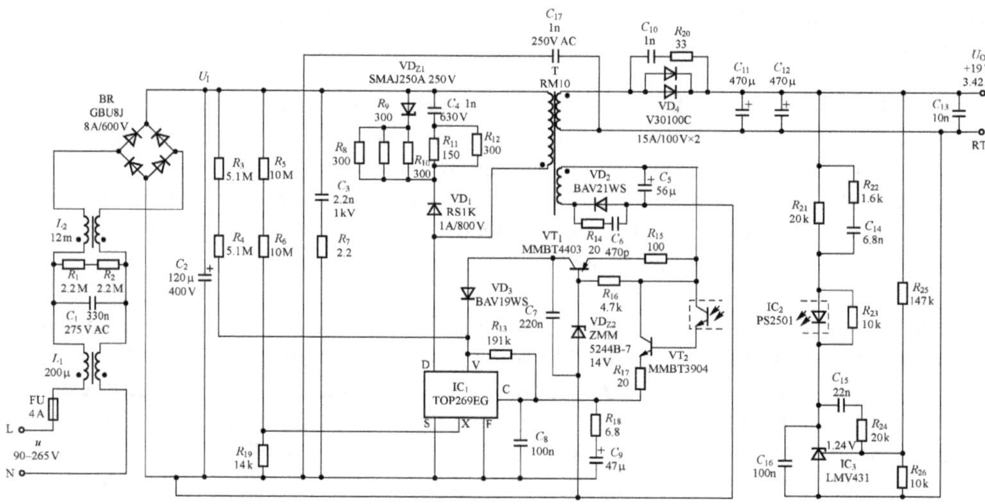

Figure 8.36 Circuit of 65 W power adapter with low no-load power consumption and high efficiency

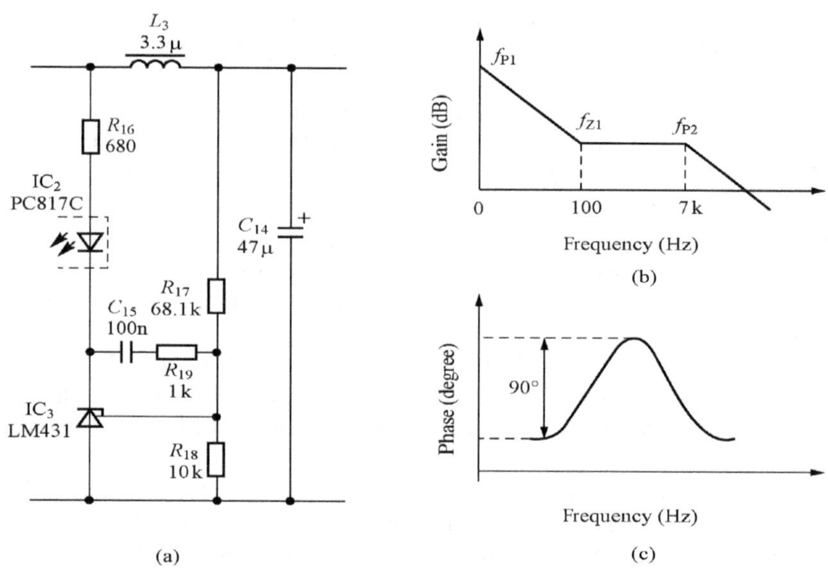

Figure 8.37 (a) Control loop before change and its (b) amplitude-frequency and (c) phase-frequency characteristic curves

See Figure 8.37(a), (b), and (c), respectively, for the control loop of Figure 8.34 and its amplitude-frequency and phase-frequency characteristic curves before change. f_{P1}–f_{P3} in the amplitude-frequency characteristic curve refer to pole frequencies. There are two poles (f_{P1} and f_{P2}, referred to as the double pole frequency) in Figure 8.37(b), wherein, the f_{P1} set by

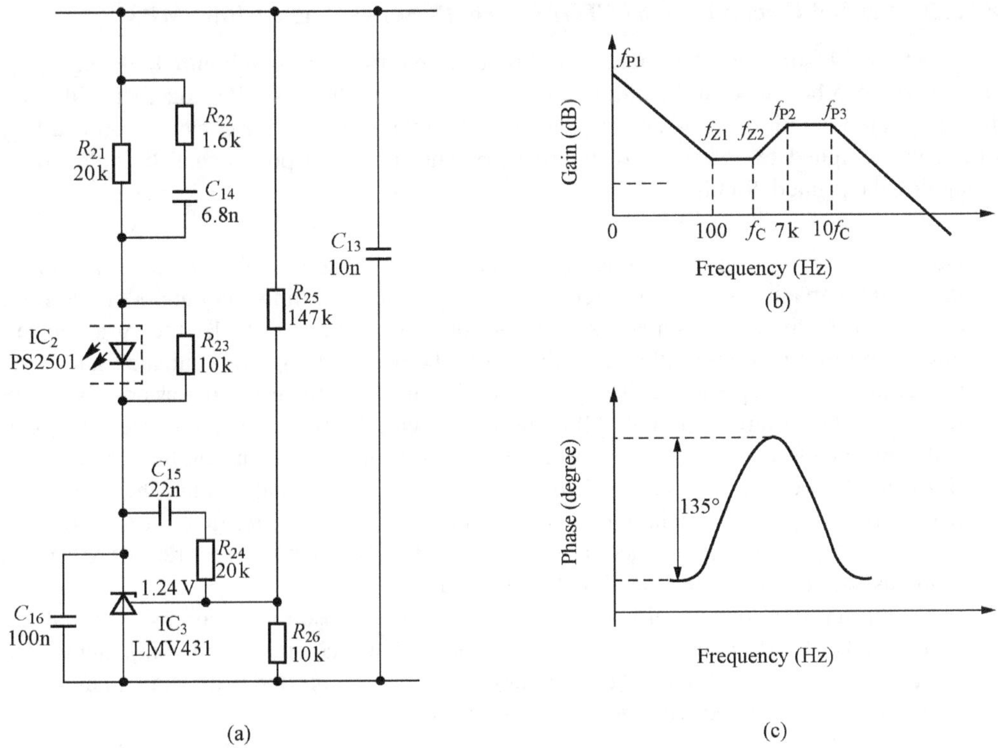

Figure 8.38 (a) Control loop after change and its (b) amplitude-frequency and (c) phase-frequency characteristic curves

C_{15} is equal to 0 Hz, the f_{Z1} set by C_{15}, R_{19}, and R_{17} is equal to 100 Hz, and the f_{P2} set by the internal resistor capacitor unit of TOP269EG is equal to 7 kHz.

See Figure 8.38(a), (b), and (c), respectively, for the control loop after change and its amplitude-frequency and phase-frequency characteristic curves. There are three poles (f_{P1}, f_{P2}, and f_{P3}, referred to as the three-pole frequency) in Figure 8.38(b). f_{Z1} and f_{Z2} refer to zero frequency, which can be one or several in number. The f_{P1} set by C_{15} is equal to 0 Hz. The f_{Z1} set by C_{15}, R_{24}, and R_{25} is equal to 100 Hz. The f_{Z2} set by R_{21}, R_{22}, and C_{14} is equal to f_C, wherein, f_C refers to the crossover frequency, indicating the corresponding frequency when the total open-loop gain is 1 (i.e., 0 dB). The crossover frequency must be less than 1/10 of the switching frequency. The f_{P2} set by TOP269EG is equal to 7 kHz. The f_{P3} set by R_{21}, R_{22} and C_{14} is equal to $10f_C$.

A bode diagram can be generated by using the amplitude-frequency and phase-frequency characteristic curves, wherein, the logarithmic scale is applied for the frequency coordinate. The bode diagram reflects the amplification capability of the control loop toward different frequency signals, capable of providing the loop gain calculation and the stability analysis with convenience

8.9.2.2 Printed Circuit Design of TOPSwitch-JX Series Single-Chip SMPS

See Figure 8.39 (some components are slightly different from those in Figure 8.36, e.g., magnetic beads can be added in the output circuit to constitute the post filter together with C_{13}.) for the printed circuit of 65 W power adapter with low no-load power consumption and high efficiency designed by referring to Figure 8.36. The following points shall be noted when designing the printed circuit:

1. During the wiring of primary side, the connection of the source pin, the negative pole of input filter capacitor C_2, and the return terminal of bias winding shall adopt single-point connection method (also known as Kelvin connection method) to let the surge current directly return to the input filter capacitor from the bias winding, thus enhancing the anti-surge ability. The bypass capacitor C_8 at control terminal shall be as close as possible to the source and the control terminal. All components with the source as the reference ground and connected to the voltage monitoring terminal (V) or external current limiting terminal (X) shall also be as close as possible to the source and corresponding pin. The source current of internal power switching tube shall return to the negative terminal of C_2 through a separate path, rather than connect to pin C, V, and X. No switching current of MOSFET shall pass through the source lead of TOP269EG.
2. The input terminal of optical coupler PS2501 shall be close to the control terminal and source of TOP269EG and bypass the wiring area of source and clamp components. The position of clamp circuit shall let the primary winding be connected to the drain to minimize the loop area constituted with the clamp capacitor.

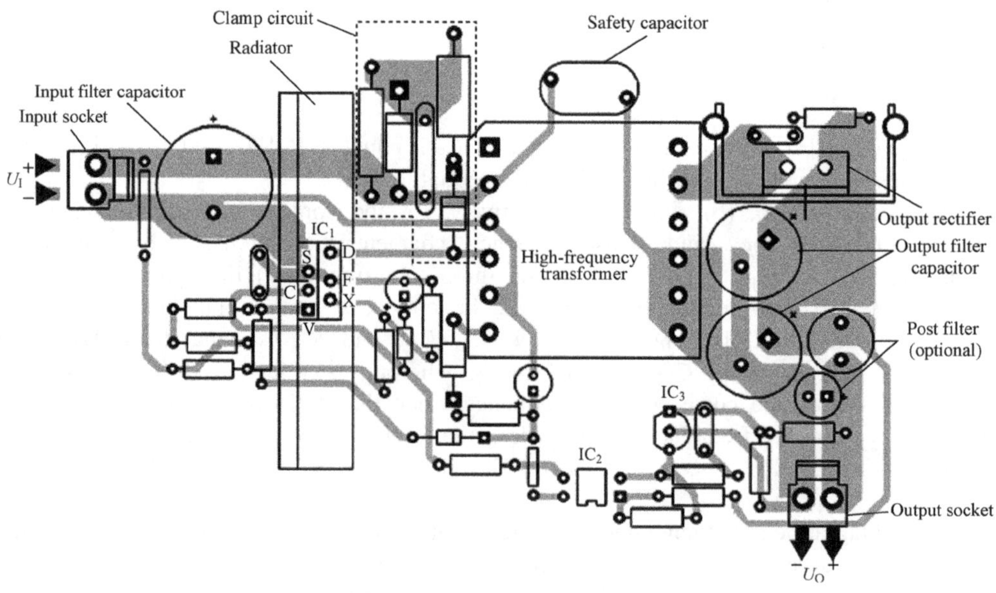

Figure 8.39 Printed circuit of 65 W power adapter with low no-load power consumption and high efficiency

3. The exposed pad of TOP269EG with E (eSIP-7C), K (eSOP-12), and V (eDIP-12) package is connected to the source in the chip. To avoid circular current, the radiator installed on the exposed pad shall not be connected to the primary ground terminal or the source. When designing the double-sided PCB, the via hole connecting the top and bottom layers can be applied to improve the heat dissipation effect. When using K package, the exposed pad can be directly soldered to a copper-clad area to constitute PCB radiator.
4. When using TOP269–TOP271 to design high-power SMPSs, it is recommended to connect in parallel an RC noise cancelling circuit between the drain and the source. Generally, take $R = 22–150\,\Omega$ and $C = 10–33\,\text{pF}$ (1 kV voltage-resistant ceramic capacitor) to not only reduce the influence of switching noise on the work of power supply but also help to reduce the electromagnetic radiation.
5. The area of loop connecting the secondary winding, the output rectifier (VD_4), and the output filter capacitor (C_{11} and C_{12}) shall be minimized. The pad area of pin VD_4 shall be large enough to facilitate heat dispersion.
6. In the case of high current output, the output ripple current will rise significantly. Even if the filter capacitor with very low ESR is adopted, the allowable value may still be exceeded. Generally, several filter capacitors may be connected in parallel, in which case, the lead length of each capacitor shall be the same to realize equal division of ripple current. If necessary, another stage of post-LC filter can be added to further reduce switching noise.
7. To reduce the leakage inductance of high-frequency transformer and PCB, it is recommended to introduce the sandwich winding method and the lead of PCB shall be as short as possible.

Finally, see Figure 8.40 for the summarized description of several key points in circuits during PCB layout and wiring for reference.

8.10 Electromagnetic Interference Waveform Analysis and Safety Code Design of SMPS

Electromagnetic compatibility is referred to as EMC. EMC is defined by the International Electrotechnic Commission (IEC) as follows: "EMC is a function of the electronic equipment, which can complete its functions without causing intolerable interference in an electromagnetic environment." This indicates that EMC has three meanings: first, the electronic equipment shall have the ability to suppress external EMI. Second, the EMI generated by the electronic equipment shall be less than the limit specified and shall not affect the normal operation of other electronic equipments in the same electromagnetic environment. Third, the EMC of any electromagnetic equipment is measurable. It was specified by the European Community (EC) as early as January 1, 1996 that EMC test must be carried out for electronic equipments (including electric appliance, electronic instrument, and devices with electric appliance or electronic parts), that is, the test of electromagnetic wave interference (CMI) and anti-interference (EMS) performance of electronic equipment. It was also specified by EC on 1 January 1997 that safety performance test must be carried out for low-voltage electronic equipment. At present, EMC and safety code have been regarded as an important indicator to test the quality of electronic products by China.

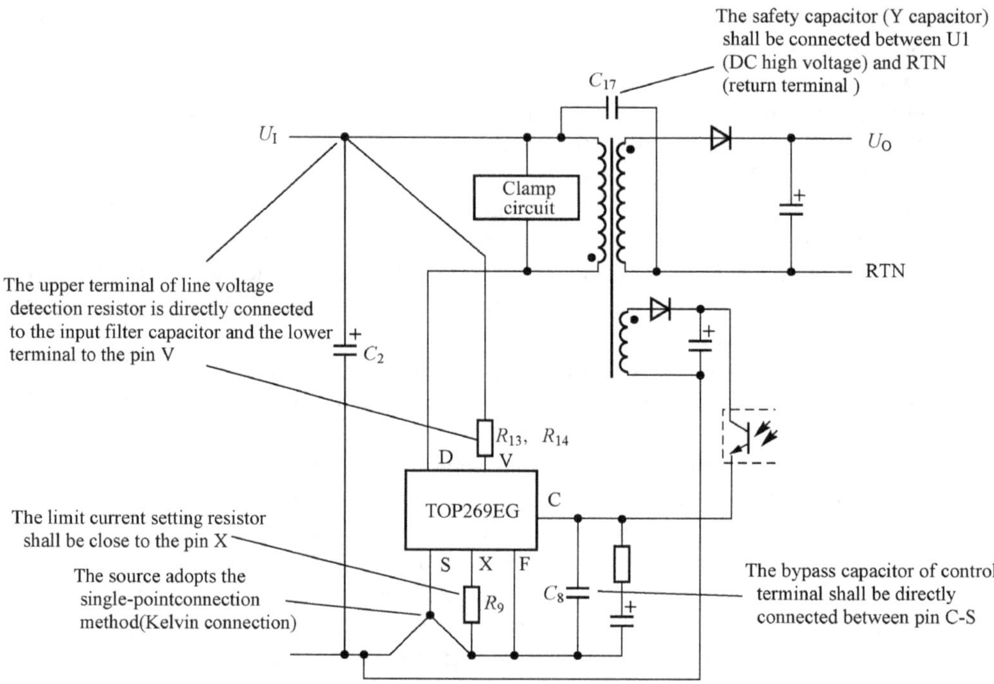

Figure 8.40 Description of several key points in circuits

8.10.1 Electromagnetic Interference Waveform Analysis of SMPS

EMI is referred to as EMI. SMPS works under high-frequency, high-voltage, and heavy-current state, so the EMI generated is divided into common-mode interference (also known as line-to-ground interference) and series-mode interference (also known as line-to-line interference), which are propagated outside through conduction or radiation. The EMC design of SMPS is intended to attenuate EMI to an allowable range, so as not to affect the normal working of the power supply itself and other electronic devices.

See Figure 8.41(a) and (b), respectively, for the simplified circuit and EMI waveform of flyback SMPS. In Figure 8.41(a), U_I is the DC input voltage, I_1 the primary current of high-frequency transformer, U_{DS} the drain-source voltage of power switching tube MOSFET (hereinafter referred to as MOSFET), and U_{D2} the voltage of output rectifier. I_2 is the secondary current and R_L is the load. Figure 8.41(b), respectively, shows the EMI waveforms of I_1, U_{DS}, I_2, and U_{D2}. The following text analyzes these four kinds of waveform.

When MOSFET is conducted, the primary current I_1 is started to be generated and rises along the slope to reach the peak I_{1P}. The value of I_{1P} is determined by the DC input voltage U_I, the primary winding inductance L_P, the switching frequency f, and the duty ratio D. The fundamental frequency of the trapezoidal current waveform is switching frequency and the harmonic wave is the interference waveform. The primary series-mode interference current passes through the primary winding, MOSFET, and U_I to form a loop. When the area of current loop is large, I_1 can radiate common-mode interference outward. The voltage waveform of U_{DS} is featured so that its voltage change rate (dU/dt) is very high and affected by

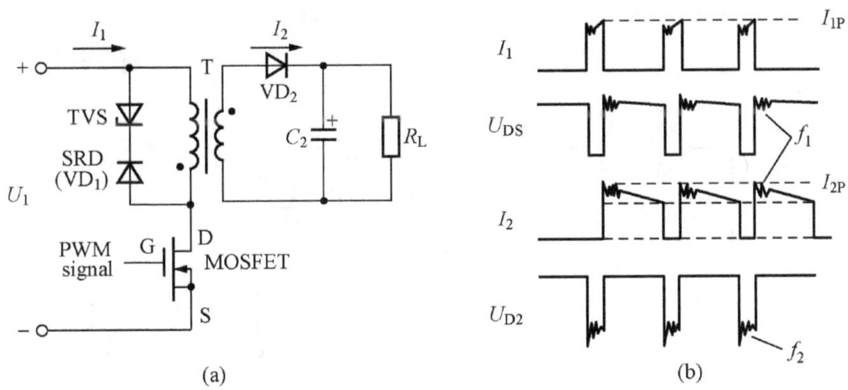

Figure 8.41 Simplified circuit and EMI waveform of flyback SMPS. (a) Simplified circuit and (b) four kinds of EMI waveform

distributed parameters such as the leakage inductance of high-frequency transformer, the distributed capacitance of high-frequency transformer, and the output capacitance of MOSFET. In addition, decaying oscillation – ringing – will be formed by U_{DS} within the frequency range of $f_1 = 3$–12 MHz. When MOSFET is turned off, I_2 current passes through the secondary side and decreases linearly from the peak I_{2P}, the decrease rate of which is determined by the secondary winding inductance L_S and the output voltage U_O. The ringing formed during the decrease process corresponds to U_{DS} in terms of time. The ringing frequency is f_1. U_{D2} is also featured in high change rate of voltage and steep rising and falling edge. The peak voltage is determined by the distributed inductances of high-frequency transformer and output rectifier. The range of ringing frequency f_2 is 20–30 MHz.

8.10.1.1 Circuit Model of Common-Mode Interference

See Figure 8.42 for the circuit model causing common-mode interference. Common-mode interference is mainly generated by the drain-source voltage U_{DS} and the output rectifier voltage VD_2.

In Figure 8.42, C_u is the coupling capacitor connected in parallel to the input terminal of AC power supply and C_{BD1}–C_{BD4} are the equivalent capacitances of four rectifiers in BR. C_{IN} is the input filter capacitor, of which the ESL and ESR are, respectively, expressed as L_{ESL} and R_{ESR}. C_{W1}–C_{W6} are the distributed capacitances of high-frequency transformer, wherein, C_{W1} and C_{W6} are, respectively, the distributed capacitances of the primary and secondary windings, and C_{W2}–C_{W5} are the distributed capacitances between the primary and secondary winding. C_{OSS} is the output capacitance of MOSFET. C_{S1} and C_{S2} are, respectively, the drain-to-ground and secondary-to-ground distributed capacitance. The above capacitances will cause five interference currents: I_{CW1}, I_{COSS}, I_{CS1}, I_{CW3}, and I_{CW4}. After the five currents are overlapped, some will be offset and the remained high-frequency current forms the common-mode interference.

Common-mode interference can be suppressed by the common-mode choke in EMI filter. The inductance of common-mode choke is usually 10–33 mH. To reduce the distributed capacitance, associated wires on the PCB shall be as short as possible.

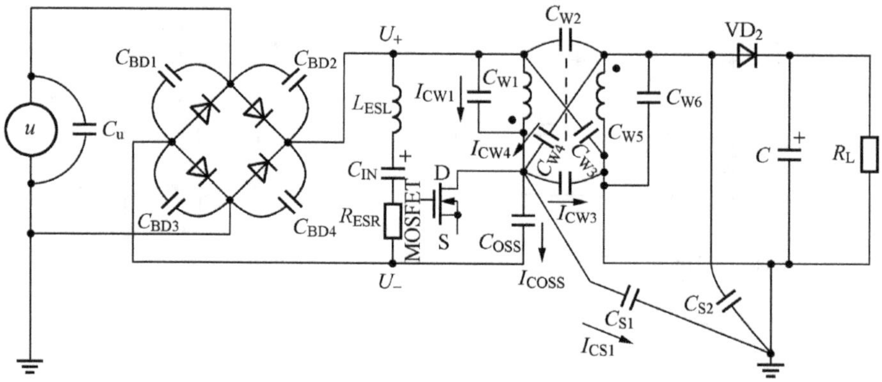

Figure 8.42 Circuit model causing common-mode interference

When measuring the common-mode inductance, one winding shall be disconnected first, and then measure the inductance of the other winding, which is the common-mode inductance of common-mode choke.

8.10.1.2 Circuit Model of Series-Mode Interference

See Figure 8.43 for the circuit model of series-mode interference. Figure (b) is the equivalent circuit. In Figure (a), C_D is the series-mode capacitance and L_D and L'_D are two series-mode chokes. R_{ESR} is the ESR of input filter capacitor C_{IN}. In Figure (b), the voltage to ground of the two power lines is expressed as U_S. The voltage polarity at the positive half cycle is shown in the figure. It is easy to see that the series-mode interference current flows in the direction from one power line to MOSFET and then flows out from the other power line. The series-mode interference filter constituted by C_D, L_D, and L'_D can suppress the series-mode interference. For example, in the 15 W SMPS, actually take $C_D = 0.1\,\mu F$, $C_{IN} = 33\,\mu F$, $R_{ESR} = 0.375\,\Omega$, and $L_D = L'_D = 74\,\mu H$. L_D and L'_D can be either discrete inductances or the ESLs separated from the common-mode interference choke. After adding the series-mode interference filter, the fundamental wave voltage of series-mode interference is 59.3 mV and the secondary harmonic voltage is reduced to 43.0 mV.

When measuring the equivalent series-mode inductance of one winding in the common-mode choke, the other winding shall be in short circuit and the inductance of L_D (or L'_D) is equal to half the measured value.

8.10.2 Safety Code of SMPS

The safety code of SMPS is referred to as "safety code." The name of IEC950 standard is "safety of information technology equipment, including commercial electric apparatus." The standard specifies detailed regulations on the design of safety equipment to prevent any damage or harm caused by some hazards. The hazards include electric shock, electric injury, fire, mechanical and thermal damage, radiation damage, and chemical damage. See Table 8.3 for the contents suitable for the SMPS safety code extracted from IEC950 standard.

Table 8.3 Safety code of SMPS

General category	Classification	Contents of safety code in IEC950 standard
Device type	Class I	Devices using the following method to obtain protection against electric shock: the electronic equipment using basic insulation must have a kind of connection device to let the conductive parts that will have dangerous voltage in the event of failure of the basic insulation be connected to the protective ground conductor in the building wiring
	Class II	Electronic devices not only depend on the basic insulation but also need additional security measures for protection against electric shock. For example, the devices using double or reinforced insulation, which relies on neither the protective earthing nor the protective measures of installation conditions
Circuit type	Primary circuit	Internal circuit directly connected to the external grid or other equivalent power supply. In the SMPS, this part of circuit includes the EMI filter, the bridge rectifier, the primary winding of high-frequency transformer, and any component directly connected to the primary side, such as the bias winding using primary feedback and the phototriode of optical coupler
	Secondary circuit	The circuit is not directly connected to the primary power terminal (except Y capacitor), the power transferred by which is from the high-frequency transformer
Voltage type	Working voltage	The maximum voltage allowed by the insulating material when the device works normally at rated voltage
	Safety extra low voltage (SELV)	The designed circuit has a protective function. The safety voltage that shall not be exceeded by the voltage between any two close parts in the secondary circuit or between one part and the protective ground terminal of class I device under normal operating or single fault conditions
Insulation type	Basic insulation	The insulation providing basic protection against electric shock
	Double insulation	The insulation constituted by the basic and additional insulation
	Reinforced insulation	A single insulation structure, the level of electric shock protection provided by which is equivalent to double insulation
	Creepage distance	The shortest distance measured along the insulator surface between two conductive parts or between the conductive part and the boundary surface of device. The creepage distance from all primary circuits to all secondary circuits in SMPS is usually 5–6 mm

Table 8.3 (*Continued*)

General category	Classification	Contents of safety code in IEC950 standard
Maximum leakage current	All devices of class II	0.25 mA
	Handheld devices of class I	0.75 mA
	Mobile devices of class I (other than handheld devices)	3.50 mA
Dielectric strength of AC insulation from the primary side to the secondary side	Basic insulation	1000 V ($u \leq 130$ V); 1500 V (130 V $\leq u \leq 250$ V)
	Additional and reinforced insulation	2000 V ($u \leq 130$ V); 3000 V (130 V $\leq u \leq 250$ V)
Safety isolating transformer	The power transformer capable of isolating the winding supplying power to SELV circuit and other windings (e.g., the primary winding and the primary bias winding). Even through the insulation suffers breakdown, no danger will be caused on the SELV winding	
Mains voltage	When determining the most unfavorable mains voltage applied to test the power supply, the following factors shall be considered: multiple rated voltages, the limit of rated voltage range, and the rated voltage tolerance specified by the manufacturer. If no tolerance is specified, the tolerance range of +6% to −10% shall be adopted	
Safe discharge	The device shall be designed to ensure that no shock hazard will be caused owing to the release of charge connected to the capacitor in the power supply circuit when the power grid is turned off. When the device has a capacitor with the rated capacity greater than 0.1 μF connected to the external power grid, all possible methods must be adopted to discharge the capacitor and the discharge time constant shall be less than 1 s. This requirement is particularly suitable for any EMI filter capacitor directly connected to the AC power grid. When the power cable is pulled out from the socket, the capacitor may cause electric shock owing to its exposed plug	
Ground leakage current	At the maximum input voltage, the maximum ground leakage current shall not exceed the specified limit. For devices of class II, the accessible conductive parts shall be tested when the output is not connected to the ground. For accessible nonconductive parts, the metal foil with the area less than 10 × 20 cm posted on the part shall be measured	
Two-wire AC input	The connection of two-wire AC input of SMPS can be constituted by one phase 1 wire and one neutral wire, wherein, the neutral wire of AC power grid is connected to the ground finally through the wiring board at this position. The safety ultra low voltage (SELV) output terminal of power supply can either be directly or indirectly connected to the ground	

Table 8.3 (*Continued*)

General category	ClassificationContents of safety code in IEC950 standard
Three-wire AC input	In the three-wire system connection method, the third wire is the ground wire applied to connect the EMI filter components, shielding, chassis, and shell. The neutral wire is regarded as the AC power cable not grounded or a separate phase wire. Therefore, safety consideration is also given to it as any AC power cable Safety codes such as IEC950, UL1950, and UL544 strictly limit the total fault current in the case of open circuit of the safety ground connection or failure of a component (e.g., Y1 capacitor). For example, it is specified in UL1950 that for the information technology equipment of class I or the three-wire (phase, neutral, and ground wire) AC 240 V and 60Hz input, the leakage current shall not exceed 3.5 mA in case of ground wire disconnection or short circuit owing to the failure of a component. Therefore, the maximum capacity of Y capacitor is limited to 39 nF. For the equipment of class II or two-wire (phase and neutral wire, without ground wire) input, the leakage current shall not exceed 250 μA when a component fails. Corresponding to the 240 V and 60 Hz input, the maximum of Y capacitor is limited to below 2800 pF. In addition, the capacitance and input voltage tolerance shall be considered

8.11 Radiator Design of Single-Chip SMPS

8.11.1 Working Principles of Radiator

After a radiator is added to the semiconductor device, the total resistance can be reduced and the distribution of thermal resistance is as shown in Figure 8.44. $R_{\theta JC}$ refers to the thermal resistance from junction to shell. $R_{\theta CS}$ is the thermal resistance from shell to the radiator surface. When $R_{\theta JS}$ is the thermal resistance from junction to radiator, $R_{\theta JS} = R_{\theta JC} + R_{\theta CS}$. $R_{\theta SA}$ is the thermal resistance from cooling plate to surrounding air, referred to as radiator thermal resistance.

See Figure 8.45 for the thermal resistance distribution of semiconductor device. It can be seen in the figure that the thermal conduction is in accordance with the order of temperature

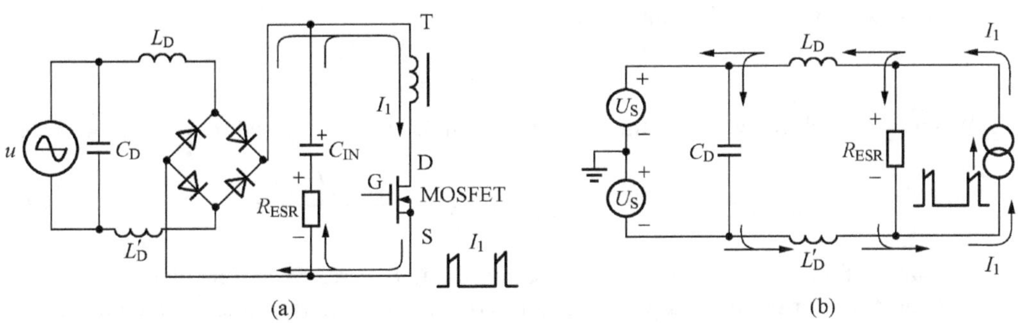

Figure 8.43 Circuit model of series-mode interference. (a) Circuit model and (b) equivalent circuit.

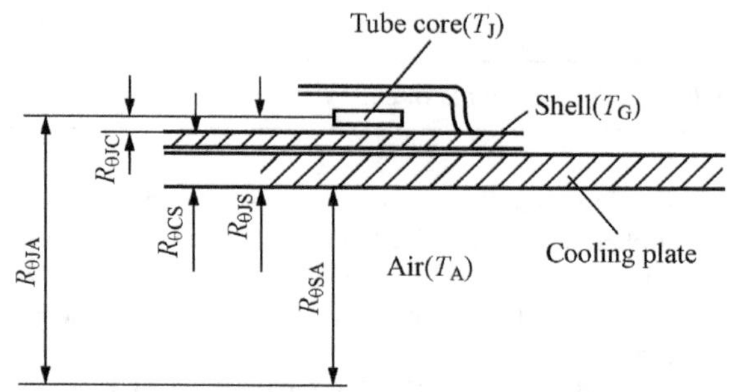

Figure 8.44 Thermal resistance of semiconductor device added with a radiator

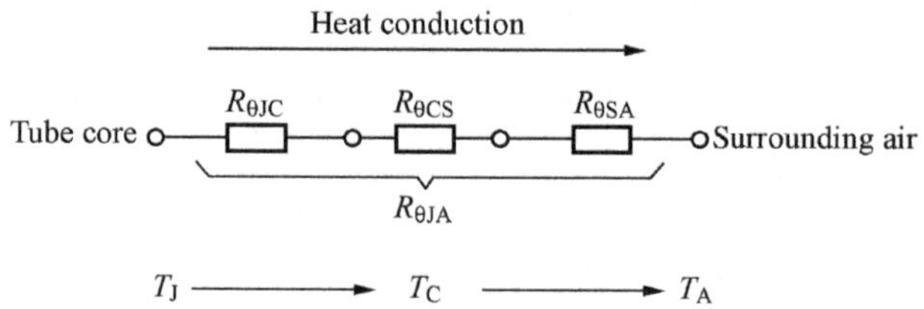

Figure 8.45 Thermal resistance distribution diagram of semiconductor device

gradient from high to low, that is, "$T_J \rightarrow T_C \rightarrow T_A$". Letting $R_{\theta JA}$ be the total thermal resistance,

$$R_{\theta JA} = R_{\theta JC} + R_{\theta CS} + R_{\theta SA} = R_{\theta JS} + \ R_{\theta SA} \tag{8.27}$$

Assuming that the total thermal resistance without cooling board is R_θ, the maximum allowable junction temperature is T_{jM}, the maximum ambient temperature is T_{AM}, the temperature rise is ΔT, and the actual power consumption after adding radiator is P_D, relevant formula is as follows:

$$P_D = \frac{\Delta T}{R_{\theta JA}} = \frac{T_{jM} - T_{AM}}{R_{\theta JS} + R_{\theta SA}} \tag{8.28}$$

Then, letting P'_D be the designed power consumption and P_{DM} be the maximum allowable power consumption, the following condition must be met:

$$P_D \leq P'_D < P_{DM} \tag{8.29}$$

When P_D is replaced with P'_D, the following formula can be obtained from Formula (8.28):

$$R_{\theta SA} = \frac{T_{jM} - T_{AM}}{P'_D} - (R_{\theta JC} + R_{\theta CS}) = \frac{T_{jM} - T_{AM}}{P'_D} = -R_{\theta JS} \tag{8.30}$$

The value of $R_{\theta SA}$ will be adopted when determining the area of cooling plate.

See Table 8.4 for the thermal parameters of common packages of linear regulator. It shall be noted that some products have a variety of packages. See Figure 8.46 for the relation curve of thermal resistance $R_{\theta SA}$ and surface area S of the cooling board, wherein, the

Table 8.4 Thermal parameters of common packages

Package		TO-220	TO-3	TO-263	SOT-223
Maximum allowable power consumption P_{DM}/W		10	20	5	2
Total thermal resistance from junction to surrounding air without radiator $R_{\theta JA}$/(°C/W)		62.5	40	40	150
Total thermal resistance from shell to radiator surface after adding the radiator $R_{\theta CS}$/(°C/W)	Direct contact with the cooling plate	7	6	3 (directly soldered on the cooling plate)	15 (directly soldered on the cooling plate)
	Applying heat conducting silicone grease	1	1	—	—
	Adding the mica sheet of 0.05 mm thick	1.8	1.8	—	—

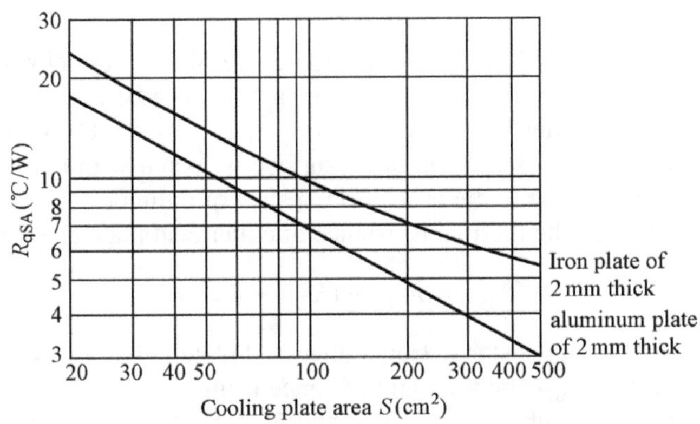

Figure 8.46 $R_{\theta SA}$-S relation curve of aluminum and iron plates

coordinates X and Y adopt the logarithmic scale. The figure, respectively, shows the curves of aluminum and iron plates with the plate thickness of both 2 mm. The use condition is that the cooling plate shall be placed vertically with the device installed in the center of cooling plate. It can be seen in the figure that the larger the cooling plate area is, the smaller will be the thermal resistance, approximately in inverse proportion. Furthermore, in the case of the same surface area and thickness, the aluminum plate has smaller thermal resistance, better than the iron plate in terms of heat dissipation performance. In addition, the aluminum plate has a density only 1/3 of that of the iron plate (2.7/7.8) and is not easy to rust.

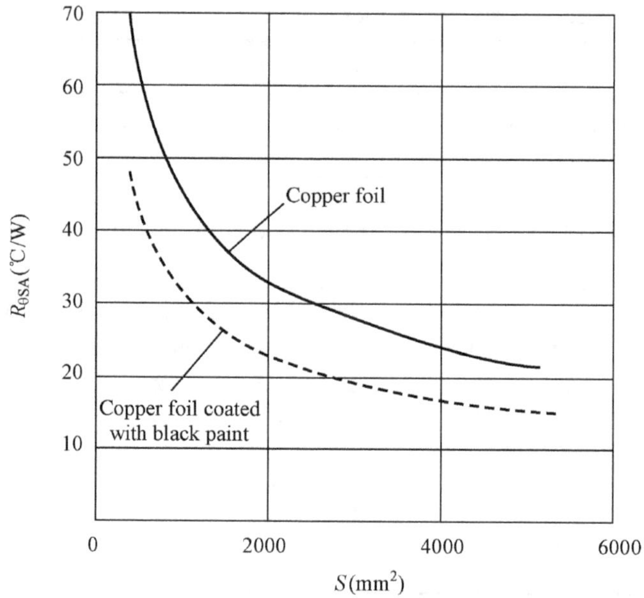

Figure 8.47 Relation curve of the thermal resistance of PCB radiator and the cooling cooper foil area

See the solid line in Figure 8.47 for the relation curve of the thermal resistance ($R_{\theta SA}$) of PCB radiator and the cooling copper foil area (S), assumed to be no airflow here. It can be seen from the figure that when $S = 1\,\text{in.}^2 = 645\,\text{mm}^2$, $R_{\theta SA} = 55\,°\text{C/W}$. The dotted line is measured under the condition that the radiator surface is coated with black paint and the air speed is 1.3 m/s, which is close to the best working state of radiator.

The formula to calculate the thermal resistance of coller clad plate is as follows:

$$R_{\theta SA} = R_{\theta JA} - (R_{\theta JC} + R_{\theta CS}) \tag{8.31}$$

wherein, $R_{\theta SA}$ is the thermal resistance from coller clad plate radiator to surrounding air (i.e., the ambient temperature), $R_{\theta JA}$ the thermal resistance from tube core to surrounding air, $R_{\theta JC}$ the thermal resistance from tube core to shell, and $R_{\theta CS}$ the thermal contact resistance from shell to coller clad plate. Two points shall be noted: first, $R_{\theta JC}$ is related to the package. Second, when the shell (or pin) is directly soldered on the coller clad plate (or soldered on the coller clad plate through the metal cooling pad at the bottom of device), $R_{\theta CS} = 0$. When the shell (or pin) is not soldered on the coller clad plate, $R_{\theta CS} = 0.5–3\,°\text{C/W}$ depending on specific contact situation. See Table 8.5 for the total thermal resistance $R_{\theta JA}$ from tube core to surrounding air of 8 kinds of typical surface-mount chip products.

8.11.2 Radiator Design Methods of Single-Chip SMPS

The power consumption of single-chip SMPS IC is mainly generated by the internal power switching tube (MOSFET). The power consumption of other unit circuits in the chip is generally negligible. In the case of low-voltage and high-current output, the power consumption

Table 8.5 Typical $R_{\theta JA}$ of eight kinds of typical surface-mount chip products

Package	SOT-23	TSSOP-8	SOT-89	μMAX-8 Micro-8	SO-8	D-PAK	D2-PAK
$R_{\theta JA}$ (°C/W)	75	45	35	35	25	3	2

of output rectifier shall be considered. The important difference between the SMPS and the linear regulated power supply is that the power switching tube works in the high-frequency switching state. The power consumption P_D varies constantly within the switching cycle, so it is difficult to accurately calculate the value of P_D. It can be known through analysis that the power consumption of single-chip SMPS mainly includes two parts: the transmission and switching loss. Transmission loss is caused by the on-state resistance $R_{DS\,(ON)}$ of MOSFET. For example, the $R_{DS\,(ON)}$ of early single-chip SMPS product TOP227Y is equal to 4.3 Ω (the typical value at 100 °C, the same below), and the $R_{DS\,(ON)}$ of new product model TOP258 with equivalent output power is reduced to 2.5 Ω. The smaller the on-state resistance is, the lower will be the transmission loss. Switching loss refers to the loss caused by the output capacitance C_{OSS} of MOSFET. The transmission loss is usually much larger than the switching loss, so the switching loss is also negligible.

The following text describes the radiator design method by calculating the average power consumption $\overline{P_D}$ of chip according to the original chart provided by the manufacturer. This method is simple and practical with application popularization value.

When designing the radiator, first calculate the average power consumption $\overline{P_D}$ of the selected product model. Letting the maximum duty ratio be D_{max}, the formula is as follows:

$$\overline{P_D} = (\overline{I_{DS\,(ON)}})^2 R_{DS\,(ON)} D_{max} = (I_{DS\,(ON)}/2)^2 R_{DS\,(ON)} D_{max} \tag{8.32}$$

Then, find out the value of $R_{\theta CS}$ according to the package and put the value of $\overline{P_D}$ and it together into the following formula to get the value of $R_{\theta SA}$:

$$R_{\theta SA} = \frac{T_{jM} - T_{AM}}{P_D} - R_{\theta JS} \tag{8.33}$$

wherein, $R_{\theta SA}$ is the thermal resistance from cooling plate to surrounding air (referred to as radiator thermal resistance), which can be applied to determine the surface area of aluminum cooling plate (or PCB radiator). Finished radiator can also be selected directly according to the value of $R_{\theta SA}$. T_{JM} is the maximum junction temperature, T_{AM} the maximum ambient temperature, P_D the power consumption of device, and $R_{\theta JS}$ the thermal resistance from junction to radiator surface.

8.11.3 Radiator Design Examples of Single-Chip SMPS

Considering the single-chip SMPS of TOPSwitch-GX (TOP242–TOP250) series for example, the thermal parameters given in its data sheet are as shown in Table 8.6. See Tables 8.7 and 8.8, respectively, for the on-state resistance (the chip junction temperature $T_J = 100\,°C$) and limit current of this series. When MOSFET is conducted, the relation curve of drain-source

Table 8.6 Thermal parameters of **TOPSwitch-GX** series

Thermal parameter	TO-220-7C (Y) or TO-262-7C (F) package	TO-263-7C (P) or SMD-8B (G) package	
Thermal resistance from junction to ambient temperature $R_{\theta JA}$ (°C/W)	80 (without radiator)	70 (the area of cooling copper foil is 232 mm^2)	60 (the area of cooling copper foil is 645 mm^2)
Thermal resistance from junction to shell $R_{\theta JC}$ (°C/W)	2 (the thermal resistance from junction to the small cooling plate on the back of device)	11 (measured at the source pin near the plastic shell surface)	
Radiator features	Using plate-type or finished radiator	Using PCB radiator, the mass per unit area of PCB being 610 g/m^2	

Table 8.7 On-state resistance of **TOPSwitch-GX** series ($T_J = 100\,°C$)

TOPSwitch-GX series Product model	On-state resistance $R_{DS\,(ON)}$ (Ω)		
	Minimum	Typical	Maximum
TOP242	—	25.7	30.0
TOP243	—	12.9	15.0
TOP244	—	8.60	10.0
TOP245	—	6.45	7.50
TOP246	—	4.30	5.00
TOP247	—	3.22	3.75
TOP248	—	2.58	3.00
TOP249	—	2.15	2.50
TOP250	—	1.85	2.15

conduction current ($I_{DS\,(ON)}$) and drain-source conduction voltage ($U_{DS\,(ON)}$) is as shown in Figure 8.48, in which case, the drain-source conduction voltage of MOSFET is generally only a few volts. When MOSFET is turned off, the relation curve of drain power consumption P_D and drain-source turn-off voltage $U_{DS\,(OFF)}$ is as shown in Figure 8.49, in which case, $U_{DS\,(OFF)}$ can be up to several hundred volts. The scale factor k in Figures 8.48 and 8.49 is related to chip models.

Design example: the TOP249Y single-chip SMPS IC with TO-220-7C package is applied to design a 70 W (19 V, 3.6 A) universal SMPS. It is known that the maximum junction temperature of TOP249Y is $T_{JM} = 150\,°C$, the maximum operating junction temperature specified by manufacture is $T_{Jmax} = 125\,°C < 150\,°C$, and the maximum ambient temperature is $T_{AM} = 40\,°C$. It is intended to adopt the aluminum sheet radiator of 2 mm thick. Try to determine the external dimensions of the radiator.

Table 8.8 Limit current of **TOPSwitch-GX** series

TOPSwitch-GX series Product model	Limit current I_{LIMIT} (A)		
	Minimum $I_{LIMIT\,(min)}$	Typical I_{LIMIT}	Maximum $I_{LIMIT\,(max)}$
TOP242P/G/Y	0.418	0.45	0.481
TOP243P/G	0.697	0.75	0.802
TOP243Y	0.837	0.90	0.963
TOP244P/G	0.930	1.00	1.070
TOP244Y	1.256	1.35	1.445
TOP245Y	1.674	1.80	1.926
TOP246Y	2.511	2.70	2.889
TOP247Y	3.348	3.60	3.852
TOP248Y	4.185	4.50	4.815
TOP249Y	5.022	5.40	5.778
TOP250Y	5.859	6.30	6.741

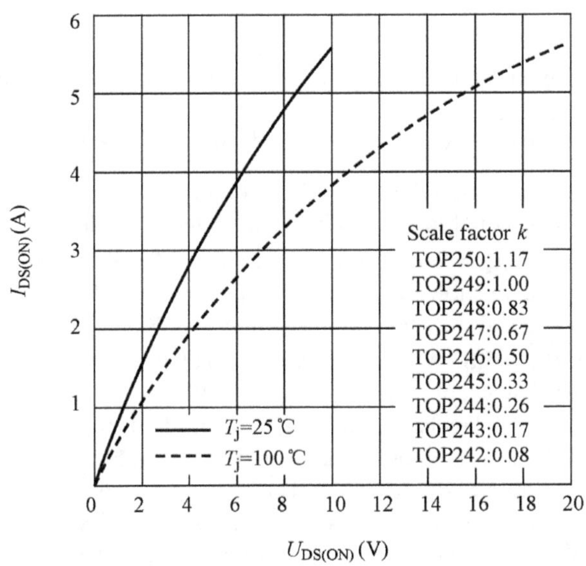

Figure 8.48 Relation curve of $I_{DS\,(ON)}$ and $U_{DS\,(ON)}$ when MOSFET is conducted

Taking into account the most unfavorable case, 100 °C can be adopted as the chip junction temperature T_J for calculation. It is found in Table 8.7 that the $R_{DS\,(ON)}$ of TOP249Y is equal to $2.15\,\Omega$ (typical value) when $T_J = 100\,°C$ and found in Table 8.8 that the limit current $I_{LIMIT} = 5.40\,A$ (typical value). The chip is always derated for use, so it is desirable to actually take $I_{DS\,(ON)} = 0.8I_{LIMIT} = 4.32\,A$. $I_{DS\,(ON)}$ rises from zero to the maximum approximately according to linear rule (see Figure 8.48), so its average value shall be taken, that is, $\overline{I_{DS\,(ON)}} = (0 + I_{DS\,(ON)})/2 = (0 + 4.32\,A)/2 = 2.16\,A$. For TOP249Y, the scale factor

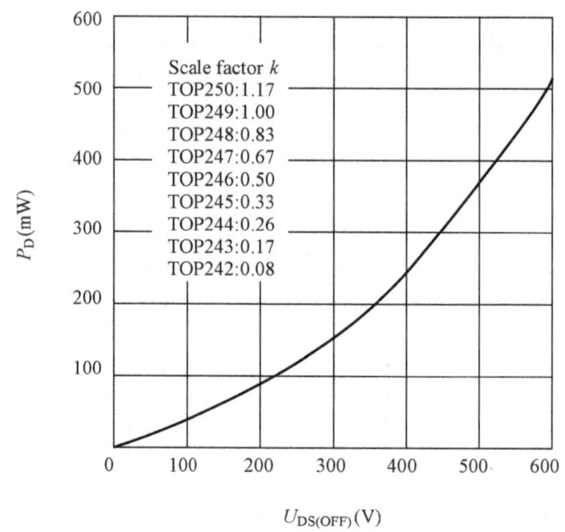

Figure 8.49 Relation curve of P_D and $U_{DS\,(OFF)}$ when MOSFET is turned off

$k = 1.00$. Letting the maximum duty ratio be $D_{max} = 60\%$, it is not difficult to calculate

$$\overline{P_D} = (\overline{I_{DS\,(ON)}})^2 R_{DS\,(ON)} D_{max} = (2.16\,\text{A})^2 \times 2.15\,\Omega \times 60\% = 6.0\,\text{W}$$

It is found in the dotted line ($T_J = 100\,°\text{C}$) of Figure 8.48 that the $U_{DS\,(ON)}$ corresponding to $\overline{I_{DS\,(ON)}} = 2.16\,\text{A}$ is equal to 4.5 V. When calculated according to the value of $U_{DS\,(ON)}$, $\overline{P_D} = \overline{I_{DS\,(ON)}} \cdot U_{DS\,(ON)} D_{max} = 2.16\,\text{A} \times 4.5\,\text{V} \times 60\% = 5.83\,\text{W}$, slightly less than 6.0 W. This is because the relation curve shown in Figure 8.48 is nonlinear, thus causing the latter's value to be slightly lower. In addition, when the switching loss $P_{DK} = 510\,\text{mW}$ (when $U_{DS\,(OFF)} = 600\,\text{V}$) is considered, $\overline{P_D} = 5.83\,\text{W} + (0.510\,\text{W}/2) = 6.09\,\text{W}$, very close to 6.0 W. The following thermal parameters are calculated as per $\overline{P_D} = 6.0\,\text{W}$.

Table 8.6 shows the thermal resistance from junction to shell $R_{\theta JC} = 2\,°\text{C/W}$ and the thermal resistance $R_{\theta CS}$ from shell to radiator surface is not provided by the manufacturer. However, it can be known from Table 8.4 that for the TO-220 (including TO-220-7C) package, $R_{\theta CS}$ is equal to $1\,°\text{C/W}$ when a layer of heat conducting silicone grease is coated between the small radiating fin of device and the external cooling plate. Put $\overline{P_D} = 6.0\,\text{W}$, $T_{JM} = 125\,°\text{C}$, $T_{AM} = 40\,°\text{C}$, $R_{\theta JC} = 2\,°\text{C/W}$, and $R_{\theta CS} = 1\,°\text{C/W}$ into Formula (8.30),

$$R_{\theta SA} = \frac{125\,°\text{C} - 40\,°\text{C}}{6.0\,\text{W}} - (2\,°\text{C/W} + 1\,°\text{C/W}) = 14.2\,°\text{C/W}$$

Finally, it is found in Figure 8.46 that the surface area of aluminum cooling plate is $S \approx 30\,\text{cm}^2$. When 1/3 margin is reserved, the actual surface area of aluminum cooling plate will be $40\,\text{cm}^2$ and the external dimension can be $8\,\text{cm} \times 5\,\text{cm}$. Generally, the ratio between the length and width of aluminum cooling plate shall not exceed 2:1. After a layer of heat conducting silicone grease is coated on the contact surface between the device and the cooling plate, $R_{\theta JS} = 1\,°\text{C/W}$. This is because with the heat conducting silicone grease coated, the cooling plate and the device are closely fit to minimize the thermal resistance of the contact surface, not only improving the heat dissipation conditions, but also significantly reducing the area of cooling plate.

8.12 Radiator Design of Power Switching Tube (MOSFET)

8.12.1 Radiator Design Methods of Power Switching Tube

SMPS can drive the power switching tube (hereinafter referred to as MOSFET) through PWM controller. MOSFET is the most important power device of SMPS, of which the power consumption (P_D) includes two parts: transmission loss and switching losses. Transmission loss P_{DR} is caused by the on-state resistance of MOSFET, also known as resistance loss. Switching losses P_{DK} refer to the loss resulted from the release of electric energy stored in the output capacitance of MOSFET at the beginning of each switching cycle.

Switching losses include the capacitance loss of MOSFET and the switching overlapping loss. The capacitance loss here said is also known as CU^2f loss, mainly caused by the distributed capacitance of MOSFET. Relevant formula is as follows:

$$P_D = P_{DR} + P_{DK} \tag{8.34}$$

The following text describes the methods and procedures to design MOSFET radiator.

1. Calculating the on-state resistance $R_{DS\,(ON)}$ of MOSFET
   The temperature coefficient of MOSFET on-state resistance is $\alpha_{TR} = (0.35\text{–}0.85\%)/°C$, which can be approximately taken as $\alpha_{TR} = 0.5\%/°C$ under normal circumstances. Let the on-state resistance at room temperature ($T_A = 25\,°C$) be $R_{DS\,(ON)\ A}$. When the junction temperature rises to T_J, the on-state resistance becomes

$$R_{DS\,(ON)} = R_{DS\,(ON)\ A}[1 + 0.5\%\ (T_J - T_A)] \tag{8.35}$$

2. Calculating the transmission loss P_{DR} of MOSFET

$$P_{DR} = I_O^2 R_{DS\,(ON)} D_{max} = I_O^2 R_{DS\,(ON)} \frac{U_O}{U_I} \tag{8.36}$$

3. Calculating the switching losses P_{DK} of MOSFET
   The distributed capacitance of MOSFET mainly includes the reverse transfer capacitance C_{RSS}, the input capacitance C_{ISS}, and the output capacitance C_{OSS}, wherein, C_{RSS} has the maximum influence on the switching losses. Therefore, the formula calculating the switching losses can be simplified to

$$P_{DK} = \frac{I_O U_I^2 f C_{RSS}}{I_{GATE}} \tag{8.37}$$

   wherein, f is the switching frequency and I_{GATE} is the gate current at the critical conduction of MOSFET. These parameters can be found in the product data sheet.
4. Calculating the total power consumption P_D of MOSFET

$$P_D = P_{DR} + P_{DK} = I_O^2 R_{DS(ON)} \cdot \frac{U_O}{U_I} + \frac{I_O U_I^2 f C_{RSS}}{I_{GATE}} \tag{8.38}$$

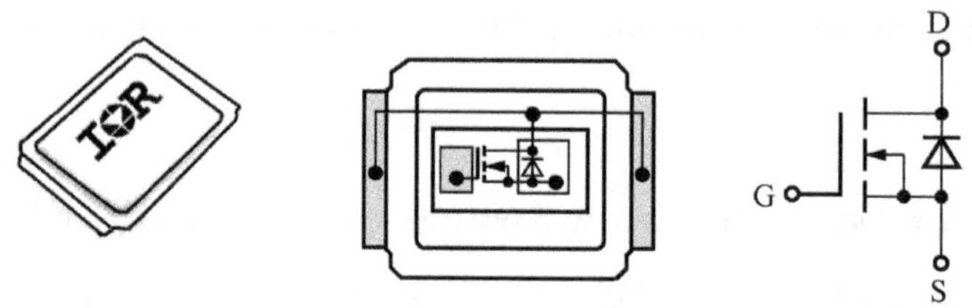

(a) (b) (c)

Figure 8.50 Outline and structure of IRF6631 N-channel MOSFET with high power. (a) Outline, (b) internal structure, and (c) circuit symbol

5. Calculating the thermal resistance $R_{\theta SA}$ of radiator

$$R_{\theta SA} = \frac{T_j - T_{AM}}{P_D} - R_{\theta JS} \tag{8.39}$$

wherein, T_{AM} is the actual maximum ambient temperature and $R_{\theta JS}$ is the thermal resistance from junction to radiator surface (findable in the product data sheet).

6. Manufacturing or purchasing a finished radiator according to the value of $R_{\theta SA}$

8.12.2 Radiator Design Example of Power Switching Tube

Assume that the input voltage of SMPS is $U_I = +10\,V$, the output is $U_O = +5\,V$, $I_O = 10\,A$, and the switching frequency is $f = 100\,kHz$. The IRF6631 N-channel MOSFET with high power produced by IR is adopted as the external power switching tube, of which the outline is shown in Figure 8.50. It has an internal protection diode and the main parameters are as follows: the on-state resistance at $T_A = 25\,°C$ is $R_{DS\,(ON)\ A} = 6.0\,m\Omega$, $C_{GD} = 1450\,pF$, $C_{GS} = 170\,pF$, $C_{DS} = 310\,pF$, and $I_{GATE} = 0.14\,A$. The maximum junction temperature of chip is $T_{JM} = 150\,°C$. For reasons of safety, the maximum operating junction temperature $T_J = 125\,°C$ is applied. The maximum ambient temperature $T_{AM} = 40\,°C$.

Radiator design steps of IRF6631 are as follows:

1. Put $R_{DS\,(ON)\ A} = 6.0\,m\Omega$, $T_J = 125\,°C$, and $T_A = 25\,°C$ into Formula (8.35) to get

$$\begin{aligned} R_{DS\,(ON)} &= R_{DS\,(ON)\ A}[1 + 0.5\%(T_J - T_A)] = 6.0\,m\Omega \times [1 + 0.5\%(125\,°C - 25\,°C)] \\ &= 9.0\,m\Omega(1) \end{aligned}$$

2. Put $I_O = 10\,A$, $R_{DS\,(ON)} = 9.0\,m\Omega$, $U_O = 5\,V$, and $U_I = 10\,V$ into Formula (8.36) to get the transmission loss of MOSFET

$$P_{DR} = I_O^2 R_{DS\,(ON)} \cdot \frac{U_O}{U_I} = (10\,A)^2 \times 9.0\,m\Omega \times \frac{5\,V}{10\,V} = 0.45\,W$$

3. Put $I_O = 10\,A$, $U_I = 10\,V$, $f = 100\,kHz$, $C_{GD} = 1450\,pF$, and $I_{GATE} = 0.14\,A$ into Formula (8.37) to get the switching losses of MOSFET

$$P_{DK} = \frac{I_O U_I^2 f C_{GD}}{I_{GATE}} = \frac{10\,A \times (10\,V)^2 \times 100\,kHz \times 1450\,pF}{0.14\,A} = 1.04\,W$$

4. The total power consumption of MOSFET is

$$P_D = P_{DR} + P_{DK} = 0.45\,W + 1.04\,W = 1.49\,W$$

5. Put $T_J = 125\,°C$, $T_{AM} = 40°C$, $P_D = 1.5\,W$, and $R_{\theta JS} = 1.4°C/W$ into Formula (8.39) to get the thermal resistance of radiator

$$R_{\theta SA} = \frac{T_j - T_{AM}}{P_D} - R_{\theta JS} = \frac{125\,°C - 40\,°C}{1.49\,W} - 1.4°C/W = 56\,°C/W$$

6. Finally, find the surface area $S = 620\,mm^2$ of PCB radiator corresponding to $R_{\theta SA} = 56\,°C$ in Figure 8.47 (roughly converted into 1 in.^{2}, 1 in.^{2} = 645 mm^2).

8.12.3 *Notes for Designing the Radiator of Power Switching Tube*

The following point shall be noted when designing the radiator of power switching tube (MOSFET):

1. The MOSFET applied in early SMPSs has a relatively high on-state resistance (is up to several or even above 10 Ω) and its switching frequency is only a few dozen kilohertz, causing the transmission loss to be far greater than the switching losses, that is, $P_{DR} \gg P_{DK}$. Therefore, the second term in Formula (8.38) is often negligible. The new type of MOSFET produced in recent years has the $R_{DS\,(ON)}$ of only several to tens of milliohms. For example, the MOSFET recently developed by IR has the $R_{DS\,(ON)}$ of only 2.6 mΩ, capable of withstanding 40–100 V voltage and 240 A current. Meanwhile, the switching frequency of SMPS is also increased to several hundred kilohertz, so the transmission loss is greatly reduced to significantly raise the share of switching losses in the total power consumption, even more than the transmission loss. In this case, the second term in Formula (8.38) cannot be ignored.
2. If the input voltage U_I is not constant, the power consumption of MOSFET shall be calculated, respectively, at the maximum input voltage $U_{I\,(max)}$ and the minimum input voltage $U_{I\,(min)}$. The maximum power consumption P_{DM} of MOSFET may occur at the minimum or maximum input voltage. This is because when $U_I = U_{I\,(min)}$, the duty ratio is the maximum and P_{DR} may reach the maximum. When $U_I = U_{I\,(max)}$, P_{DR} may reach the maximum under the influence of "U_I^2" in Formula (8.38). The reasonable design shall make P_{DM} basically constant in the two extreme cases of $U_{I(max)}$ and $U_{I\,(min)}$.
3. When P_{DM} significantly rises when $U_I = U_{I\,(min)}$, the transmission loss plays a dominant role. In this case, the MOSFET with higher power can be applied to reduce $R_{DS\,(ON)}$. When the power consumption is increased significantly when $U_I = U_{I\,(max)}$, the MOSFET with lower power shall be adopted to raise its switching speed.
4. If necessary, the switching frequency can also be reduced to diminish the switching losses.

Table 8.9 General troubleshooting

Failure phenomenon	Failure reason	Troubleshooting
Failing to work after power-on	The circuit is latched by the trigger pulse of switching current, so the device does not work	The single-point grounding method is introduced to directly connect the input filter capacitor, the bypass capacitor at control terminal, and the filter capacitor of bias winding to the source (S)
		For the single-chip SMPS with TO-220 package, the pin lead must be cut to the shortest length
		Improve the design of printed circuit and insulate the radiator with the circuit
	The reverse breakdown voltage U_B of drain clamping diode VD_{Z1} of the high-frequency transformer is too low	The 5 W transient voltage suppressor with relatively high U_B is adopted. Letting the primary induced voltage be U_{OR}, the selection principle is $U_B \approx 1.5U_{OR}$. Typically, $U_{OR} = 135$ V, P6KE200 TVS can be adopted, in which case, $U_B = 200$ V
	The polarity of bias winding is reverse	Change the polarity
	The output rectifier circuit is damaged	Replace the rectifier or filter capacitor
	The drain clamp protection circuit is damaged	Replace the clamping diode VD_{Z1}. VD_1 shall be SRD with the reverse recovery time t_{rr} less than 75 ns and the reverse withstand voltage of generally 600 V
		Replace the blocking diode VD_1
	During the initial trigger pulse latching period, the output current of optical coupler is too high	Increase the LED current limiting resistance of optical coupler
		The linear optical coupler with current transfer ratio of CTR = 50–200% is applied. Too low CTR will reduce the adjustment sensitivity, and two high CTR will be easy to cause incorrect shutdown during start-up or load mutation. 4N × × (e.g., 4 N25 and 4 N26) series optical couplers with switching characteristics shall not be adopted
		Connect a 270–620 Ω resistor in series at the emitter of optical coupler
	The shunt resistance at control terminal is too high, greater than 15 Ω	The shunt resistance can be 6.2 or 6.8 Ω. However, it sometimes can be increased to 15–100 Ω to eliminate the influence of auto-restart capacitor on the control loop. When the shunt resistance is greater than 15 Ω, a 0.1 μF bypass capacitor must be connected in parallel between the source and the control terminal

The device bursts at start-up or overload	The drain-source voltage is too high owing to insufficient clamp voltage or too high leakage inductance of high-frequency transformer, so the internal MOSFET suffers breakdown	Replace the transient voltage suppressor and blocking diode of appropriate model to ensure the drain voltage is clamped to below the safety voltage Redesign and manufacture the high-frequency transformer by adopting the multistrand parallel winding method for the secondary winding to reduce the leakage inductance. Appropriately reduce the primary induced voltage U_{OR} Replace the device
The device works at 50 kHz or lower harmonic	The clamping action is too strong	Weld the positive pole of VD_{Z1} directly on the primary winding
	The primary or secondary damp is too large, so the amplitude and frequency of switching signal are reduced	Reduce the primary or secondary damp, including the reduction of resistance or capacitance to reduce the time constant $\tau = RC$
	The turn-to-turn capacitance of primary winding of the high-frequency transformer is too high	Rewind the primary winding and add insulating layers between the layers to reduce the turn-to-turn capacitance. In addition, properly reduce the turns of primary winding
The transient voltage suppressor VD_{Z1} suffers overheating	The U_B of VD_{Z1} is too low	Replace the clamping diode to let $U_B = 1.5U_{OR}$
	The heat dispersion of VD_{Z1} is poor	Reduce the pin length of VD_{Z1} (cut off the excess part) Properly increase the width of printed conductor
	The switching speed of blocking diode VD_2 is too low	Replace the FRD VD_2 with SRD
	The wattage of VD_{Z1} is too small	A 5 W TVS is adopted and a 0.01 μF/200 V capacitor is connected in parallel to VD_{Z1}
	The leakage inductance of high-frequency transformer is relatively high	Rewind the high-frequency transformer to reduce the leakage inductance

Table 8.9 (*Continued*)

Failure phenomenon	Failure reason	Troubleshooting
The device suffers overheating	The inductance of primary winding of the high-frequency transformer is too small, so I_{RMS} is increased	Increase the inductance of primary winding Simultaneously increase the turn ratio and the inductance of primary winding
	The heat dispersion of the device is poor	Increase the radiator size and coat a layer of heat conducting silicone grease between the small cooling plate of device and the external radiator to reduce the thermal resistance
	The output current and power of the device are too low	The device with higher power is applied
When the AC input voltage rises or the load becomes lighter, the output voltage is increased	The output voltage of bias winding is too low	Increase the output voltage of bias winding to 30 V (RMS)
	No dummy load is connected to the output terminal	Connect dummy load (i.e., the minimum load resistance) to the output terminal to reduce the no-load voltage

Table 8.10 Troubleshooting of basic feedback circuit

Failure phenomenon	Failure reason	Troubleshooting
The output voltage is unstable	The gain/phase margin is insufficient	The device with higher power is introduced to improve the output capability Increase the capacitance of auto-restart capacitor
The output voltage changes too fast or exceeds the limit	The regulating speed of control loop is too slow	Adjust the components of R and C in the control loop to increase the bandwidth of control loop Connect a 22 μF soft-start capacitor in parallel to the regulator in the feedback circuit
The load regulation ability is poor	Spike pulse is generated owing to the leakage inductance of bias winding	Add the RC filter in the bias winding to filter out spike pulses Change the winding order of high-frequency transformer to wind the primary winding N_P first, then the secondary winding N_S, and finally the bias winding N_B. Wind N_B with multistrand parallel thick enameled wire to completely cover the winding surface, thus minimizing the leakage inductance
	The device works in uncontinuous mode	Redesign the high-frequency transformer by properly increasing the inductance of primary winding to let the device work in continuous mode

8.13 Common Troubleshooting Methods of SMPS

SMPS failures are mostly resulted from improper circuit design, unreasonable component selection, error welding or installation, serious EMI, unreasonable PCB layout, and poor heat dispersion, which relate to both the technical reasons and workmanship problems. Table 8.9 comprehensively and thoroughly analyzes the phenomena, reasons, and remedies of common single-chip SMPS failures. In order to facilitate the description, the common troubleshooting of three-terminal single-chip SMPS (hereinafter referred to as the device) is divided into three categories: (i) general troubleshooting (see Table 8.9), (ii) troubleshooting of basic feedback circuit (see Table 8.10), and (iii) troubleshooting of optocoupler feedback circuit with TL431 (see Table 8.11).

Table 8.11 Troubleshooting of optocoupler feedback circuit

Failure phenomenon	Failure reason	Troubleshooting
The output voltage U_O is unstable	The output-stage pi filter causes phase shift, resulting in insufficient gain/phase margin	Directly connect the optical coupler to the input terminal of pi filter to reduce the regulation lag
	The design of compensation circuit is unreasonable	Increase the resistance of LED current limiting resistor Increase the capacitance of bypass capacitor
	The CTR of optical coupler is too low or too high and the linearity is poor	Replace the optical coupler of 4N × × series with the linear optical coupler and CTR = 50–200% is adopted
The output voltage of secondary winding changes too fast or exceeds the limit	The control loop has too long response time and too slow regulating speed	Increase the bandwidth of the control loop Properly reduce the capacitance of soft-start capacitor. The capacitor is connected in parallel between the cathode and anode of TL431 with the capacitance of generally 4.7–47 μF, capable of not only completing the soft-start function but also suppressing the spike voltage in the optocoupler feedback circuit
The output voltage ripple of secondary winding is too large	The bandwidth of control loop is too narrow	Increase the bandwidth of the control loop
	The capacity of filter capacitor is too small	Increase the capacity of the filter capacitor

9

SMPS Testing Technology

9.1 Parameter Testing of SMPS

9.1.1 Testing Methods of Main SMPS Parameters

See Figure 9.1 for the test circuit of SMPS, wherein, *T* is the autotransformer, *S* the switch used for no-load test, and R_L the adjustable load. The circuit uses one standard AC voltmeter, DC voltmeter, and DC ammeter (amperemeter or milliammeter), respectively. To improve the measurement accuracy, the above-mentioned meters can also be replaced with the calibrated digital voltmeter (DVM) and ammeter.

9.1.1.1 Measuring the Output Voltage Accuracy

The accuracy of output voltage is also known as precision, which indicates the relative error between the actual output voltage and the nominal output voltage. Apply the nominal input voltage and rated load to SMPS and measure the actual output voltage U'_O with the DC voltmeter. Compare U'_O with the nominal output voltage U_O and then calculate the accuracy of the output voltage according to the following formula:

$$\gamma_V = \frac{U'_O - U_O}{U_O} \times 100\% \tag{9.1}$$

9.1.1.2 Measuring the Voltage Regulation

Voltage regulation (S_V) is also known as linear regulation, which is generally expressed as a percentage. It refers to the change in the rate of output voltage when the input voltage changes within the specified range. The method to measure the voltage regulation is as follows: apply the rated load to SMPS to measure the output voltage U'_O at the nominal input voltage. Then, adjust the AC input voltage *u* continuously to let it vary from the specified maximum value (u_{min}) to the maximum value (u_{max}). Write down the maximum deviation $\Delta U'_O$ between the output voltage and the nominal value and finally put it into the following formula:

Optimal Design of Switching Power Supply, First Edition. Zhanyou Sha, Xiaojun Wang, Yanpeng Wang, and Hongtao Ma.

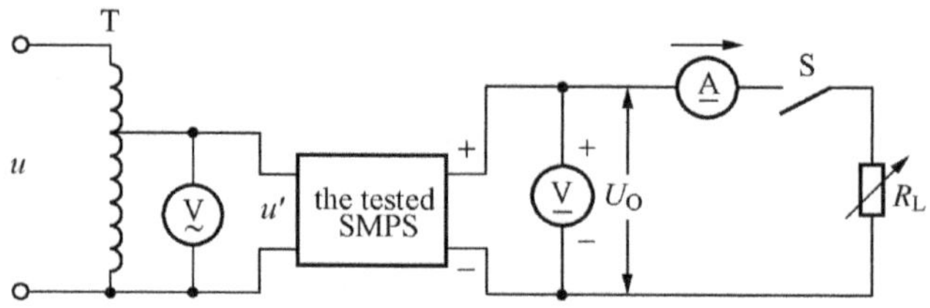

Figure 9.1 Test circuit of SMPS

$$S_V = \frac{\Delta U'_O}{U'_O} \times 100\% \tag{9.2}$$

9.1.1.3 Measuring the Load Regulation

Load regulation (S_I) is also known as current regulation, which is generally expressed as a percentage. It indicates the ability of SMPS to keep the output voltage constant when the load current varies. The method to measure the load regulation is as follows: adjust the input voltage to the nominal value to measure the output voltages U_1 and U_2 of SMPS at full load and no load, respectively. Then, put them into the following formula to calculate the load regulation:

$$S_I = \frac{U_2 - U_1}{U_O} \times 100\% \tag{9.3}$$

It shall be noted that the load regulation of SMPS is usually measured when I_O varies from 10% to 100% of the full load. In this case, U_2 in Formula (9.3) shall be replaced with the output voltage at $I_O = 10\% I_{OM}$.

9.1.1.4 Measuring the Output Resistance

The output resistance of SMPS is also known as equivalent internal resistance. It is equal to the ratio (absolute value) between the variance ΔU_O of the output voltage and the variance ΔI_L of the load current at the rated mains voltage. The relevant formula is as follows:

$$R_O = |\Delta U_O / \Delta I_L| \tag{9.4}$$

9.1.1.5 Measuring the Output Ripple

Output Ripple Voltage

It refers to the ripple voltage between the output terminals and is synchronized with the grid and switching frequency. The output ripple voltage of SMPS is usually expressed as the peak-to-peak value instead of the average. It belongs to the high-frequency narrow pulse, of which the average may be only a few millivolts when the peak-to-peak value is relatively high (e.g., ±60 mV), so the peak-to-peak value is more representative.

Output Noise Voltage
It refers to the noise voltage between the output terminals and varies randomly and is also expressed as the peak-to-peak value.

Output Ripple Noise Voltage
It refers to the sum of the output ripple voltage and the noise voltage at the rated output voltage and the load current, also known as the maximum ripple voltage. When measuring the ripple voltage containing high-frequency components, the oscilloscope of 20 MHz bandwidth is recommended to observe the peak-to-peak value. In order to prevent the radiated noise of SMPS from being introduced from the ground clamp of oscilloprobe, it is recommended to use shielding or twisted pair line as the intermediate connection wire to let the oscilloscope be far away from the SMPS.

Ripple Factor
It refers to the percentage ratio between the RMS U_{RMS} of the output ripple voltage and the output DC voltage U_O at the rated load current. The relevant formula is as follows:

$$\gamma = \frac{U_{RMS}}{U_O} \times 100\% \tag{9.5}$$

Ripple Rejection Ratio
It refers to the ratio (expressed by common logarithm) between the peak-to-peak value U_{RI} of the input ripple voltage and the peak-to-peak value U_{RO} of the output ripple voltage at the specified ripple frequency (e.g., 50 Hz). The relevant formula is as follows:

$$\text{PSRR} = 20\ \lg(U_{RI}/U_{RO}) \tag{9.6}$$

9.1.2 Power Measurement Technology

In order to calculate and analyze the efficiency of SMPS, various power parameters must be measured accurately, including the AC input power and the power losses and the total power consumption of the components. There are mainly three methods to measure these power parameters: the method of direct power measurement with wattmeter, the power calculation method by measuring the voltage and current, and the DC heat equivalent method, which are described, respectively, in the following.

9.1.2.1 Direct Power Measurement Method

Ordinary AC RMS meters are not suitable for measuring the power parameters of SMPS. These meters are only suitable for measuring undistorted sine wave signal, but the SMPS has a variety of high-frequency nonsine waves and transient interferences, such as the 100 kHz square wave, sawtooth wave, switching distorted waveform, AC ripple, higher harmonics, and electromagnetic interference (EMI) signals of spike voltage, ringing voltage, audio noise, transient voltage from the grid, and so on. The crest factors of nonsine waves are greater than 1.414 (crest factor of sine wave).

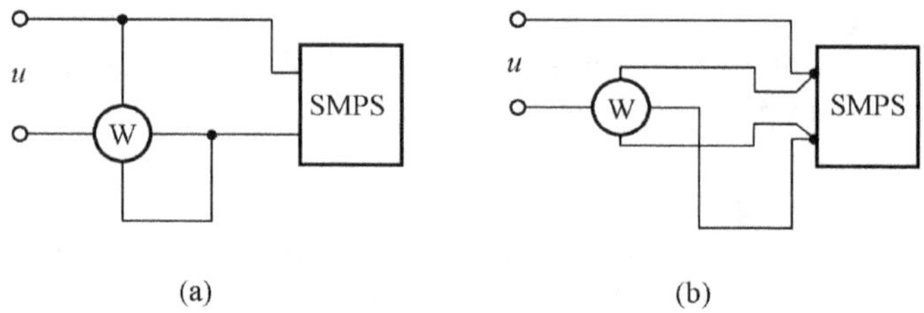

Figure 9.2 Measuring the input power with wattmeter. (a) Wrong connection and (b) right connection

Crest factor (K_P) is equal to the ratio between the peak voltage (U_P) and the RMS voltage (U_{RMS}). The relevant formula is as follows:

$$K_P = \frac{U_P}{U_{RMS}} \tag{9.7}$$

Wattmeter can be used to directly measure the AC input power and various power losses. For nonsine waves with $K_P \geq 3$ (e.g., narrow pulse), the wattmeter is also suitable.

The input filter capacitor of SMPS will cause severe distortion of AC input current waveform because the crest factor is relatively high and the power factor is relatively low. The typical power factor of SMPS is $\cos\phi = 0.6$–0.8, which depends on the impedance and AC input voltage of the AC line.

When measuring the input power with wattmeter, the circuit shall be connected as per Figure 9.2(b). When measuring the voltage, try to use the cross-connection for it at AC input terminal of SMPS, or 1–2% measurement error will be caused by the voltage drop of the power lead as shown in Figure 9.2(a). If there is no wattmeter, the AC input voltage u can also be replaced with the DC high voltage, directly applied to the AC input terminal. In this way, ordinary DC voltmeter and ammeter can be used to measure the DC input power. Most SMPSs can work properly under the two input modes of AC and DC. However, if electric fan and mains transformer are connected in parallel on the AC power line, such electric fan and mains transformer must be disconnected first and then the DC high voltage applied. Otherwise, when the DC high voltage is applied, very high short-circuit current will form in the winding of the electric fan or the transformer, and it will be easy to burn the electric fan and high-frequency transformer. In addition, when the AC input is replaced with the DC input, the power value measured will be higher by 1–2%. The first reason is that the voltage drop of the component is relatively low in case of DC input. The second reason is that the filter capacitor now has no grid frequency fluctuation with the power consumption lower than the normal value. This indicates that the DC input method can only obtain the approximate value of the power supply efficiency.

When no ready DC high voltage is available at hand, simple bridge rectification and filter method can also be used to convert the alternating current into DC high voltage. For example, approximately 300 V DC high voltage can be obtained after rectifying and filtering the 220 V alternating current. Two points shall be noted during the measurement: first, the fluctuation of alternating current shall be as small as possible. Second, this method does not adopt any line frequency transformer or grid isolation, so attention must be paid to the safety.

9.1.2.2 Calculation Method

The power loss of components can also be obtained by measuring the voltage and the current. This method is suitable for measuring the power loss of the following components: the input filter capacitor C_1, the output filter capacitor C_2, the output filter inductor L_1, the high-frequency transformer T, the feedback winding, and the feedback circuit.

Digital storage oscilloscope is capable of functional operation, which can directly calculate and display the average power from the voltage and current waveforms. However, it shall be noted that some oscilloprobes have 50 ns delay time, which will cause error in power loss measurement. In this case, it is recommended to use the DC heat equivalent method.

9.1.2.3 DC Heat Equivalent Method

This method is very useful in obtaining the approximate value of power loss, especially for the power devices such as power switching tube and output rectifier, which have similar transmission loss, reverse recovery time, switching loss, and other issues. When using this method, first measure the temperature rise of the device at normal work. Then, let DC current flow through the device to generate the same temperature rise. Finally, calculate the average power loss of the given device according to the measured DC voltage and DC current. The measurement principle of DC heat equivalent method is using the direct current to simulate the thermal effect of alternating current to convert the AC measurement into the measurement of thermodynamic temperature and pure DC parameters. It is an indirect measurement.

The DC heat equivalent method is suitable for measuring the power loss of the bridge rectifier, TOPSwitch chip, blocking diode VD_1 of clamp protection circuit, output rectifier VD_2, regulator VD_Z, and common-mode choke L_2 of EMI filter.

9.2 Performance Testing of SMPS

Performance testing of SMPS is the important basis for the evaluation of its technical level and advanced level of process and quality. The following section describes the conventional testing methods of SMPS and the electrical performance testing methods of high-frequency transformer.

9.2.1 Testing Methods of Main SMPS Parameters

The instruments required for the routine testing of small- and medium-power SMPS mainly include

1. One 3½ digit AC and DC DVM (e.g., TD 1915) or one 3½ digit digital multimeter.
2. One 3½ digit digital multimeter (e.g., VC890D and VC9808A+).
3. One pointer multimeter, for example, the 500-type multimeter.
4. One Level 0.5 DC ammeter, for example, C31-A.
5. 0–250 V and 0–2 A wattmeter, used to measure the AC input power.
6. One 0–250 V and 0.5 kVA auto voltage regulator, for example, TDGC2-0.5.

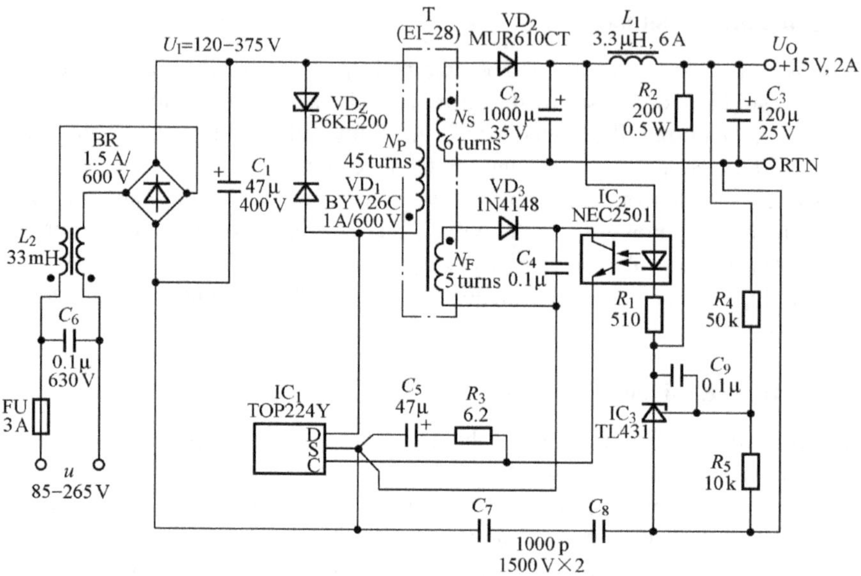

Figure 9.3 Circuit of 15 V and 30 W SMPS

7. One piece of 1 kW and 100 Ω resistance wire, used as the dummy load R_L of SMPS. It can be replaced with the electric furnace heating wire only under amateur conditions, but the measurement time shall be as short as possible to avoid the resistance of electric furnace heating wire changes owing to heating.

The following text illustrates the testing methods of SMPS regulation performance with examples.

See Figure 9.3 for the circuit of TOP224Y 15 V and 30 W SMPS, wherein the maximum output current $I_{OM} = 2$ A. It uses one TOP224Y SMPS, which is configured with TL431 adjustable precision shunt regulator and NEC2501 linear optical coupler to constitute the optocoupler feedback circuit of external error amplifier to let its voltage regulation $S_V = \pm 0.2\%$, the load regulation ($I_O = 10$–$100\% I_{OM}$) $S_I = \pm 0.65\%$, and the power supply efficiency be 84%. 120–375 V DC high voltage will be generated after the rectification and filtering of 85–265 V AC input voltage. The drain clamp protection circuit constituted by VD_Z and VD_1 can suppress the spike voltage caused by the leakage inductance. The output voltage of the secondary winding becomes the DC voltage after the rectification of VD_2 and the filtering of C_2, of which the high-frequency ripple voltage will be filtered out by L_1 and C_3. R_2 can improve the load regulation at light loads. The bias voltage of TOP224Y will be generated after the rectification and filtering of the feedback winding voltage, respectively, by VD_3 and C_4. The sampling voltage obtained through the voltage division of output voltage U_O by R_4 and R_5 will be compared with the internal 2.50 V reference voltage of TL431 to provide the error voltage signal output, which flows into the control terminal of TOP224Y through NEC2501 to directly control the output duty ratio, thus obtaining stable voltage output. The nominal value of output voltage can be adjusted by changing the voltage division ratio between resistors R_4 and R_5. R_3 and C_5 are used to improve the frequency response of

Table 9.1 Measured input characteristic data

u (V, AC)	60	85	100	120	150	180	200	220	245
U_O (V)	15.10	15.22	15.38	15.39	15.48	15.48	15.48	15.48	15.48
I_O (A)	2.01	2.02	2.05	2.06	2.06	2.06	2.06	2.06	2.06
P_O (W)	30.3	30.7	31.5	31.5	31.9	31.9	31.9	31.9	31.9
S_V (%)	−2.4	−1.6	−0.65	−0.58	< 0.1	< 0.1	< 0.1	< 0.1	< 0.1

Table 9.2 Measured output characteristic data

R_L (Ω)	Open circuit	97.4	36.2	24.7	19.5	14.8	8.6	8.1	6.0
U_O (V)	15.59	15.59	15.57	15.56	15.56	15.54	15.50	15.49	15.46
I_O (A)	—	0.16	0.43	0.63	0.80	1.05	1.80	1.92	2.57
P_O (W)	—	2.49	6.70	9.80	12.45	16.32	27.90	29.74	39.73
S_I (%)	—	+0.65	+0.52	+0.45	+0.45	+0.32	+0.06	0	−0.19

the control loop. C_7 and C_8 are safety capacitors, which can filter out the common-mode interference caused by the coupling capacitance of the primary and secondary windings after connected in series.

See Tables 9.1 and 9.2, respectively, for the measured input and output characteristics of SMPS. In addition, it is measured that the output ripple voltage is 37 mV. It can be known by analyzing the measured data that $S_V < 1.6\%$ within the wide range of $u = 85$–245 V (AC). When $u = 150$–245 V (AC), the calculated value of the voltage regulation has been reduced to zero. Taking into account the accuracy and the resolution limit of the measured instrument, it can be actually regarded that $S_V < 0.1\%$. When the load current changes from 10 to 100% I_{OM}, $S_I < 0.65\%$ and the actual maximum output current $I_{OM}' = 2.57$ A > 2 A. It is also measured that the power supply efficiency, $\eta = 85\%$. The measured results show that the SMPS fits the design indexes.

9.2.2 *Electrical Performance Testing Methods of High-Frequency Transformer*

9.2.2.1 Voltage Resistance Test

Apply 3000 V and 50 Hz high voltage among the primary, feedback, and the secondary windings of high-frequency transformer for 1 min without any arc discharge or breakdown.

9.2.2.2 Measuring the Inductance of the Primary Winding

Let the secondary and the feedback windings be in open circuit to measure the inductance L_P at both ends of the primary winding with digital inductance meter, with 10% error allowed. The efficiency of SMPS can be improved by properly increasing the primary inductance.

9.2.2.3 Measuring the Leakage Inductance of the Primary Winding

Let the secondary winding be in short circuit to measure the leakage inductance L_{P0} at both ends of the primary winding, which shall be less than a few dozen microhenries.

9.2.2.4 Measuring the Secondary Leakage Inductance and Total Primary Leakage Inductance of PCB

The secondary leakage inductance (L_{S0}) of PCB is also known as the secondary trace inductance with the inductance of about 20–40 nH. Although L_{S0} is very small, it will increase the total primary leakage inductance as follows after reflected to the primary circuit according to the relationship of the square of turn ratio (n^2):

$$L_{PS0} = L_{P0} + n^2 L_{S0} \tag{9.8}$$

To measure the total primary leakage inductance L_{PS0} of SMPS, the high-frequency transformer must be welded on the PCB for online measurement. In this case, let the output rectifier (VD_2) and the output filter capacitor (C_2) be in short circuit, respectively, with two thick wires rather than directly let the secondary winding be in short circuit. Then, measure the leakage inductance at both the ends of the primary winding, which is the real total online equivalent leakage inductance L_{PS0}. The following text illustrates the method to calculate L_{S0} with an example.

Use TOP245 to design an 85–265 V AC SMPS with the input and output of 5 V and 45 W. It is known that the turn ratio of the primary and secondary windings is $n = 25$. The L_{P0} of the high-frequency transformer measured separately is equal to 17 μH. It is measured that $L_{PS0} = 36.7$ μH by welding the high-frequency transformer on PCB and letting VD_2 and C_2 be in short circuit. According to Formula (9.8),

$$L_{S0} = \frac{L_{PS0} - L_{P0}}{n^2} \tag{9.9}$$

Put the data into Formula (9.9) to get $L_{S0} = 31.5\text{nH}$.

9.3 SMPS Measurement Skills

The following section describes the skills to measure the input current, output ripple voltage, switching regulator efficiency, and the load power of SMPS (including the switching regulator), respectively.

9.3.1 Using the Noncontact Method to Measure the Input Current of SMPS

The AC input current of SMPS has pulse waveform owing to the influence of the input rectifier filter, so the ordinary AC DVM (including the AC voltage gear of digital multimeter) cannot be used, which belongs to the mean instrument, designed based on the undistorted sine wave. The true RMS digital ammeter has the advantage incomparable for the mean DVM, which can accurately measure the RMS of various current waveforms in real time without

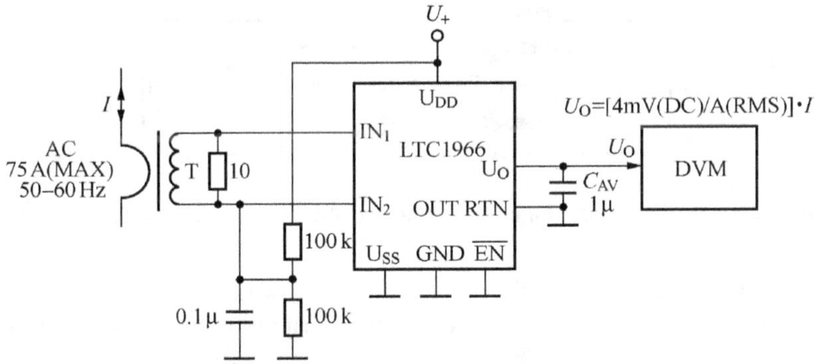

Figure 9.4 Circuit using +5 V single power supply and differential-input TRMS digital ammeter

considering the waveform type and the distortion degree. The following text describes the noncontact method measuring the input current of the SMPS with the AC transformer and the true RMS/DC (TRMS/DC) converter.

LTC1966 is the precision broadband TRMS/DC converter developed by LT. It uses the brand-new Δ-$\sum$ computing technology to further improve the accuracy and the linearity indicators. Its conversion gain error within the frequency range of 50 Hz to 1 kHz is 0.1% and the total error is 0.25%. Its linearity is up to 0.02%. It has the rail-to-rail output, with the output voltage amplitude equal to the mains voltage. It has ultra-low power consumption and uses +2.7 to 5.5 V single power supply or ±5.5 V dual power supply. The typical value of the supply current is only 155 μA with the maximum below 170 μA. The supply current under standby mode is as low as 0.1 μA. The ranges of the single and the dual power supplies required by the similar product AD536A are +5 to +36 V and ±3 to ±18 V, respectively. LTC1966 can complete TRMS/DC conversion just with an external capacitor, suitable for measuring various AC RMS voltages with the crest factor of $K_P \leq 4$.

See Figure 9.4 for the circuit using +5 V single power supply and differential-input TRMS digital ammeter. The circuit uses the CR8348-2500-N current transformer produced by CR MAGNETICS, which belongs to the PCB-mount current transformer with high transformation ratio, suitable for the noncontact measurement method of SMPS input current. The maximum RMS current is 75 A. The measured AC current frequency range is 50–400 Hz and the operating temperature range is −25 to +66 °C. The output voltage sensitivity of the instrument is 4 mV (DC)/A (RMS), so the measured current value can be directly displayed with DVM.

9.3.2 Methods to Accurately Measure the Output Ripple Voltage

Output ripple voltage is an important indicator for SMPS. However, it is not easy to accurately measure the output ripple voltage. This is because the high-speed switching circuit in SMPS generates the spike pulse voltage and this common-mode interference signal will generate a lot of energy after overlapped on the output ripple voltage. Therefore, this kind of noise must be eliminated. The following text describes the measurement method that can eliminate common-mode noise interference.

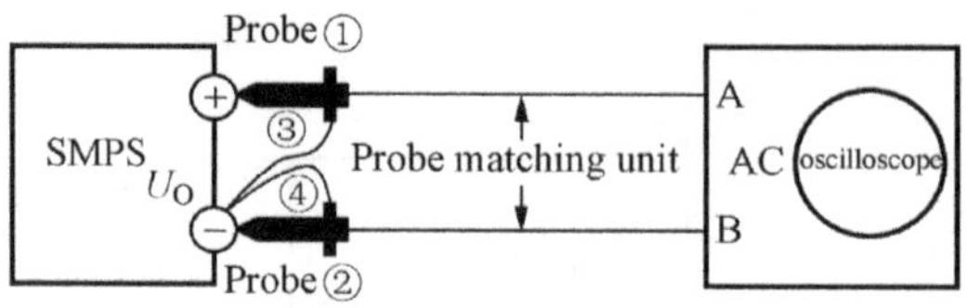

Figure 9.5 Using differential-input method to accurately measure the output ripple voltage

When observing the waveform of the output voltage U_O with oscilloscope, differential-input method can be used to accurately measure the output ripple voltage, the circuit of which is shown in Figure 9.5. The two AC input channels A and B of oscilloscope are connected to probe ① and ②, respectively, through the probe matching unit. Probe ① is connected to the positive terminal of U_O and probe ② to the negative terminal of U_O. The ground wires of the two probes are expressed as ③ and ④, respectively.

When using this method, probes ① and ② are used to pick up the ripple voltage, and ground wires ③ and ④ can, respectively, obtain the spike pulse voltages. In view of the differential-input method, the spike pulse voltages obtained in ③ and ④ will be equal with opposite phase as long as the two probes are matched properly. After the offset of the spike pulse voltages, the waveform observed from the oscilloscope is the output ripple voltage waveform with the common-mode interference eliminated.

When the RMS of the output ripple voltage has to be measured, an AC DVM can be connected at the input terminal of the oscilloscope. In order to deduct the noise voltage from the measurement, first connect the positive input terminal of AC DVM to the negative terminal of U_O and write down the reading 1. Then, connect to the positive terminal of U_O and write down the reading 2. The difference between the two readings is the RMS of output ripple voltage.

The following point shall be noted during measurement:

1. The waveform of the output ripple voltage is nonsine wave, so the TRMS DVM must be used rather than the mean-response AC DVM. The AC voltage gear of the ordinary digital multimeter belongs to the mean instrument, which is designed according to the exact relationship ($U_{RMS} = 1.111\overline{U}$) between the RMS and the average of the sine wave. Therefore, it is only suitable for measuring the undistorted sine wave, incapable of measuring the severely distorted sine wave, letting alone nonsine waves such as the square, rectangular, sawtooth, and noise wave. RMS voltages of various waveforms can be accurately measured in real time with the TRMS meter. The so-called TRMS refers to the true root mean square. It is calculated based on the RMS definition formula, so it is called TRMS. TRMS DVM is featured by high accuracy, fast response, and wide measuring surface.
2. The −3 dB bandwidth of AC DVM shall be at least three times of the measured signal bandwidth.

9.3.3 Method to Measure the Efficiency of Switching Regulator

The total power P of the switching regulator (i.e., DC/DC converter) is equal to the sum of the output power P_O and the total power consumption P_D:

$$P = P_O + P_D \tag{9.10}$$

Therefore, the conversion efficiency of switching regulator is

$$\eta = \frac{P_O}{P_O + P_D} = \frac{P_O}{P} \times 100\% \tag{9.11}$$

To determine the conversion efficiency, the output power and the total power shall be measured. The output voltage and the output current are DC, so the method to measure the output power (i.e., load power) is very simple. The formula is

$$P_O = U_O I_O \tag{9.12}$$

If the DC input voltage and the input current are U_I and I_I, respectively, the total power will be

$$P = U_I I_I \tag{9.13}$$

It shall be noted that I_I in Formula (9.13) is neither pure DC current nor sine wave current, instead it refers to the average current. Therefore, it is not easy to accurately measure the input power. Although the input voltage is DC, the operating current of switching regulator is DC pulsating current, which will cause relatively large error if measured with DC ammeter. However, it is not the sine wave AC current, so it cannot be measured with clip-on ammeter, which will deliver meaningless measurement result.

9.3.3.1 Input Power Estimation Method

The input power estimation method is drawing a horizontal line in the input current waveform to estimate the average input current. First, connect the bandwidth probe of the oscilloscope in series at the input terminal of switching regulator to display the input current waveform on the oscilloscope. Then, draw a horizontal line in the current waveform diagram to let the waveform area (A) above the horizontal line be equal to the "missing" waveform area (B) below the horizontal line as shown in Figure 9.6. The current "averaged" is equal to the DC current capable of generating the same input power.

9.3.3.2 Accurate Measurement Method of Input Power

To accurately measure the input power, the LC filter constituted by L_1, C_1, L_2, and C_2 shall be inserted between the DC power supply and the switching regulator as shown in Figure 9.7. There is no high-frequency switching device between the DC power supply and the switching regulator, so the DC current from the DC power supply is the average input current I_I of the switching regulator as long as the filter is appropriate, which can be accurately measured with DC ammeter. Connect the DVM to the input terminal of the switching regulator to measure the

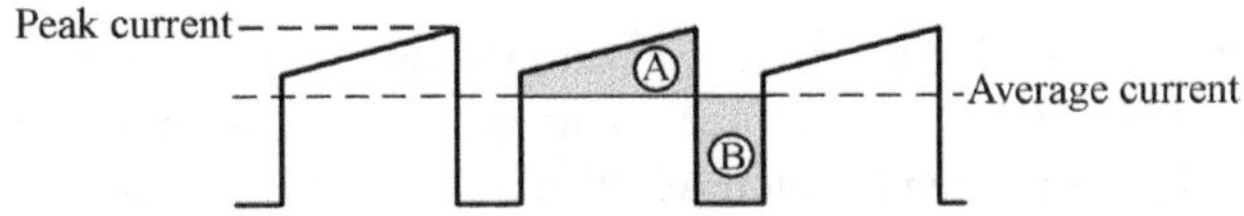

Figure 9.6 Averaging method of input current waveform

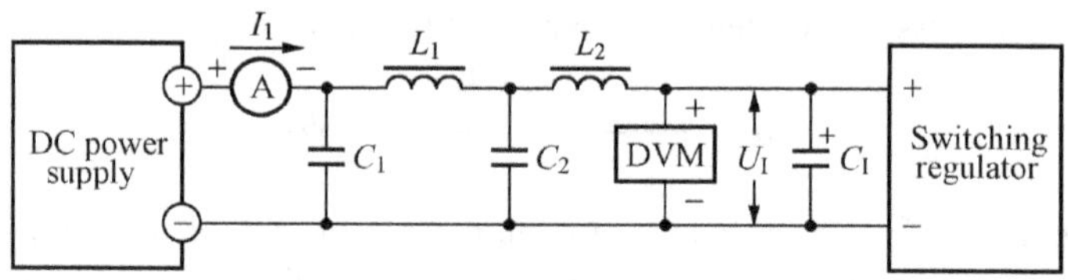

Figure 9.7 Inserting LC filter between the DC power supply and the switching regulator

value of U_I, thus calculating the input power and the conversion efficiency. It shall be noted that from the perspective of the switching regulator, the LC filter shows high impedance for the switching current at the input terminal of switching regulator, so the switching current is not directly gotten from the output terminal of the DC power supply, but provided by C_I. C_I shall be the capacitor with large capacity and low ESR.

9.3.4 Simple Method to Measure the Input Power of Isolated AC SMPS

It is very difficult to accurately measure the total power of the AC power supply owing to the influence of the power factor. In an AC circuit, the power factor is defined as λ, which refers to the cosine of the phase difference between the voltage and the current, equal to the ratio between the active power P and the apparent power S, that is,

$$\lambda = \cos\phi = P/S \tag{9.14}$$

In the isolated AC SMPS (i.e., isolated AC/DC converter), the input filter capacitor C_I is the reason for nonsine wave of the input current. Therefore, the power factor shall be redefined. For converter with input filter capacitor, the power factor is defined as

$$\lambda = P_1/S \tag{9.15}$$

wherein P_1 is the actual power, which is equal to the sum of the load power and the total power consumption in the power supply (with the unit of W). The apparent power S is equal to the input current RMS multiplied with the input voltage RMS (with the unit of VA). Therefore, the formula to calculate the power factor becomes

$$\lambda = \frac{P_1}{U_I I_I} \tag{9.16}$$

wherein U_I and I_I in the formula are RMSs. The typical power factors in case of single-phase AC power supply and three-phase power supply are, respectively, about 0.6 and 0.9. For the given actual power, the AC grid must increase the provided input current RMS owing to the reduction of power factor.

See Figure 9.8 for the simple method to measure the input power of the isolated AC SMPS. It uses the bridge rectifier and the input filter capacitor to get the unregulated DC voltage, so as to measure the actual power of the isolated AC SMPS. The input power is measured behind the input bridge rectifier and the input filter capacitor and the current from C_I is pulsating direct current (expressed as I_{DC} in the figure), so it is easy to measure the average current with DC

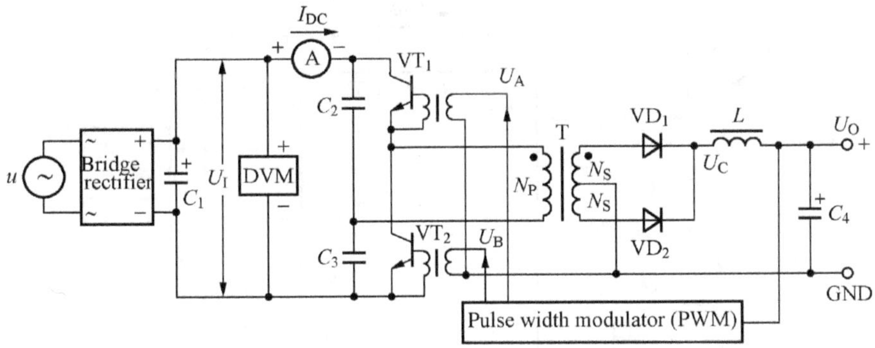

Figure 9.8 A simple method to measure the input power of isolated AC SMPS

current. U_I can be measured with DVM. The total power of the isolated AC SMPS is equal to the sum of the power ($U_I I_{DC}$) provided by C_I and the power consumption (P_Q) of the bridge rectifier. The formula is as follows:

$$P_1 = U_I I_{DC} + P_Q \tag{9.17}$$

The power consumption of the bridge rectifier accounts for a small proportion, so it is generally negligible.

9.3.5 Method to Measure the Load Power of SMPS

The load power of SMPS is equal to the load voltage multiplied with the load current. The following text describes the digital DC power meter constituted by the DC power detector MAX4211, suitable for measuring the load power and power level, constituting the intelligent battery pack/charger and the peripheral controller of intelligent computer, and so on.

MAX4211 uses the precision current-sense amplifier to measure the load current and then calculates the power with analog multiplier, so the ground path of the load is not affected, especially suitable for measuring the power and current of battery-powered system. Max4211 integrates a high-side current-sense amplifier, the 1.21 V band-gap reference voltage source, and two open-drain output comparators. The maximum errors of the detection power and current are both less than ±1.5%, and the frequency bandwidth is 220 kHz. The range of the measured load voltage is +4 to 28 V. The full-scale voltage during current sense is 100 or 150 mV. The range of the mains voltage is +2.7 to 5.5 V, and the operating current is 670 μA (typical). The operating temperature range is −40 to +85 °C and the chip surface area is 4 mm × 4 mm only.

The resistance divider in MAX4211A/B/C is connected to the terminal RS+ and the input terminal of analog amplifier. This design can accurately measure the source power and protect the power supply (e.g., battery). See Figure 9.9 for the typical application circuit of MAX4211A/B/C, wherein U_{PULLUP} is the pull-up power supply of comparators 1 and 2. R_{SENSE} is the current-sense resistor. In ideal cases, the maximum load current generates the full-scale detection voltage at both ends of R_{SENSE}. Select proper gain to let the current-sense amplifier obtain the maximum output voltage without saturation. When calculating the

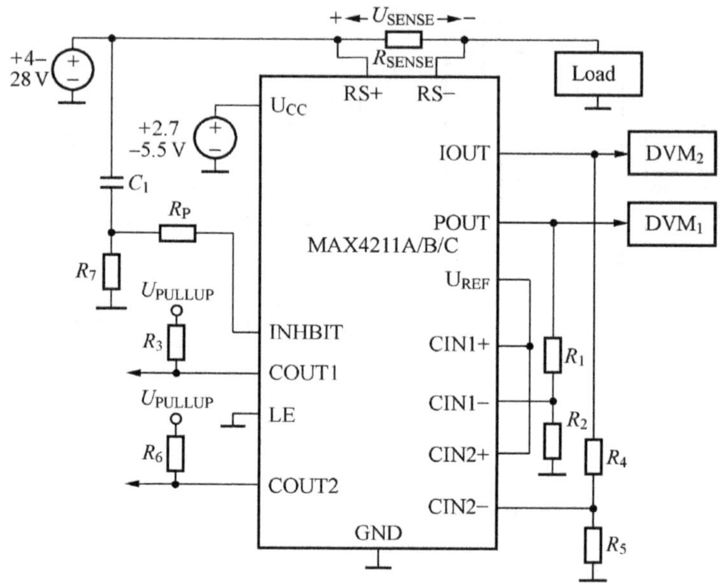

Figure 9.9 Typical application circuit of MAX4211A/B/C

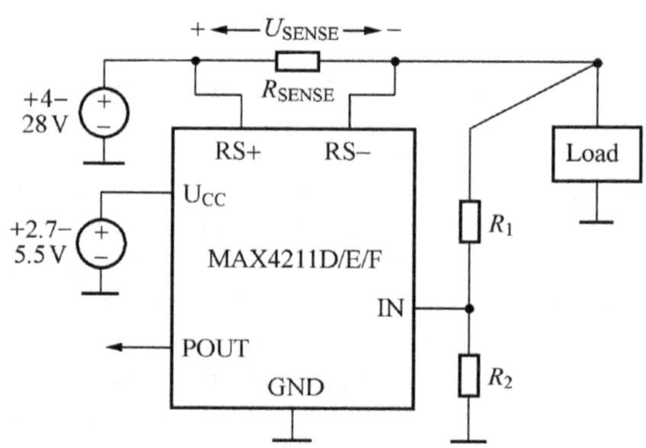

Figure 9.10 Circuit measuring the load power with external resistance dividers

maximum R_{SENSE}, the differential voltage between the terminals RS+ and RS– shall not exceed the full-scale detection voltage. Properly increase the resistance of R_{SENSE} to raise U_{SENSE}, which is helpful to reduce the output error. However, when detecting a high current, it is recommended to use a dedicated current-sense resistor and consider the power consumption of R_{SENSE} itself. When the allowable power consumption of the resistor is exceeded, the resistance will significantly rise with temperature and even cause the resistor to be burned. The digital voltmeter DVM_1 is used to display the measured power. DVM_2 can display the input current value.

To accurately measure the load power, the MAX4211D/E/F with the external resistance divider can also be used to connect the resistance dividers R_1 and R_2 directly to the load as shown in the circuit of Figure 9.10. This design can reduce the power consumption influence of R_{SENSE} to improve the measurement accuracy of load power.

9.4 Accurate Measurement Method of Duty Ratio

Duty ratio (D) refers to the percentage ratio between the pulse width (high level duration t of signal) and the period (T). The formula is as follows:

$$D = \frac{t}{T} \times 100\% \tag{9.18}$$

When measuring the pulse width modulation (PWM) SMPS and frequency control system, it is often necessary to measure the duty ratio of the impulse signal. At present, the method is generally using the oscilloscope to observe the waveform and then using the time scale to calculate the duty ratio, which is not only cumbersome but also inaccurate. The pulse duty ratio can be measured quickly and accurately by connecting a DVM in the duty ratio detection circuit. The measurement range is $D = 0$–100%. The accuracy is up to ±0.2% to ±2%. The input pulse amplitude range is 0.6–10 V and the frequency range is 20 Hz to 1 MHz, fully meeting the requirements of the routine measurements.

See Figure 9.11 for the circuit measuring the duty ratio. The measurement circuit is constituted by the input terminal protection circuit, voltage amplifier, resistance divider, and calibration circuit, configured with the 200 mV DVM constituted by ICL7106. The input pulse signal is expressed as U_{IN} (f_i) with the allowable duty ratio of 0–100%. R_1 is the current limiting resistor. VD_1 and VD_2 use high-speed silicon switching diode 1N4148 to constitute a bidirectional over-voltage protection limiter circuit. D_1 and D_2 are CMOS inverters, which can use the same CD4069 with the mains voltage provided by the +2.8 V reference voltage source E_0 in ICL7106. D_1 can be biased in the linear amplification area by R_f. After the pulse signal is amplified by D_1 and D_2, the amplitude $U_P \approx 2.8$ V, which will be reduced to U'_P after the voltage division of R_2 and RP. The average amplitude is $\overline{U_{IN}}$. RP is the duty ratio calibration potentiometer. It can be realized that $U'_P = 100$ mV by adjusting RP. The amplitude values of U_P and U'_P can be measured with oscilloscope or peak voltmeter. R_3 and C constitute the high-frequency filter at the analog input terminal. The input from DVM is the pulse voltage, which reflects the

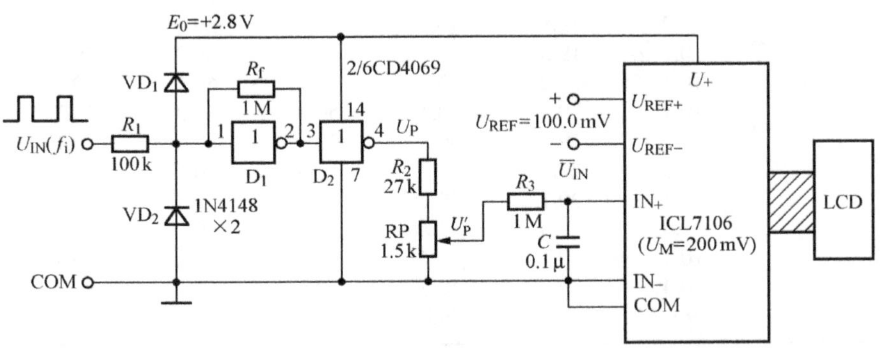

Figure 9.11 Circuit measuring the duty ratio

average pulse voltage $\overline{U}_{IN}$. In addition, $\overline{U}_{IN}$ is related to the duty ratio D, that is,

$$\overline{U}_{IN} = DU'_P \tag{9.19}$$

According to the measurement principle of ICL7106, the displayed value shall be

$$N = \frac{1000}{U_{REF}} \cdot \overline{U}_{IN} = \frac{1000}{100.0} \cdot \overline{U}_{IN} = 10\overline{U}_{IN}$$

that is,

$$\overline{U}_{IN} = 0.1N \tag{9.20}$$

Put Formula (9.19) into Formula (9.20). $U'_P = 100\,\text{mV}$, so $\overline{U}_{IN} = DU'_P = D \times 100\,\text{mV} = 100D\,(\text{mV}) = 0.1N$, that is,

$$D = 0.001N = 0.1N(\%) \tag{9.21}$$

Obviously, the pulse duty ratio can be read directly with 200 mV DVM.

To configure a 2-V DVM, R_2 and R_{RP} shall be 1.5 and 1 kΩ, respectively, and RP shall be adjusted to let $U'_P = 1$ V.

Use 1632-type function pulse generator to obtain the 20 Hz to 1 MHz pulse waveform with the amplitude of 4 V and $D = 10$–90%. Use the above duty ratio meter to display the duty ratio. See Table 9.3 for all the measured data. It can be seen from the table that the measurement error of duty ratio generally does not exceed ±2%, fully meeting the requirements of the routine measurements.

Table 9.3 Measured duty ratio data*

Pulse frequency f (Hz)	Measured duty ratio D (%)						
	10	20	40	50	60	80	90
	Readings of 200 mV digital voltmeter (%)						
20	10.1	19.7	40.2	50.1	—	—	—
100	10.0	19.8	40.1	50.2	60.2	—	—
200	9.8	19.9	40.3	50.0	60.1	79.8	89.6
500	10.2	19.8	40.1	49.7	59.8	79.6	89.9
1 k	10.3	19.8	39.7	49.7	59.3	79.9	89.7
2 k	10.1	20.2	39.9	50.1	60.1	79.9	90.2
5 k	10.2	20.4	40.1	50.0	60.3	79.7	90.1
10 k	9.9	19.8	39.8	50.3	60.0	80.1	89.3
20 k	10.3	20.2	39.2	50.5	60.4	79.7	90.5
50 k	10.1	20.1	40.2	50.1	59.9	80.6	89.5
100 k	9.9	19.6	39.8	50.0	60.2	80.2	89.4
200 k	10.1	20.1	39.7	50.1	59.7	79.9	90.1
500 k	10.4	20.5	40.5	49.8	60.5	80.3	89.9
1 M	9.8	19.7	39.0	—	—	—	—

*Some data in the table is vacant owing to the duty ratio adjustment capability constraints of 1632-type function pulse generator.

To configure a 2-V DVM, R_2 and R_{RP} shall be 1.5 and 1 kΩ, respectively, and RP shall be adjusted to let $U'_P = 1$ V.

9.5 Method to Detect the Magnetic Saturation of High-Frequency Transformer with Oscilloscope

High-frequency transformer is an important part in the isolated SMPS, and the design of high-frequency transformer is also a key technology for SMPS. In practical applications, SMPS is often damaged by unreasonable design of the high-frequency transformer or poor production process. The magnetic saturation of the high-frequency transformer is an important cause of failure. The following section describes the simple method to detect the magnetic saturation of the high-frequency transformer with oscilloscope for your reference.

9.5.1 Magnetic Saturation Characteristic of High-Frequency Transformer and Its Harmfulness to SMPS

9.5.1.1 Magnetic Saturation Characteristic of High-Frequency Transformer

During the magnetization of ferromagnetic material, the magnetic induction B first rises continuously with the external magnetic field intensity H. However, when H exceeds a certain value, the magnetic induction B approaches to a certain fixed value to enter into magnetic saturation state. See Figure 9.12 for the typical magnetization curve. It enters into the critical saturation region when $B \approx B_P$ and into the magnetic saturation region when $B \approx B_O$. For SMPS, the state when the magnetic flux ($\phi = BS$) in the high-frequency transformer does not change significantly with the rise of external magnetic field intensity is called magnetic saturation. The change of magnetic induction B is very small when the magnetic field intensity changes H, so the magnetic permeability decreases significantly and the magnetic permeability $\mu = \Delta B / \Delta H$. In this case, the inductance L_P of primary winding is also significantly reduced. It can be seen from Figure 9.12 that the magnetic permeability is equal to the slope of the magnetization curve. However, the magnetization curve is nonlinear, so μ is not a constant.

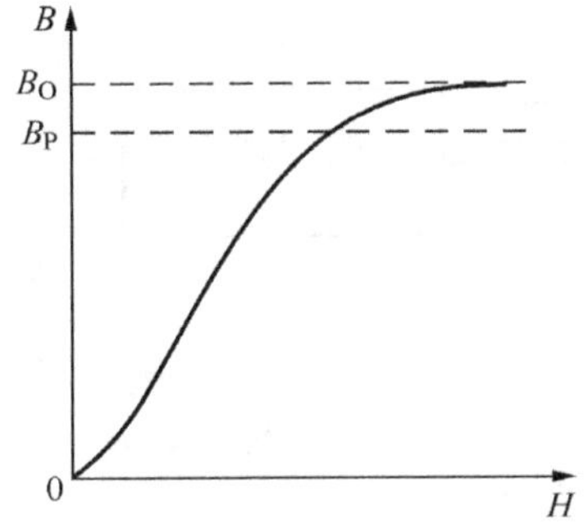

Figure 9.12 Magnetization curve of ferromagnetic material

9.5.1.2 Harmfulness of Magnetic Saturation to SMPS

Once the magnetic saturation occurs, the harmfulness to SMPS is great, ranging from component overheating to the component damage. In the case of magnetic saturation, the inductance of the primary winding L_P decreases significantly to rapidly increase the DC resistance (copper resistance) of the primary winding and the power consumption of the internal power switching tube MOSFET, thus resulting in the sharp increase of the primary current. It is possible that MOSFET has been damaged before the protection provided by the current limiting circuit in TOPSwitch. The main phenomena in the case of magnetic saturation are as follows: (i) the high-frequency transformer is very hot and TOPSwitch chip is overheated. (ii) When the load is increased, the output voltage drops quickly, failing to reach the designed output power.

There are many methods to prevent the magnetic saturation of the high-frequency transformer. The main method is properly reducing the turns of the primary winding. In addition, try to use the magnetic core with relatively large size and reserve a certain air gap width δ for the magnetic core, which can also prevent the magnetic core from entering into the magnetic saturation state.

9.5.2 Method to Detect the Magnetic Saturation of High-Frequency Transformer with Oscilloscope

In amateur conditions, it is relatively difficult to detect the magnetic saturation of the high-frequency transformer. The author has concluded a simple and effective method in practice, namely, measuring the current slope mutation of primary winding. If mutation exists, it is proved that the magnetic saturation occurs. See Figure 9.13 for the magnetic saturation detection method. First, generate the 1–3 kHz square wave signal through the square wave signal generator, which then delivers the power signal output within ±10 to 20 V and ±10 A after passing through the AC power amplifier with over-current protection. After that, the power signal is applied to the sampling resistor R_0 through the primary winding. Finally, observe the voltage waveform of R_0 with the oscilloscope. R_0 can be 0.1 Ω and 2 W precision wire wound resistor.

For an ideal inductor, when a fixed DC voltage is applied, its waveform of current I changes with time t as shown in Figure 9.14(a). In the case of a small current, the change of I can be considered linear. Figure 9.14(b) refers to the current waveform of inductor corresponding to

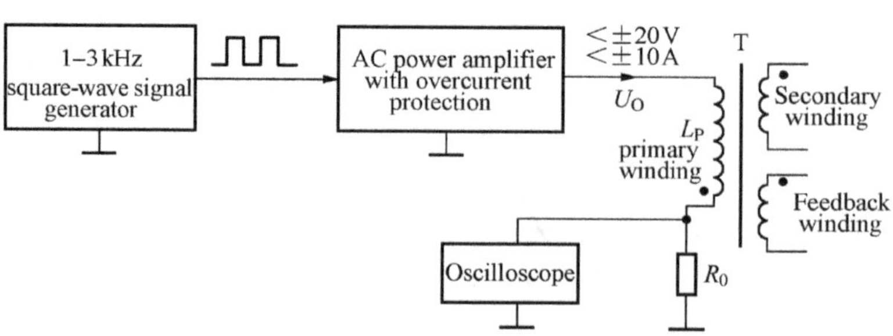

Figure 9.13 Magnetic saturation detection method

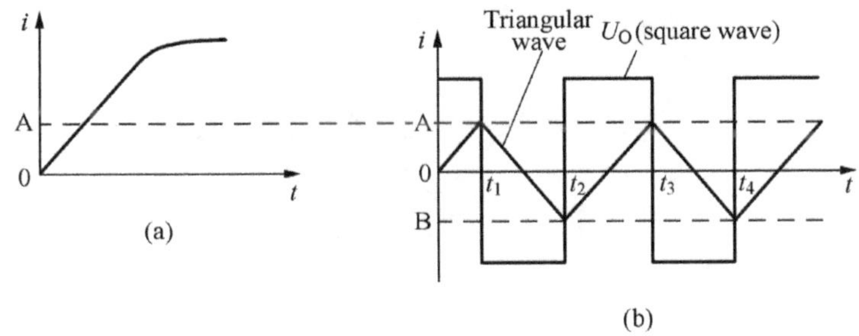

Figure 9.14 Corresponding relation of two waveforms. (a) Current waveform of ideal inductor when applying a fixed DC voltage and (b) current waveform of inductor when applying a square wave voltage

the square wave voltage U_O. At the positive half cycle of square wave output (e.g., at the stage of $t_2 \rightarrow t_3$, corresponding to the turn-on period of power switching tube), the inductor current linearly rises to point A. At the negative half cycle of square wave output (e.g., at the stage of $t_3 \rightarrow t_4$, corresponding to the turn-off period of power switching tube), the inductor current linearly decreases to point B. The slopes of the current waveform during the rise and decrease are the same, so eventually a symmetrical triangular wave is formed.

When there is no magnetic saturation, the voltage waveform U_{R0} of R_0 observed from the oscilloscope shall be triangular wave voltage. When a very small spike voltage at the top of U_{R0} waveform is observed, it is proved that the current slope of the primary winding starts mutation, which indicates that the high-frequency transformer has reached the critical magnetic saturation regions. When the spike voltage is relatively high, it means that the current slope has obvious mutation, which indicates that the high-frequency transformer has entered into the magnetic saturation region. See Figure 9.15 for the waveform comparison of the nonmagnetic saturation, critical magnetic saturation, and the magnetic saturation.

The above method has the following features: (i) It can determine whether the magnetic saturation occurs in the high-frequency transformer through simulation. (ii) It uses the low voltage and high current to detect the critical magnetic saturation point, and the power amplifier output can automatically define the maximum output power. (iii) The high-frequency transformer does not need any peripheral device, delivering easy operation and good safety. (iv) The rise rate of the primary current *i* is relatively low to facilitate observation and operation.

The author have measured the critical magnetic saturation current of SMPS with the test data as shown in Table 9.4. The following rules can be concluded: firstly, when using the same core model, the fewer the turns of primary winding are, the smaller will be the inductance and higher the critical magnetic saturation current. This is because the magnetic field intensity (H) is proportional to the product ($N_P \cdot I_P$) of the turns of the primary winding and the primary peak current. Therefore, when I_P is constant, $N_P\downarrow \rightarrow H\downarrow$, which is not likely to cause magnetic core saturation. Secondly, higher critical magnetic saturation current can be obtained by using the magnetic core with larger size at the same output power.

Finally, it shall be noted that the critical magnetic saturation current of the high-frequency transformer shall be greater than the limit current I_{LIMIT} of SMPS to prevent the high-frequency transformer from entering into the magnetic saturation state before the over-current protection of SMPS.

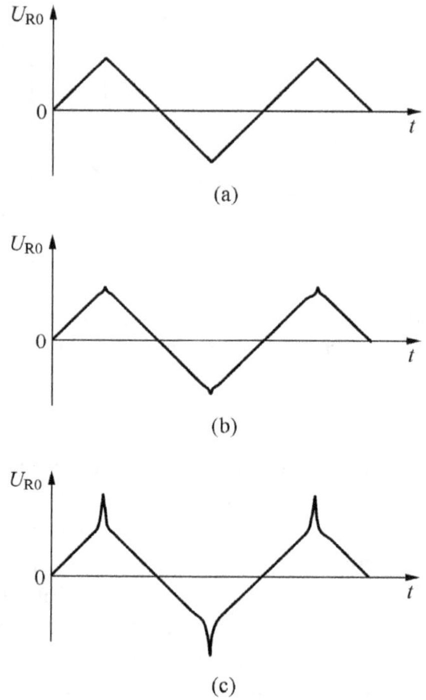

Figure 9.15 Comparison of three waveforms. (a) Nonmagnetic saturation waveform, (b) critical magnetic saturation waveform, and (c) magnetic saturation waveform

Table 9.4 Measurement data of critical magnetic saturation current

Core model	E30		E33		E40	EI25	EI40
Turns of the primary winding (T)	65	45	56	33	51	177	34
Critical magnetic saturation current (A)	1.92	2.9	2.25	4.0	5.21	1.04	7.5

9.6 Digital Online Current/Resistance Meter

Online measurement technology is also known as equipotential isolation technology. It can accurately measure the parameters such as the current, resistance, and transistors in the printed circuit of SMPS without destroying the circuit integrity. When the current is measured by traditional measurement method, the printed wire shall be cut off to connect an ammeter in series. In the case of resistance measurement by traditional method, a pin must be cut off through welding for measurement, which not only wastes time but also is easy to damage the PCB and components. In addition to the simple operation, the online measurement technology can not only measure the current distribution of PCB but also diagnose the failures of unit circuit and components. Therefore, it has high practical value, providing a new test method for SMPS maintenance. The following section describes the circuit designs and application examples of online DC current and resistance measurement, respectively.

9.6.1 Principle and Application of Online DC Current Measurement

9.6.1.1 Measurement Principle

See Figure 9.16(a) for the basic principle of online DC current measurement. It is featured that the four-wire system is used to reduce the contact resistance, and two sets of double probes and operational amplifier are used to isolate the circuit. In addition, after the *I*/*U* conversion of the measured online current, the current value is measured directly by the milliammeter. The milliammeter can also be replaced with a DVM, which is different in terms of the connection method. Assume that the measured current flowing through the printed wire is I_x, the resistance between points a and b on the printed wire is R_0, A is the operational amplifier, the open-loop voltage gain is A_{VD}, and the actual voltage amplification factor during closed-loop application is K. R_I is the internal resistance of DC ammeter, of which the reading is I_1. Points a and b are, respectively, connected to the inverting and noninverting input terminals of operational amplifier. Operational amplifier A here is used for buffering and isolation.

Point b is grounded. Point a is "virtual ground." The distance between a and b is very short and the resistance of R_0 is very small, so

$$U_O = -KU_I = -KI_xR_0 = I_1R_I$$

that is,

$$I_1 = -\frac{U_O}{R_I} = -\frac{KI_xR_0}{R_I} = -\frac{KR_0}{R_I}\cdot I_x \tag{9.22}$$

Put $K = -R_I/R_0$ into the above formula to get

$$I_1 = I_X \tag{9.23}$$

The measured current can be determined just by the reading of milliammeter. That is the basic principle of online current measurement.

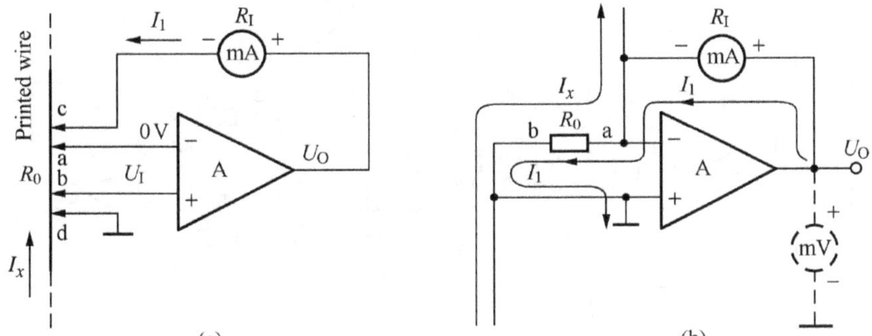

Figure 9.16 Principle of online current measurement. (a) Basic principle diagram and (b) equivalent circuit

According to the equivalent circuit shown in Figure 9.16(b), the measurement principle can be further analyzed. I_1 is used to offset the current I_x of R_0 to let the voltage drop of R_0 be zero, thus realizing equipotential between points a and b. R_0 is very small, so it will not affect I_x. This method is also known as local short-circuit measurement. In addition, point a is virtual ground, so the ammeter is equivalent to being connected across the output terminal of operational amplifier and the ground. In this case, the ammeter becomes a voltmeter with the full-scale voltage of $U_M = I_M R_I$, wherein I_M refers to the full-scale current of ammeter. It shall be noted that the ground connected by point d in Figure 9.16(a) is independent, which shall not be connected to the common ground wire of PCB.

9.6.1.2 Circuit Design of Online Current Meter

See Figure 9.17 for the circuit of online current meter with three ranges. The three current ranges are 20 mA, 200 mA, and 2 A. The left arrow in the figure refers to the four probes of the meter. S is the range switch. The circuit adopts the four-stage operational amplifier (referred to as operational amplifier). In order to improve the measurement accuracy and reduce the zero drift, A_1 is ICL7650 chopper-stabilized zero precision operational amplifier (IC_1). A_2 and A_3 use the same LM358 general-purpose operational amplifier (IC_2) of low power consumption together. A_4 uses a LM358 (IC_3, now using one set of operational amplifier only) independently. A_1 uses ±5 V dual power supply. To improve the current output ability, the one-stage emitter follower is constituted by NPN transistor JE9013 (VT). R_I (R_{I1}–R_{I3}) is equivalent to the internal resistance of ammeter in Figure 9.16. When R_I is 1.000, 10.00, and 100.0 Ω, the current range is 2 A, 200 mA, and 20 mA, respectively. R_I and A_2–A_4 can be used to convert I_1 into a voltage signal, which is sent to the DVM (can be replaced with the 2VDC gear of digital multimeter) with the range of 2 V to display the online current. A_2 and A_3 are buffers. A_4 is a voltage amplifier. Let the output voltages of A_2 and A_3 be U_1 and U_2, respectively. Assume that the voltages at the inverting and noninverting input terminals of A_4 are U_- and U_+, respectively, and the input currents at the two terminals are I_- and I_+, respectively. According to the basic principle of difference ratio operational circuit, the currents at the noninverting and

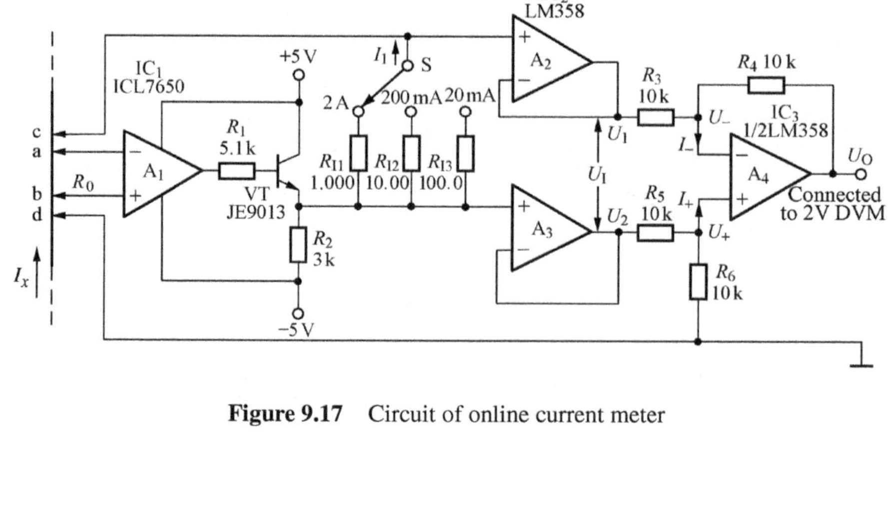

Figure 9.17 Circuit of online current meter

inverting input terminals of A_4 are both zero, that is, $I_+ = I_-$, which is called "virtual open circuit." However, the potentials at the two terminals are equal, that is, $U_+ = U_-$, which is called "virtual short circuit." Therefore,

$$U_+ = \left(\frac{R_6}{R_5 + R_6}\right) \cdot U_2$$

$$U_- = \frac{R_4}{R_3 + R_4} \cdot U_1 + \frac{R_3}{R_3 + R_4} . U_O$$

$$\frac{R_6}{R_5 + R_6} \cdot U_2 = \frac{R_4}{R_3 + R_4} \cdot U_1 + \frac{R_3}{R_3 + R_4} \cdot U_O$$

After finishing, it can be obtained that

$$U_O = \frac{R_6(R_3 + R_4)}{R_3(R_5 + R_6)} \cdot U_2 - U_1 \tag{9.24}$$

$R_3 = R_4 = R_5 = R_6 = 10\,\text{k}\Omega$ during circuit design, so Formula (9.24) can be simplified to

$$U_O = U_2 - U_1 \tag{9.25}$$

$(U_2 - U_1)$ is the input signal voltage U_I of A_4, so

$$U_O = U_I \tag{9.26}$$

This indicates that the voltage amplification factor K of A_4 is equal to 1.

R_{I1} to R_{I3} in Figure 9.17 shall be precision metal film resistors with error of ±0.1 to ±0.5%. The offset voltage of ICL7650 is less than 1 μV, the long-term drift is only 10 μV/month, and $A_{VD} > 120\,\text{dB}$ (equivalent to 10^6 times). When considering A_1 only, the relative error of online measurement current is

$$\gamma = \pm\frac{\Delta I_x}{I_x} = \pm\frac{R_I}{A_{VD} R_0} \times 100\% \tag{9.27}$$

Assume that $R_I = 10.00\,\Omega$, $A_{VD} = 1 \times 10^6$, and $R_0 = 0.01\,\Omega$, put them into Formula (9.27) to get $\gamma = 0.1\%$. In fact, A_1 and VT constitute a closed-loop amplification circuit, so the error is about ±1% to ±3%.

9.6.1.3 Typical Application of Online Current Meter

See Table 9.5 for the measurement data at gear 200 mA of the online current meter. The exact current in the table is measured by connecting the VC890D digital multimeter in series to the printed circuit. The online current is measured by switching to gear 2 V (DC) of VC890D and connecting VC890D to the voltage output terminal of online current meter (moving the decimal point to the right by one digit, i.e., expanding the reading by ten times, with the unit of mA). It can be seen that the measurement error is generally not more than ±3%. During measurement, the distance between a and b shall be greater than 5 mm.

Table 9.5 Online current measurement data

Online current I_x' (mA)	10.5	20.0	30.6	46.2	60.2	70.4	83.2	94.2	113.0	146.5	168.4
Exact current I_x (mA)	10.8	20.3	31.3	46.7	60.8	71.1	83.9	95.1	113.9	147.7	169.4
Relative error γ (%)	−2.8	−1.5	−1.6	−1.1	−1.0	−0.4	−1.0	−1.0	−0.8	−1.2	−0.6

9.6.2 Principle and Application of Online Resistance Measurement

9.6.2.1 Measurement Principle

The basic principle of the online resistance R_x measurement is that regardless of the complexity of circuit, the components connected in parallel to R_x can always be equivalent to two resistors R_1 and R_2 connected in series to constitute a triangular resistor network. See Figure 9.18 for the basic principle diagram. As long as the two ends of R_1 (or R_2, the same below) are equipotential, $U_{R1} = 0$. In this case, R_1 is equivalent to being in an open circuit and R_2 becomes the load resistor of the operational amplifier. Therefore, R_1 and R_2 have no shunting effect, which helps to directly measure the resistance of R_x. E is a test voltage. I_S is a test current. Assume that the currents flowing through R_x and R_1 are I_x and I_1, respectively, and the input bias current of operational amplifier A is I_{IB}.

$$I_S = I_x + I_1 + I_{IB} \tag{9.28}$$

According to Kirchhoff's law, it can be known that I_{IB} is very small, which is negligible. According to the "virtual ground" principle, $U_{cd} = I_1R_1 = 0$, so $I_1 \approx 0$, which is also negligible. Therefore,

$$I_S = I_x \tag{9.29}$$

Take into account the grounding point c and the virtual ground point d, so

$$I_S = E/R_0 \tag{9.30}$$

Thus, it can be deduced that

$$U_x = I_xR_x = I_SR_x = \frac{E}{R_0} \cdot R_x \tag{9.31}$$

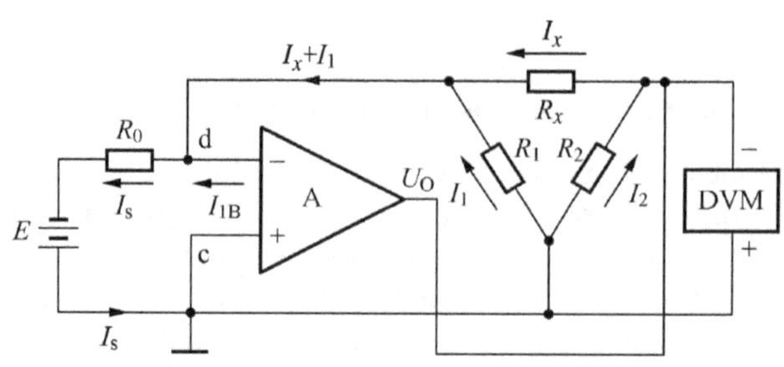

Figure 9.18 Basic principle diagram of online resistance measurement

Obviously, the value of R_x can be calculated as long as the voltage drop U_x across the two ends of R_x is measured with DVM. Further, point d is the virtual ground, so the DVM can be directly connected to the two ends of the load resistor R_2 (i.e., between U_O and the ground).

It shall be noted that the I_{IB} in Formula (9.28) is divided into positive and negative values. The direction of the input-stage bias current of the operational amplifier depends on the internal circuit structure. For the operational amplifier with PNP differential pair tube as the input stage (e.g., LM358 and LM324), the I_{IB} is outgoing (Figure 9.18), which is expressed as $+I_{IB}$. For the operational amplifier with NPN differential pair tube as the input stage (e.g., μA741), the I_{IB} is ingoing, which is expressed as $-I_{IB}$. However, I_{IB} is very small (typically only 5–80 nA) regardless of its current direction, so it is always negligible in Formula (9.28).

9.6.2.2 Circuit Design of Online Resistance Meter

See Figure 9.19 for the four-range online resistance meter. The four ranges are 200 Ω, 2 kΩ, 20 kΩ, and 200 kΩ with test currents of 10 mA, 1 mA, 100 μA, and 10 μA, respectively. The diagram within the dashed box refers to the measured resistor network. R_{01}–R_{04} are the range-setting resistors. S is the range switch. A_1 and A_2 share the same LM358 dual operational amplifier. To improve the output capacity, A_2 is connected as a buffer with the output terminal connected to a 2-V-range DVM. The selected test voltage $E = 3$ V. Take 200 Ω, for example, $I_S = E/R_{01} = 3\text{ V}/300\,\Omega = 10\text{ mA} = 0.01$ A and $U_x = 0.01R_x$ (V), that is, $R_x = 100U_x$ (with the unit of Ω). Obviously, the online resistance can be directly read just by moving the decimal point of 3 ½ DVM to the right by two digits to become ×××.× (Ω). The measurement range of gear 200 Ω of the meter is 0.1–200.0 Ω.

When it is not required to directly read the resistance, R_{01} can also be 150 Ω to let $I_S = 20$ mA. $U_x = 0.02R_x$ (V), so $R_x = 100U_x/2$ (Ω). The actual resistance shall be half of the reading, so the upper measurement limit is reduced to 100.0 Ω.

The error of the online resistance measurement is caused by the input bias current I_{IB} of the operational amplifier and the current of R_1. I_1 is formed owing to $U_{cd} \neq 0$. U_{cd} is related to the offset voltage U_{IO} of operational amplifier and the open-loop voltage gain A_{VD}. Therefore, the U_{IO} of the used operational amplifier shall be as low as possible and A_{VD} shall be

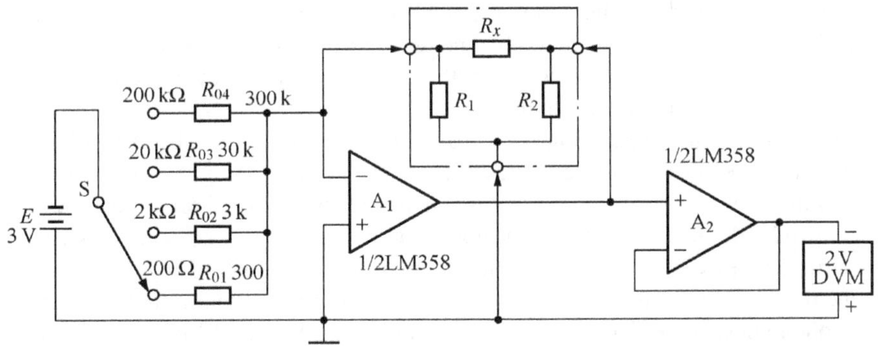

Figure 9.19 Circuit of online resistance meter

Table 9.6 Online resistance measurement data

Measured online resistance (Ω)	1788	1420	1044	169
Exact value of the measured resistor (Ω)	1784	1419	1050	176
Measurement error γ (%)	−0.2	−0.07	+0.5	+4.1

large enough. LM358 high-gain operational amplifier is used here, of which $U_{IO} = \pm 2\,\text{mV}$ (typical value) and $A_{VD} = 100\,\text{dB}$ (equivalent to 10^5 times), meeting the general requirements.

9.6.2.3 Typical Application of Online Resistance Meter

Measure the resistances of the four online resistors with the gear $2\,\text{k}\Omega$ of the meter. Then, respectively, cut off one pin of each resistor through welding. Finally, measure the exact value of the offline resistor with VC890D digital multimeter, of which the measurement data are shown in Table 9.6. It can be seen from the table that the error will increase when measuring low resistance, which is generally not more than ±5%. If ICL7650 or LF412/LF412A high-gain operational amplifier with low offset voltage is used, the error can be less than ±1%.

9.7 Electromagnetic Compatibility Measurement of SMPS

Electromagnetic compatibility is referred to as EMC. This concept began in the 1940s and was not developed into an independent new science until the1970s. With the rapid development of electronic, electrical, communications, and computer technology, the importance of EMC in industrial, scientific research, civilian, and military fields is increasingly recognized by people.

9.7.1 Research Fields of EMC

9.7.1.1 Concept of EMC

The EMI and the anti-interference ability of electronic equipment are not only a pair of inextricably deep-seated antagonists but also an interrelated paradox. On the one hand, EMI is increasingly serious owing to the popularization of the electronic devices. On the other hand, EMI puts forward higher requirements for the design of electronic devices. The core issue is how to ensure that electronic devices can both work properly and meet the design indicators in a complex electromagnetic environment.

EMC is defined by the International Electro-technical Commission (IEC) as follows: "EMC is a function of the electronic equipment, which can be completed in an electromagnetic environment without any interference intolerable." In the generic "EMC" standards recently issued by China, EMC is defined as follows: "the ability of the equipment or system to work normally in its electromagnetic environment without causing intolerable EMI to any other matters in the environment." The electromagnetic environment said here refers to the sum of all the electromagnetic phenomena in a given situation. This indicates that the EMC has three meanings: firstly, the electronic equipment shall have the ability to suppress external EMI. Secondly, the EMI generated by the electronic equipment shall be less than the limit specified and shall not

Table 9.7 Pulse energy capable of damaging electronic components

Component	Single pulse energy (μJ)	Continuous pulse energy (μJ)
0.25 W resistor	10^4	10^2
Electrolytic capacitor	60–1000	0.6–10
Relay	10^3–10^5	10–10^3
Diode (point contact type)	10^{-2}–10	10^{-4}–10^{-1}
Low-power transistor	20–10^3	0.2–10
High-power transistor	10^3	10

affect the normal operation of other electronic devices in the same electromagnetic environment. Thirdly, the EMC of any electronic equipment is measurable. Obviously, EMC means much more than the usually said "anti-interference ability."

EMC is important in the military. For example, in the Falklands (Islas Malvinas called by Argentina) war in 1982, the Sheffield missile destroyer of British had to temporarily turn off the radar system to ensure that the remote communication is not interfered owing to the failure in the EMC problem settlement, so it was hit by the Exocet missile of Argentina, causing destroyer crash and death. During 1991–2003, from the Gulf War and Kosovo War to the recent Iraq War, US forces used the "new concept weapon" of electromagnetic pulse (EMP) bomb to make the enemy radar communications network completely paralyzed to fully gain the air mastery.

EMC is very harmful. See Table 9.7 for the energy of single and continuous pulse capable of damaging electronic components. In addition, electromagnetic radiation can not only detonate electric initiating devices and ammunition depots but also cause harm to humans. It is shown by experiments that when the microwave irradiation power density is 10 mW/cm^2, the person's body temperature rises by about 1 °C. If a dog is under the radiation of 1200 MHz and 330 mW/cm^{2}microwave, it will be killed within 15 min. The human safety margin of microwave radiation developed by China is 0.025–0.05 mW/cm^2. Furthermore, if a mobile phone is used on the plane, it will interfere with the navigation system to threaten flight safety.

At present, EMC has been a quality indicator of household appliances in China. For example, a few years ago, the State Bureau of Technical Supervision and the Ministry of Information Industry carried out special spot checks for 29 kinds of domestic and imported large-screen color TV, of which three kinds of imported color TV sets were exposed owing to EMC incompetence.

It was specified by the European Community on 1 January 1996 that EMC test must be carried out for electronic devices (including electric appliance, electronic equipment, and devices with electric appliance or electronic parts), that is, the electromagnetic wave interference (CMI) and anti-interference (EMS) performance test of electronic equipment. It was also specified by the European Community on 1 January 1997 that safety performance test must be carried out for low-voltage electronic equipment. Only the electronic equipment passing the inspection of competent certification authority accredited by the European Community can obtain CE certification, have CE mark on its products, and enter into the European market for sale.

9.7.1.2 Research Fields of EMC

It mainly includes the generation and transmission of EMI, EMC design (including the development of standards), EMI diagnosis and suppression, and EMC test (including tests). Taking EMC, for example, the objects under research are as follows:

- EMI source
  - Natural interference source
    - Atmospheric noise source (e.g., lightning and the corona discharge caused by sandblasting)
    - Sky noise source (e.g., solar and cosmic noise)
    - Thermal noise(e.g., resistance thermal noise)
    - Grid interference source (50 Hz industry interference)
    - Electrical brush interference source (caused by motor)
  - Artificial interference source
    - Ignition system interference source(car ignition device)
    - Household appliance interference source (e.g., daylight lamp interference)
    - Radio-frequency interference source (e.g., interference emitted by video computer games and mobile phones)
    - Interference source made intentionally (e.g., electronic countermeasures battle and electromagnetic pulse bomb)

China started late in this field but has also achieved remarkable results. For example, it has completed many major scientific research projects such as the EMI measurement and analysis of Beijing city, the EMC study of electrified railways, the asymmetric transverse electromagnetic cell research, and the electronic countermeasures battle.

9.7.1.3 EMC Standards

There have been more than 100 EMC standards. The representative standards include the "Terminology for Electromagnetic Compatibility" (IEC50-161), the "Term Definition and Unit System of Electromagnetic Interference and Electromagnetic Compatibility" (MIL-STD-463A), the "System Requirements for Electromagnetic Compatibility" (MIL-E-6051D), the "Measurement of Electromagnetic Interference Characteristics" (MIL-STD-462), and the "Radio Interference and Immunity Measurement Apparatus and Method Specification" (CISRR 16-1-1993) developed by the International Special Committee on Radio Interference. China also formulates relevant national and national military standards, such as the "Radio Interference Terminology" (GB4363-84), the "Electromagnetic Interference and Electromagnetic Compatibility Terminology" (GJB72-85), the "Radio Interference and Immunity Measurement Apparatus Specification" (GB/T6113-1995), the "Measurement Methods and Allowable Values of Radio Interference Characteristics of Electric Tools, Household Appliances and Similar Appliances" (GB4343-84), and the "Electromagnetic Compatibility Term" (GB/T4365-1995). For example, the allowable continuous interference voltages of household appliances specified in GB4343 are shown in Table 9.8. The promulgation and implementation of these standards lay the foundation for EMC improvement.

Table 9.8 Allowable continuous interference voltage of household appliances

Frequency range (MHz)	0.15–0.20	0.20–0.50	0.50–5.0	5.0–30
Allowable value (mV)	3	2	1	2

9.7.2 *EMC Measurement*

Measurement work is an important basis to evaluate the EMC design of equipment. The test items mainly include the emission testing (measuring the electromagnetic energy, frequency, etc. radiated or conducted by the electronic equipment during work) and the sensitivity test (measuring the ability of electronic equipment to suppress external EMI). The electromagnetic shielded room can be used to prevent EMI and purify the test site of electromagnetic environment. See Table 9.9 for the EMC measurement devices and their functions. The total frequency range of the measured radio interference voltage and interference current and field is 9 kHz to 18 GHz.

The high-frequency noise simulator is mainly used to superimpose interference pulse on the operational power supply to implement the power line interference resistance test of equipment. The "high frequency" here means that the frequency of the harmonic component in the output waveform is very high. The following text describes the typical application of ENS-24XA high-frequency noise simulative generator produced by Shanghai Sanki Electronic Industries Co. Ltd.

The power line interference resistance test of equipment shall be carried out through a dedicated power line coupling/decoupling network as shown in Figure 9.20. The coupling/decoupling network adopts 50 Ω coaxial cable with the cable length of 2 m.

Common-mode interference refers to the interference between the power line and the ground line or between the neutral line and the ground line. For the three-phase alternating current, the common-mode interference exists between any phase and the ground line. See Figure 9.21 for the features of common-mode interference, wherein U_{CM} refers to the common-mode voltage between the current carrying conductor and the ground and I_{CM} refers to the common-mode current. Series-mode interference exists between the phase and the neutral lines of power supply (for the three-phase alternating current, also exists between any two phases) as shown in Figure 9.21(b), wherein, U_{DM} refers to the series-mode voltage between current carrying conductor and I_{DM} refers to series-mode current.

During the common-mode test with high-frequency noise simulator, the two cores of the coupling/decoupling network output cable are, respectively, connected to the power input terminal of the tested equipment, and the shielding layer of the output cable is connected to the reference ground plate (*note*: it shall be low-impedance connection and the connecting wire shall be as short as possible). The ground terminal of the tested equipment shall be connected to the reference ground plate with short and thick wire in a low-impedance way. Connect the shell of the high-frequency noise simulator to the ground. Pulse injection lines A and B are, respectively, connected to the power line through coupling capacitors C_{S1} and C_{S2}. In addition, the shielding layer of the injection line is connected to the shell. Therefore, the pulse applied to the power line is actually relative to the shell, that is, to the ground, so the circuit is used for common-mode test.

The interference between the lines shall be measured during differential mode test. Therefore, the shielding layer of the coaxial cable can be connected to the line of lines A and B

Table 9.9 EMC measurement devices and their functions

Classification	Name	Measurement function
Measurement device	Automatic EMI test system	Measuring various parameters of EMI and waveform analysis
	EMI analyzer	Automatically analyzing the amplitude, incidence rate, and duration of EMI
	Spectrum analyzer and scanning receiver	Measuring and analyzing the interference within the frequency range of 1 kHz–18 GHz
	Peak measuring receiver	Measuring the peak of 9 kHz–1 GHz pulse interference
	Average receiver	Measuring the average of 9 kHz–1 GHz narrow-band interference
	RMS receiver	Measuring the RMS of 9 kHz–1 GHz of interference voltage
	Audio interference voltmeter	Measuring the noise immunity of 20 Hz–20 kHz audio system
Auxiliary device	Artificial mains network	Coupling the RF interference voltage to the measurement receiver
	Current probe	Using the clamp converter to measure 30 Hz–1 GHz interference current of the wire with the noncontact method
	Voltage probe	Measuring the RF interference voltage between the grid power line and the ground
	Absorbing clamp	Measuring 30–100 MHz interference power of lead radiation
	Antenna	Receiving 150 kHz–30 MHz radio interference signal
	Coupling network	Measuring the interference immunity of experimental equipment toward 15 kHz–150 MHz conduction current
	Transverse electromagnetic cell	Testing laboratory dedicated to radiated power measurement
	Reverberant field	Testing laboratory dedicated to total radiated power measurement
	Test field	Open area test site used to measure the radio interference field strength within the frequency range of 30 MHz–1 GHz

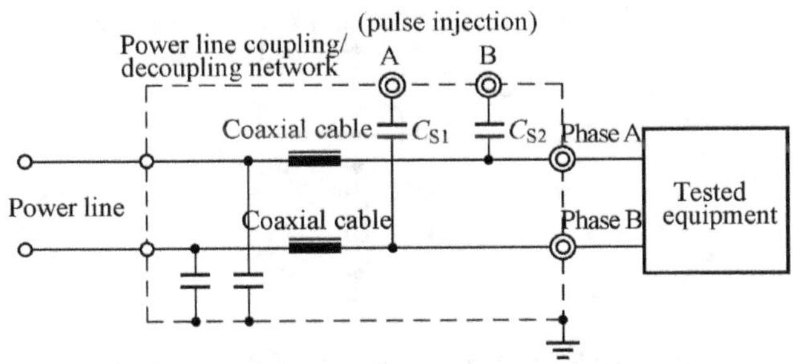

Figure 9.20 Power line coupling/decoupling network

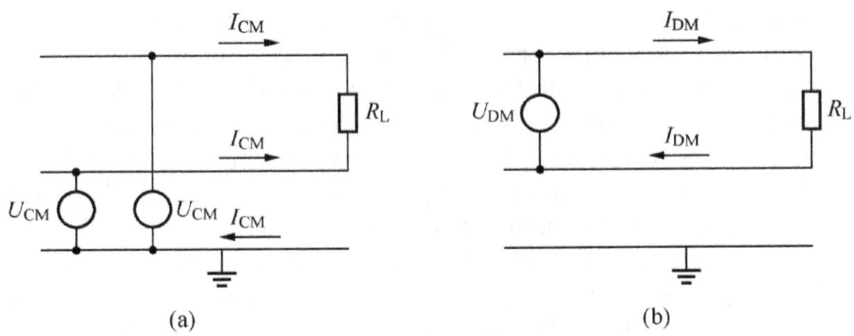

Figure 9.21 (a) Common- and (b) series-mode interference

without pulse injection. For example, when the pulse is injected to line A during test, the shielding layer of the coaxial cable must be connected to line B. It shall be noted that it is prohibited to connect the shielding layer to line A, which will cause the pulse to be in short circuit. It is preferable that the connected cable is the neutral line of power line, so as to prevent the test personnel from being injured owing to the charged shell of coupling/decoupling network. For these reasons, the interference pulse shall be injected into the phase line of the power supply during differential test. In this way, the cable-shielding layer can be connected to the ground wire of the power supply to prevent any danger during the test. If the live wire of power supply cannot be determined, the shell of the coupling/decoupling network shall be floating. Otherwise, if the shielding layer is connected to the live wire and the coupling/decoupling network is connected to the ground, then the power supply will be in short circuit. The safest way is to add an isolation transformer between the mains and the coupling/decoupling network.

9.8 Waveform Test and Analysis of SMPS

9.8.1 Key Waveform Test Method of PWM Controller

PWM controller is the core part of SMPS. During power supply commissioning, in addition to measuring the voltages at various test points in the control circuit with voltmeter, the most

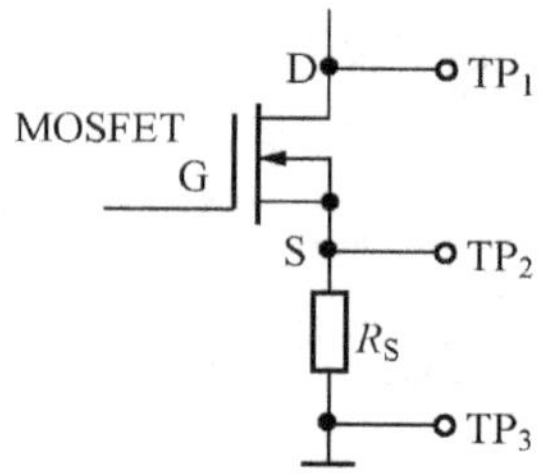

Figure 9.22 Test point selection of PWM controller

important thing is to observe the voltage and current waveform of PWM controller with oscilloscope, so as to timely determine whether the power supply works normally. See Figure 9.22 for the test point selection of the PWM controller. TP_1 is the drain of power switching tube (MOSFET). TP_2 is the source of power switching tube. R_S is the current sampling resistor. The two test points can be connected to the two channels (CH1 and CH2) of dual trace oscilloscope to simultaneously observe the waveforms of the two points. In this case, the ground terminals of the two probes shall be both connected to the negative terminal of primary DC high-voltage input, that is, position TP_3 in the figure. During the actual measurement, the ground clamp of the probe can be directly clamped on the ground pin of R_S.

The drain voltage waveform of power switching tube can be seen at TP_1, which can reflect the drain spike voltage, the input DC high voltage, the conduction voltage drop, turn-on time of power switching tube, the turn-off time, and other information. In single-end flyback SMPS, the drain voltage waveform of power switching tube is shown in Figure 9.23.

The source voltage waveform of power switching tube can be seen at TP_2, which is the voltage waveform of sampling resistor R_S and can reflect the drain current, the turn-on and turn-off time, and other information. See Figure 9.24 for the drain current waveform of power switching tube, which reflects that the SMPS works in continuous current mode. In each cycle, when the switching tube is turned on, the drain current rises from a relatively small initial current. Before the switching tube is turned off, the drain current reaches its peak.

TP_1 and TP_2 are the two key test points, which can basically reflect the working and failure conditions of SMPS. During commissioning, special attention shall be paid to the waveforms

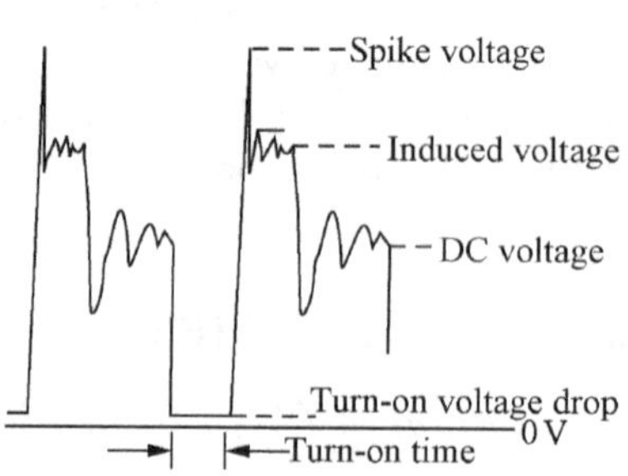

Figure 9.23 Drain voltage waveform of power switching tube

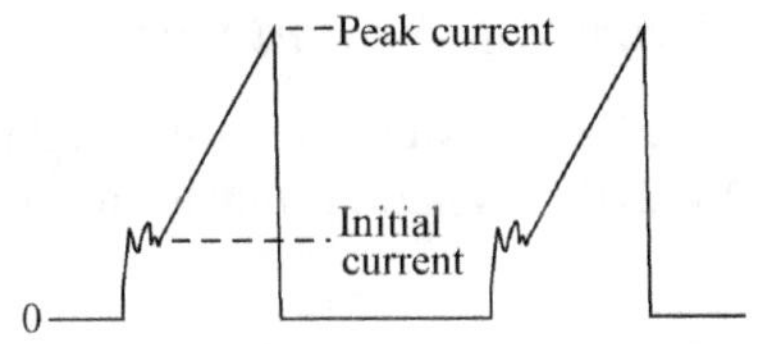

Figure 9.24 Drain current waveform of power switching tube

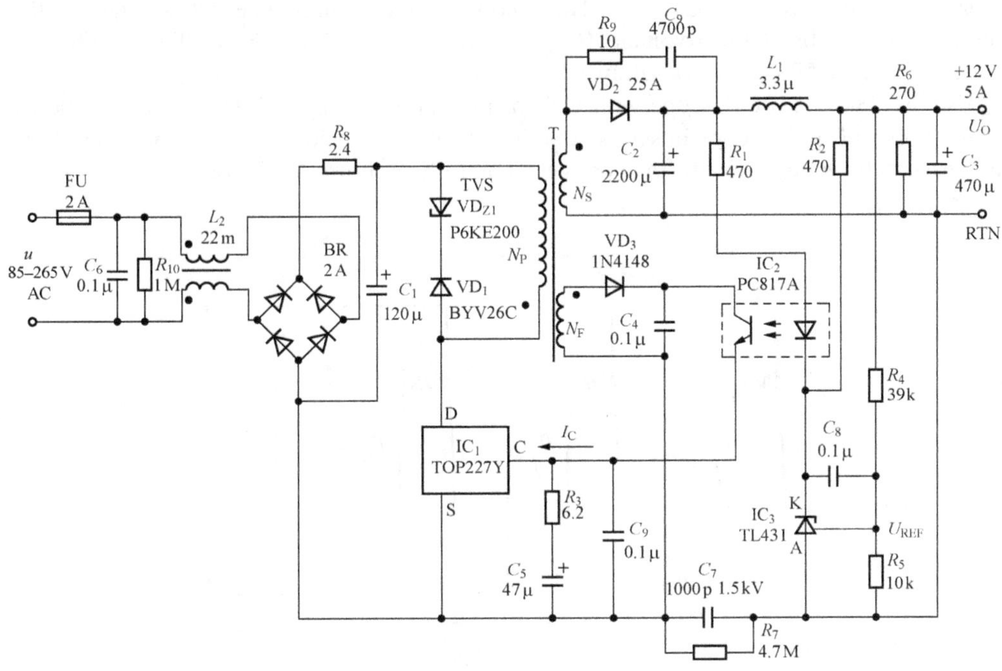

Figure 9.25 Typical circuit of the tested single-chip SMPS

of these two test points. When the input voltage is gradually increased, once it is found that the peak voltage or peak current exceeds the design limit, the power supply shall be turned off to prevent the power switching tube from being damaged.

The following text takes the 12 V and 60 W SMPS constituted by TOP227Y, for example, to introduce the waveform test and analysis methods of single-chip SMPS. See Figure 9.25 for the typical circuit of the tested single-chip SMPS, of which the AC input voltage is 85–265 V, the output is +12 V and 5 A, and the rated output power is 60 W. It belongs to the optocoupler feedback precision SMPS. There are three ICs in the circuit: IC_1 (TOP227Y single-chip SMPS IC), IC_2 (PC817A linear optical coupler), and IC_3 (TL431 adjustable precision shunt regulator).

9.8.2 *Measuring the Primary Voltage and Current Waveforms*

To ensure the test safety, an isolation transformer must be added between the tested SMPS and the grid. 200 VA and 220 V/220 V isolation transformer is used in the actual measurements.

9.8.2.1 Measuring the Primary Voltage Waveform

When $u = 150$ V and $I_O = 1.5$ A, respectively, connect the two input probes of the oscilloscope between the drain (D) of TOP227Y and the primary common terminal. See Figure 9.26 for the measured primary voltage waveform. The maximum voltage amplitude is 435 V, far less than the drain-source breakdown voltage $U_{(BR)DS}$ ($U_{(BR)DS} \geq 700$ V) of MOSFET in TOP227Y. Therefore, the SMPS can work safely.

Use the SS-7802 oscilloscope with CRT display produced by IWATSU. Connect the two input probes of oscilloscope in series at both ends of the transient voltage suppressor (TVS) to measure the voltage waveform at both ends of TVS as shown in Figure 9.27.

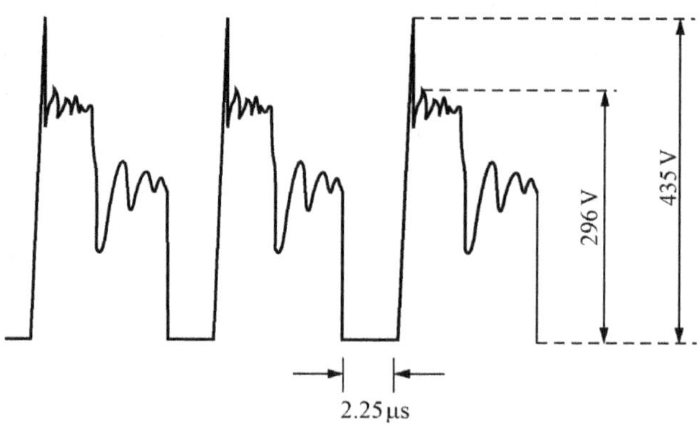

Figure 9.26 Primary voltage waveform

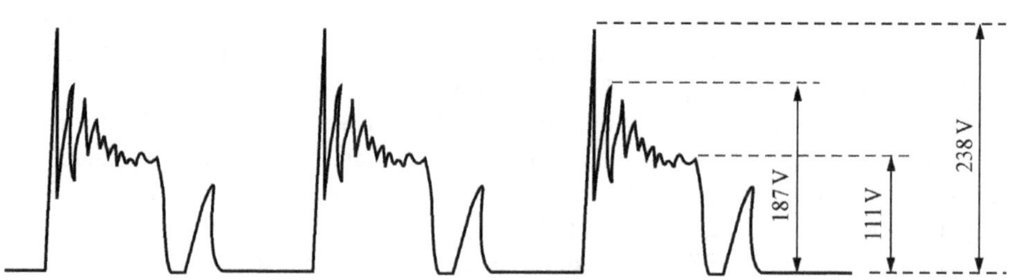

Figure 9.27 Voltage waveform at both ends of TVS

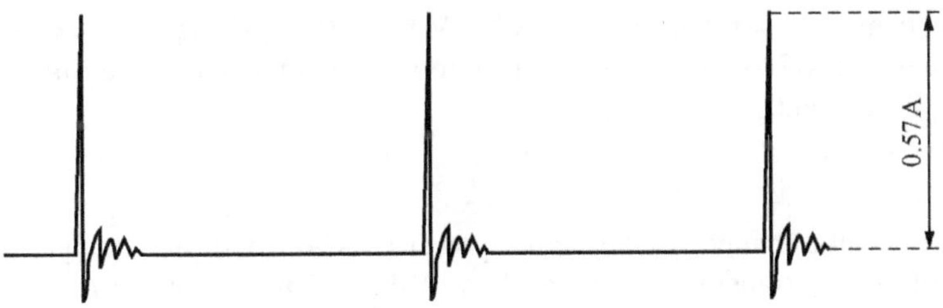

Figure 9.28 Primary spike current waveform

9.8.2.2 Measuring the Spike Current Waveform of Primary Clamp Circuit

The method to measure the spike current in the primary clamp circuit is as follows: firstly, connect a 1 Ω and 1 W sampling resistor R_{S1} to VD_{Z1} in series. Secondly, measure the voltage drop U_{R1} at both ends of the sampling resistor. Finally, calculate the spike current according to Ohm's law $I_J = U_{R1}/R_{S1}$. When $u = 150$ V and $I_O = 1.15$ A, $U_{RI} = 0.57$ V. Therefore, $I_J = 0.57$ A. See Figure 9.28 for the spike current waveform measured with the oscilloscope. The experiment also shows that when u decreases, the spike current will rise and when u rises, the spike current will decrease. It shall be noted that the primary impedance is relatively high, so the measurement will not be affected even if resistance of R_{S1} is slightly greater.

9.8.2.3 Measuring the Primary Peak Current Waveform in Noncontinuous Mode

The method to measure the primary peak current is as follows: connect in series a 0.5 Ω and 1 W sampling resistor R_{S2} at the inlet terminal of DC high voltage U_I and then calculate the primary peak current by measuring the voltage drop of the sampling resistor. When $u = 150$ V and $I_O = 1.16$ A, connect the two input probes of oscilloscope to the two ends of R_{S2}. See Figure 9.29 for the measured primary peak current waveform. During the turn-on period of MOSFET (the turn-on time $t_{ON} = 1.74$ μs), $\Delta U_1 = 0.386$ V and $I_{P1} = 0.772$ A. During the turn-off period of MOSFET (the turn-off time $t_{OFF} = 9.76$ μs), $\Delta U_2 = 0.169$ V and $I_{P2} = 0.338$ A. It can be seen from the figure that the switching current starts from zero in each switching cycle, and the current is in noncontinuous state, which indicates that the SMPS works in noncontinuous mode under the above conditions.

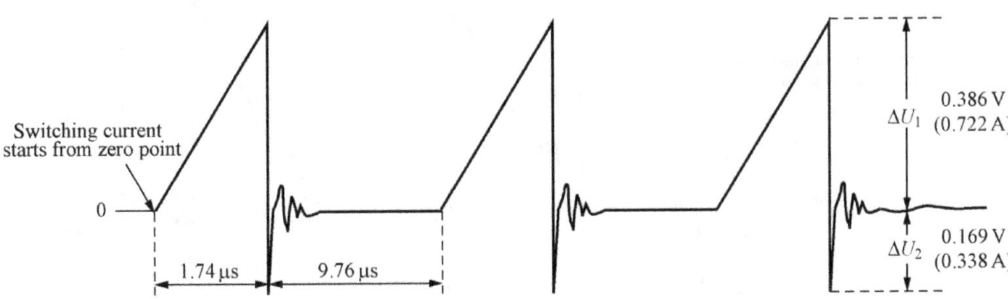

Figure 9.29 Primary peak current waveform (noncontinuous mode)

The experiment also shows that when $u = 260$ V, $t_{ON} = 0.96$ µs, and when $u = 60$ V, t_{ON} varies within the range of 4.74–5.62 µs, at which time the output voltage is unstable. Only when $u > 80$ V, U_O is constant.

Test result analysis:

1. The load does not change, so the output power of SMPS will not change. From the perspective of energy transmission, regardless of the AC input voltage u, the energy value ($E = I_P^2 L_P$) finally stored by the high-frequency transformer is the same as long as I_P does not change.
2. The lower the u is, the smaller will be the rise rate of primary current, the longer the time to reach the rated peak current, the higher the output duty ratio of TOP227Y, and vice versa.

9.8.2.4 Measuring the Primary Peak Current Waveform in Continuous Mode

Adjust the load resistance to realize that the measured $\Delta U_1 = 0.91$ V, $I_{P1} = 1.82$ A, $\Delta U_2 = 0.558$ V, and $I_{P2} = 1.12$ A when $I_O = 6.24$ A. See Figure 9.30 for the primary peak current waveform measured with oscilloscope. The switching current in each switching cycle starts from a nonzero value, so it can be determined that the SMPS currently works in continuous mode. It can be known through further observation that when $u > 230$ V, the SMPS enters into the noncontinuous mode.

Test result analysis:

1. The work mode can be changed when the AC input voltage does not change and the load current changes significantly.
2. The work mode can be changed when the load does not change and the AC input voltage changes significantly.
3. The continuous mode is caused by the incomplete release of the energy stored in the high-frequency transformer during flyback. The slope of energy release is basically unchanged, but the discharge time is shortened obviously to let the primary current fail to pass the point zero, thus causing part of the energy failing in release.

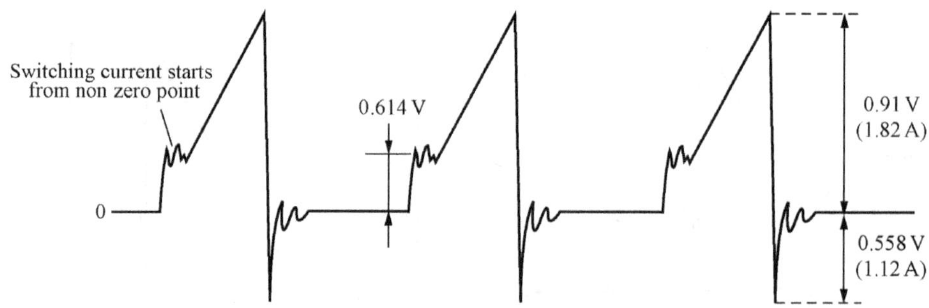

Figure 9.30 Primary peak current waveform (continuous mode)

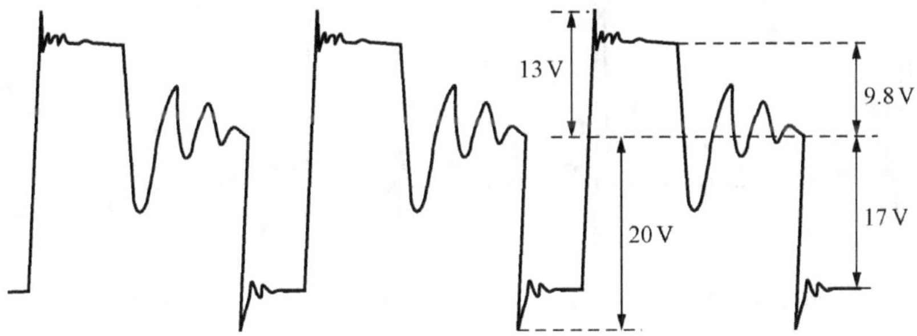

Figure 9.31 Secondary voltage waveform

9.8.3 Measuring the Secondary Voltage and Current Waveforms

9.8.3.1 Measuring the Secondary Voltage Waveform

See Figure 9.31 for the measured secondary voltage waveform.

9.8.3.2 Measuring the Secondary Current Waveform

The method to measure the secondary current is as follows: connect in series a 0.15 Ω and 8 W sampling resistor R_{S3} to the output rectifier VD_2 and then calculate the secondary current I_S by measuring the voltage drop of the sampling resistor. When $u = 150$ V and $I_O = 1.54$ A, do not connect R_9 and C_9 and measure the secondary current waveform with oscilloscope as shown in Figure 9.32. When R_9 and C_9 are connected, the secondary current waveform is as shown in Figure 9.33. It can be known that after R_9 and C_9 are connected, the secondary high-frequency oscillation can be suppressed obviously. It is also observed that the primary spike voltage can be reduced after R_9 and C_9 are connected. It is shown by the experiment that the secondary current waveform does not vary with the AC input voltage. However, when u is reduced to about 60 V, audio howl starts to appear and the secondary current change is relatively large, which indicates that the work status of SMPS has been close to the continuous mode. It shall be noted that the secondary impedance is very low and the resistance of R_2 shall be as small as possible. This is because if the secondary sampling resistance is too high, not

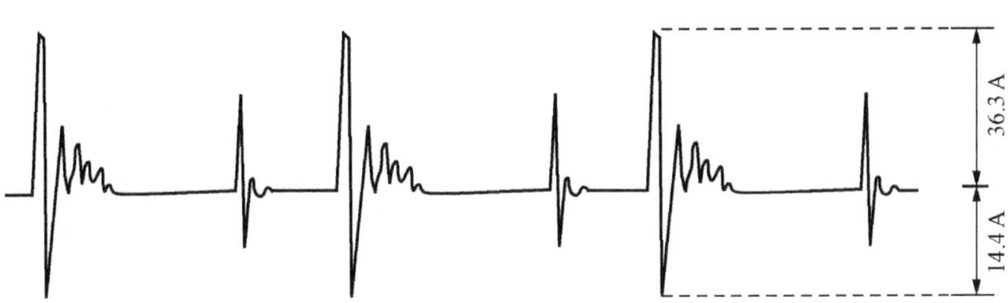

Figure 9.32 Secondary current waveform (R_9 and C_9 not connected)

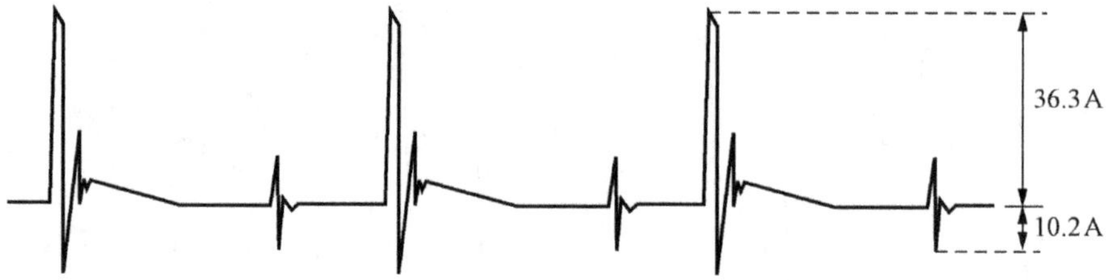

Figure 9.33 Secondary current waveform (R_9 and C_9 connected)

only the secondary power consumption will be increased but also the primary peak inverse voltage will be raised. In addition, the loss of the clamp circuit will be increased.

10

Protection and Monitoring Circuit Design of SMPS

10.1 Design of Drain Clamp Protection Circuit

For the flyback switching-mode power supply (SMPS), whenever the MOSFET is converted from turn-on into turn-off, peak and induced voltages will be generated in the primary winding of SMPS. The said peak voltage is formed due to the leakage inductance (i.e., the self-inductance generated by the magnetic leakage) of high-frequency transformer, which is superimposed with DC high voltage U_{I} and induced voltage U_{OR} on the drain of the MOSFET, and that can easily damage the MOSFET. Therefore, a drain clamp protection circuit must be added to clamp or absorb the spike voltage.

10.1.1 Potential Distribution of Voltage Parameters on the Drain

This section describes in detail the potential distribution of the six voltage parameters of the maximum input DC voltage U_{Imax}, the induced voltage U_{OR} in the primary winding, the clamp voltages U_{B} and U_{BM}, the maximum drain voltage U_{Dmax}, and the drain-source breakdown voltage $U_{\mathrm{(BR)DS}}$, so that readers can have a quantitative concept.

For the TOPSwitch-××series single-chip SMPS, the drain-source breakdown voltage $U_{\mathrm{(BR)DS}}$ of its power switching tube is no less than 700 V. Now consider the lower limit value 700 V. The induced voltage $U_{\mathrm{OR}} = 135\,\mathrm{V}$. Originally, when the clamp voltage U_{B} of clamping diode is 135 V, it can be superimposed on U_{OR} to be absorbed by the spike voltage caused by the leakage inductance. However, it is actually not the case. The parameter value of U_{B} given in the manual only refers to the value working at room temperature and small current. In fact, the clamping diode [i.e., the transient voltage suppressor (TVS)] also has a positive temperature coefficient, of which the clamp voltage at high temperature and high current is much higher than U_{B}. Experiments show that they have the following relationship:

$$U_{\mathrm{BM}} \approx 1.4\,U_{\mathrm{B}} \tag{10.1}$$

Optimal Design of Switching Power Supply, First Edition. Zhanyou Sha, Xiaojun Wang, Yanpeng Wang, and Hongtao Ma.

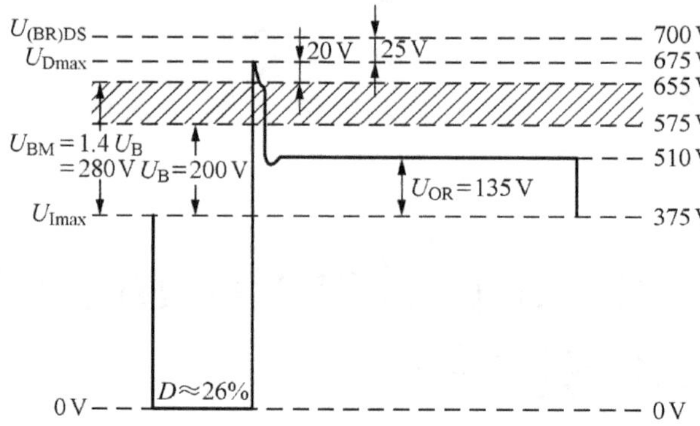

Figure 10.1 Potential distribution of voltage parameters on the drain of MOSFET

This indicates that U_{BM} is higher than U_B by about 40%. To prevent the primary induced voltage U_{OR} from being subject to the clamping action of clamping diode, the clamp voltage of the used TVS shall be calculated as per the following formula:

$$U_B = 1.5\,U_{OR} \tag{10.2}$$

In addition, the effect of the blocking diode VD_1 connected in series with the clamping diode shall be considered. Superfast recovery diode (SRD) is generally adopted for VD_1, which is featured by short reverse recovery time (t_{rr}). However, there is the forward recovery time (t_{fr}) between the reverse turn-off and the forward turn-on, so 20-V voltage margin shall be reserved.

After considering the above factors, empirical formula to calculate the maximum drain-source voltage of TOPSwitch-×× shall be

$$U_{Dmax} = U_{Imax} + 1.4 \times 1.5\,U_{OR} + 20\,\text{V} \tag{10.3}$$

When TOPSwitch-×× series single-chip SMPS is at the 230-V AC fixed input, the potential distribution of voltage parameters on the drain of MOSFET is shown in Figure 10.1 and the duty ratio is $D \approx 26\%$. In this case, $u = 230 \pm 35$ V, that is, $u_{max} = 265$ V, $U_{Imax} = \sqrt{2}\,u_{max} \approx 375$ V, $U_{OR} = 135$ V, $U_B = 1.5\,U_{OR} \approx 200$ V, $U_{BM} = 1.4\,U_B = 280$ V, $U_{Dmax} = 675$ V, and 25-V voltage margin is reserved. Therefore, $U_{(BR)DS} = 700$ V. In fact, $U_{(BR)DS}$ also has a positive temperature coefficient. $U_{(BR)DS}$ rises with the ambient temperature, so that the above-mentioned design provides the chip withstand voltage with extra margin.

10.1.2 Basic Types of Drain Clamp Protection Circuit

The following are the five main types of drain clamp protection circuits (see Figure 10.2 for the circuit):

1. See Figure 10.2(a), for the TVS- and VD-type clamping circuits constituted by the TVS (P6KE200) and the blocking diode (SRD UF4005), wherein, N_P, N_S, and N_B refer to the primary, secondary, and bias winding, respectively. However, the bias winding N_B is replaced with the feedback winding N_F in some SMPSs.

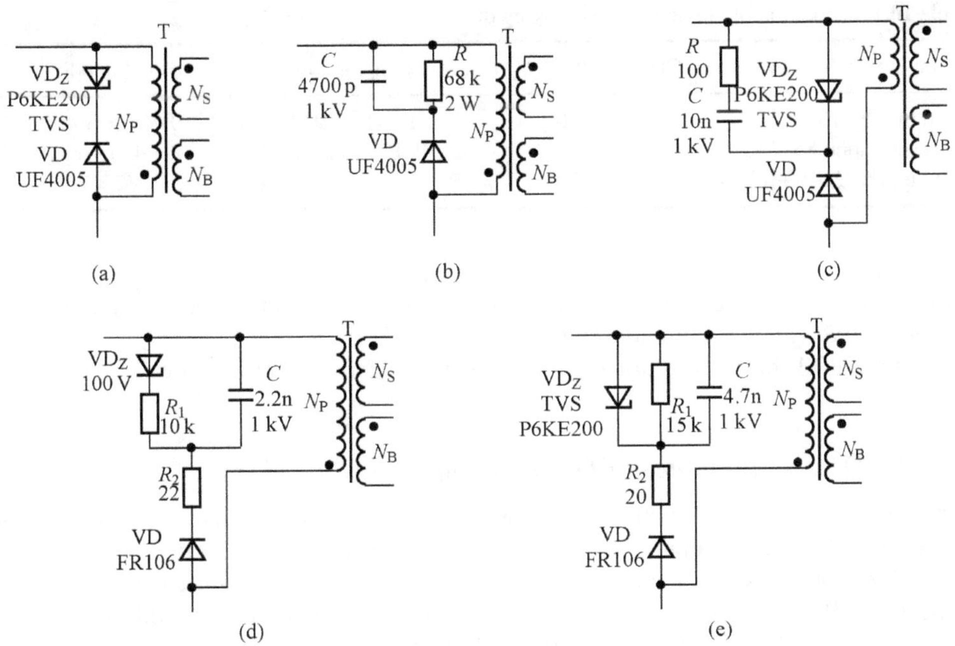

Figure 10.2 Five types of drain clamp protection circuit. (a) TVS- and VD-type clamping circuits; (b) R-, C-, and VD-type clamping circuits; (c) R-, C-, TVS-, and VD-type clamping circuits; (d) VD_Z-, R-, C-, and VD-type clamping circuits; and (e) TVS-, R-, C-, and VD-type clamping circuits

2. See Figure 10.2(b), for the R-, C-, and VD-type clamping circuits constituted by resistance-capacitance components and blocking diodes.
3. See Figure 10.2(c), for the R-, C-, TVS-, and VD-type clamping circuits constituted by resistance-capacitance components, TVS, and blocking diodes.
4. See Figure 10.2(d), for the VD_Z-, R-, C-, and VD-type clamping circuits constituted by the voltage regulator tube (VD_Z), resistance-capacitance component, and blocking diode (FRD).
5. See Figure 10.2(e), for the TVS-, R-, C-, and VD-type clamping circuits constituted by TVS, resistance-capacitance component, damping resistor, and blocking diode (FRD).

Program (5) has the best protective effect, which can give full play to the advantages of TVS such as fast response and transient high-energy pulse resistance and adds the RC absorption circuit. In view of the relatively severe discreteness of the nominal breakdown voltage ($U_{1\,\mathrm{mA}}$) and the response speed slower than that of TVS, voltage-sensitive resistor (VSR) is generally not employed to constitute the drain clamp protection circuit in SMPS.

The following two points shall be noted: firstly, FRD or SRD can generally be applied for the blocking diode. However, glass passivated rectifier VD (1N4005GP) with relatively long reverse recovery time is specially applied to recover the leakage inductance energy, thus improving the power supply efficiency. The reverse recovery time of glass passivated rectifier is between those of FRD and ordinary silicon rectifier, but 1N4005GP shall not be replaced

Table 10.1 Selection of clamping and blocking diode

AC input voltage u (V)	Clamp voltage U_B (V)	Clamping diode	Blocking diode
Fixed input: 110	90	P6KE91(91 V/5 W)	BYV26B(400 V/1A)
Universal input: 85–265	200	P6KE200(200 V/5 W)	UF4005(600 V/1A)
Fixed input: 230(1 ± 15%)	200		BYV26C(600 V/1A)

with ordinary silicon rectifier 1N4005. Secondly, clamping circuit is usually required by the AC/DC SMPS with the continuous output power less than 1.5 W.

See Table 10.1 for the selection of common clamping and blocking diodes.

10.1.3 Design Examples of Drain Clamp Protection Circuit

TOPSwitch-HX series TOP258P chip is adopted. The switching frequency is $f = 132\,\text{kHz}$ and $u = 85$–$265\,\text{V}$. The two outputs are U_{O1} (+12 V, 2 A) and U_{O2}(+5 V, 2.2 A), respectively, and $P_O = 35\,\text{W}$. The peak drain current is $I_P = I_{LIMIT} = 1.65\,\text{A}$. The measured primary leakage inductance of high-frequency transformer $L_0 = 20\,\mu\text{H}$. P6KE200 TVS is introduced. Consider $U_{Q(max)} = U_B = 200\,\text{V}$. It is proposed to apply the drain clamp protection circuit as shown in Figure 10.2(e).

The calculation steps are as follows:

Maximum allowable drain voltage: $U_{D(max)} = \sqrt{2}\ u_{max} + U_{Q(max)} \leq 700 - 50\,\text{V} = 650\,\text{V}$.

Ripple voltage of the clamping circuit: $U_{RI} = 0.1\,U_{Q(max)} = 0.1\,U_B = 0.1 \times 200\,\text{V} = 20\,\text{V}$.

Minimum clamp voltage: $U_{Q(min)} = U_{Q(max)} - U_{RI} = U_B - 0.1\,U_B = 90\%\,U_B = 180\,\text{V}$.

Average voltage of the clamping circuit: $\overline{U_Q} = U_{Q(max)} - 0.5\,U_{RI} = U_B - 0.5 \times 0.1\,U_B = 0.95\,U_B = 190\,\text{V}$.

Energy stored in the primary leakage inductance: $E_{L0} = \frac{1}{2}I_P^2L_0 = \frac{1}{2} \times (1.65\,\text{A})^2 \times 20\,\mu\text{H} = 27.2\,\mu\text{J}$.

Calculating the energy absorbed by the clamping circuit: when $P_O = 35 < 50\,\text{W}$, $E_Q = 0.8E_{L0} = 0.8 \times 27.2\,\mu\text{J} = 21.8\,\mu\text{J}$. When $P_O > 50\,\text{W}$, $E_Q = E_{L0}$.

The clamp resistance R_1 is

$$R_1 = \frac{\overline{U_Q}^2}{E_Q f} = \frac{(190\,\text{V})^2}{21.8\,\mu\text{J} \times 132\,\text{kHz}} = 12.5\,\text{k}\Omega$$

The clamp capacitance C is

$$C = \frac{E_Q}{(U_{Q(max)}^2 - U_{Q(min)}^2)/2} = \frac{2 \times 21.8}{200^2 - 180^2} = 5.7\,\text{nF}$$

Let the time constant determined by R_1 and C be τ

$$\tau = R_1C = \frac{\overline{U_Q}^2}{E_Q f} \cdot \frac{E_Q}{(U_{Q(max)}^2 - U_{Q(min)}^2)/2} = \frac{2\overline{U_Q}^2}{(U_{Q(max)}^2 - U_{Q(min)}^2)f} \tag{10.4}$$

Put $U_{Q(max)} = U_B$, $U_{Q(min)} = 90\% \, U_B$, $\overline{U_Q} = 0.95 \, U_B$ and $f = 132\,\text{kHz}$ into Formula (10.4) to get $\tau = R_1C = 9.47/f = 9.47\,T(\mu s)$. When $f = 132\,\text{kHz}$, the switching period $T = 7.5\,\mu s$ and $\tau = 9.47 \times 7.5\,\mu s = 71.0\,\mu s$. This indicates that the time constant of R_1 and C is related to the switching period and equal to 9.47 times of the switching period. When the error of resistance-capacitance component is considered, τ can be $10T$ when estimating the time constant. Actually consider the clamp resistance $R_1 = 15\,\text{k}\Omega$ and the clamp capacitance $C = 4.7\,\text{nF}$. In this case, $\tau = 70.5\,\mu s$.

The power consumption of R_1 is

$$P_{R1} = \frac{\overline{U_Q}^2}{R_1} = \frac{(190\,\text{V})^2}{15\,\text{k}\Omega} = 2.4\,\text{W}$$

Considering that the clamp protection circuit works only in the half cycle corresponding to the turn-off of power switching tube and the actual power consumption of R_1 is about 1.2 W (assume that the duty ratio is 50%),the resistor with the rated power of 2 W can be adopted. The primary DC high voltage $U_C > 1.5\,U_{Q(max)} + U_{I(max)} = 1.5 \times 200 + 265\,\text{V} \times \sqrt{2} = 674\,\text{V}$, so consider 1 kV as the actual withstand voltage.

The reverse withstand voltage of blocking diode VD $U_{BR} \geq 1.5\,U_{Q(max)} = 300\,\text{V}$, so the FRD FR106 (1A/800 V, the forward peak current up to 30 A) is applied. Its forward peak current is required to be much higher than I_P (here $30 \gg 1.65\,\text{A}$).

Note that FRD is employed for VD instead of SRD to cooperate with the damping resistance R_2 to transfer part of the leakage inductance energy to the secondary side, thereby improving the power supply efficiency.

The damping resistance shall meet the following condition:

$$\frac{20\,\text{V}}{0.8\,I_P} \leq R_2 \leq 100\,\Omega \tag{10.5}$$

Finally, calculate the damping resistance R_2 as per Formula (10.5):

$$\frac{20\,\text{V}}{0.8 \times 1.65\,\text{A}} = 15 < 100\,\Omega$$

Actually consider the nominal resistance of $20\,\Omega/2\,\text{W}$.

10.2 Overvoltage Protection Circuit Constituted by Discrete Components

Overvoltage protection is referred to as OVP. When the input or output voltage is too high, it can be applied to prevent the SMPS from being damaged by overvoltage failure.

10.2.1 Input/Output Overvoltage Protection Circuit Constituted by Silicon Controlled Rectifier (SCR)

See Figure 10.3(a) and (b) respectively for the input/output overvoltage protection circuits constituted by silicon controlled rectifier (SCR).

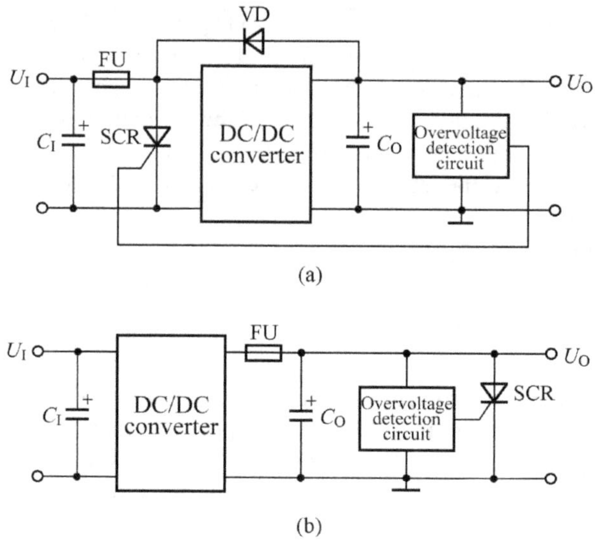

Figure 10.3 Basic principle of SCR overvoltage protection circuit. (a) SCR connected in parallel to the input terminal of switching regulator and (b) SCR connected in parallel to the output terminal of switching regulator

SCR is connected in parallel to the input terminal of DC/DC converter in Figure 10.3(a) and to the output terminal in Figure 10.3(b). U_I and U_O refer to DC input and output voltages, respectively. C_I and C_O refer to input and output capacitors, respectively. Once overvoltage failure occurs, the signal generated by the overvoltage detection circuit will trigger the SCR for turn-on. At the same time, C_I (or C_O) is discharged to rapidly reduce U_I (or U_O) for overvoltage protection. VD refers to the protection diode, which is employed to prevent C_O from discharging through the internal circuit of DC/DC converter during the turn-on of SCR. The fuse is connected in series to the inlet terminal of DC/DC converter in Figure 10.3(a) and to the output terminal of DC/DC converter in Figure 10.3(b). When the DC/DC converter itself has the overcurrent protection function, the fuse in Figure 10.3(b) can be omitted. This kind of protection circuit is also known as crowbar overvoltage protection circuit and the SCR employed here is also called crowbar SCR. The crowbar here is an action term, which means that the protection circuit can be triggered just by a small voltage increment.

The following section describes the key design points of SCR overvoltage protection circuit.

10.2.1.1 Overvoltage Detection Circuit

The overvoltage detection circuit is introduced to trigger the SCR for turn-on once the overvoltage failure is detected. The overvoltage detection circuit is required to have strong anti-interference ability to avoid malfunction. See Figure 10.4 for the overvoltage detection circuit constituted by voltage regulator tube, which is featured by simple circuit and low cost. Its working principle is as follows: when the output voltage U_O of switching regulator exceeds the reverse breakdown voltage U_Z of regulator tube, the regulator tube suffers

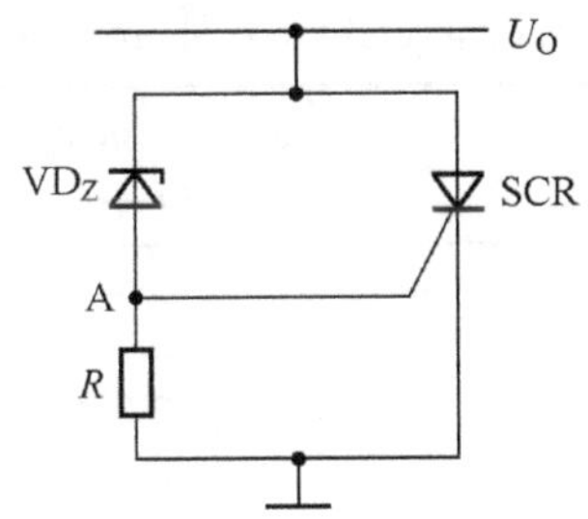

Figure 10.4 Overvoltage detection circuit constituted by voltage regulator tube

breakdown and enters into the voltage regulation area to let the potential at point A exceed the gate trigger voltage U_{GT} (typically 0.55–0.70 V) of SCR, so that the SCR is turned on. R is the bias resistance of voltage regulator tube. The disadvantages of this circuit are that the ability to drive the SCR gate is poor and the rise rate of trigger voltage is relatively low.

In addition, the overvoltage detection integrated circuit MC3423 can also be adopted as shown in Figure 10.4.

10.2.1.2 Selection of Crowbar SCR

When the overvoltage protection circuit works, the crowbar SCR is easy to be damaged or suffer performance reduction due to the surge current impact from the input or output terminal. In this case, the surge current waveform of crowbar SCR is shown in Figure 10.5, wherein, I_P is the peak current and the critical rate of rise of on-state current $\mathrm{d}i/\mathrm{d}t \approx \Delta i/\Delta t$. Table 10.2 shows the main technical indicators of several kinds of crowbar SCR for selection. They all belong to the ON Semiconductor products. The peak non-repetitive surge current (I_{TSM}) in the table refers to the maximum non-repetitive forward overload current caused by abnormal circuit conditions and letting the working temperature of SCR to be no more than the maximum junction temperature.

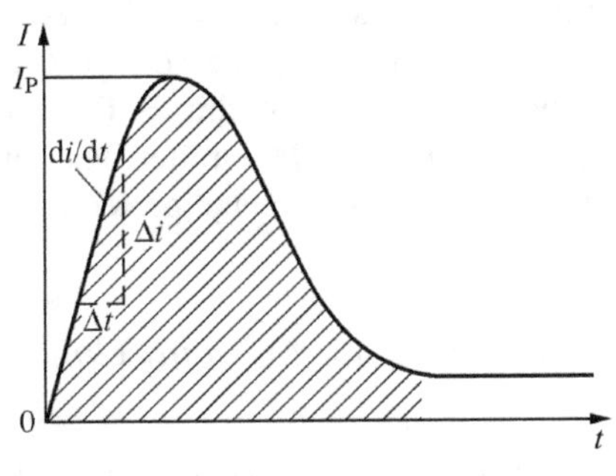

Figure 10.5 Surge current waveform of crowbar SCR

Table 10.2 Main technical indicators of several kinds of crowbar SCR

SCR Model	Peak non-repetitive surge current I_{TSM} (A)	Average on-state current $I_{T\,(AV)}$ (A)	Critical rate of rise of on-state current $\mathrm{d}i/\mathrm{d}t$ (A/μs)
MCR68-2	100	8	75
MCR69-2	100	8	75
MCR703A	25	2.6	100
MCR716	25	2.6	100

10.2.1.3 Critical Rate of Rise of On-State Current d*i*/d*t*

Critical rate of rise of on-state current ($\mathrm{d}i/\mathrm{d}t$) is the maximum rate of rise of on-state current with no adverse impact on the SCR when the SCR transfers into turn-on from turn-off under specified conditions. $\mathrm{d}i/\mathrm{d}t$ reflects the rapid conduction capability of components toward high current, and is related to the factors of components such as the turn-on speed, the turn-on consumption, the thermal parameter, and the rate of rise of trigger pulse. When driving the gate area of SCR, a period of time is needed to expand the area, which is very small at the beginning and then gradually expands. When a relatively high anode current flows through the area, instant high temperature will be generated at some part due to the high current density, which may damage the SCR or affect the service life of components. Therefore, in any case $\mathrm{d}i/\mathrm{d}t$ shall not exceed the specified value of component and a certain margin shall be reserved. At high frequencies, it is also necessary to properly reduce the on-state current and improve the heat dissipation conditions.

The surge current and $\mathrm{d}i/\mathrm{d}t$ can be reduced by connecting a small resistor in series to the anode of SCR.

10.2.1.4 Output Overvoltage Protection Circuit Constituted by Discrete SCR

See Figure 10.6 for the output overvoltage protection circuit constituted by discrete SCR.

Two PNP and NPN transistors VT_1 and VT_2 are employed here to constitute the discrete SCR, of which the three electrodes are respectively the anode A, cathode K, and gate (also known as control electrode) G. The feedback voltage U_{FB} provides the gate voltage U_G after the voltage division by regulator tube VD_{Z2} and resistor R_1. Under normal cases, U_G is relatively low and SCR is turned off. When overvoltage occurs at the secondary side, $U_O\uparrow\rightarrow U_{FB}\uparrow\rightarrow U_G\uparrow$ to trigger SCR for turn-on, thereby letting the voltage U_C at control terminal become low level to turn off the TOPSwitch series single-chip SMPS for protection. The sum of the regulated voltage of regulator tube VD_{Z2} and the emitter junction voltage of VT_2 is equal to $(U_{Z2}+U_{BE2})$. When $U_{FB}>U_{Z2}+U_{BE2}$, overvoltage protection will be implemented.

10.2.2 Output Overvoltage Protection Circuit Constituted by Bidirectional Trigger Diode

Bidirectional trigger diode is also known as DIAC, which belongs to the symmetrical double-ended component and can be equivalent to the NPN transistor with the base in

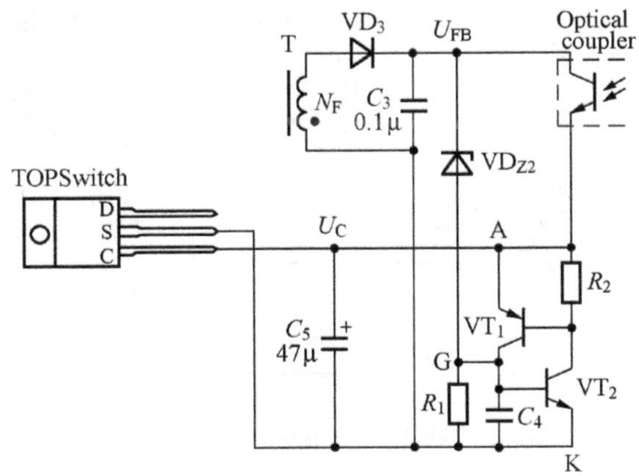

Figure 10.6 Output overvoltage protection circuit constituted by discrete SCR

open circuit and the emitter and the collector completely symmetrical. The positive and negative volt-ampere characteristics are completely symmetrical, so the component is in high impedance state when the voltage U across DIAC is less than the forward break-over voltage $U_{(BO)}$ and DIAC is turned on when $U > U_{(BO)}$. Similarly, when U exceeds the reverse breakover voltage $U_{(BR)}$, the diode can also be turned on. The symmetry of forward and reverse breakover voltages can be expressed as $\Delta U_{(B)}$. Generally, it is required that $\Delta U_{(B)} = U_{(BO)} - U_{(BR)} < 2\,V$. The bidirectional trigger diode has a simple structure and low cost, so it is often applied to constitute overvoltage protection circuit and suitable for triggering TRIAC.

See Figure 10.7 for the output overvoltage protection circuit constituted by bidirectional trigger diode. Once the breakover voltage of DIAC is exceeded by U_{FB} due to the overvoltage output, DIAC will be turned on to clamp the voltage U_{CE} of phototriode, so U_C will be reduced to turn off TOPSwitch. An MBS4991 bidirectional trigger diode is adopted in this figure, of which the forward and reverse breakover voltages are 10 V. The maximum on-state current is 2 A. The power consumption is 0.5W. R is the current limiting resistor.

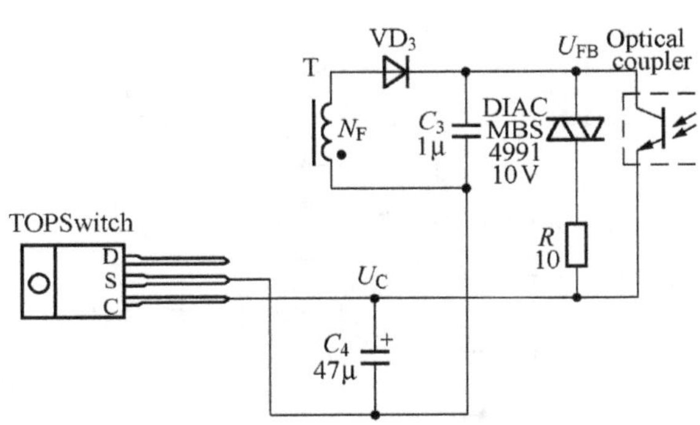

Figure 10.7 Output overvoltage protection circuit constituted by bidirectional trigger diode

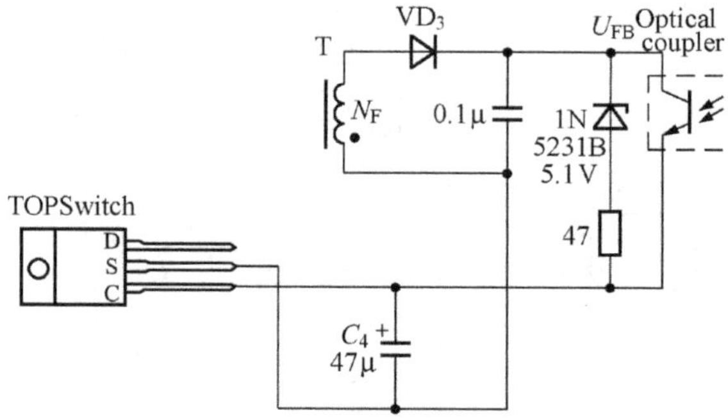

Figure 10.8 Output overvoltage protection circuit I constituted by regulator tube

10.2.3 Output Overvoltage Protection Circuit Constituted by Regulator Tube

Figure 10.8 illustrates the output overvoltage protection circuit I constituted by regulator tube. A 1N5231B-type 5.1V and 20-mA regulator tube is introduced to limit the output voltage here, which can also realize the protection action when the phototriode is damaged or the secondary winding is in open circuit. 1N5231B can be replaced with the domestic regulator tube 2CW340.

Figure 10.9 depicts the output overvoltage protection circuit II constituted by regulator tube. Its working principle is that when overvoltage failure occurs at the output terminal, the bias winding voltage will rise to let the 9.1-V regulator tube VD_Z (BZX79-B9V1) suffer reverse breakdown, thus rapidly increasing the current at terminal BP/M. When the current exceeds 7 mA, TNY377P will be turned off until the voltage at terminal BP/M is reduced to below 4.8 V after turning off the AC input, thus realizing the function of output overvoltage protection.

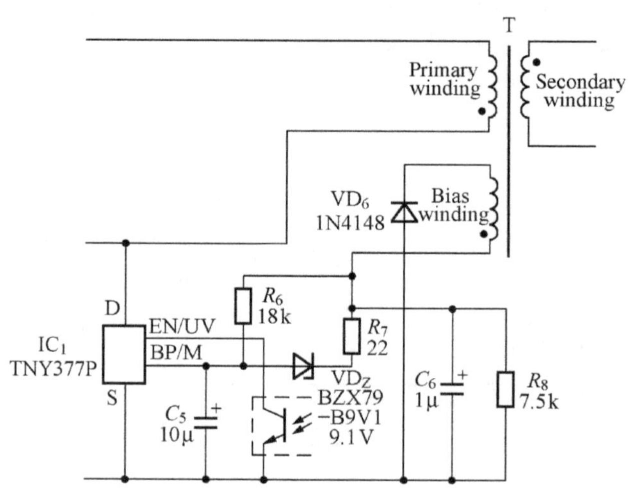

Figure 10.9 Output overvoltage protection circuit II constituted by regulator tube

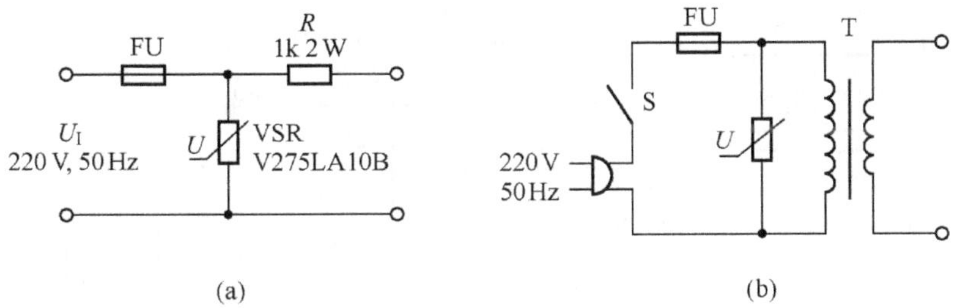

Figure 10.10 Typical application circuit of voltage-sensitive resistor. (a) Surge voltage protection circuit and (b) lightning protection circuit

10.2.4 Overvoltage Protection Circuit Constituted by VSR

Figure 10.10(a) and (b) shows the protection circuits constituted by VSR. Figure 10.10(a) is the input circuit of AC/DC converter with surge voltage protection function. V275LA10B-type VSR with the nominal voltage of 275 V (AC RMS) and the current of 2500 A is adopted for the surge voltage protection component, which can absorb up to 6 kV surge voltage for duration of 60 μs. R is the 1 kΩ and 2 W input resistor. When the AC input voltage is 120 V, V130LA10B-type VSR with the nominal voltage of 130 V (AC RMS) can be applied and R_3 can be omitted. Figure 10.10(b) is the lightning protection circuit added at the inlet terminal of user power supply in the 1 + 1 carrier telephone equipment, which can absorb the overvoltage caused by induction lightning to ensure personal and equipment safety.

10.3 Application of Integrated Overvoltage Protector

In view of the requirement of modern power technology for overvoltage protection improvement, the overvoltage protector is now developed toward high performance and integration.

10.3.1 Overvoltage Protection Circuit Constituted by NCP345

NCP345 is the new overvoltage protection IC launched by ON Semiconductor, which can be widely applied in mobile phones, digital cameras, laptop computers, personal digital assistants (PDA), portable CD players, car spare chargers, and other low-voltage power supply systems. NCP345 shall be connected between the AC/DC power adapter (or battery charger) and load, wherein, the battery charger can be Li-Ion and NiMH battery chargers. It has the overvoltage power-off and under-voltage lockout functions, capable of detecting overvoltage condition and quickly turn off the input power supply to prevent the electronic equipment from being damaged by overvoltage or power adapter failure.

Currently, many portable electronic products are equipped with AC/DC power adapter to convert AC voltage into DC voltage, charging the internal storage battery. Once the overvoltage occurs because of the failures of power adapter such as self-oscillation, sensitive electronic components will be damaged. In addition, when the battery is suddenly disconnected by the user during the charging process, a transient voltage with relatively high amplitude will be generated, which may cause damage to the product. In response to those problems,

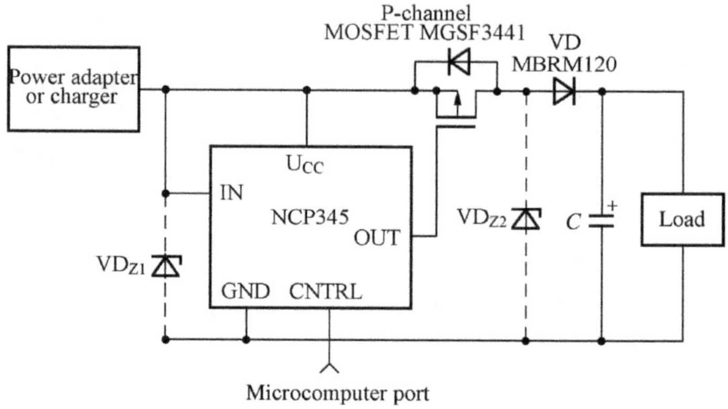

Figure 10.11 Circuit of overvoltage protector constituted by NCP345

NCP345 and P-channel MOSFET can be introduced to constitute the overvoltage protector of which the circuit is shown in Figure 10.11. P-channel MOSFET is employed as a switch. The MGSF3441-type MOSFET with internal protection diode can be adopted as a switching device. The main parameters are as follows: the drain-source voltage $U_{DS} = 20\,V$, the gate-source voltage $U_{GS} = 8.0\,V$, the drain current $I_D = 1\,A$, the maximum drain current $I_{DM} = 20\,A$, the maximum power consumption $P_{DM} = 950\,mW$, and the on-state resistance $R_{DS(ON)} = 78\,m\Omega$. VD adopts the MBRM120 (1 A/20 V) Schottky diode with low voltage drop, of which the conduction voltage drop is only 0.34 V when $I_F = 1\,A$ and $T_A = 25\,°C$. The said Schottky diode is connected in series with MOSFET to prevent the short circuit of battery. NCP345 can be applied to monitor the input voltage, allowing the turn-on of MOSFET only under safe conditions. Zener diodes VD_{Z1} and VD_{Z2} are connected in parallel to the input and load terminals, respectively, for secondary overvoltage protection.

10.3.2 Overvoltage Protection Circuit Constituted by MAX4843

MAX4843 series is the integrated overvoltage protector produced by Maxim, which includes the four models of MAX4843, MAX4844, MAX4845, and MAX4846. The input voltage range is +1.2–28 V. The overvoltage thresholds (U_{OVLO}) are 7.4, 6.35, 5.8, and 4.65 V. When the input voltage $U_{IN} > U_{OVLO}$, the internal charge pump driver is adopted to turn off the external N-channel MOSFET to prevent the protection device from being damaged. No external capacitor is needed by the internal charge pump. When the input voltage is less than the undervoltage threshold (U_{UVLO}), the chip enters into the low-current standby mode with the supply current of 10 μA only. The undervoltage thresholds of MAX4843/4844/4845 and MAX4846 are 4.15 and 2.5 V, respectively. MAX4843 series has a failure alarm output terminal, which will deliver a high-level output in the case of overvoltage or undervoltage failure. The working condition of chip can be monitored through the on-chip state machine.

See Figure 10.12(a) and (b) for the typical application circuit of *MAX*4843–*MAX*4846 series. The monitored input voltage range is +1.2–28 V. *C* is the bypass capacitor at input terminal, which is recommended to introduce 1 μF ceramic capacitor. FLAG is the failure alarm output terminal (open drain output). *R* is the pull-up resistor. In the case of overvoltage

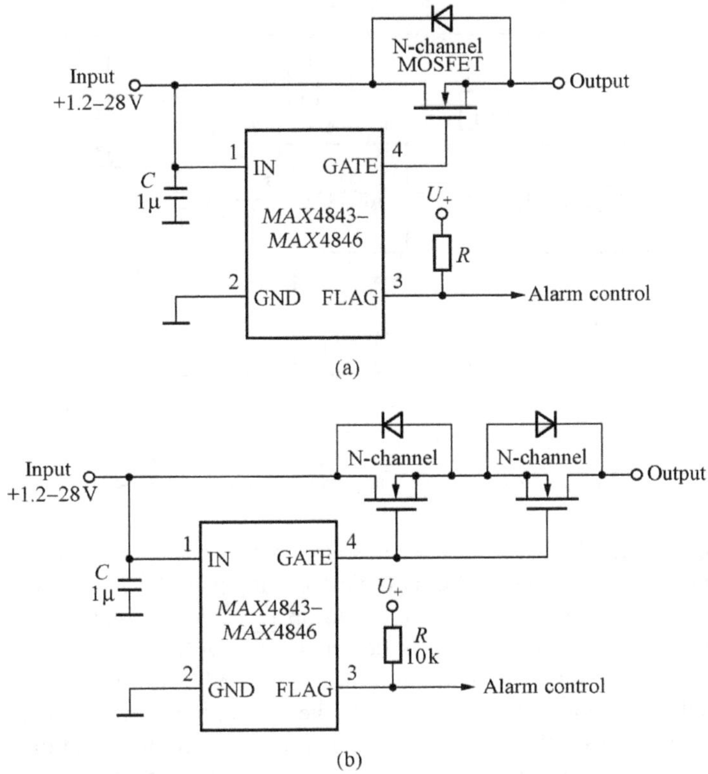

Figure 10.12 Typical application circuit of MAX4843–MAX4846 series in SMPS. (a) Driving one N-channel MOSFET and (b) driving two N-channel MOSFETs

or under-voltage lockout, terminal FLAG will be triggered to become high level after 50 ms delay time, thus delivering an alarm signal to the host system. GATE is the gate drive terminal (charge pump output), which will become high level to turn on the external N-channel MOSFET when $U_{UVLO} < U_{IN} < U_{OVLO}$. Terminal GATE can drive not only a single N-channel MOSFET, but also two back-to-back N-channel MOSFETs, of which the circuit is shown in Figure 10.12(b). In addition, the overvoltage and undervoltage thresholds of the host system can be changed by adding a resistance divider at the input terminal.

10.3.3 Overvoltage Protection Circuit Constituted by MC3423

MC3423 is an overvoltage detection IC produced by ON Semiconductor dedicated to SCR driving, which is featured by programmable overvoltage threshold, programmable trigger delay time, indicator output terminal, remote on/off control, strong anti-interference ability, and so on. The mains voltage range of MC3423 is +4.5–40 V. The output current is up to 300 mA. The rate of rise is 400 mA/μs.

Figure 10.13 illustrates the typical application circuit of MC3423. PDIP-8 or SOIC-8 package is adopted for MC3423. U_{CC} and U_{EE} are connected to the positive and negative terminals of mains voltage, respectively. Sense 1 and Sense 2 are two overvoltage detection

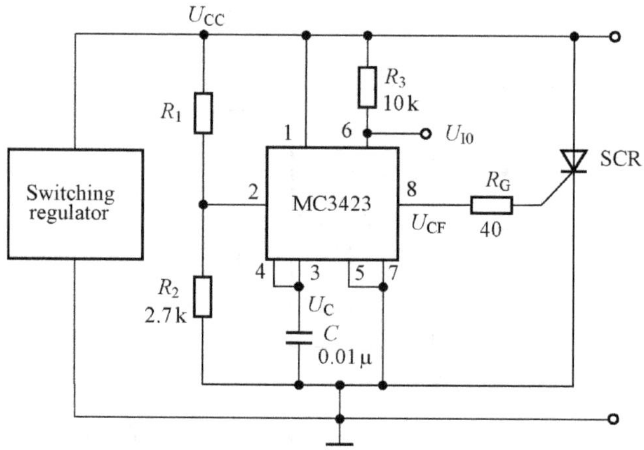

Figure 10.13 Typical application circuit of MC3423

input terminals. Current Source is the current source output terminal. OUT is the drive output terminal, connected to the gate of SCR. Indicator Output is the indicator output terminal (open collector output, connected with the pull-up resistor externally), which can deliver low-level output when an overvoltage is detected. $\overline{\text{ON}}$/OFF is the remote on/off control terminal (also known as remote activation terminal). MC3423 can work normally when the terminal $\overline{\text{ON}}$/OFF is connected to high level. In the case of high-level connection, the output of MC3423 will be turned off and the overvoltage protection circuit is out of work.

The monitored voltage is connected to Sense1 (pin 2) after the voltage division of R_1 and R_2. Sense 2 (pin 3) shall be shorted with the current source output terminal (pin 4) and then be grounded through capacitor C, which is a delay capacitor. When the remote on/off control terminal (pin 5) is not applied, it shall be connected to U_{EE} (pin 7). R_3 is the pull-up resistor of indicator output terminal (pin 6). The drive signal from the drive output terminal (pin 8) is connected to the gate of SCR through R_G.

MC3423 also has a remote on/off control function. This function can be employed to turn off the regulators in the specified order when the system power supply fails. See Figure 10.14

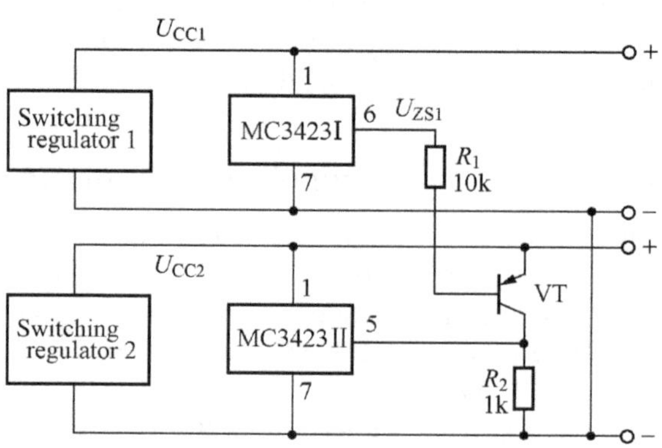

Figure 10.14 Remote ON/OFF control circuit

for the remote on/off control circuit. Two pieces of MC3423 (MC3423 I and MC3423 II) are adopted here respectively for overvoltage protection of the switching regulators 1 and 2, in which case, the indicator output signal U_{ZS1} of MC3423 I is introduced to control the turn-on and turn-off of MC3423 II. The features of the circuit are as follows: when U_{ZS1} is low level, VT is turned on to pull the terminal $\overline{\text{ON}}/\text{OFF}$ of MC3423 II to high level, thus letting the protection function of MC3423 II be out of work. When U_{ZS1} is high level, VT is turned off to pull the terminal $\overline{\text{ON}}/\text{OFF}$ of MC3423 II to low level through R_2, thus activating the protection function of MC3423 II.

10.4 Design of Undervoltage Protection Circuit

Undervoltage protection is briefly known as UVP. When the input voltage is too low, the UVP circuit can be employed to prevent the switching power supply IC from being damaged.

10.4.1 Input UVP Circuit Constituted by Optical Coupler

Figure 10.15 shows the input UVP circuit constituted by optical coupler. When the DC input voltage U_I is less than the lower limit, the base potential of VT $U_B < 4.7\,\text{V}$ after the voltage division of R_1 and R_2, so VT and VD_4 are turned on and the voltage U_C of control terminal is less than 4.7 V to immediately turn off TOPSwitch. It is easy to see that

$$U_B = \frac{U_I R_2}{R_1 + R_2} \tag{10.6}$$

Therefore,

$$R_2 = \frac{U_B}{U_I - U_B} \cdot R_1 \tag{10.7}$$

Assume that $U_I = 100\,\text{V}$ in the case of undervoltage. Consider $R_1 = 1\,\text{M}\Omega$ and $U_B = 4.4\,\text{V}$ and put them into Formula (10.7) to get $R_2 = 46.4\,\text{k}\Omega$. To reduce the power consumption of

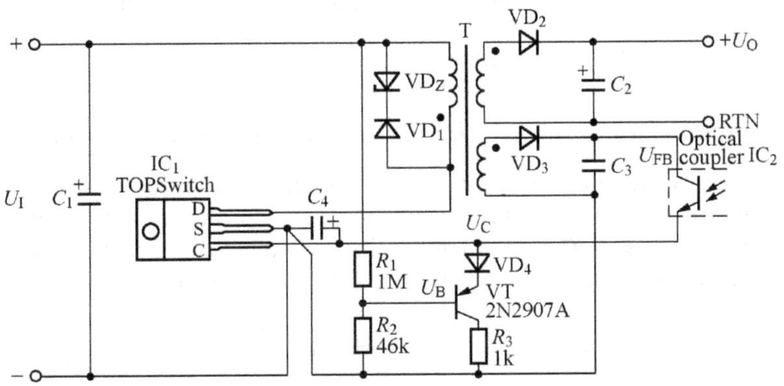

Figure 10.15 Input undervoltage protection circuit constituted by optical coupler

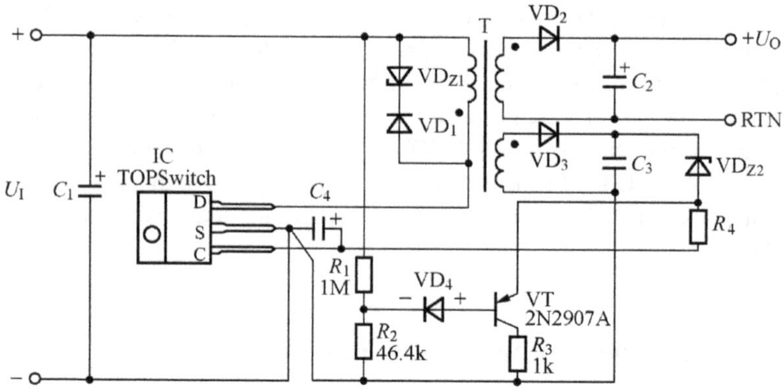

Figure 10.16 Input undervoltage protection circuit constituted by bias winding

the protection circuit, the feedback voltage U_{FB} shall be designed to be 12 V. The PNP-type transistor 2N2907A can also be replaced with JE9015.

When the AC voltage u is suddenly powered down, U_I will decrease with the discharge of C_1 to reduce U_O. Once U_O is reduced to a value outside the automatic voltage regulation range, C_4 will start discharge, which can also turn off the TOPSwitch.

10.4.2 Input UVP Circuit Constituted by Bias Winding

TOPSwitch series single-chip SMPS has no internal undervoltage protection circuit, but the input UVP function can be added just through the peripheral circuit. See Figure 10.16 for the input UVP circuit constituted by bias winding. When U_I is undervoltage, the transistor VT is turned on and U_C is low level to turn off the TOPSwitch. When U_I becomes normal again, VD_4 and VT are turned off and TOPSwitch turns to normal work. The circuit can also avoid accidental start of TOPSwitch and allows restart only when U_I is higher than the undervoltage. VD_4 can prevent the emitter junction reverse voltage of VT from being too high. Similarly, when the AC power supply is suddenly powered down, the circuit can also play a protective role.

10.4.3 Switching Regulator with Under-Voltage Lockout Function

See Figure 10.17 for the under-voltage lockout circuit of LM2576-××. It is featured that when the input voltage is less than the specified threshold, the regulator is in the off state to play a protective role. Let the regulated voltage of regulator tube VD_Z be U_Z, and the emitter junction voltage of transistor VT be U_{BE}. Therefore, the undervoltage threshold is set to be

$$U_{UV} = U_Z + U_{BE} \tag{10.8}$$

When $U_I < U_{UV}$, the regulator tube does not work. The base of transistor is grounded through R_2 to turn off VT and the collector is high level. The on/off control terminal

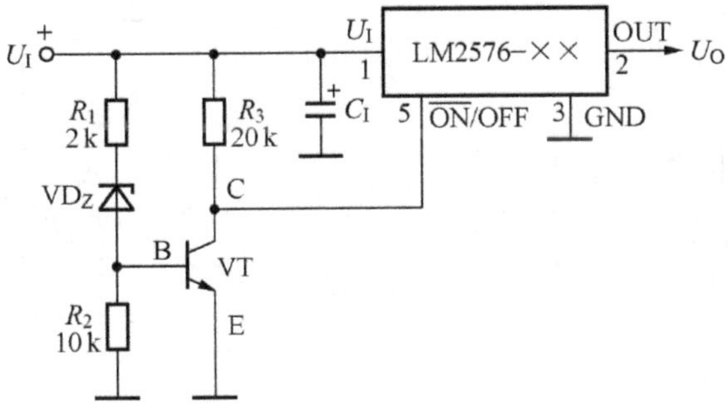

Figure 10.17 Switching regulator circuit with under-voltage lockout function

$\overline{ON}/OFF = 1$ (high level), so the output of LM2576- ×× is turned off to play a protective role. Only when $U_I > U_{UV}$, the regulator tube suffers breakdown and enters into the regulated area to let VT be turned on, in which case, $\overline{ON}/OFF = 0$ and LM2576 can work normally.

10.4.4 External Drive Circuit Realizing Overvoltage and Undervoltage Control

See Figure 10.18(a)–(f) respectively for the six kinds of external drive circuit realizing overvoltage and undervoltage control. Figure 10.18(a) shows the drive circuit constituted y two transistors 2N1711. When the overvoltage signal $\overline{OV} = 0$ (low level), VT_1 is turned off to turn on VT_2, thus driving the external control circuit.

Figure 10.18(b) shows the drive circuit constituted by TIP-110 Darlington tube and relay. TIP-110 can control the 2 A load current and drive high-power relays. In fact, the Darlington tube itself can be applied as a switching device and it plays the role similar to that of MOSFET and IGBT. Its working principle is as follows: when $\overline{OV} = 0$ (low level), VT_1 is turned off, VT_2 is turned on and the relay is closed through absorption to switch on the normally open contact (J) for overvoltage protection. VD is the freewheeling diode, which can provide the reverse potential generated by the relay at the moment of power-off with a bleed-off circuit to prevent the Darlington tube from being damaged. On the basis of Figure 10.18(b), an overvoltage display and alarm circuit constituted by a 3-V battery pack, current limiting resistor, and LED can be added.

The undervoltage signal $\overline{UV}$ is employed for Figure 10.18(c) and (d) to drive the heating elements respectively through P-channel and N-channel MOSFETs. When the external power supply $U_+ = 12\,V$, R is $2.4\,k\Omega$. When $U_+ = 6\,V$, $R = 1.2\,k\Omega$. The resistance error of ±5% is allowed.

Figure 10.18(e) shows the circuit adopting the undervoltage signal $\overline{UV}$ to control the motor or compressor through the IRGBC40S-type insulated gate FET-bipolar transistor (IGBT).

In Figure 10.18(f), the undervoltage signal $\overline{UV}$ drives the AC load through the MOC3011-type optical coupler (containing the 0.1A/400 V bidirectional trigger diode) and 4A/400 V bidirectional SCR 2N6073A.

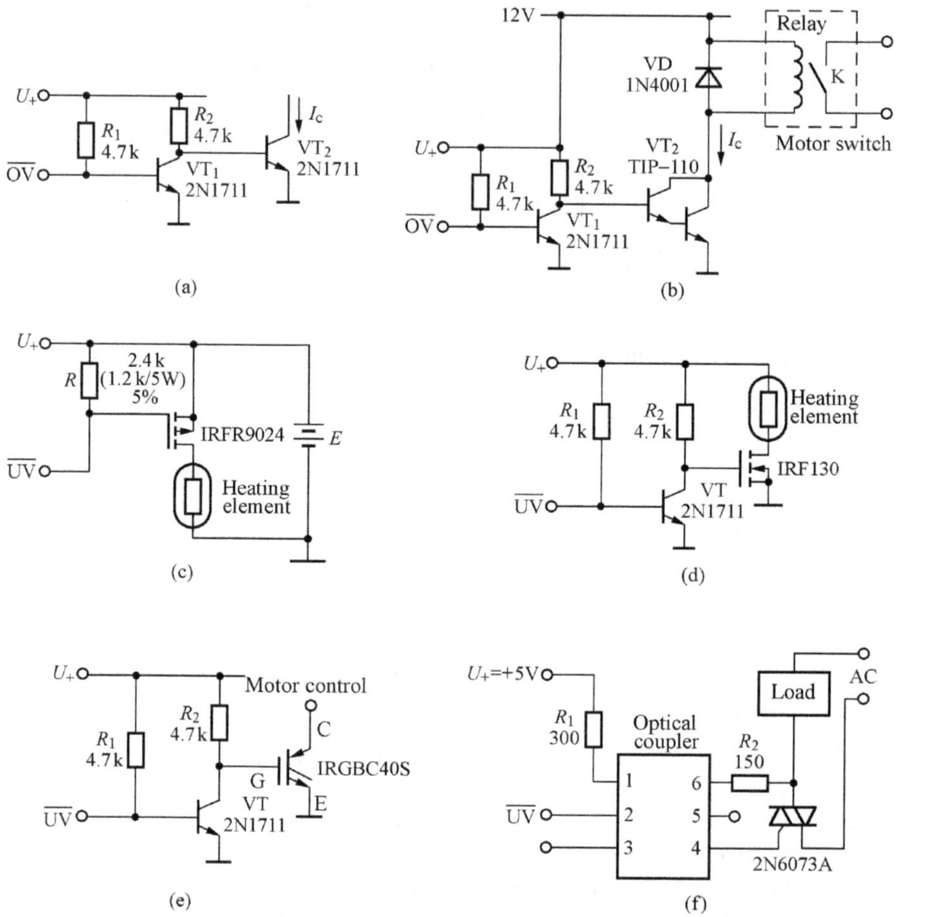

Figure 10.18 Six kinds of drive circuit realizing overvoltage and undervoltage control. (a) Driving with two power transistors, (b) driving the relay with Darlington tube, (c) driving with P-channel MOSFET (d) driving with N-channel MOSFET, (e) driving with IGBT, and (f) driving with optical coupler and bidirectional SCR

10.5 Design of Overcurrent and Overpower Protection Circuit

Overcurrent protection (OCP) circuit can be adopted to prevent SMPS overload. Over power protection circuit can be implemented to protect the load.

10.5.1 Current Limiting Protection Circuit Constituted by Power Thermistor

Power thermistor belongs to the negative temperature coefficient resistor (NTCR). Its resistance decreases with the rise of temperature within the operating temperature range. The resistance temperature coefficient α_T is generally $-(1–6)\%/°C$. When the temperature rises

Table 10.3 Typical power thermistor products

Model	Nominal resistance (Ω)	Rated current (A)	Rated power (W)
8-101	10	1	10
5-052	5	2	20
10-103	10	3	90
13-056	5	6	180
15-204	20	4	320
15-473	47	3	423

substantially, the resistance can be reduced by three to five orders of magnitude. The power thermistor is featured in low nominal resistance (only 1–47 Ω), high rated power (10–500 W), high working current (1–10 A) and so on, especially suitable for being applied as the start protection component of various power supplies.

See Table 10.3 for the typical power thermistor products, wherein, the number in the front of model refers to diameter (mm), the number in the middle refers to the nominal resistance, and the number at the end refers to the rated current (A).

See Figure 10.19 for the current limiting protection circuit of SMPS constituted by power thermistor, which is employed for current limiting protection here. The 220 V AC first becomes the pulsating direct current after passing through the bridge rectifier, and then passes through the 5-052-type power thermistor and filter capacitor respectively to generate about +300 V DC high voltage, which is sent to the pulse width modulator (PWM). At the beginning of power-on, the voltage drop of filter capacitor C cannot change suddenly and the capacitive reactance tends to be zero, so the instantaneous charging current is very high, and easy to damage the 200 μF/400 V high-voltage electrolytic capacitor. To solve this problem, the "hard start" approach is usually applied, that is, connecting in series a low-resistance current limiting resistor R of several ohms in the circuit. However, the resistance of ordinary resistor is basically constant, so the power consumption of R will inevitably reduce the power supply efficiency after the SMPS enters into normal work. The better solution is to replace the ordinary current limiting resistor with power thermistor and the current limiting value can be slightly higher. Its work feature is that at the beginning of power-on, the resistance of R is relatively high with good instantaneous current limiting effect and with the heat emitted by the current passing through, its resistance decreases rapidly to significantly reduce the power consumption, truly achieving two aims.

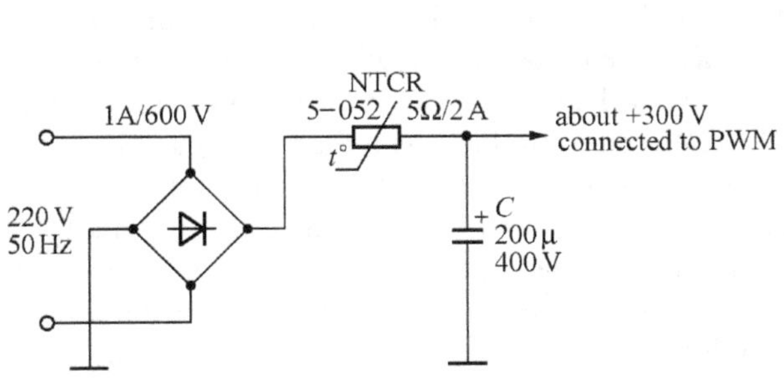

Figure 10.19 Current limiting protection circuit of SMPS constituted by power thermistor

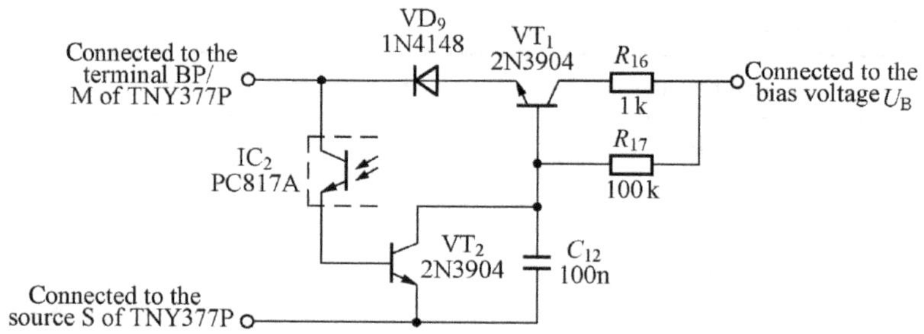

Figure 10.20 Overcurrent protection circuit constituted by transistor

10.5.2 Overcurrent Protection Circuit Constituted by Transistor

Figure 10.20 shows the overcurrent protection circuit constituted by transistor. Two 2N3904-type NPN transistors VT_1 and VT_2 are adopted for the circuit, a diode VD_9 (1N4148), R_{16} and R_{17}. Once overload occurs, the output voltage will decrease. The phototransistor in optical coupler PC817A will cause VT_2 to be turned off due to lack of feedback current. Capacitor C_{12} is charged by the bias voltage U_B to turn on VT_1. In this case, the current flowing into terminal BP/M of TNY377P through VD_9 exceeds 7 mA, which quickly turns on TNY377P to realize the overcurrent protection function.

10.5.3 Overcurrent Protection Circuit Constituted by Resettable Fuse

Resettable fuse is referred to as RF. It is a new type of overcurrent protective device launched in 1990s. Traditional fuses are one-off overcurrent protectors, not convenient for application. The resettable fuse made from a polymer mixed with conductor can satisfactorily resolve these problems. The resettable fuse has the advantages such as small size, complete specifications, good switching characteristics, automatic recovery, reuse, and maintenance elimination, which can be widely applied in SMPSs, electronic instruments, household appliances, computers, and communications equipment for automatic protection.

The overcurrent protection function can be realized just by connecting the resettable fuse in series to the DC input terminal of switching regulator, of which the circuit is shown in Figure 10.21. In this case, the fuse at the AC input terminal can be omitted. The switching regulators of L4960, L4970A and other series are suitable for the circuit. It shall be noted that the resettable fuse is generally employed for low-voltage overcurrent protection, which cannot be connected to 220-V AC voltage. The DC input voltage range of L4960 is $U_I = 9$–46 V, which can employ the RXE series products with the operating voltage less than 60 V.

A coaxial cable is usually introduced for satellite television receiving equipment to transfer television signals. When overcurrent occurs at the user's interface of coaxial cable due to short-circuit failure, it is easy to damage the television components. Therefore, a resettable fuse can be connected in series to the power supply of television as shown in Figure 10.22, which will become high impedance state in the case of overcurrent failure to effectively protect the television. After the failure is removed, the resettable fuse can automatically recover without affecting the normal work of television.

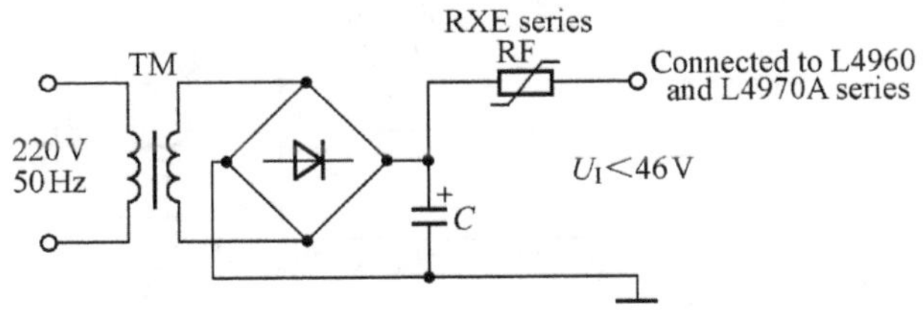

Figure 10.21 Overcurrent protection circuit constituted by resettable fuse

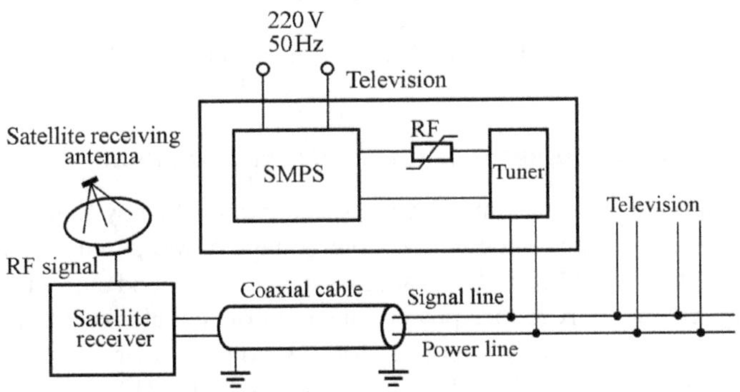

Figure 10.22 Satellite television receiver protection circuit constituted by resettable fuse

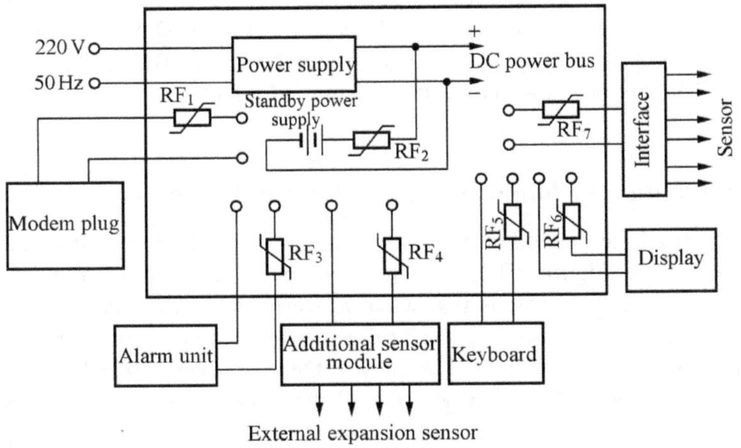

Figure 10.23 Automatic alarm system protection circuit constituted by resettable fuse

Figure 10.23 shows the automatic alarm system protection circuit constituted by resettable fuse. The circuit can prevent any unit from being damaged due to overcurrent or overheating. This design is also in compliance with the security requirements of US UL864 certification.

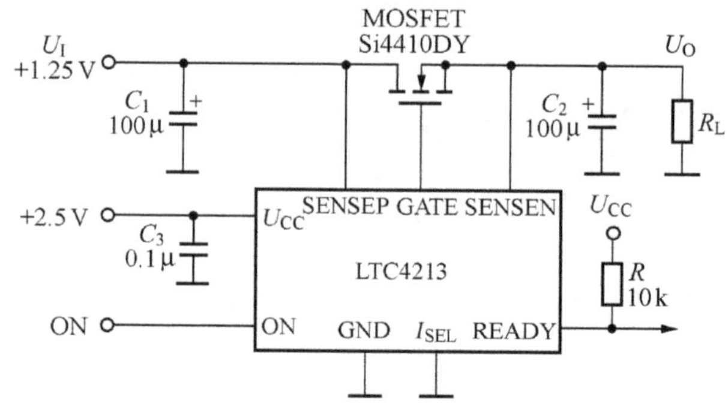

Figure 10.24 Typical application circuit of LTC4213

10.5.4 Application of Integrated Overcurrent Protector

LTC4213 is the integrated overcurrent protector recently launched by Linear Technology, suitable for the overcurrent protection devices of low-voltage power supply system. The on-state resistance $R_{DS(ON)}$ of external MOSFET is applied to detect the load current, so no sense resistor is needed. This can not only reduce power consumption, but also reduce costs and simplify circuit design, which is particularly important for low-voltage power supply system.

Figure 10.24 shows the typical application circuit of LTC4213. U_I is connected to load R_L through MOSFET. C_1 and C_2 are bypass capacitors at the input and output terminals, respectively. When ON is connected to high level, LTC4213 works normally. Si4410DY-type FET and $R_{DS(ON)} = 0.015\,\Omega$ (typical value) are adopted for the circuit. When terminal I_{SEL} is connected to GND, $U_{DS} = 25\,\text{mV}$. It is not difficult to calculate that the current threshold at mild overload is $I_{LIMIT} = U_{DS}/R_{DS(ON)} = 25\,\text{mV}/0.015\,\Omega = 1.67\,\text{A}$. At severe overload, the current threshold is $I'_{LIMIT} = U_{DS\ (FAST)}/R_{DS\ (ON)} = 100\,\text{mV}/0.015\,\Omega = 6.67\,\text{A}$. The normal load current is 1A. R is the pull-up resistor at terminal READY.

10.5.5 Application of Integrated Over-Power Protector

Chapter 9 introduced the method to measure the load power of SMPS with MAX4211A/B/C (see Figure 9.9). See Figure 10.25 for the overpower protection circuit constituted by MAX4211. Its working principle is that when a overpower fault is detected, the protection circuit will turn off the load current. The circuit can effectively prevent the battery from being damaged due to short circuit fault or overpower. Once an over power fault is detected, the P-channel MOSFET (V) will be turned off until the manual reset button is pressed. At the same time, the input power will let pin LE become low level to relieve the latching of OUT1 of the comparator 1 and reset the protection circuit.

During power-on or when load characteristics are changed, the load may generate surge current to form a relatively high voltage at terminal POUT, thus letting the voltage at terminal CIN+ exceed the reference voltage at CIN−. In this case, terminal COUT1 will become high level to trigger the protection circuit, causing malfunction. To avoid this situation, an RC network (R_4 and C_1) can be connected to deliver high-level input to the terminal INHBIT of comparator 1. In this period, comparator 1 suspends work. The suspension time is determined

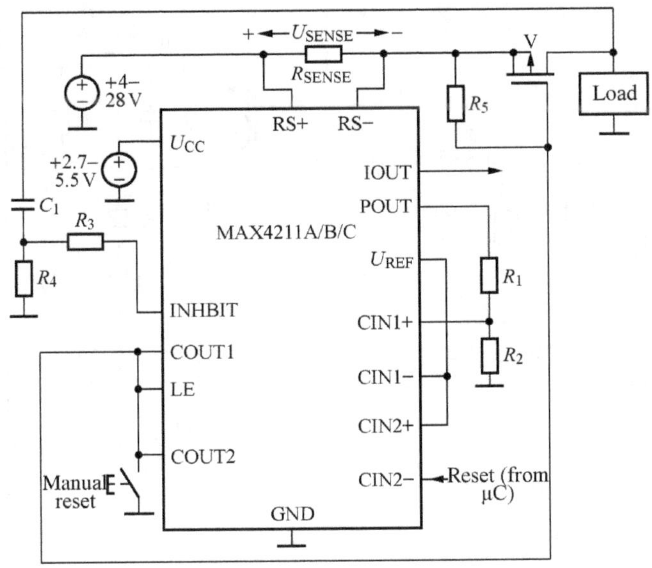

Figure 10.25 Overpower protection circuit constituted by MAX4211

by the following formula:

$$t = R_4 C_1 \ln(\Delta U/0.6) \tag{10.9}$$

wherein, ΔU is the variance of load voltage. It shall be noted that the RC network does not affect the protection of long-term overpower (too high load power or short circuit of load). R_3 is the current limiting resistor at terminal INHIBIT, typically 10 kΩ. Overcurrent fault can be detected by following the above circuit. The specific method is changing the connection position of resistance dividers R_1–R_2 from terminal POUT to terminal IOUT.

10.6 Design of Soft-Start Circuit

The so-called soft-start means gradually increasing the output voltage of SMPS to the rated voltage at the beginning of power-on to protect the SMPS.

10.6.1 Soft-Start Circuit

10.6.1.1 Optocoupler Feedback Soft-Start Circuit

The soft-start capacitor C_{SS} can be added to eliminate the circuit impact at the moment of power-on, thus letting the output voltage rise smoothly. See Figure 10.26(a) and (b) respectively for the two kinds of optocoupler feedback soft-start unit circuit. Figure 10.26(a) shows adding a soft-start capacitor to the optocoupler feedback circuit with regulator tube. Figure 10.26(b) shows adding a soft-start capacitor to the precision optocoupler feedback circuit. C_{SS} can limit the spike current generated during the turn-on of LED in the optocoupler,

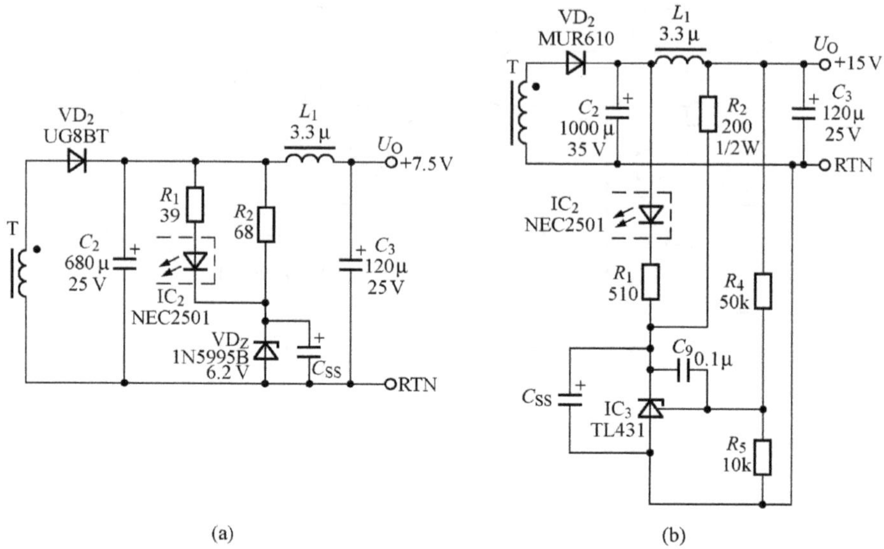

Figure 10.26 Two kinds of optocoupler feedback soft-start circuit. (a) Ordinary optocoupler feedback circuit and (b) precision optocoupler feedback circuit

thereby limiting the duty ratio. C_{SS} does not work during normal operation. After power-off, C_{SS} is discharged through R_2. 4.7–47 μF electrolytic capacitors can be introduced for the soft-start capacitor.

10.6.1.2 Basic Feedback Soft-Start Circuit

Soft-start capacitor can be applied to eliminate the turn-on spike voltage of basic feedback single-chip SMPS. See Figure 10.27 for the basic feedback soft-start circuit. At the moment the TOPSwitch is turned on, the soft-start capacitor C_{SS} can increase the current I_C at the control terminal to limit the duty ratio, letting the output voltage tend to be stable. When the power supply is turned off, C_{SS} is discharged through the resistor R_D.

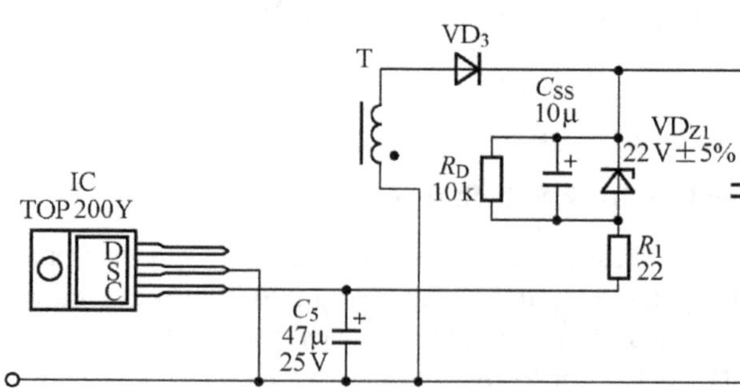

Figure 10.27 Basic feedback soft-start circuit

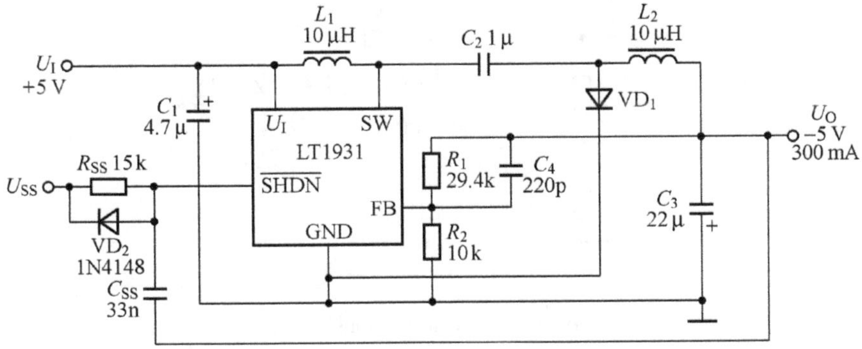

Figure 10.28 Circuit of +5/ − 5 V power supply converter with soft-start function

10.6.2 +5/−5 V Power Supply Converter with Soft-Start Function

The soft-start circuit can be added to let the output voltage rise to the rated value after a period of delay time, avoiding overload at the output terminal at the beginning of startup. See Figure 10.28 for the circuit of +5/ − 5 V power supply converter with soft-start function. R_{SS} and C_{SS} are respectively the soft-start resistance and capacitor. When $R_{SS} = 15\,\text{k}\Omega$ and $C_{SS} = 33\,\text{nF}$, the soft-start time is about 1 ms. This indicates that U_O reaches −5 V after 1 ms. VD_2 is employed to rapidly discharge C_{SS} when LT1931 is turned off. The soft-start time can be reset just by changing the time constant $\tau = R_{SS}C_{SS}$.

10.6.3 Switching Regulator with Delayed Start Function

Figure 10.29 shows the delayed start circuit designed for LM2576- × ×. The pin $\overline{\text{ON}}$/OFF can be adopted to realize the delayed start function. At the beginning of power-on, the voltage drop across C_D cannot be changed suddenly, so $\overline{\text{ON}}$/OFF = 1 and LM2576- × × has no output. With rapid charging of C_D, pin $\overline{\text{ON}}$/OFF becomes low level, so the regulator enters into normal work. When the input voltage is 20 V, the start time of LM2576- × × will be delayed by about 10 ms. The start time can be extended by increasing the time constant ($\tau = R_DC_D$) of R_D and C_D. However, if the time constant is too large, the ripple above 50 Hz or 100 Hz will be generated at the input terminal after the coupling of pin $\overline{\text{ON}}$/OFF.

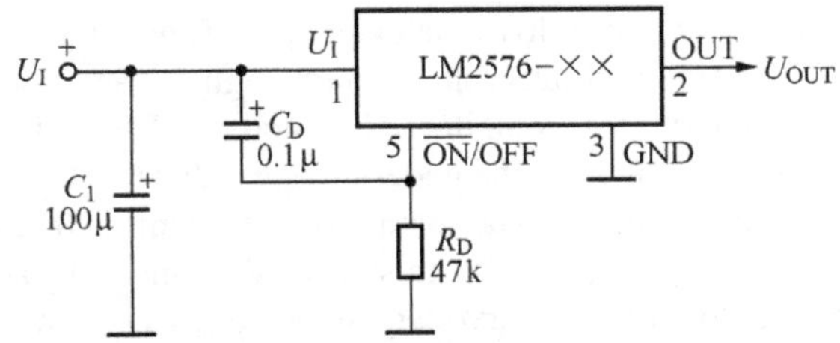

Figure 10.29 Circuit of switching regulator with delayed start function

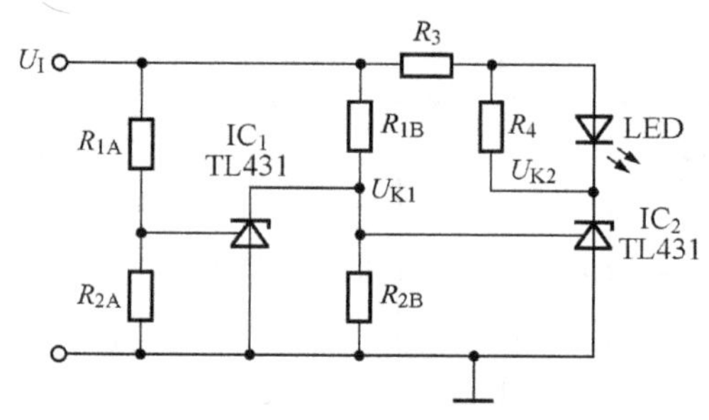

Figure 10.30 Voltage monitor circuit

10.7 Mains Voltage Monitor

10.7.1 Voltage Monitor Constituted by TL431

See Figure 10.30 for the voltage monitor circuit constituted by two programmable precision shunt linear regulators TL431 (IC_1 and IC_2). Now LED is introduced as the mains voltage indicator under normal state. The upper voltage limit (U_H) and lower voltage limit (U_L) are respectively determined by the following formulae:

$$U_H = \left(1 + \frac{R_{1A}}{R_{2A}}\right) U_{REF} \tag{10.10}$$

$$U_L = \left(1 + \frac{R_{1B}}{R_{2B}}\right) U_{REF} \tag{10.11}$$

R_3 is the current limiting resistor of LED. The resistance of R_4 shall make the cathode current of IC_2 to be greater than 1 mA. IC_1 and IC_2 can be equivalent to two paralleling switches. Only when the mains voltage is normal, that is, $U_H > U_I > U_L$, can LED lights up to indicate that the monitored voltage U_I meets requirements. Once $U_I > U_H$, that is, overvoltage, IC_1 will be turned on to cause $U_{K1} \downarrow$, which makes IC_2 be turned off to cause $U_{K2} \uparrow$, so LED goes out owing to its cathode connecting to high potential. When $U_I < U_L$, that is, undervoltage fault, IC_1 and IC_2 will be turned off simultaneously to make the LED go out.

10.7.2 Undervoltage and Overvoltage Monitor Constituted by LM3914

The LED driver of LM3914 can also drive the external TTL gate circuit (e.g., 74LS04 HEX inverter) to obtain undervoltage and overvoltage alarm signals. See Figure 10.31 for the circuit of undervoltage and overvoltage monitor. The anode of LED_2–LED_{10} is connected to U_{CC} through R_9 and LED_1 is applied as the undervoltage indicator. C_1 is a decoupling capacitor. C_2 is a noise cancelling capacitor. The feature of the circuit is that when $U_{CC} < 4.51\,V$, LED_2–LED_{10} are turned off (going out). In this case, VT_1 and VT_2 are turned off and the output of VT_2 is high level to let the undervoltage indicator LED_1 light up.

When $U_{CC} = 4.51$–$5.40\,V$, there must be an LED in LED_2–LED_{10} lights up (turned on) to make VT_2 turned on and deliver low-level output, thus letting LED_1 go out.

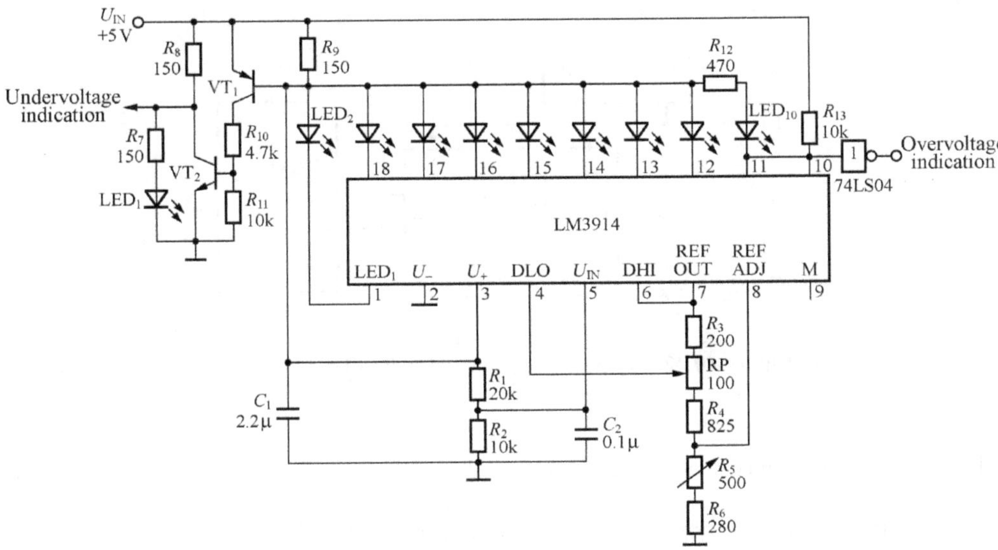

Figure 10.31 Undervoltage and overvoltage monitor

When $U_{CC} > 5.41$ V, the output of 74LS04 is high level, indicating overvoltage fault. The low-level output of TTL circuit is set by R_8, and the high level is provided by the pull-up resistor R_9. The undervoltage and overvoltage signals can be adopted to realize the on/off control of linear regulator. The output of VT_2 can also drive other TTL circuits.

10.7.3 Mains Voltage Monitor Constituted by HYM705/706

HYM705 and HYM706 are the low-cost microprocessor (μP) and microcontroller (MCU) monitoring circuits with watchdog produced by Wuhan Hao Yu Microelectronics Co., Ltd. It is suitable for the fields such as CNC power supply, battery-powered system, intelligent instrument, personal digital assistant (PDA), communications system, and automotive electronics, capable of significantly improving system reliability.

HYM705/706 has the following four functions: (i) When the power supply is turned on, turned off or fluctuates, a reset signal will be generated. (ii) It has independent watchdog output. When the watchdog output terminal is not triggered within 1.6 s, the watchdog output will become low level. (iii) There is a voltage detector internally with the power failure threshold of 1.25 V, which can realize the functions such as power failure alarm, low battery detection and monitoring power supplies out of +5 V. (iv) There is a manual reset input terminal with the reset pulse width of 200 ms (active low) and the lagging voltage of 40 mV. The reset threshold voltage of HYM705 is 4.65 V and that of HYM706 IS 4.40 V. There are the mains voltage monitoring comparator and power failure comparator in the chip. Power failure comparator is employed to monitor other voltages even negative, of which the internal threshold voltage is 1.25 V. The external threshold voltage can be set just by connecting a resistance divider to its input terminal. When the power failure time exceeds 1.6 s, the watchdog delivers high-level output to reset the system. The manual reset signal is compatible with the

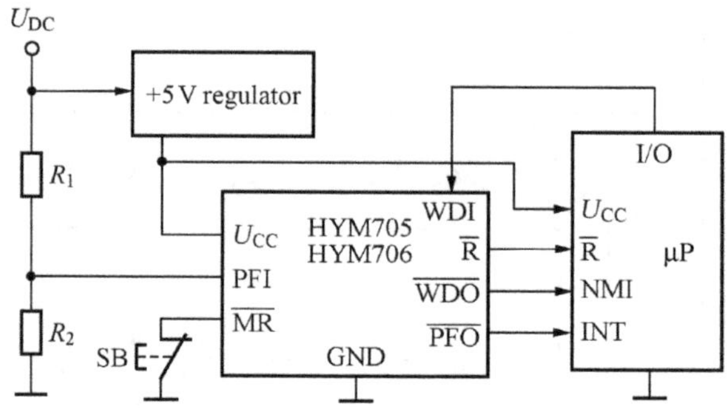

Figure 10.32 Typical application circuit of HYM705/706

TTL/CMOS level. The typical mains voltage of HYM705/706 is +5 V and the supply current is 150 μA.

Figure 10.32 shows the typical application circuit of HYM705/706. U_{DC} is a DC voltage not regulated. SB is a manual reset button. $\overline{R}$ is connected to the reset terminal of microprocessor. When the watchdog of HYM705/706 is applied to monitor the work of microprocessor, the microprocessor will send a signal at a regular interval (less than 1.6 s) under normal circumstances to trigger the watchdog input terminal (WDI) once, making the terminal $\overline{WDO}$ remain high level. When μP is out of control, the terminal $\overline{WDO}$ will deliver low-level output to issue an NMI (Non Maskable Interrupt) request to the microprocessor μP, preventing μP from continuing to run in the wrong state.

The inverting input terminal of power failure comparator is connected to the internal 1.25-V reference voltage source and the non-inverting input terminal (PFI) is connected to the resistance divider. By adjusting the resistance dividers R_1 and R_2, it can be realized that when the mains voltage decreases to a certain threshold, the terminal $\overline{PFO}$ issues a maskable interrupt request INT to μP to let μP store the data before the power supply is turned off.

Figure 10.33 shows the circuit simultaneously monitoring +5 V and +12 V mains voltages. When HYM705 is introduced, the reset threshold voltage is 4.65 V. Assume that the divider resistances $R_1 = 1\,\text{M}\Omega$ and $R_2 = 130\,\text{k}\Omega$, and the set 12-V mains voltage threshold is about 11 V. Therefore, the feature of the circuit is that as long as the 5-V mains voltage is less than

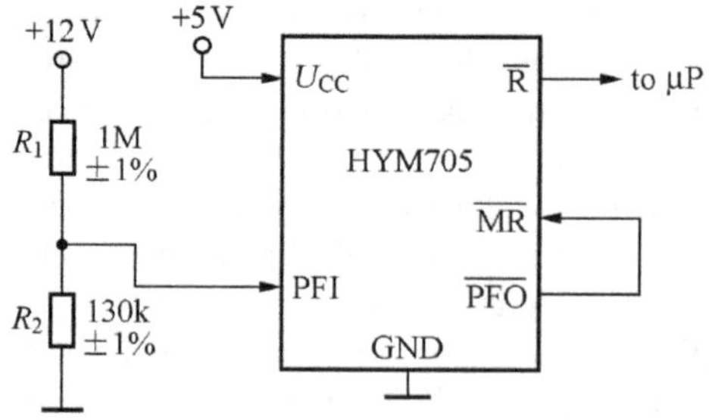

Figure 10.33 Circuit simultaneously monitoring +5 V and +12 V mains voltages

4.65 V or the 12-V mains voltage is less than 11 V, terminal $\overline{R}$ will deliver low-level output to reset μP immediately, thus realizing protection.

10.7.4 *Mains Voltage Monitor Constituted by MCP1316 Series*

MCP1316 series is the mains voltage monitor chip developed by Microchip Technology Inc., which is suitable for SCM. This series has the functions of automatic reset, manual reset, and watchdog to ensure the reliable operation of SCM system. The series includes 10 models of MCP1316–MCP1322, MCP1316M, MCP1318M, and MCP1319M. See Table 10.4 for the classification and main features of MCP1316 series.

See Figures 10.34 and 10.35 respectively for the typical application circuit of MCP1316 series. Figure 10.34(a) is the battery voltage monitor constituted by MCP1316. Figure 10.35(b) is the switching regulator voltage monitor constituted by MCP1317. Figure 10.35(a) is the SCM +5 V voltage monitor constituted by MCP1321. The power supplies of MCP1321 and SCM are both derived from the output voltage of +5 V switching regulator. *C* is the bypass capacitor. MCP1321 is connected to the terminal $\overline{RST}$ of SCM. When the +5 V voltage is lower than +2.9 V or higher than +5.8 V, SCM will be reset by MCP1321 immediately. The I/O port of SCM constantly emits negative pulse to the terminal WDI of MCP1321 to reset the

Table 10.4 Classification and main features of **MCP1316** series

Model	$\overline{RST}$ Type	$\overline{RST}$ Pull-up resistor	$\overline{RST}$ Active level	RST Type	RST Pull-up resistor	RST Active level	WDI	$\overline{MR}$
MCP1316	Push-pull output	—	Active low	—	—	—	Yes	Yes
MCP1316M	Open-drain output	Self-owned internally	Active low	—	—	—	Yes	Yes
MCP1317	Push-pull output	—	Active high	—	—	—	Yes	Yes
MCP1318	Push-pull output	—	Active low	Push-pull output	—	Active high	Yes	No
MCP1318M	Open-drain output	Self-owned internally	Active low	Push-pull output	—	Active high	Yes	No
MCP1319	Push-pull output	—	Active low	Push-pull output	—	Active high	No	Yes
MCP1319M	Open-drain output	Self-owned internally	Active low	Push-pull output	—	Active high	No	Yes
MCP1320	Open-drain output	Connected externally	Active low	—	—	—	Yes	Yes
MCP1321	Open-drain output	Connected externally	Active low	Push-pull output	—	Active high	Yes	No
MCP1322	Open-drain output	Connected externally	Active low	Push-pull output	—	Active high	No	Yes

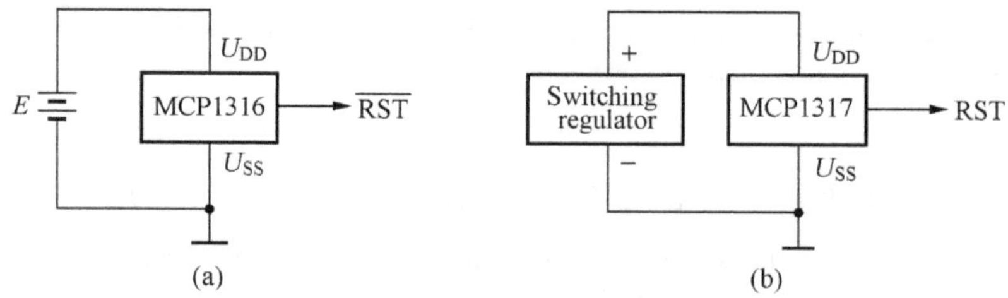

Figure 10.34 Typical application circuit I of MCP1316 series. (a) Battery voltage monitor and (b) switching regulator voltage monitor

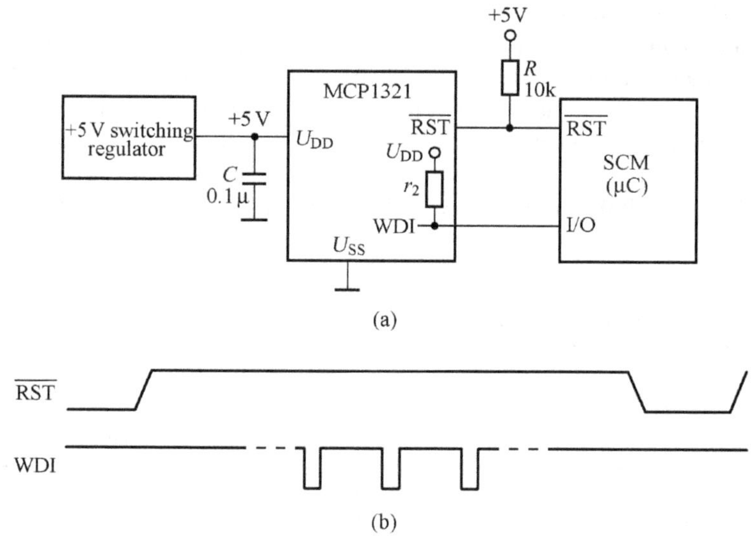

Figure 10.35 Typical application circuit II of MCP1316 Series. (a) SCM +5 V voltage monitor and (b) $\overline{RST}$ and WDI signal waveforms

watchdog timer regularly. Once the program runs out, the $\overline{RST}$ signal from MCP1321 will reset the SCM. r_2 is an internal pull-up resistor at terminal WDI. R is an external pull-up resistor at terminal $\overline{RST}$. Figure 10.35(b) shows the $\overline{RST}$ and WDI signal waveforms.

10.8 Transient Interference and Audio Noise Suppression Technology of SMPS

10.8.1 Transient Interference Suppression Method

Transient interference refers to the instantaneous interference signals generated in the AC grid such as surge voltage, ringing voltage, and spark discharge, which is featured by extremely short action duration, high voltage amplitude, and high transient energy. Transient interference

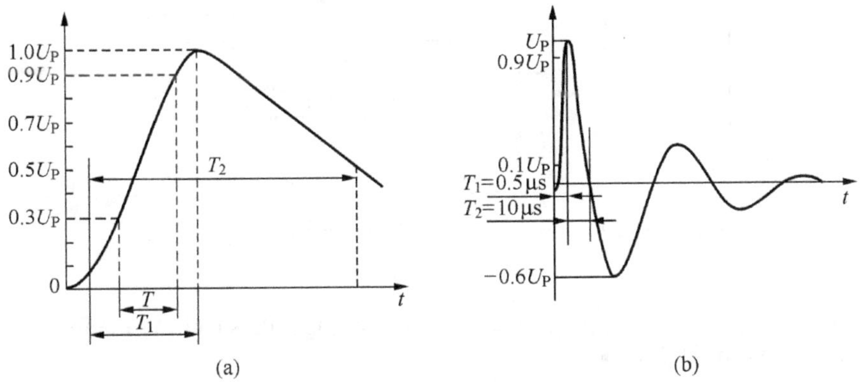

Figure 10.36 Two typical waveforms of transient voltage. (a) Surge voltage and (b) ringing voltage

will cause fluctuations in the output voltage of SMPS. When the transient voltage is superimposed on U_I to make $U_I > U_{(BR)DS}$, the TOPSwitch chip will also be damaged, so measures must be taken to suppress transient interference.

10.8.1.1 Transient Voltage Features

See Figure 10.36(a) and (b) respectively for the two typical waveforms of transient voltage. Figure 10.36(a) shows the typical surge voltage waveform of IEC1000-4-5 standard compiled by the IEC, wherein, U_P is the peak surge voltage. The test voltage of $U_P = 3000\,\text{V}$ is usually adopted. T is the time interval of the surge voltage rising from $0.3\,U_P$ to $0.9\,U_P$. T_1 is the rise time and $T_1 = 1.67\,T = 1.2 \pm 0.36\,\mu\text{s}$. The time duration of the surge voltage decreasing to $0.5\,U_P$ is T_2, which is equal to $50\,\mu\text{s}$. Figure 10.36(b) shows the typical ringing voltage waveform of IEEE-587 standard, of which the peak is also 3000 V (typical value). During the first cycle, the forward pulse rise time $T_1 = 0.5\,\mu\text{s}$, the time duration $T_2 = 10\,\mu\text{s}$, and the peak of negative pulse has decayed to $-0.6\,U_P$.

10.8.1.2 Transient Interference Suppression Method

1. Change the AC two-wire input into AC three-wire input and terminal G must be grounded.
2. Adopt two-stage EMI filter. To prevent the interference signals of two EMI filters from being superimposed mutually during resonance, L_1 and L_2 shall be no more than 10 mH and $L_2 \geq 2L_1$.
3. Wind three to five layers of polyester insulation tape of 0.05 mm thick between the primary and secondary windings to reduce the distributed capacitance of high-frequency transformer.
4. The radiator connected outside of the TOPSwitch shall be connected with the small cooling plate on the chip. When an insulating pad is placed between them and the connection position between the radiator and the circuit is inappropriate, the distributed capacitance between the radiator and the cooling plate will have resonance with the inductance of

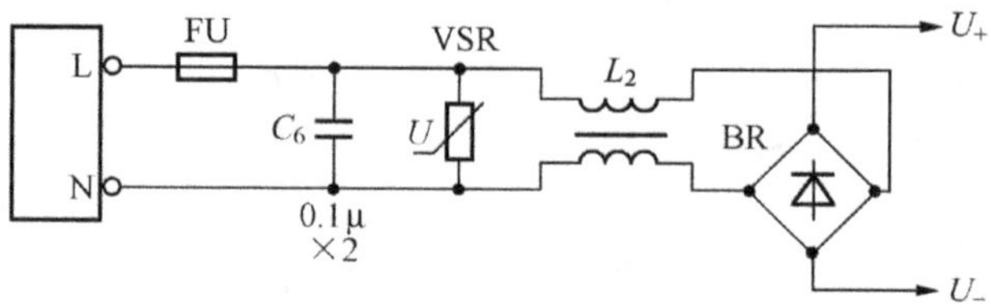

Figure 10.37 Circuit using VSR to clamp the surge voltage

the circuit to generate a high-frequency ringing voltage, thus causing malfunction of the turn-off trigger in TOPSwitch.

5. Increase the withstand voltage of bridge rectifier and properly increase the capacity of the input filter capacitor C_1.
6. Connect a VSR in parallel to the AC inlet terminal to clamp the surge voltage as shown in Figure 10.37.

10.8.2 Audio Noise Suppression Method

The audio noise of SMPS can be heard by human ears, which is mainly generated by capacitors and high-frequency transformers.

10.8.2.1 Noise of Capacitor

The dielectric material has piezoelectric effect and its deformation size is related to the electric field force with either linear or nonlinear relationship. Some nonlinear dielectric has the piezoelectric effect at room temperature. For example, the high-voltage resistant ceramic capacitor employed in the RC absorption circuit of TinySiwtch drain is sintered with nonlinear dielectric barium titanate and other materials, which will produce relatively large audio noise due to the distortion of dielectric under the effect of cyclical spike voltage. The high-voltage resistant polyester film capacitor can be applied to reduce the noise of capacitor.

10.8.2.2 Noise of High-Frequency Transformer

The attractive force between the EE- or EI-type magnetic cores of high-frequency transformer can cause displacement of the two magnetic cores. The attractive or repulsive force between the winding current can also cause displacement of coils. In addition, the high-frequency transformer may suffer periodical deformation in the case of mechanical shock. The factors discussed earlier will make the high-frequency transformer produce audio noise during work. The frequency range of audio noise of SMPS below 10 W is about 10 Hz–20 kHz.

To prevent relative displacement between magnetic cores, epoxy resin is usually employed as adhesive to conglutinate the three contact surfaces of the two magnetic cores (including the center column). However, the effect of this rigid joint approach is not ideal. Because it cannot minimize the audio noise and the magnetic core is easy to break under mechanical stress if the adhesive is excessive. Good effect can be obtained by using a special “glass beads” adhesive

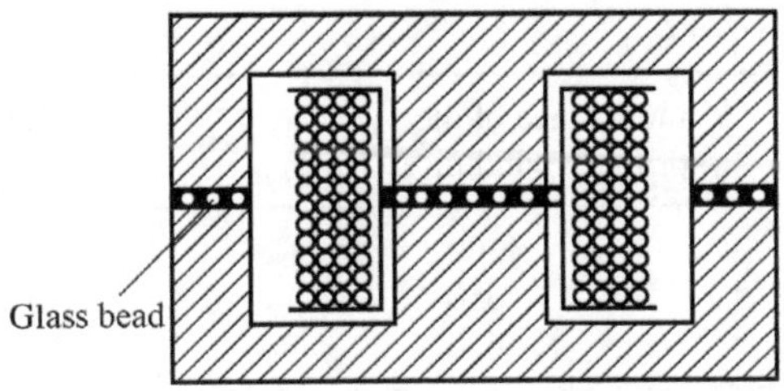

Figure 10.38 Internal structure of the high-frequency transformer

to bond ferrite cores of EE, EI, and other types. This kind of adhesive is made by mixing the glass beads and agglutinative matters according to the ratio of 1:9, which can solidifies after staying in the environment above 100 °C for 1 h. It shares similar role with ball bearing. After solidification, each magnetic core can still suffer deformation or displacement independently within a small range with the overall position unchanged, which suppresses the deformation. See Figure 10.38 for the internal structure of the high-frequency transformer conglutinated with glass beads adhesive. This process can reduce the audio noise by 5 dB.

10.8.3 Methods to Suppress Other Interferences

10.8.3.1 Suppressing Spike Voltage

The spike voltage generated by leakage inductance can be absorbed by connecting in parallel a clamping circuit between the primary inlet terminal of high-frequency transformer and the drain of TOPSwitch, thus providing the drain with clamp protection.

10.8.3.2 Ringing Voltage

The primary series mode interference forms a loop after passing through the primary winding, TOPSwitch and U_I. When the current loop area is relatively large, I_1 can radiate common mode interference. Under the influence of parameters such as the leakage inductance of transformer, the output capacitance of TOPSwitch and the distributed capacitance of transformer, U_{DS} will form a ringing voltage within the frequency range of $f_1 = 3$–$12\,\text{MHz}$. When TOPSwitch is turned off, current I_2 flows through the secondary side and decreases linearly from the peak I_{2P}, of which the rate of descent depends on the inductance L_S of secondary winding and the output voltage U_O. Ringing will also be formed during the above descent. The peak of U_{D2} depends on the distributed capacitance of transformer and output rectifier, of which the ringing frequency $f_2 = 20$–$30\,\text{MHz}$.

10.8.3.3 Other Internal Interferences

The secondary spike voltage generated by the leakage inductance of high-frequency transformer can recharge the filter capacitor at the output terminal to the peak voltage (also known as peak charging effect) to cause unstable output voltage.

Table 10.5 Allowable range of L_{P0}

Output power of SMPS P_O (W)	Inductance of the primary winding L_P (μH)	Leakage inductance of the primary winding L_{P0} (μH)
8	4300	≤ 50
20	650	≤ 35
30	620	≤ 11

For the SMPS with high voltage and low current output (e.g., 30 V, 20 mA), in order to reduce the peak charging effect of the output filter capacitor, a small resistor of 10 Ω can be connected in series to the output rectifier, which forms a low-pass filter together with the output filter capacitor C_2 to filter out the spike voltage generated by leakage inductance, thus prevent C_2 from being charged to the peak.

10.8.3.4 Reducing the Leakage Inductance of High-Frequency Transformer

Table 10.5 shows for the allowable range of leakage inductance (L_{P0}) of high-frequency transformers. L_{P0} is proportional to the square of the turns of the primary winding (${N_P}^2$), so the leakage inductance can be reduced effectively by properly reducing the turns of the primary winding. In addition, it also helps to reduce the leakage inductance by selecting low-loss core material and proper shape (i.e., the aspect ratio shall be as high as possible), improving winding method, reducing the thickness of the insulating layer between windings, and increasing the degree of coupling. When designing the multi-output SMPS, the main output winding shall be close to the primary winding to reduce leakage inductance. When conditions permit, the stacking winding shall be applied. It shall be noted that the secondary side shall be disconnected when measuring the inductance of the primary winding. When measuring the leakage inductance of the primary winding, the digital inductance meter shall be connected in parallel at both ends of the primary winding and the secondary and bias windings shall be in short circuit.

10.9 Design of Overheating Protection Component and Cooling Control System

The following first describes the basic principles of the overheating protection circuit of SMPS and then introduces the principles and applications of two kinds of overheating protection component – temperature fuse tube and temperature protector.

10.9.1 Basic Principles of the Overheating Protection Circuit of SMPS

Figure 10.39 shows the basic principles of the overheating protection circuit of SMPS. The reverse breakdown voltage of silicon transistor emitter junction (E-B) is adopted for the regulator tube VD_Z here as the reference voltage U_{REF}. This method can get 5.8–7 V reference

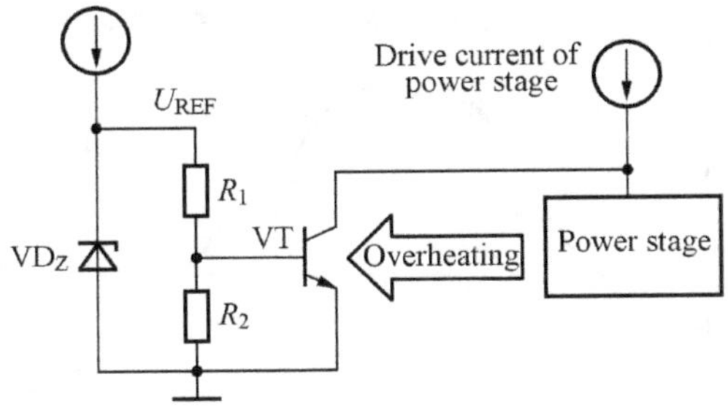

Figure 10.39 Basic principles of overheating protection circuit of SMPS

voltage, which has a positive temperature drift. The temperature coefficient of reverse breakdown voltage of the emitter junction is $\beta_T \approx +3.5\,\text{mV}/°\text{C}$, which means that U_{REF} can be increased by about 3.5 mV per 1 °C rise of ambient temperature. NPN-type transistor VT is employed as a temperature sensor. R_1 and R_2 are the base bias resistances. VT is placed close to the power stage (i.e., the regulator) to sense the temperature of regulator. The emitter junction voltage U_{BE} of NPN-type transistor has a negative temperature coefficient $\alpha_T \approx -2.1\,\text{mV}/°\text{C}$, which means U_{BE} can be reduced by 2.1 mV per 1 °C rise of ambient temperature. At room temperature, U_{BE} is far lower than the turn-on voltage of NPN transistor, so VT is turned off. When the chip temperature rises to the maximum junction temperature (T_{jM}) for some reason (overload or rise of ambient temperature), VT will be turned on to divide the drive current of power stage, which will reduce or even fully switch off the load current to achieve the purpose of overheating protection.

To prevent frequent action of the overheating protection circuit due to the fluctuation of junction temperature when $T = T_{jM}$ to improve the reliability of the protection circuit, the overheating protection circuit generally has the thermal hysteresis characteristics as shown in Figure 10.40. Once the chip temperature T reaches or exceeds the maximum junction temperature T_{jM}, the SMPS will be turned off. However, when the chip temperature is just lower than T_{jM}, the SMPS is still in turn-off state, which will not be restarted until the chip temperature is reduced to the safe temperature T_j. T_j is generally lower than T_{jM} by 20–75 °C, of which the specific value depends on the chip model. This indicates that the overheating protection circuit has hysteresis temperature (also known as thermal shutdown hysteresis temperature), which is expressed as T_{HYST}. $T_{HYST} = T_{jM} - T_j$. For example, the thermal shutdown temperature of the fifth generation single-chip SMPS of TOPSwitch-HX series is $T_{jM} = 142\,°\text{C}$ (typical value) and the hysteresis temperature $T_{SYHT} = 75\,°\text{C}$. The above-mentioned hysteresis characteristic is similar to that of Schmitt trigger in digital circuit.

10.9.2 Principles and Applications of Two Kinds of Overheating Protection Component

10.9.2.1 Principle and Application of the Temperature Fuse Tube

Temperature fuse tube (TF) is a new kind of overheating protection component. It is featured by extremely small cold resistance, fast response to temperature, accurate fusing temperature

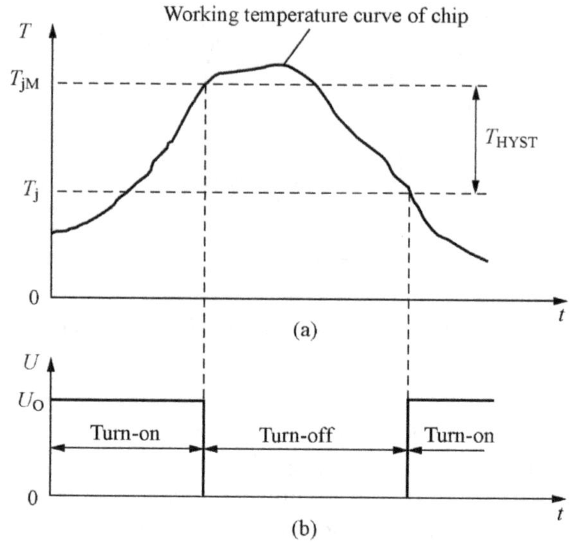

Figure 10.40 Thermal hysteresis characteristics

Table 10.6 Product classification of TF

Series	Nominal fusing temperature T_F (°C)	Current capacity (A)	External dimension (mm)
RY01	65,70,76,83,92,96,100,110,120,130,142,150,165,169, 185,190,195,200,210,225,230,250,280,320	10	$\phi 4 \times 13.5$
RY02	65,75,85,95,100,110,115,125	3	$\phi 9.5 \times 26$
RH01	83,95,100,110,115,125,130,135,145,150	1	$\phi 5.3 \times 6.5$

(T_F), small size, low price, and easy replacement , which can be widely applied in overheating protection devices and temperature control circuits. It can also be introduced together with the thermopaint to indicate the surface temperature of the test sample with the color of the thermopaint on the one hand, and control the temperature with the TF on the other hand. TF is also a one-off component like ordinary fuse. Once damaged, it must be replaced with the product of the same specification.

The typical products of TF include RY-1, RY02, RH01 and other series. Each series includes a variety of specifications. See Table 10.6 for the product classification. Take RY01 series TF for example. See Figure 10.41 for its outline and symbol. Figure 10.42 shows the application of TF in the DC heater. Connect RY01 and the load R_L (e.g., electric stove wire) in series and then put them in the monitored temperature environment. Once the limit temperature is exceeded owing to overheating ($T > T_F = 65\,°C$), RY01 will be fused quickly to turn off the power supply, thus realizing protection.

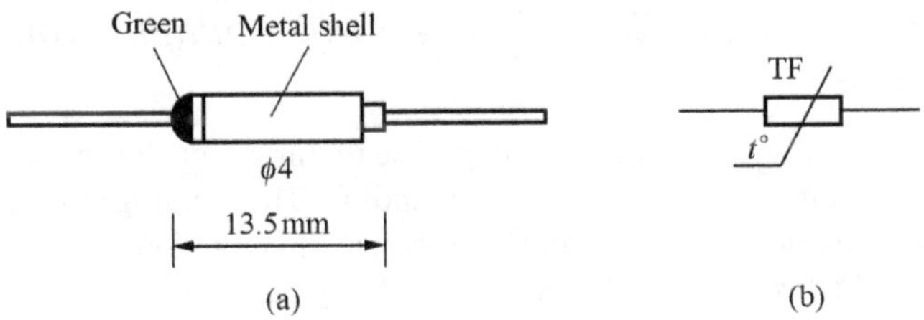

Figure 10.41 RY01 series TF. (a) Outline and (b) symbol

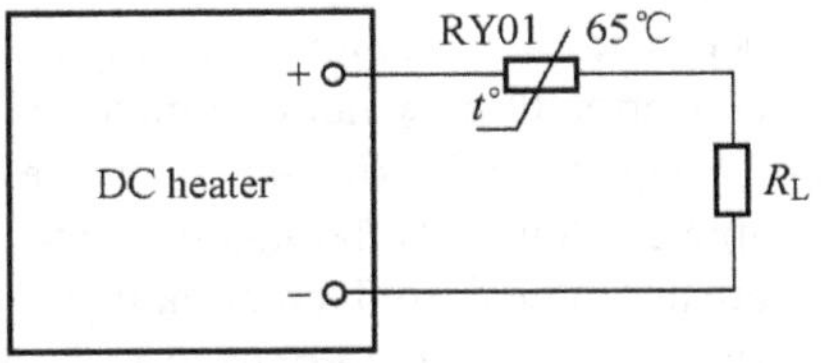

Figure 10.42 Typical application of TF

10.9.2.2 Principle and Application of the Temperature Protector

Temperature protector is a kind of overheating protection device that can be adopted for multiple times. The contact is disconnected under normal conditions and connected in the case of overheating. It can be employed repeatedly. See Figure 10.43 for the outline and contact structure of RS01 series temperature protector. The current capacity of the device is 8 A and the external dimension is 6 × 33 mm. The product specifications are 75, 80, 85, 90, 95, 100, 105, 110, 115, 120, and 130 °C. It can be seen from Figure 10.43(b) that when $T > T_F$, the temperature contact is turned on to switch off the current through the intermediate relay. With the decrease of temperature, the contact is disconnected again to automatically complete temperature control.

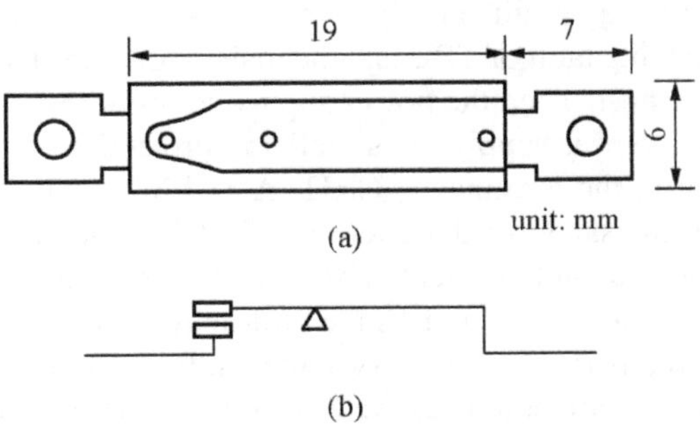

Figure 10.43 RS01 series temperature protector. (a) Outline and (b) contact structure

10.9.3 Design of Cooling Control System with Multiple Protection Functions

With the development of science and technology, the requirements for the safety and reliability of cooling control system also become more demanding. This cooling system must have sound multiple protection functions. Once the detected object has any kind of overheating fault, multiple measures shall be taken for protection. Even though the power supply controller fails to work, the main power supply of microcomputer system can also be turned off through auxiliary shutdown circuit. At present, it is called the cooling protection system with "Advanced Configuration and Power Interface" (ACPI) internationally.

The LM76 type produced by NSC can be applied as the intelligent temperature sensor suitable for constituting the cooling protection system. LM76 is the intelligent temperature sensor based on I^2C bus interface. The chip contains a band-gap temperature sensor, a 13-digit (including the sign digit) A/D converter, three window comparators (T_CRIT, T_H, and T_L), two output stages (T_CRIT_A and INT), an I^2C series bus interface, and seven registers. It takes 400 ms (typical value) to complete a temperature/data conversion. The output data is in the form of twos complement. There are three digital temperature window comparators internally, which are respectively the critical limit temperature excess alarm comparator (T_CRIT), the upper limit temperature comparator (T_H), and the lower limit temperature comparator (T_L). They correspond to the temperature control points of t_{CRIT}, t_H, and t_L, respectively. It also has the programmable temperature hysteresis characteristics with the temperature hysteresis quantity of t_{HYST}. Fault queue counter can be adopted to effectively prevent noise interference from causing the false triggering of output terminal. The above structure simplifies the design of the temperature control system with ACPI. When the measured temperature exceeds the programmable window range (t_H–t_L), the interrupt output (INT) of open drain is effective. The limit temperature excess alarm unit works only when $t > t_{CRIT}$. The temperature measurement range of LM76 is −25 to +125 °C. The temperature measurement accuracy is ±0.5 °C at +25 °C, and ±1 °C within the range of −10 to +100 °C. The resolution is up to 0.0625 °C.

See Figure 10.44 for the circuit block diagram of cooling control system constituted by LM76 with multiple protection functions. The system consists of four parts: (i) the intelligent temperature sensor LM76, adopted to measure the temperature of CPU or μP; (ii) the system power supply, including the +3.3 V main power supply (dedicated to CPU and the main circuit), and the +3.3 V auxiliary power supply (dedicated to LM76) and the main power supply controller; (iii) the power supply controller of CPU and all main circuits; and (iv) the independent power shutdown circuit. The system has sound overheating protection function. Regardless of the overheating fault of CPU and the main circuit, the terminal INT of LM76 can immediately suspend the host. Then the power supply controller will send a signal to quickly turn off the +3.3 V main power supply, thus realizing protection. In addition, in the case of critical limit excess alarm, the terminal T_CRIT_A of LM76 can also directly turn off the main power supply. For the sake of safety, terminal T_CRIT_A can also turn off the +3.3 V main power supply through an independent hardware shutdown circuit to guard against failure of the main power supply control. R_1 and R_2 are pull-up resistors.

The above-discussed temperature control system is fully in conformity with ACPI technical specification. Whenever the temperature exceeds the set window, LM76 will send interrupt signals. In addition, the user can also program for the window on the site according to the desired temperature interval and reset each temperature threshold during temperature measurement to

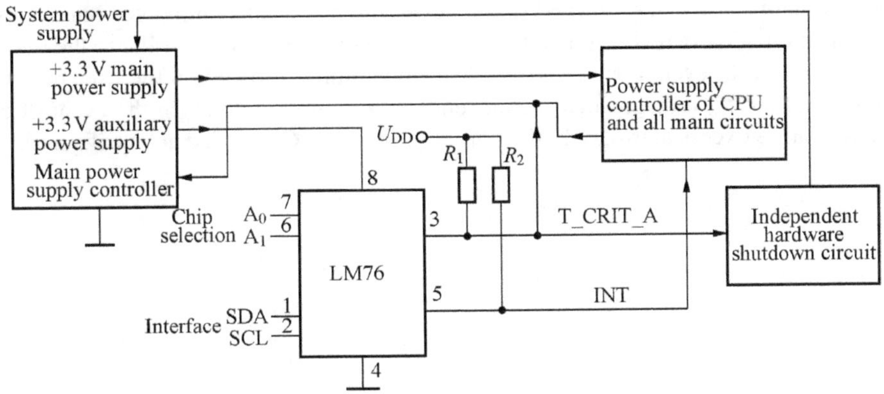

Figure 10.44 Circuit block diagram of microcomputer cooling control system with multiple protection functions

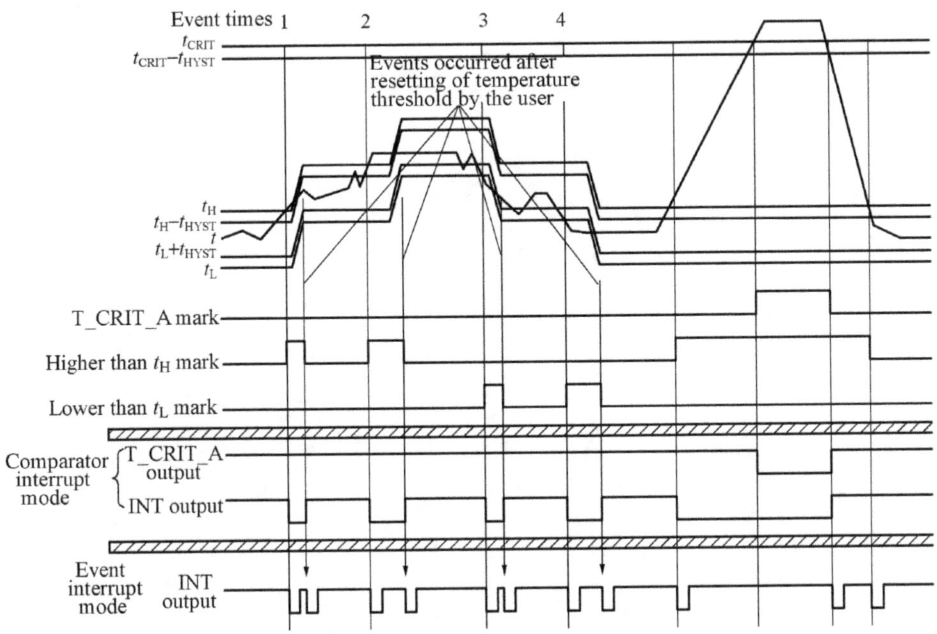

Figure 10.45 Temperature response curve of LM76

create a new setting window. See Figure 10.45 for the temperature response curve of LM76. The figure lists the various events generated during temperature measurement. The first event is enabling INT when $t > t_H$, so that the system response is interrupted. Then it is known that the temperature exceeds the upper limit by querying the status of LM76, which means that the temperature is rising. After that, the user resets an upper limit t_H, higher than the original one by a temperature interval. There will be an interrupt during the reprogramming of LM76, because the temperature has returned to the new window. In the comparator interrupt mode, LM76 can automatically revoke this interruption. The second event is similar to the first event. The third and fourth events occur during the temperature drop. When the user receives the information of $t < t_L$, it means that the current temperature is falling. In the event of critical limit temperature excess alarm, only the output terminal T_CRIT_A can be triggered.

References

[1] Sha Zhanyou, Wang Yanpeng, Ma Hongtao, et al., Optimal Design of SMPS, (2nd edition) Beijing: China Electric Power Press, 2013 (1)

[2] Sha Zhanyou, Pang Zhifeng, Zhou Wanzhen, et al., Design Entry of SMPS, Beijing: China Electric Power Press, 2013 (1)

[3] Ma Hongtao, Sha Zhanyou, et al., Production and Commission of SMPS, Beijing: China Electric Power Press, 2010 (7)

[4] Sha Zhanyou, Wang Yanpeng, Ma Hongtao, 500 Questions About the Practical Technology of SMPS, Beijing: China Electric Power Press, 2012 (2)

[5] Sha Zhanyou, Design and Application of New-Type Single-Chip SMP, Beijing: Electronic Industry Press, 2001 (6)

[6] Sha Zhanyou, Latest Application Technology of Single-Chip SMPS (2nd edition), Beijing: China Machine Press, 2006 (1)

[7] Sha Zhanyou, Design and Application Technology of Single-Chip SMPS, Taipei, China: Chuan Hwa Science & Technology Book Co., Ltd., 2006 (12)

[8] Sha Zhanyou, Ma Hongtao, Wang Yanpeng, Design and Application of Special Integrated Power Supply, Beijing: China Electric Power Press, 2007 (1)

[9] Sha Zhanyou, Pang Zhifeng, Peripheral Component Selection and Testing of SMPS, Beijing: China Electric Power Press, 2009 (7)

[10] Sha Zhanyou, Wang Li, Ma Honttao, Optimal Design of LED Lighting Drive Power Supply, Beijing: China Electric Power Press, 2011 (2)

[11] Sha Zhanyou, Wang Yanpeng, Ma Hongtao, Design Entry of LED Drive Power Supply, Beijing: China Electric Power Press, 2012 (1)

[12] Abraham Pressman, Keith Billings, Taylor Morey, Switching Power Supply Design, (3rd edition), McGraw-Hill Professional, 2009

[13] Marty Brown, Power Supply Cookbook, Butterworth-Heinemann, 1994

[14] Sanjaya Maniktala, Switching Power Supplies A to Z, Wolgan Valley, Australia: Newnes, 2009

[15] Sanjaya Maniktala, Switching Power Supply Design & Optimization, New York: McGraw-Hill, c2005, 2004

[16] Sanjaya Maniktala, Troubleshooting Switching Power Converters, Newnes, 2007

[17] Colonel William T. McLyman, Transformer and Inductor Design Handbook, London: CRC Press Inc.; 4th Revised edition, 2011

[18] Sha Zhanyou, Principle and Application of Single-Chip Switching Integrated Regulator, Electrical Measurement & Instrumentation, 1990 (6), 3–4

[19] Sha Zhanyou, Application of a Power Noise Filter, Automation Instrumentation, 1991 (9), 72–74

Optimal Design of Switching Power Supply, First Edition. Zhanyou Sha, Xiaojun Wang, Yanpeng Wang, and Hongtao Ma.

[20] Sha Zhanyou, Overview on Single-Chip TRMS/DC Converter, Electrical Measurement & Instrumentation, 1994 (3), 209–211
[21] Sha Zhanyou, Circuit Design of SMPS, ElectricAutomation, 1997 (3), 160–162
[22] Sha Zhanyou, Design and Measurement of Electromagnetic Compatibility, Electronic Measurement Technology, 1997 (4), 219–222
[23] Sha Zhanyou, Development and Its Application of Single-Chip SMPS, Application of Electronic Technology, 2000 (1), 4–8
[24] Sha Zhanyhou, Ma Hongtao, Protection Circuit Design of Single-Chip SMPS, Electrotechnical Technology, 2000 (12), 230–233
[25] Sha Zhanyou, Sui Bingdong, Wang Yanpeng, Online Measurement Technology of Digital Instrument, Electrotechnical Technology, 1999 (11), 208
[26] Sha Zhanyou, Zhou Wanzhen, Wang Yanpeng, Circuit Design of Special Integrated SMPS, Power Supply Technologies and Applications, 2001 (5), 179–184
[27] Sha Zhanyou, Design Principle of EMI Filter, Application of Electronic Technique, 2001 (5), 74–77
[28] Sha Zhanyou, Key Component Selection of Single-Chip SMPS, Power Supply Technologies and Applications, 2002 (5), 87–89
[29] Sha Zhanyou, Wang Yanpeng, New Technology and Its Application of SMPS, Power Electronic technology, 2003 (3), 6–8
[30] Sha Zhanyou, Protection Component and Circuit Design at the Input Terminal of SMPS, Power Supply Technologies and Applications, 2009 (11), 237–239
[31] Sha Zhanyou, Design of Input Rectifier Filter and Clamp Protection Circuit, Power Supply Technologies and Applications, 2009 (12), 227–230
[32] Sha Zhanyou, An Guochen, Radiator Design of SMPS, Power Supply Technologies and Applications, 2010 (1), 227–230
[33] Sha Zhanyou, Ma Hongtao, Component Selection and Circuit Design of Output Rectifier Filter, Power Supply Technologies and Applications, 2010 (2), 68–69
[34] Sha Zhanyou, Optical Coupler and Optocoupler Feedback Circuit Design, Power Supply Technologies and Applications, 2010 (4), 91–93
[35] Sha Zhanyou, Key Design Points of SMPS Clamp Protection Circuit and Radiator, Power Supply Technologies and Applications, 2010 (6), 196–198
[36] Sha Zhanyou, Wang Yanpeng, Key Design Points of SMPS, Power Supply Technologies and Applications, 2010 (combined issue of 1 and 2), 162–165
[37] Sha Zhanyou, Ma Hongtao, Derivation and Verification of Formula for Selecting High-Frequency Transformer Core based on AP Method, Power Supply Technologies and Applications, 2011 (11), 130–134
[38] Sha Zhanyou, Meng Zhiyong, Method for Improving the Efficiency of SMPS, Power Supply Technologies and Applications, 2012 (3), 165–170
[39] Sha Zhanyou, Ma Hongtao, Radiator Design of Power Switch Tube and LED Driver Chip, Power Supply Technologies and Applications, 2012 (6), 198–200
[40] Sha Zhanyou, Ma Hongtao, Stability Design of Optocoupler Feedback Control Loop of SMPS, Power Supply Technologies and Applications, 2013 (5), 171–175

Index

Optimal Design of Switching Power Supply, First Edition. Zhanyou Sha, Xiaojun Wang, Yanpeng Wang, and Hongtao Ma.

www.ingramcontent.com/pod-product-compliance
Lightning Source LLC
LaVergne TN
LVHW062204200926
848484LV00006B/1316